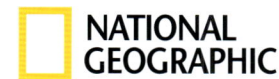

NATIONAL GEOGRAPHIC

INVISIBLE
WONDERS

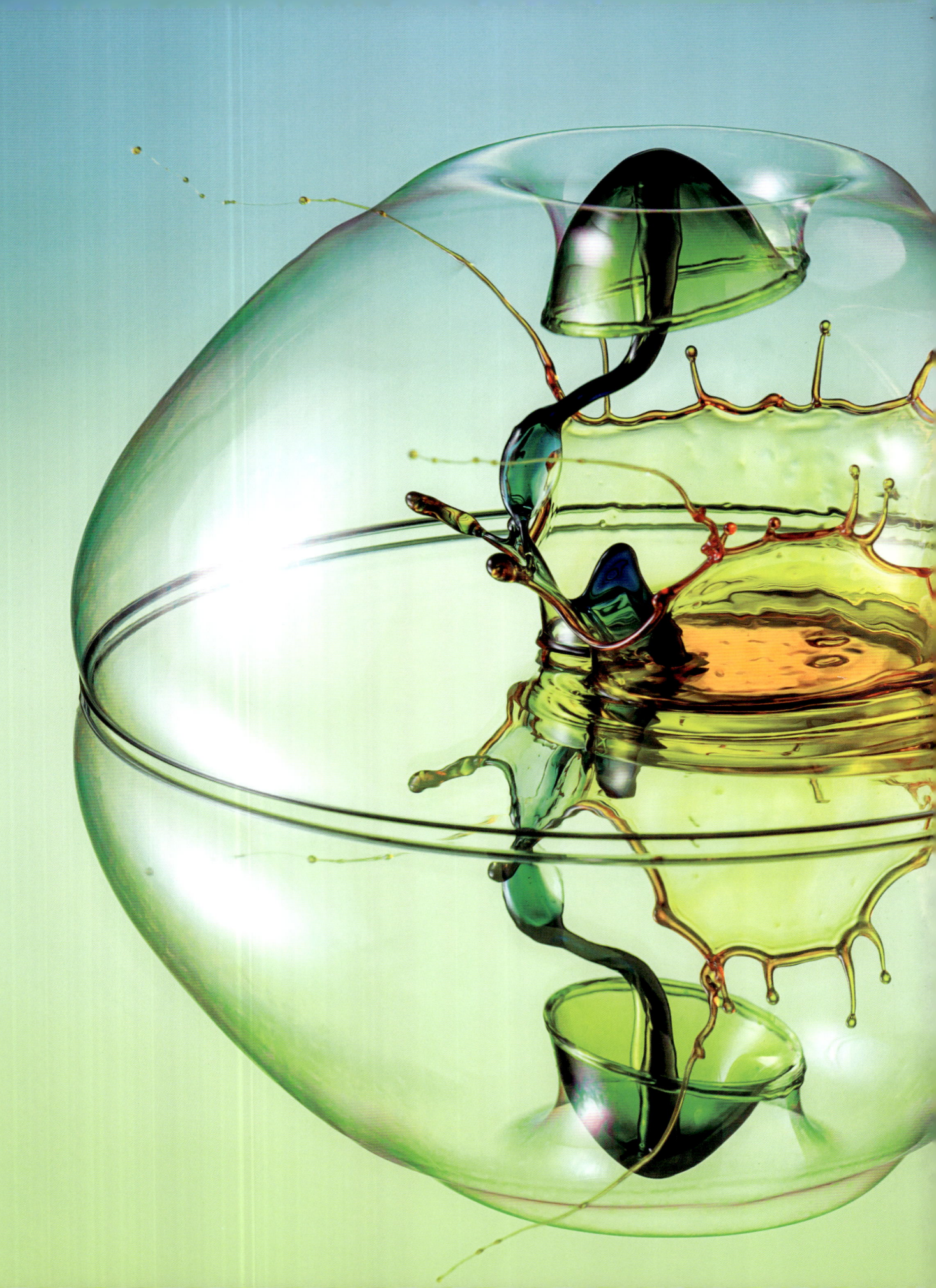

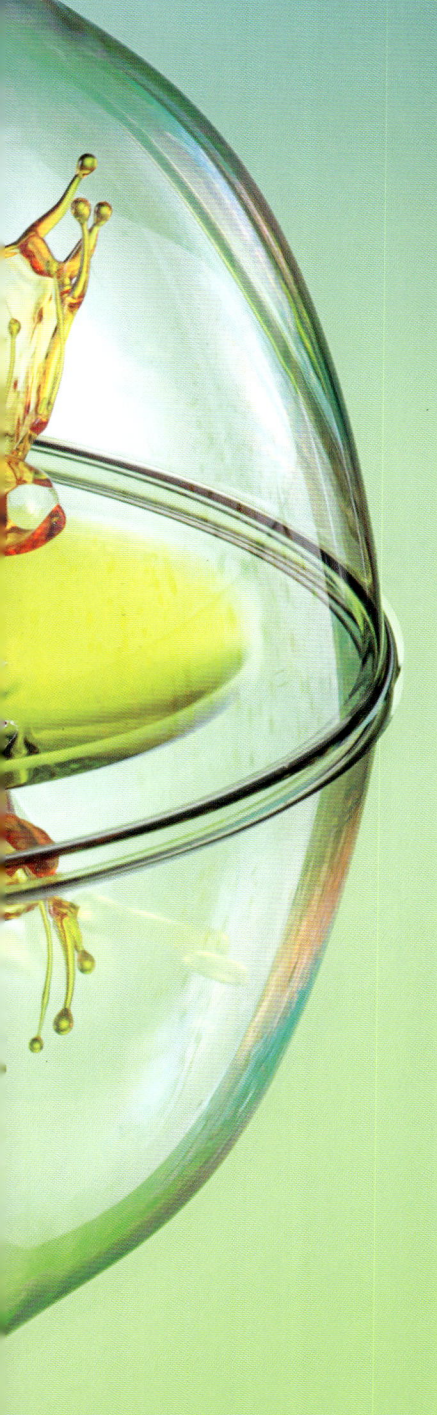

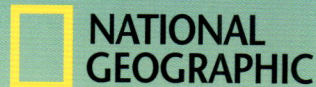

NATIONAL
GEOGRAPHIC

INVISIBLE

PHOTOGRAPHS OF THE HIDDEN WORLD

WONDERS

ANAND VARMA

NATIONAL GEOGRAPHIC

WASHINGTON, D.C.

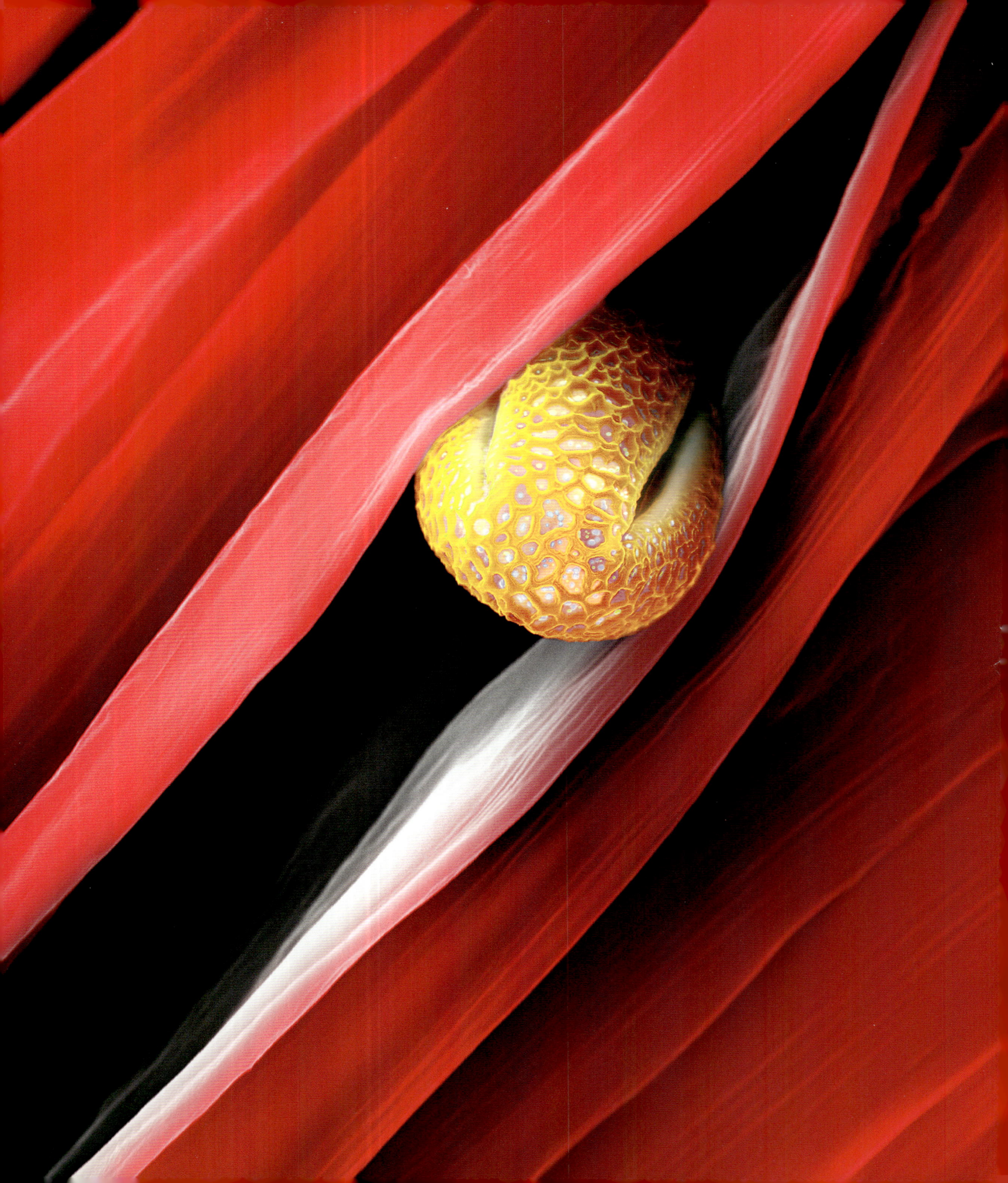

CONTENTS

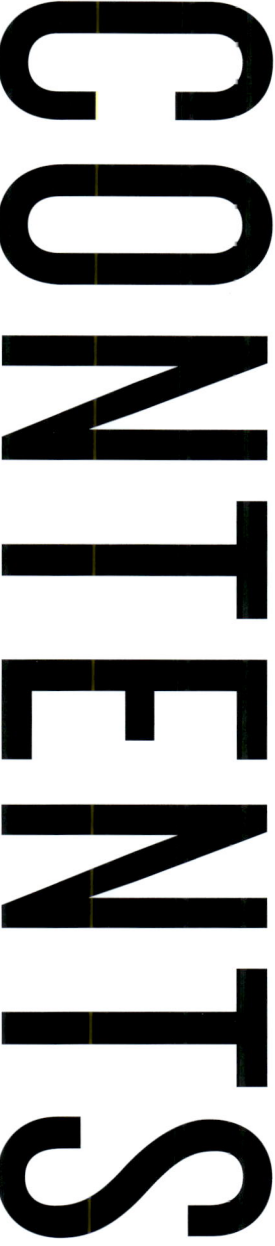

Martin Oeggerli / A single grain of pussy willow pollen: Lined up end to end, 1,400 of these would measure one inch.

FRONT AND BACK COVER: Robert Berdan / Crystalized amino acids
PAGES 2-3: Markus Reugels / Droplets of colored water bounce and collide at high speed.
PAGES 6-7: Richard Mosse / A multispectral aerial image shows the aftereffects of fire in Brazil's Pantanal.
PAGES 8-9: Anand Varma / Water vapor reveals the airflow around an Anna's hummingbird's wings in flight.
PAGES 10-11: Amy Gulick / Wild Alaska salmon hatchlings with yolk sacs attached
PAGES 12-13: Anand Varma / Artificial insemination of a queen honeybee

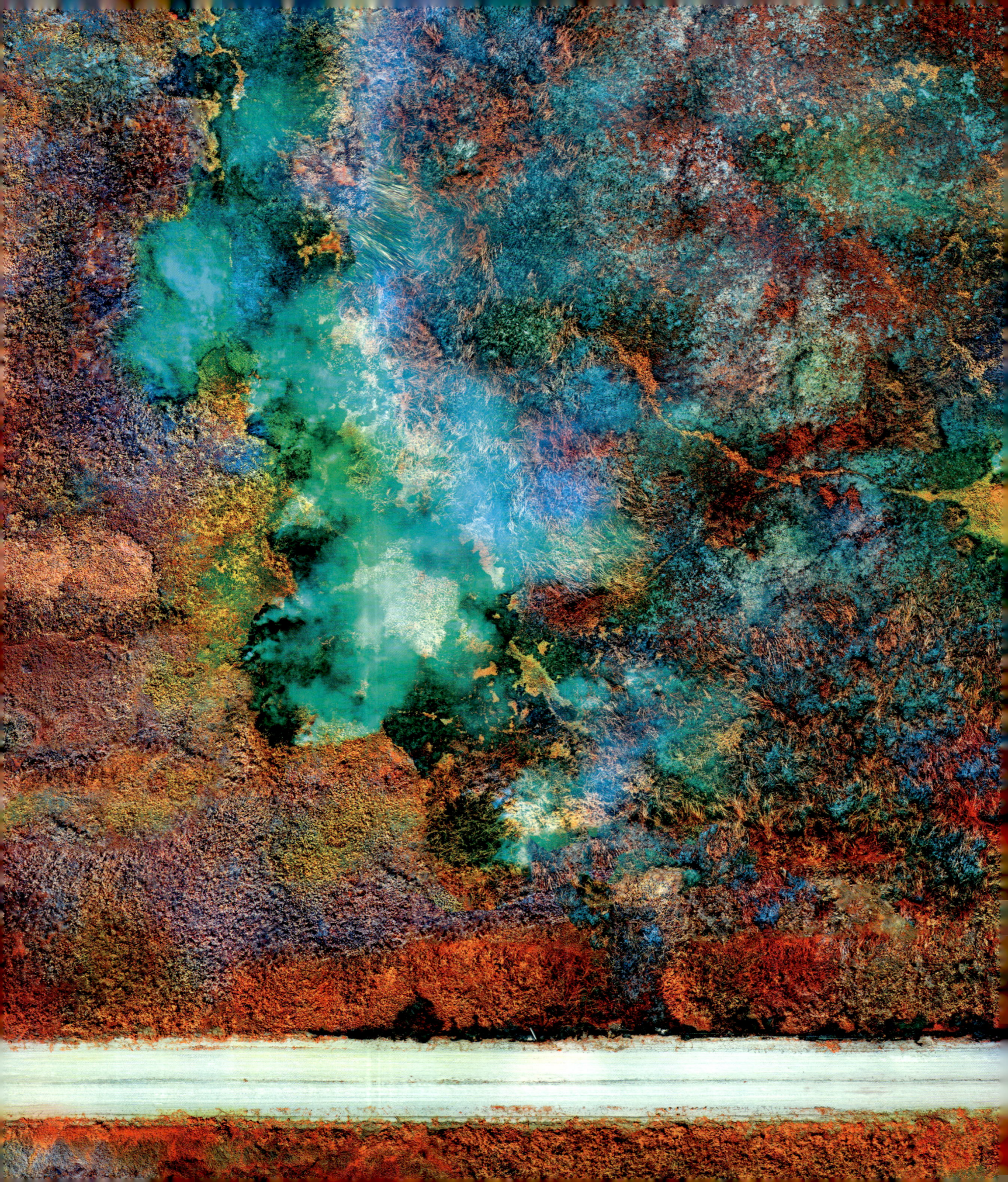

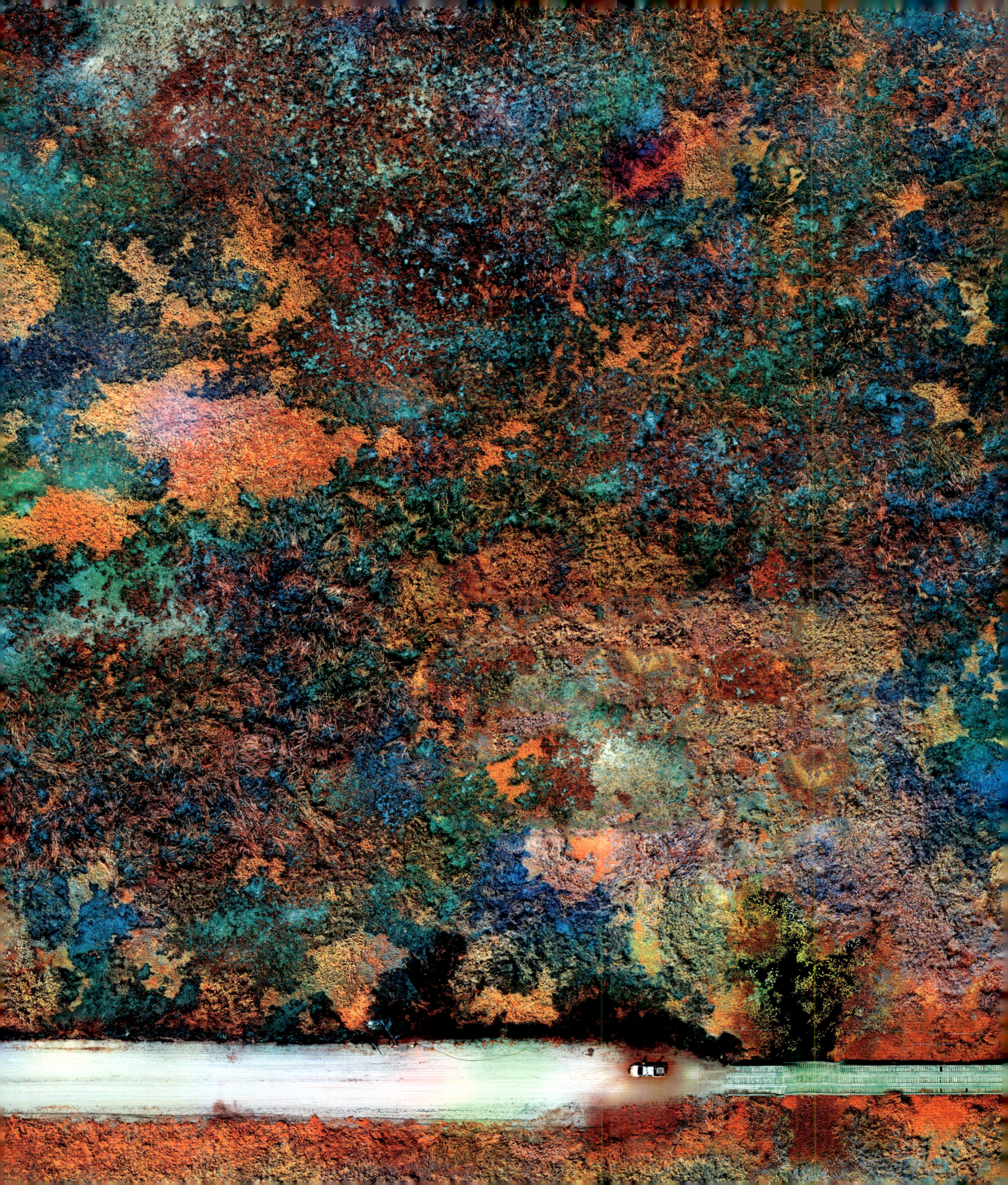

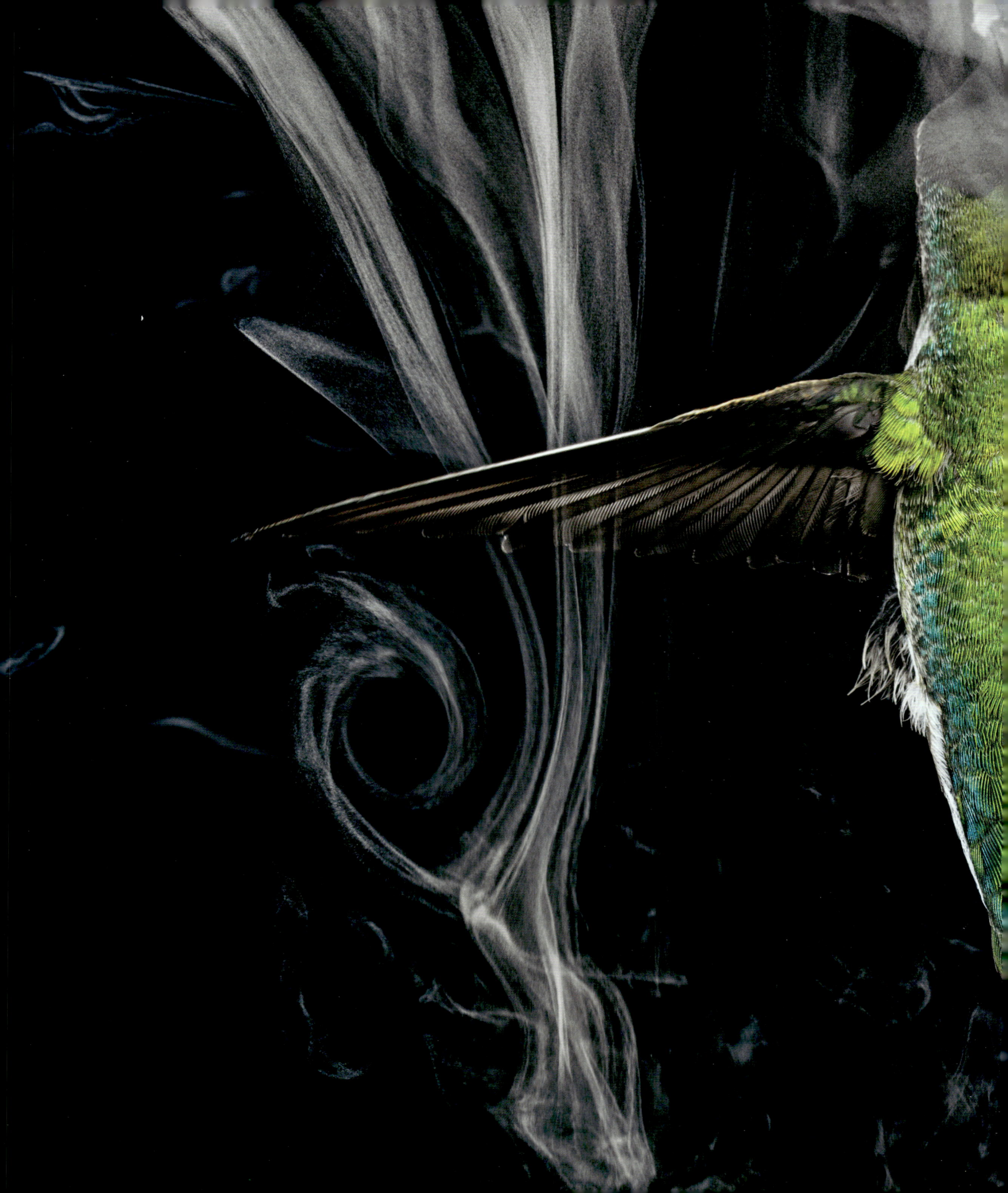

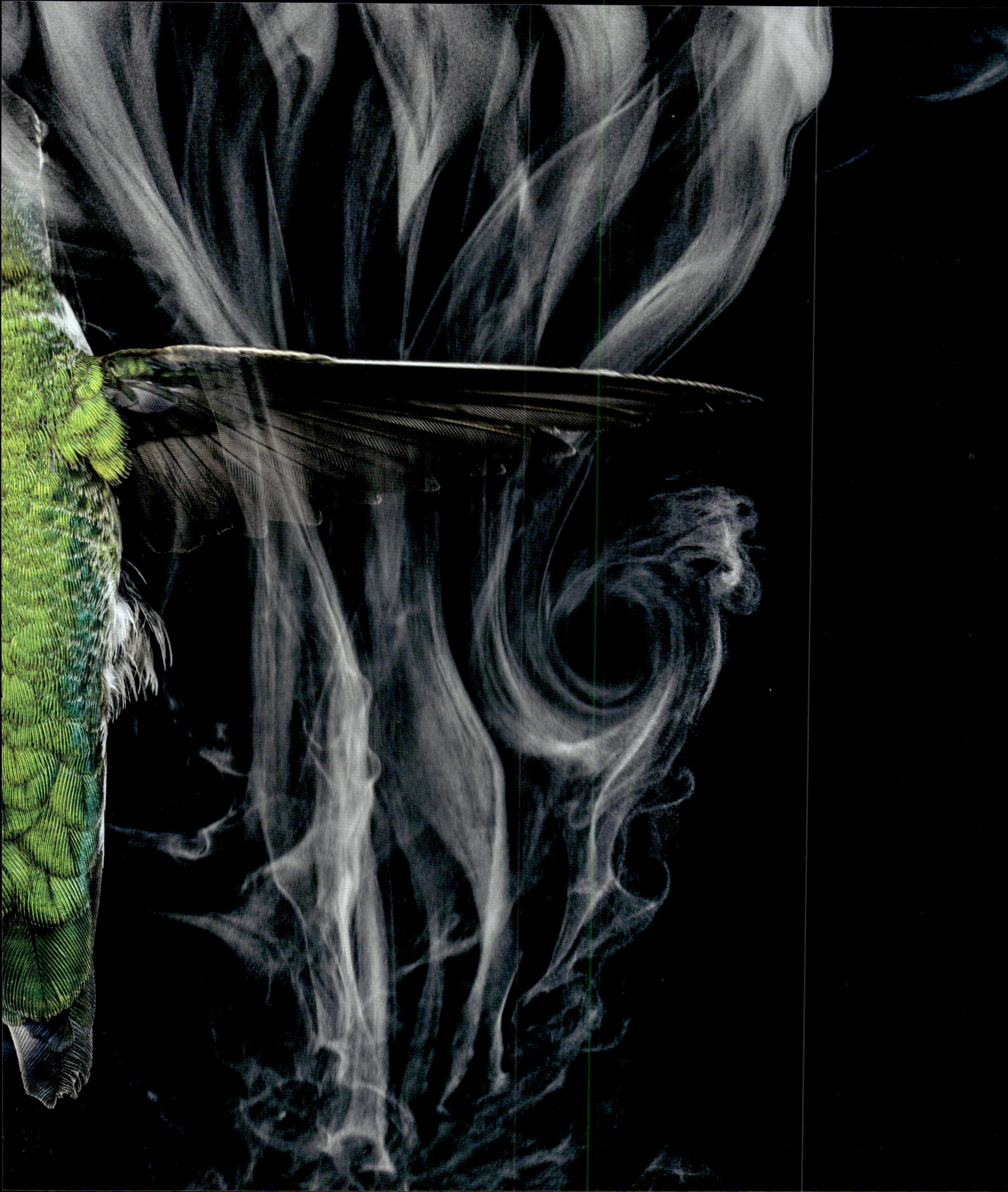

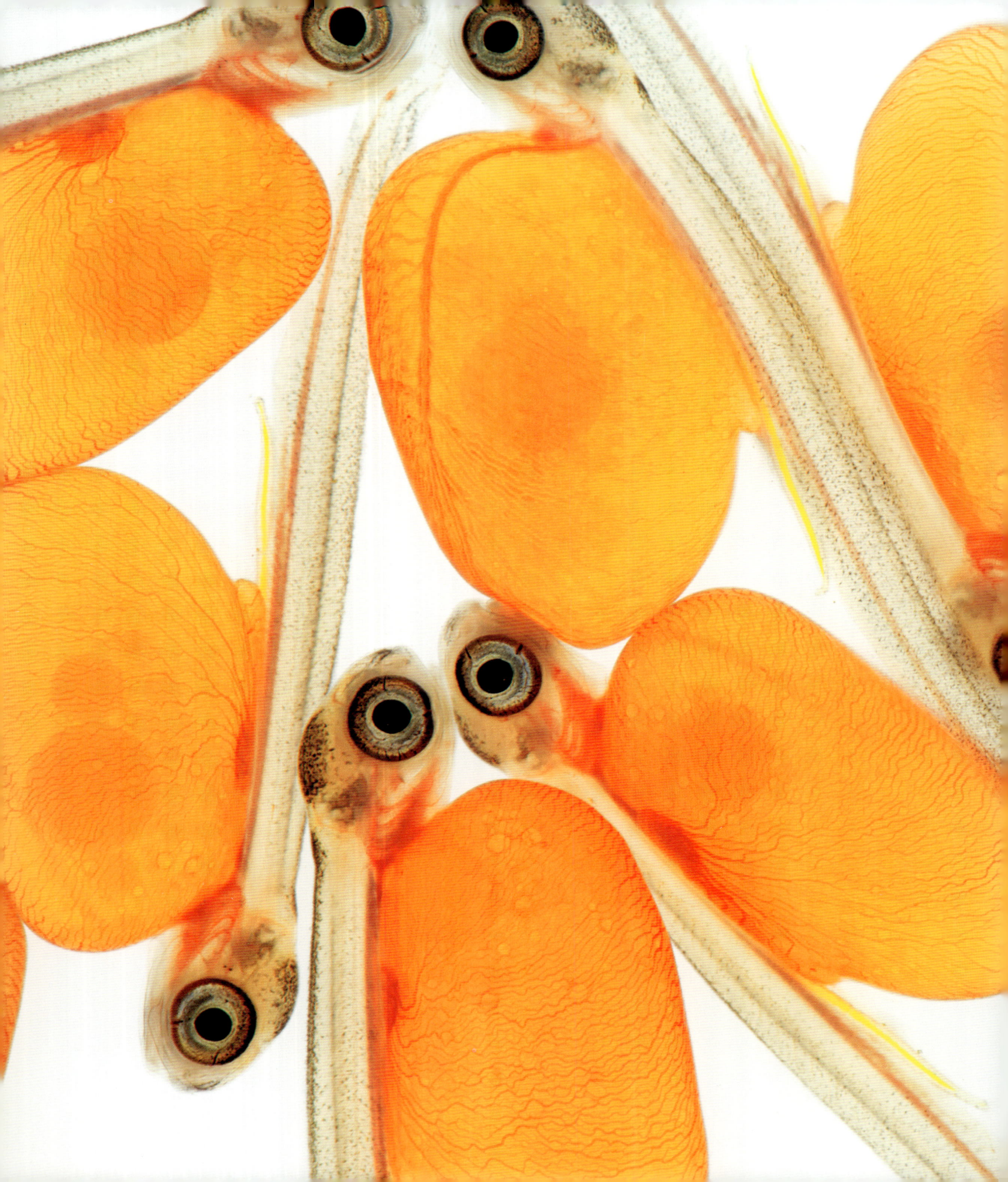

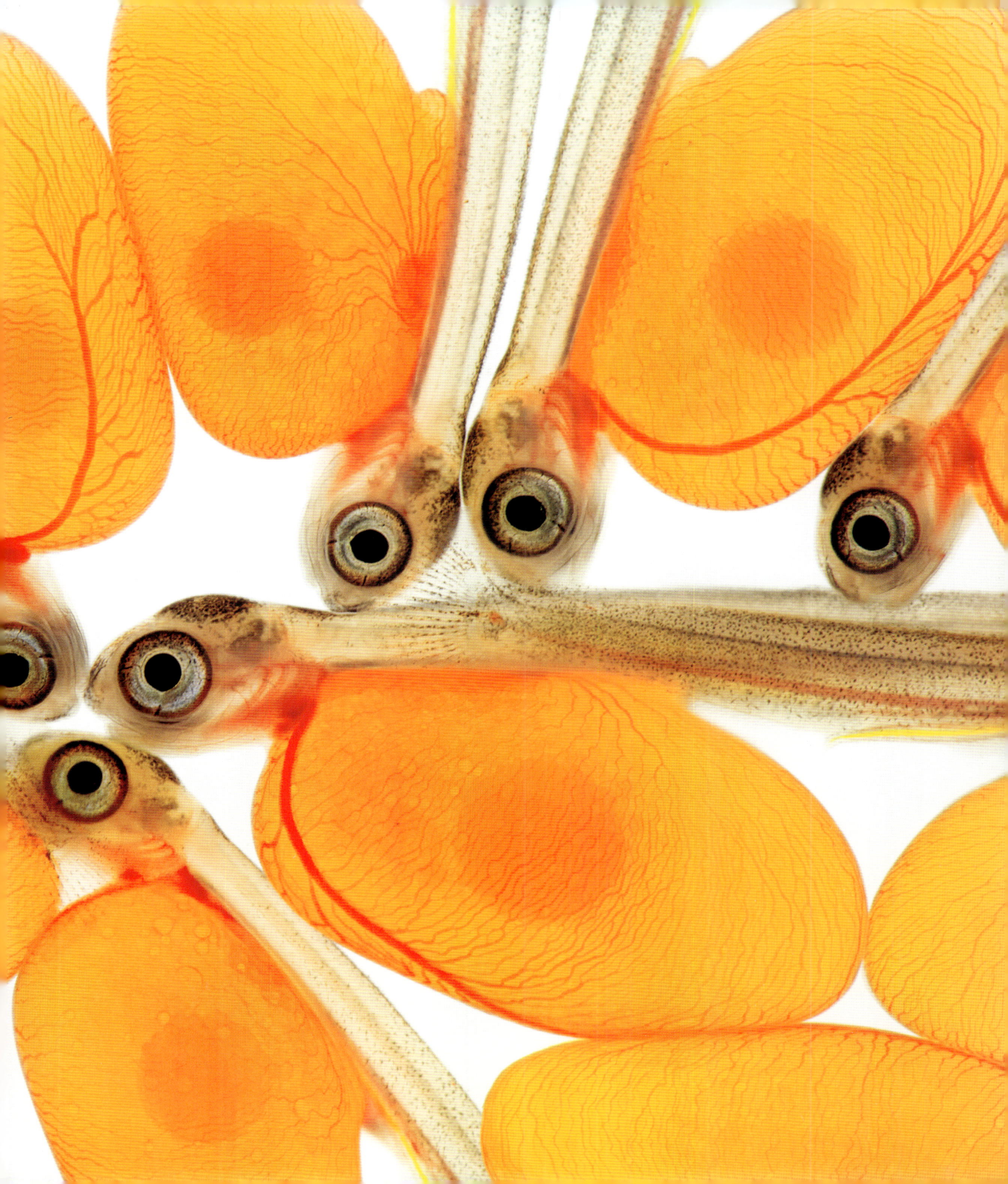

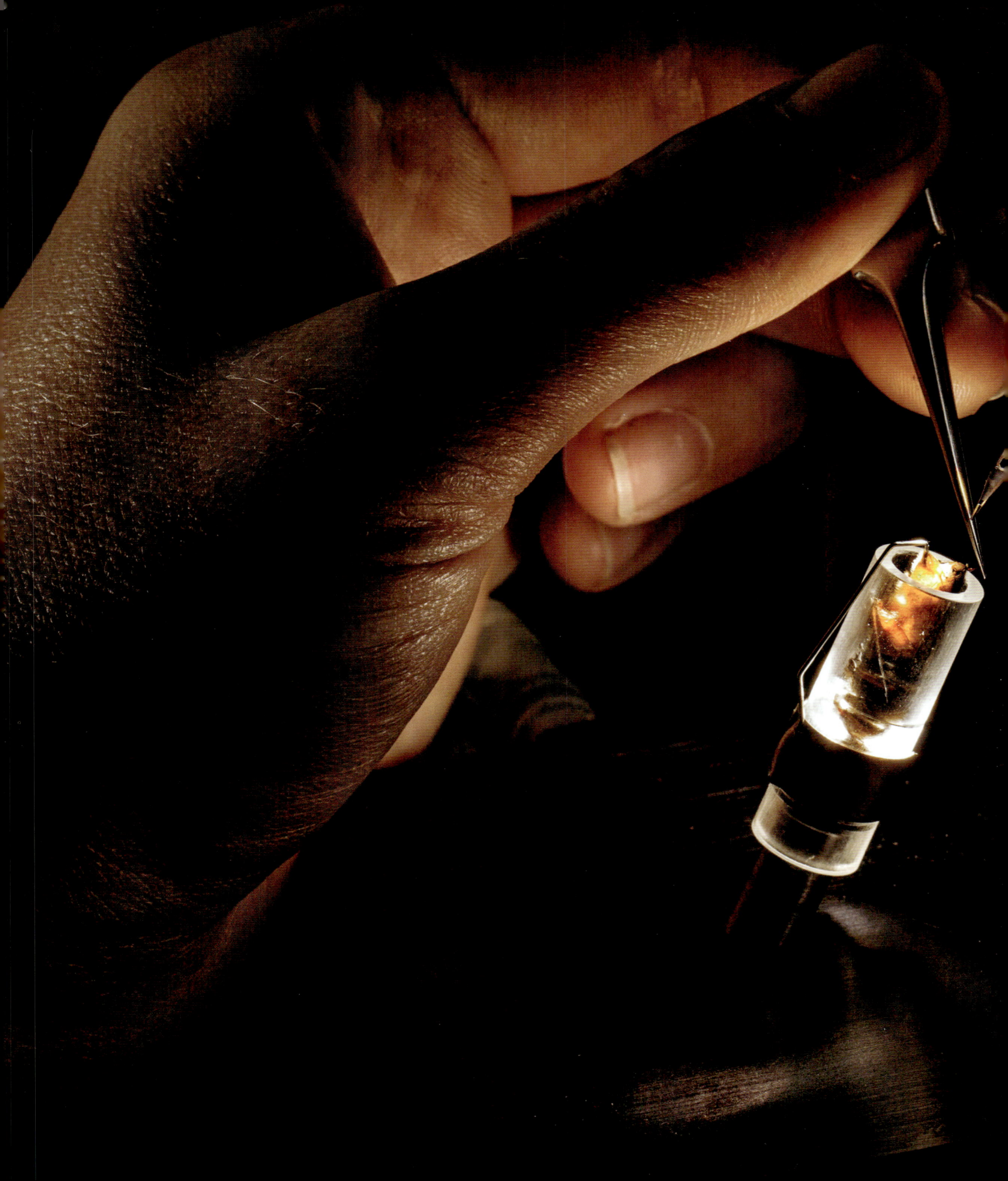

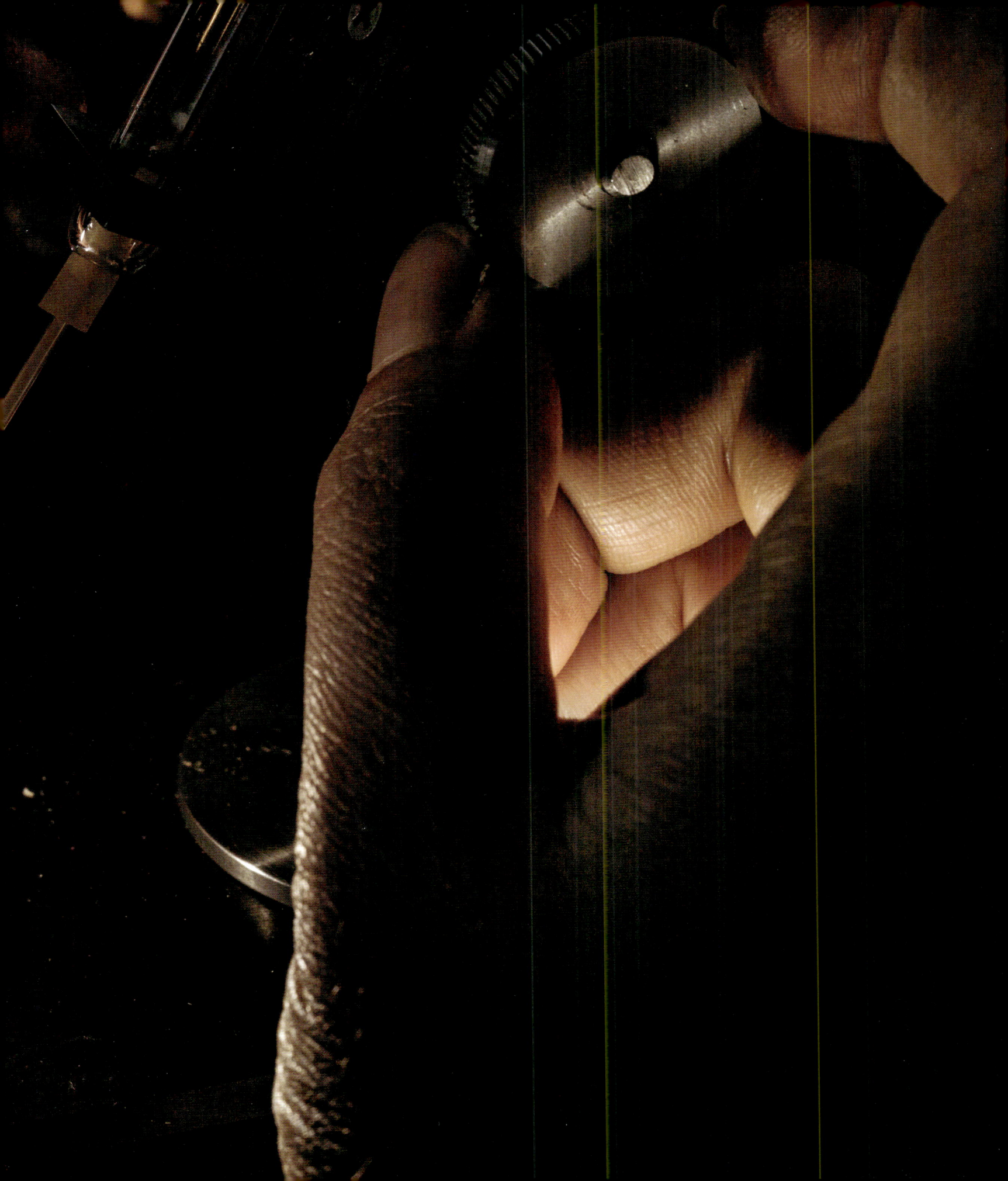

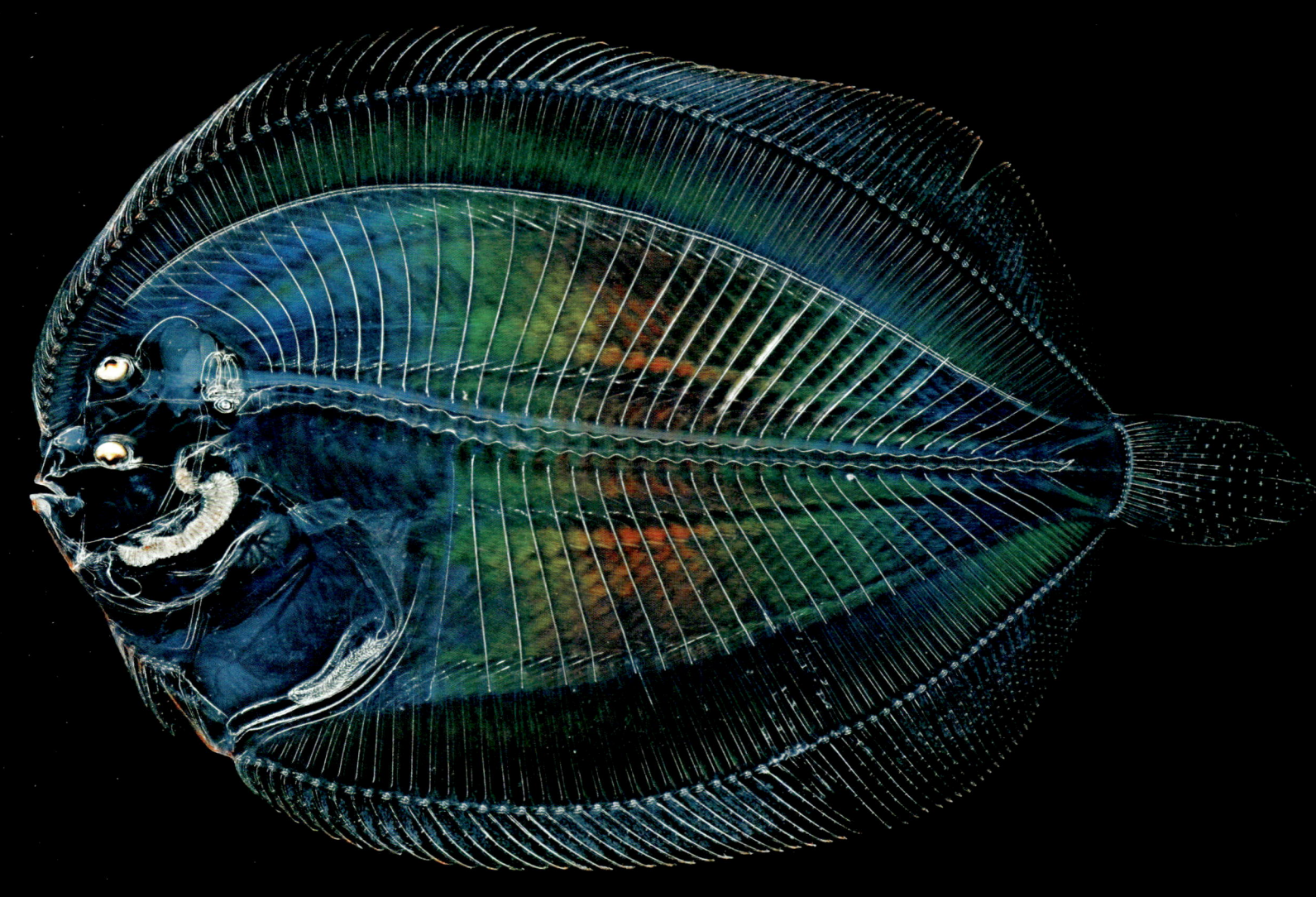

INTRODUCTION

As a kid, I dreamed of becoming a marine biologist and living my life by the sea. Since I grew up in a landlocked suburb of Atlanta, I lived out this fantasy by setting up aquariums at home. At 14, I started working at my neighborhood aquarium shop. By 16, I had seven fish tanks spread throughout my house. Then, at 20, I was introduced to photographer David Liittschwager, who hired me to help him with a *National Geographic* magazine assignment on marine life.

We spent 10 days aboard the *Oscar Elton Sette,* a 224-foot NOAA research vessel sailing off the Kona coast of Hawaii. David's assignment was to document the astounding biodiversity found at the surface of the ocean. My role was to collect the specimens for him to photograph. Every night, after the *Sette* had completed its scientific mission, I would cast a floating lamp off the port side of the ship. Like moths drawn to a flame, mysterious creatures would emerge from the depths in search of this light. Shimmering baby eels, tiny transparent crabs, sparkling squid. I'd carefully select an ambassador for each species and set up aquariums to house them as they waited for David to take their glamour shot.

Those nights aboard the *Sette* made me feel as if I were on another planet. At the end of each night, I'd sit and marvel at my dazzling collection. I had never imagined such bizarre life-forms could exist in our oceans. My eyes were glued to the tanks, my attention absorbed by these alien beings.

But I didn't grasp the true magic of what was in front of me until I saw the photographs David took of these creatures I had been collecting. The biggest surprise was his image of a baby flounder. I caught this fish by accident, scooping it up as I was chasing some more obvious target. Only later, when I inspected the contents of my collecting jar, did I notice its two tiny eyeballs staring back at me. The only other feature I could discern was the faintly wriggling outline of its transparent body.

But David's photograph of this flounder revealed a universe of detail that even my own eager eyes had missed. His macro lens magnified its delicately articulated ribs. The lightning-fast exposure froze its motion, providing a crisply

OPPOSITE:
David Liittschwager
A young flounder
caught off the coast
of Kona, Hawaii

defined view. A precisely aimed light released the rainbow hidden in the flounder's skin. And the black background eliminated all distractions to focus our attention on the quiet beauty at hand. Size, time, light, focus. These are the features of David's photograph that expanded my understanding of that little creature—and these are the tools that photography gives us to witness the invisible wonders of our world.

In this book, I have curated images to illustrate these concepts of size, time, light, and focus. In chapter 1, photographs transport us into tiny realms, immersing us in the details of feathers, crystals, fungi, and more. Chapter 2 demonstrates how cameras compress and expand time to reveal concealed patterns of motion. Chapter 3 teaches us to learn the language of light. Chapter 4 trains us to narrow our focus and search for those secrets hidden in plain sight. Not every photograph in this book shows something invisible in the sense of being too small or too quick for our eyes to see. But every image reveals something we would not have noticed without a photographer thoughtfully directing the flow of our attention.

Years after that project in Hawaii, I was snorkeling at night on a shallow reef in French Polynesia. Out of the darkness, another baby flounder emerged and settled on my mask. This time, I knew what to look for. I angled my flashlight toward the little fish in front of me and saw with my own eyes the shimmering colors and delicate bones that David's image had first revealed. Before working for David, I had assumed the goal of photography was simply to reproduce an observation so that others could share the same experience. It had never occurred to me that photography could expand our visual perception and thereby teach us to see the world anew. I learned that by revealing beauty in unexpected places, a compelling photograph can change what we pay attention to and what we choose to value. It reminds us of how much astonishing complexity in our lives goes by unnoticed. And it hints at how much surprise and delight there is yet to discover.

David's portrait of a humble flounder expanded my universe. I hope you find a photograph in this book that expands yours. ∎

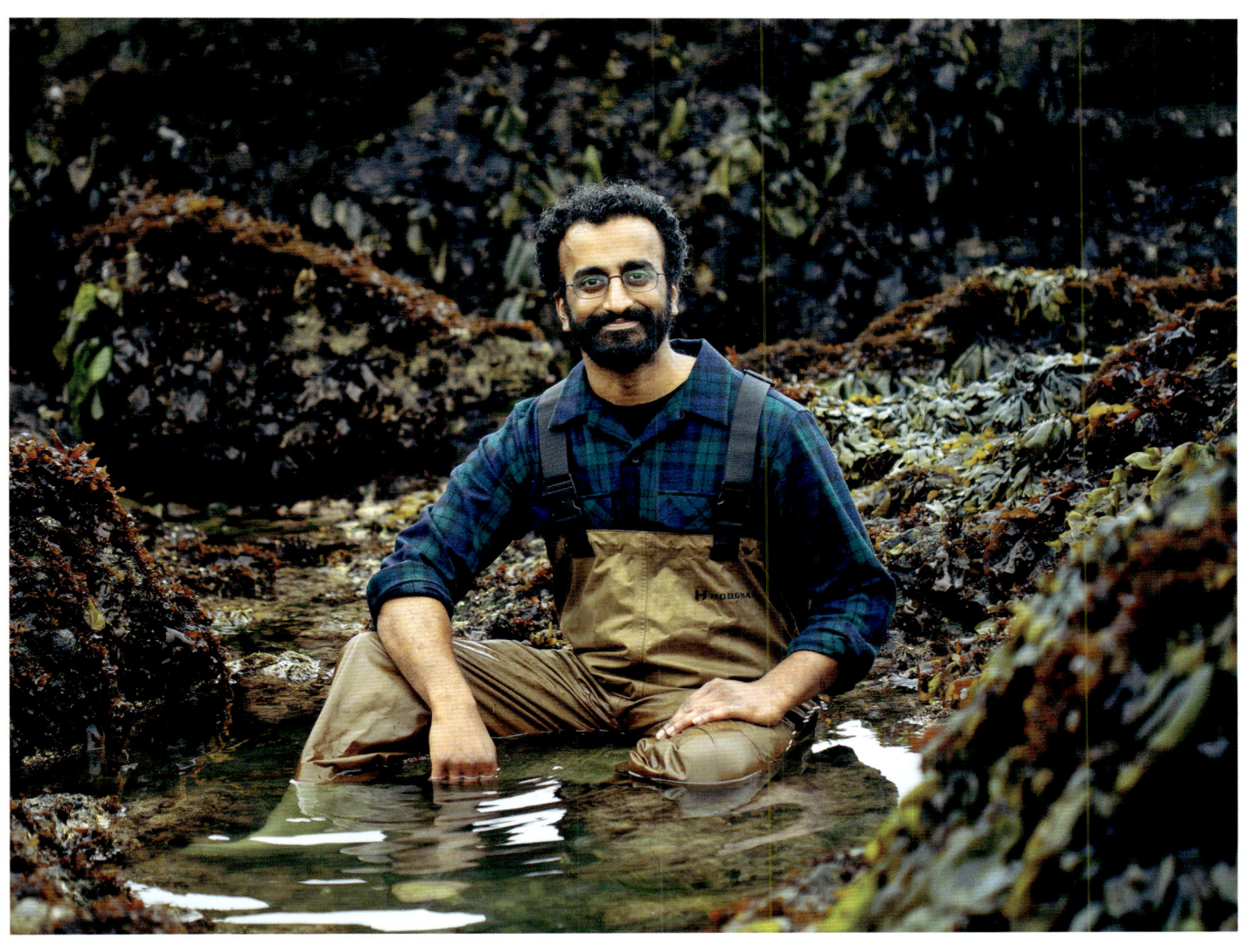

Vijay Varma
Anand Varma exploring one of his favorite places, the tide pools of Pigeon Point, California

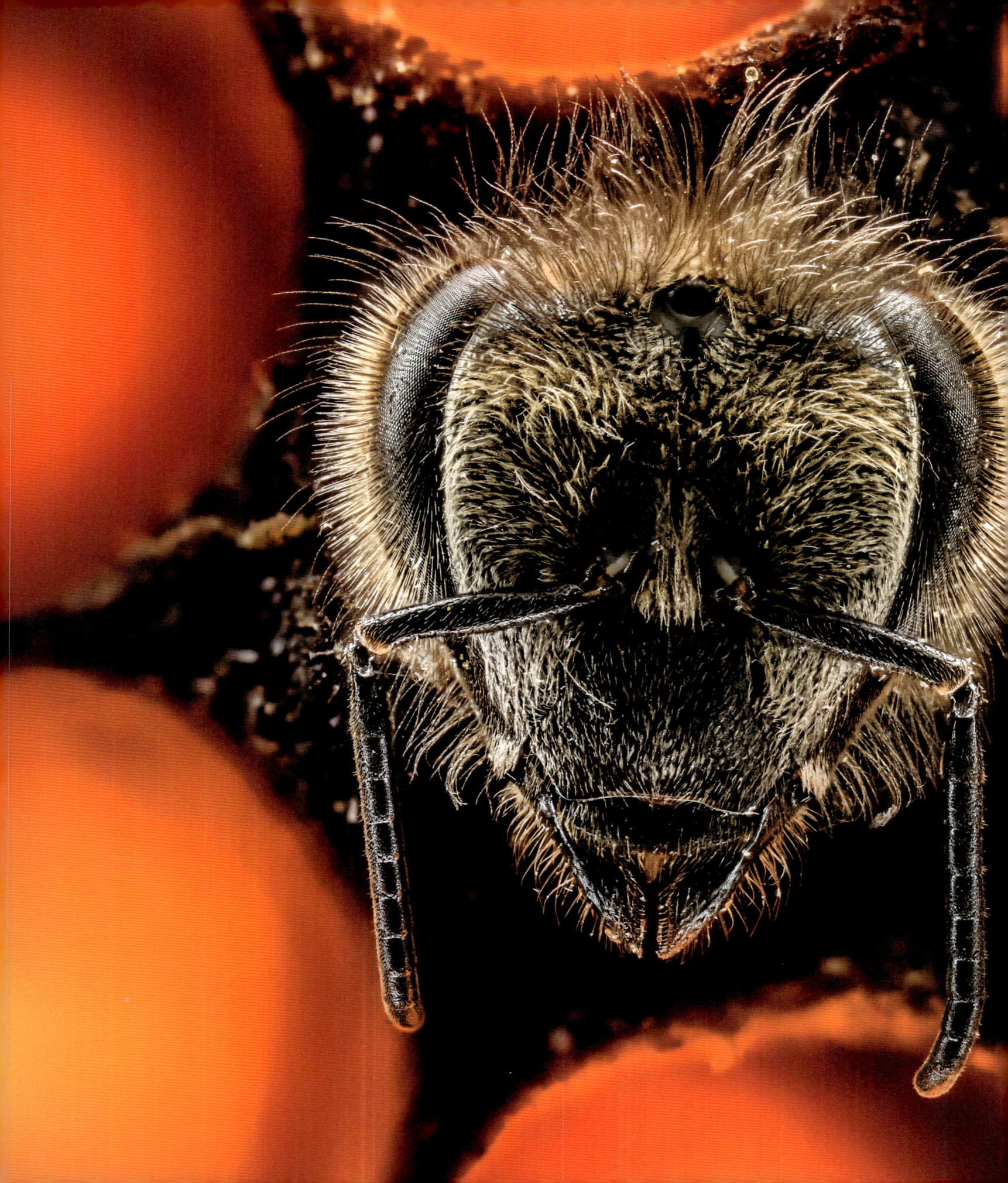

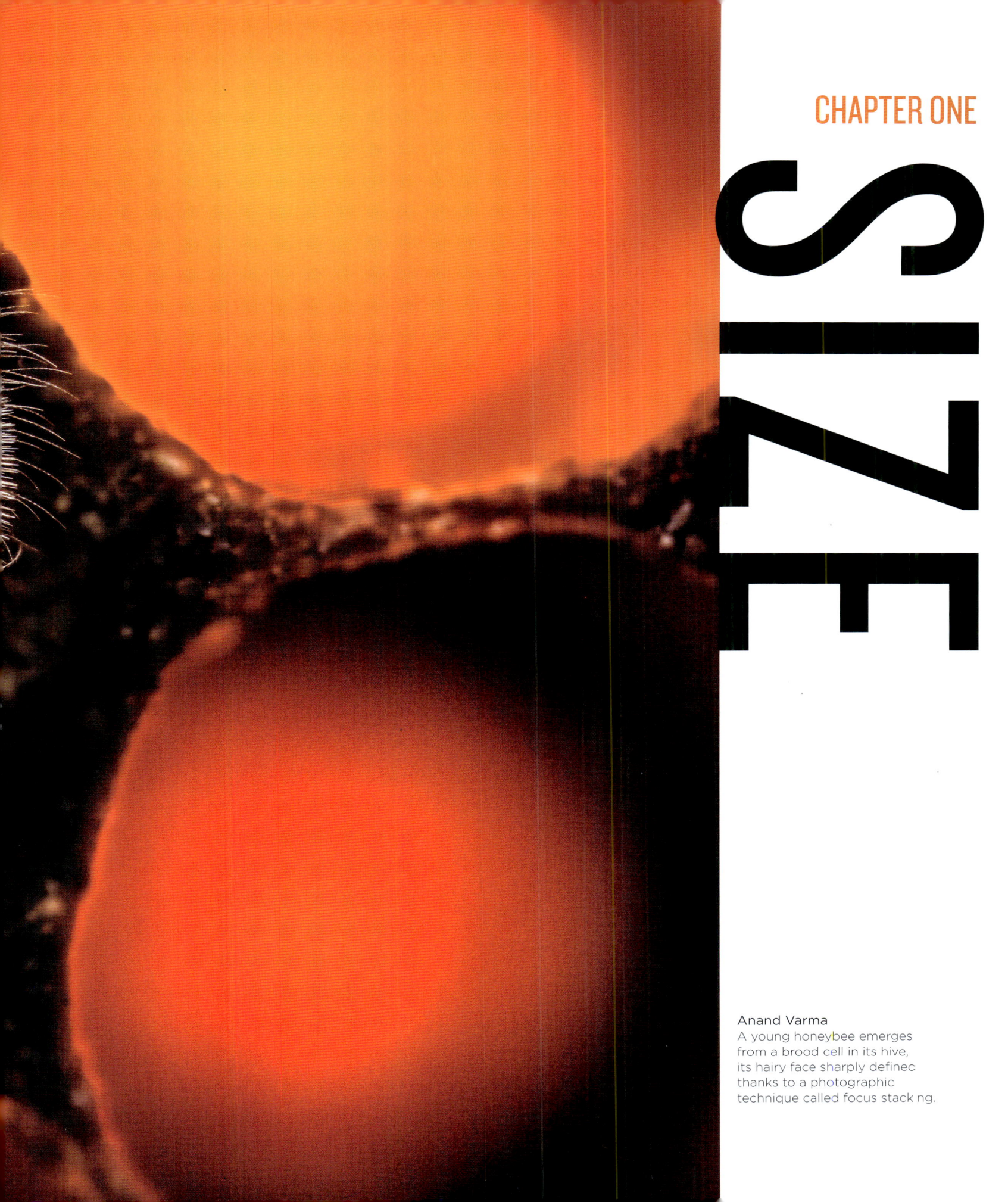

CHAPTER ONE

SIZE

Anand Varma
A young honeybee emerges
from a brood cell in its hive,
its hairy face sharply definec
thanks to a photographic
technique called focus stack ng.

WHEN EDITORS at *National Geographic* magazine asked me to

photograph a story on honeybees, I was not excited to take on the assignment. Honeybees had already garnered so much media attention that I couldn't imagine how I could contribute anything new. But as an insecure young photographer, how could I say no to *National Geographic*? So I faked some enthusiasm and landed a job I had no idea how to accomplish.

I started by learning how to keep bees in my backyard in Berkeley, California. I hoped that with enough time to study them, I might discover some new way of portraying their lives. I tried a variety of approaches, photographing them in orchards, in flowers, in petri dishes. I tried specialty lighting techniques I had developed on previous projects. But the results never quite hit the mark and often just mimicked images that had already been taken.

One day I noticed something out of place. A young bee had gotten trapped as it was emerging from the hive. Using a trick I'd learned from an entomologist, I plucked one of my eyelashes and used its thin, flexible tip to brush a bit of debris away from the bee's face. I positioned a light behind the hive to make the surrounding wax glow. The bee's position allowed me to bring my camera up close and capture features on its head I had never noticed before, like its jointed antennae and furry face. Despite having spent a year tending to my hive, I had never experienced a bee this way before. Once it was brought eye to eye, the intimacy inspired new questions. How does this bee perceive its environment? What do I look like to this creature? Why all the hair?

We tend to relate best to beings our own size, or at least to creatures we can see with an unaided eye. Photography can break this barrier by magnifying the subject and creating perspectives we aren't able to experience on our own. The next time we peer into a flower, we can imagine the structure of the pollen hidden inside, thanks to Martin Oeggerli's brilliant micrograph (page 4). Alison Pollack brings us into the miniature forest growing on the surface of a leaf with her photograph of an iridescent slime mold (page 34). Javier Aznar's frog portraits (pages 44 and 45) are far from microscopic, but with proximity, he inspires an unexpected affection for these comical creatures. Even familiar objects are given new life with a close-up lens, as in Roni Hendrawan's image that reveals the intricate cracks in a grain of rice (page 22).

WE TEND TO RELATE BEST TO BEINGS OUR OWN SIZE. PHOTOGRAPHY CAN BREAK THIS BARRIER BY MAGNIFYING THE SUBJECT AND CREATING NEW PERSPECTIVES.

By enlarging these alien realms to a size we can comprehend, these photographs invite our eyes to linger and explore. Rich in texture and detail, they reward the curiosity of the careful observer. We are reminded that our world holds so many wonderful surprises if we learn to slow down and pay careful attention. ■

FOLLOWING PAGES:
Roni Hendrawan
The cracks and fissures of a single grain of rice become visible with the help of colored lights and focus stacking.

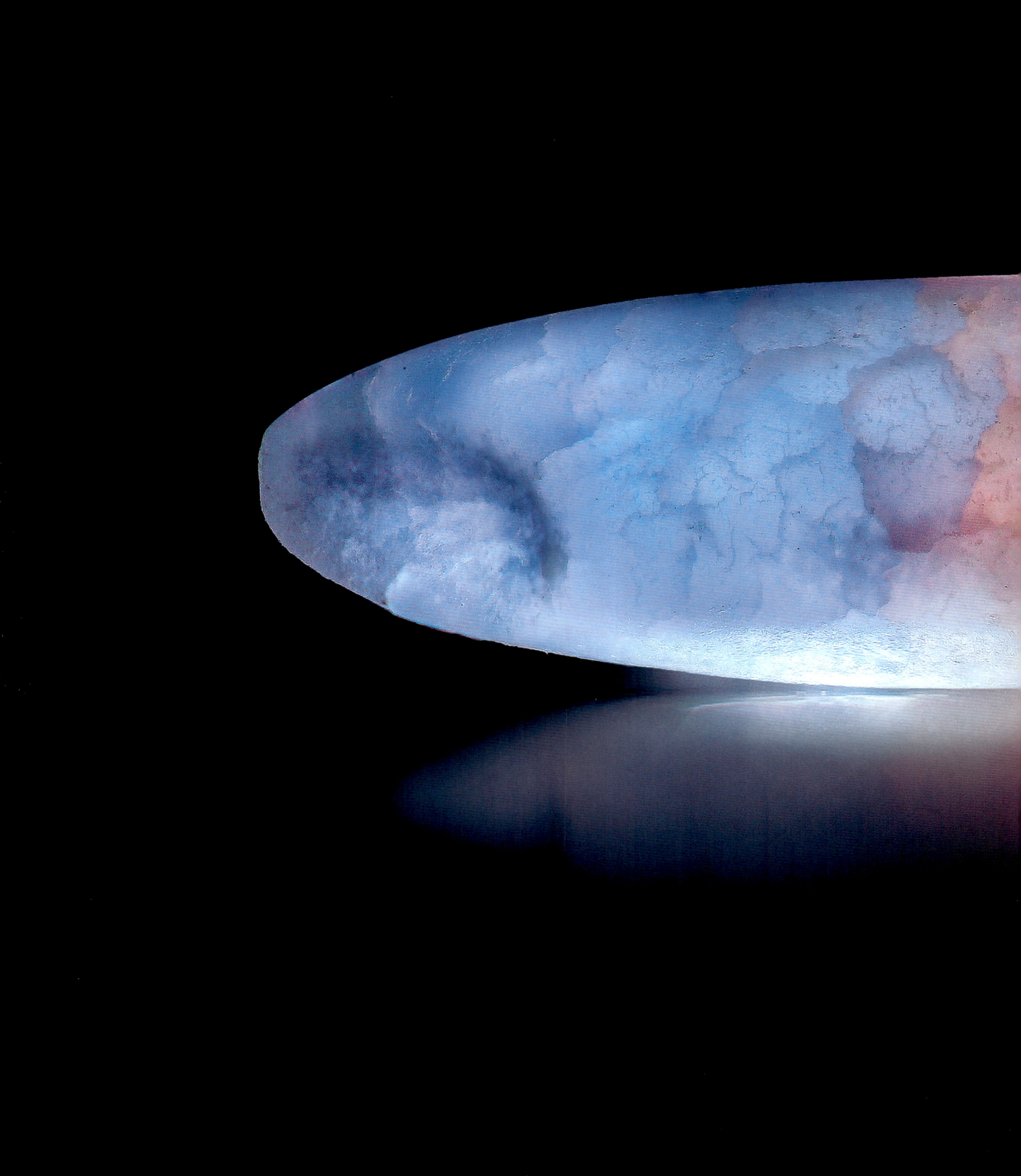

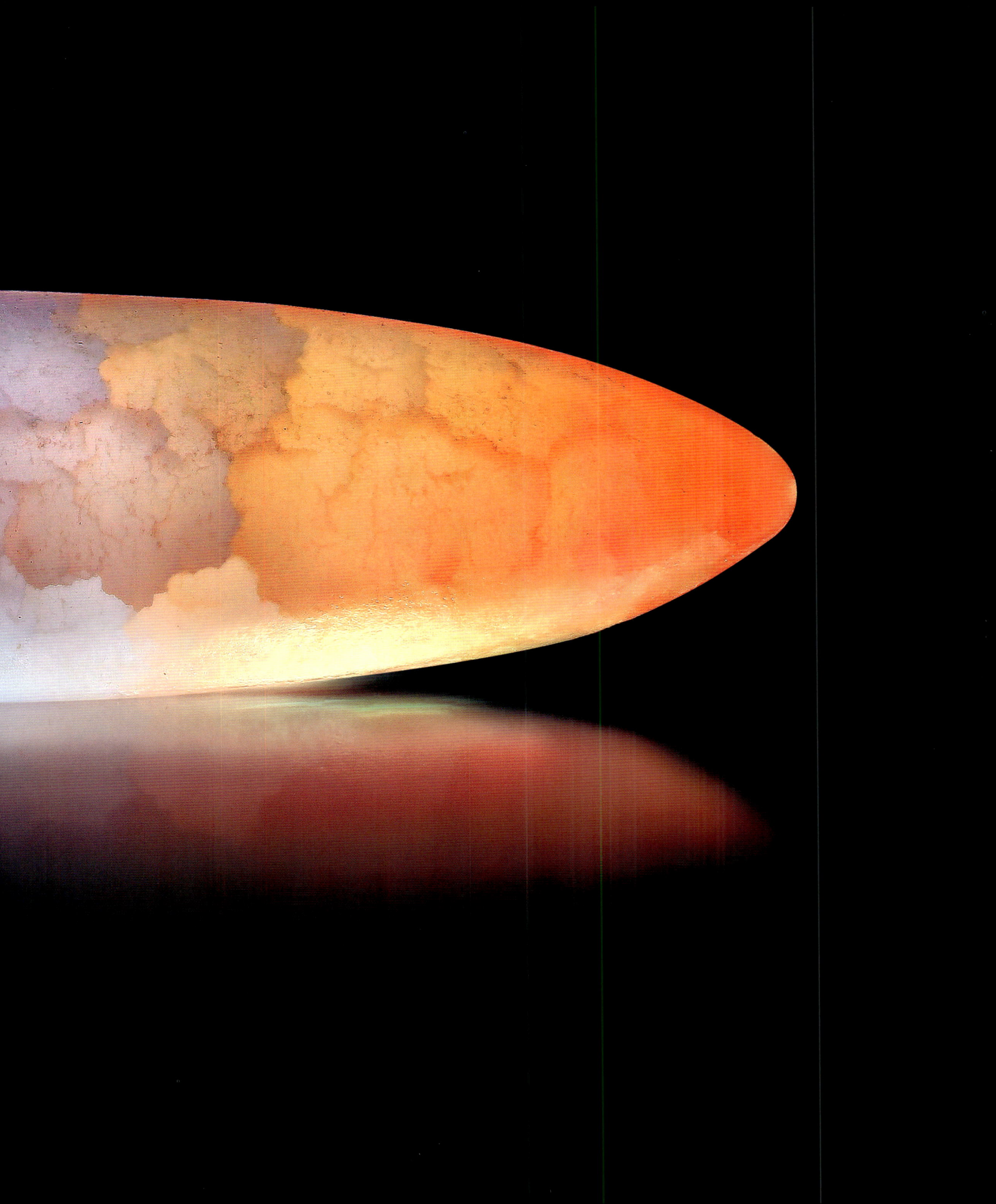

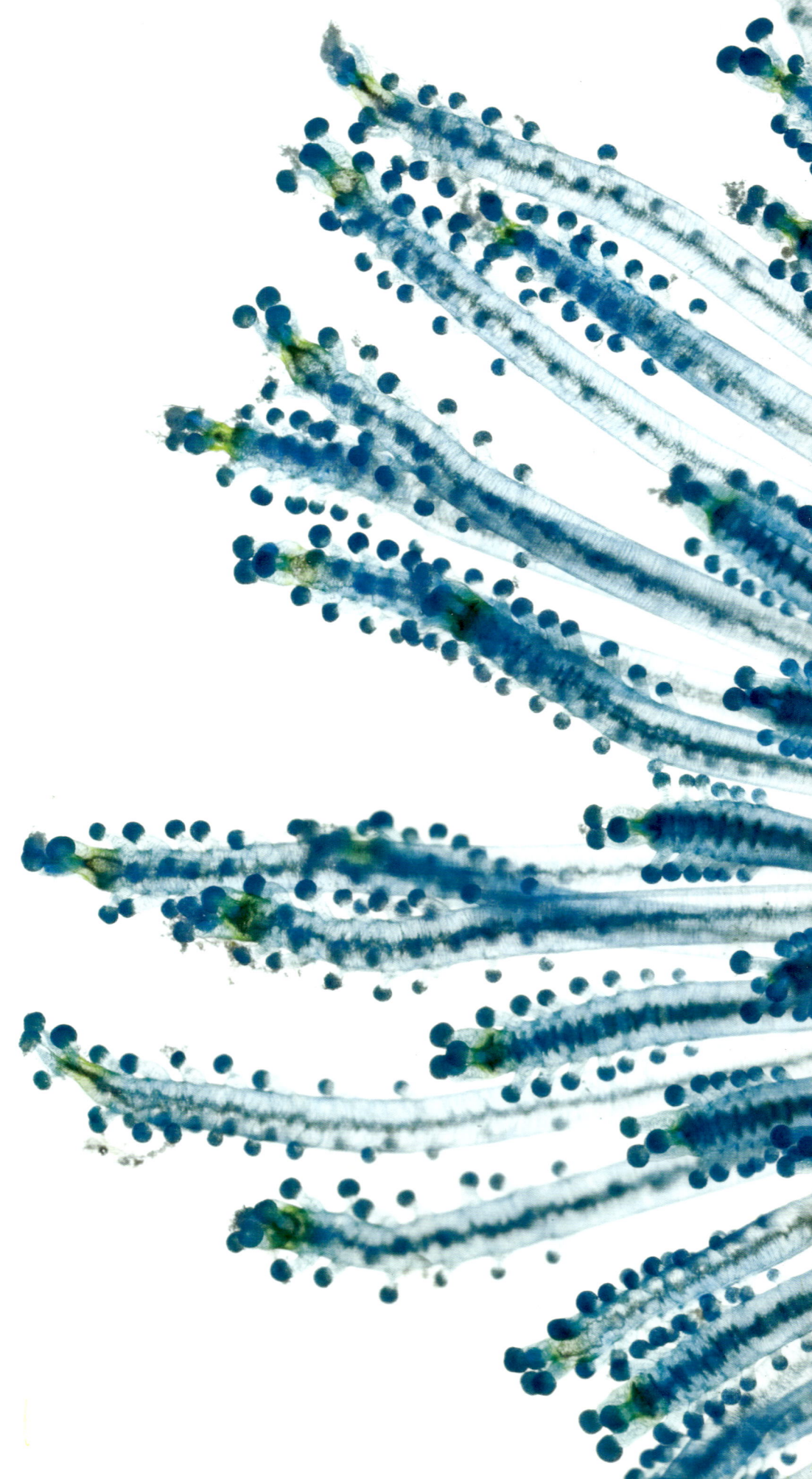

David Liittschwager
A blue button, a tiny relative of the jellyfish. This creature's blue pigment is thought to block ultraviolet rays.

FOLLOWING PAGES:
Angel Fitor
An underside view of the frilly arms of a barrel jellyfish

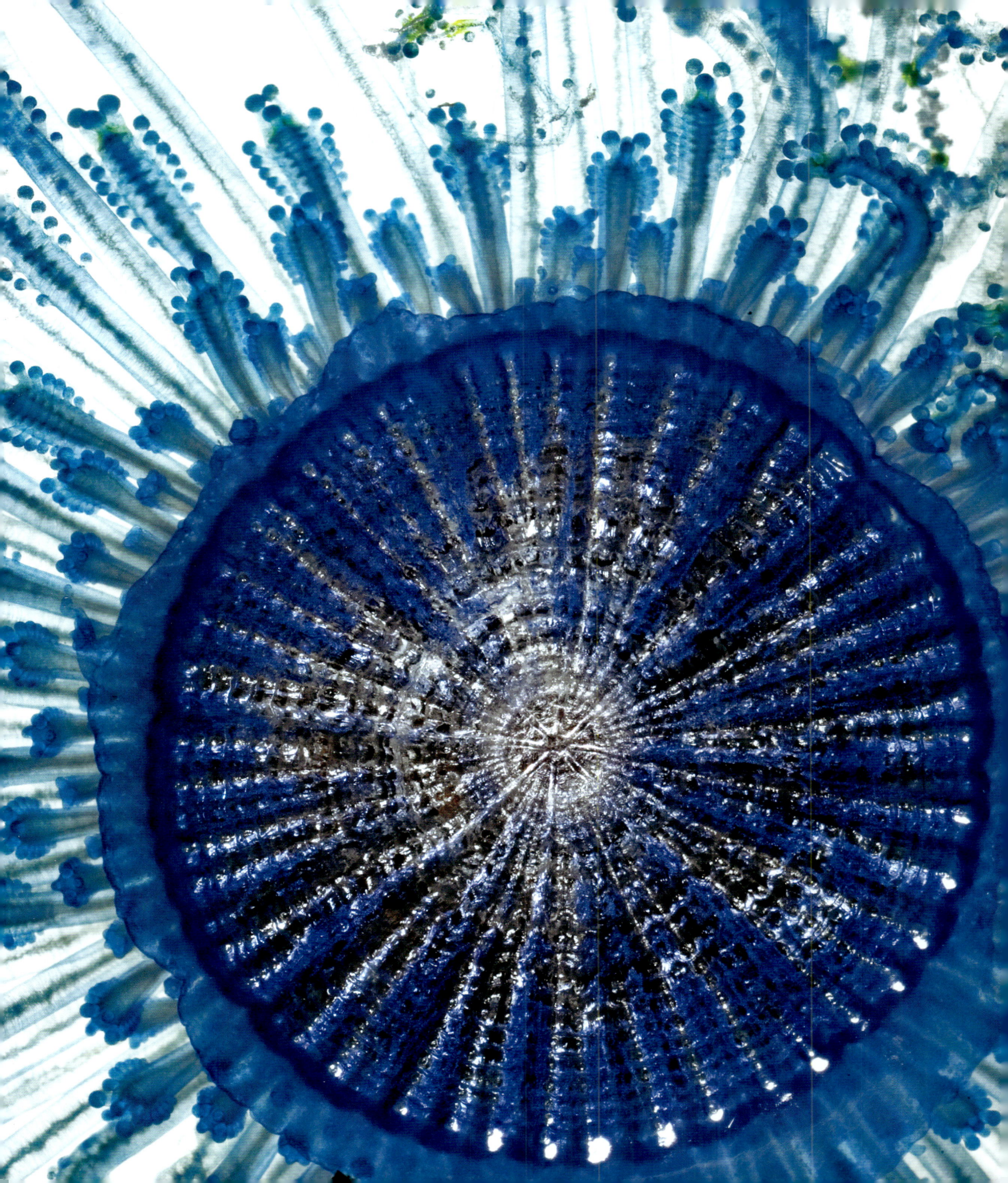

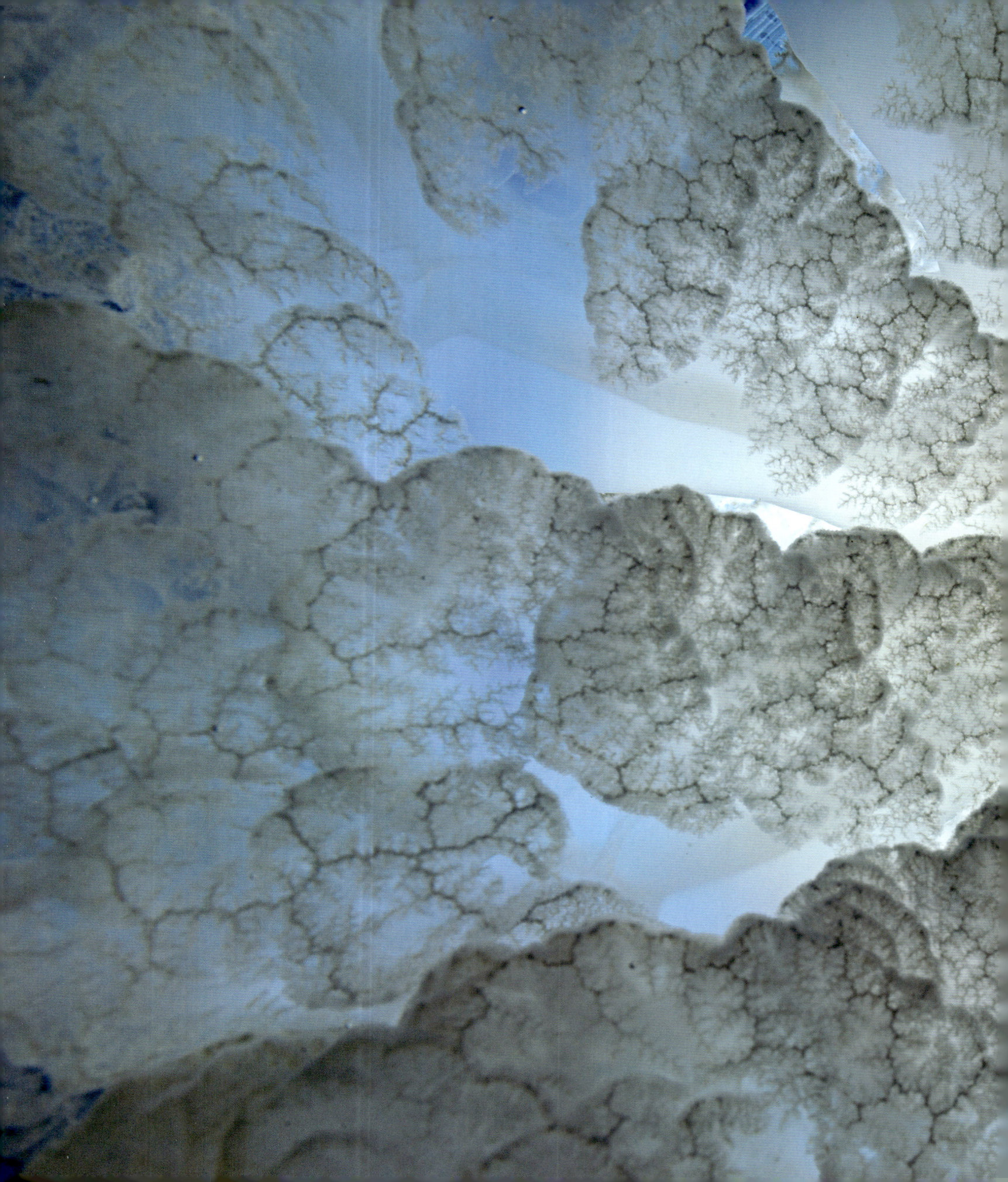

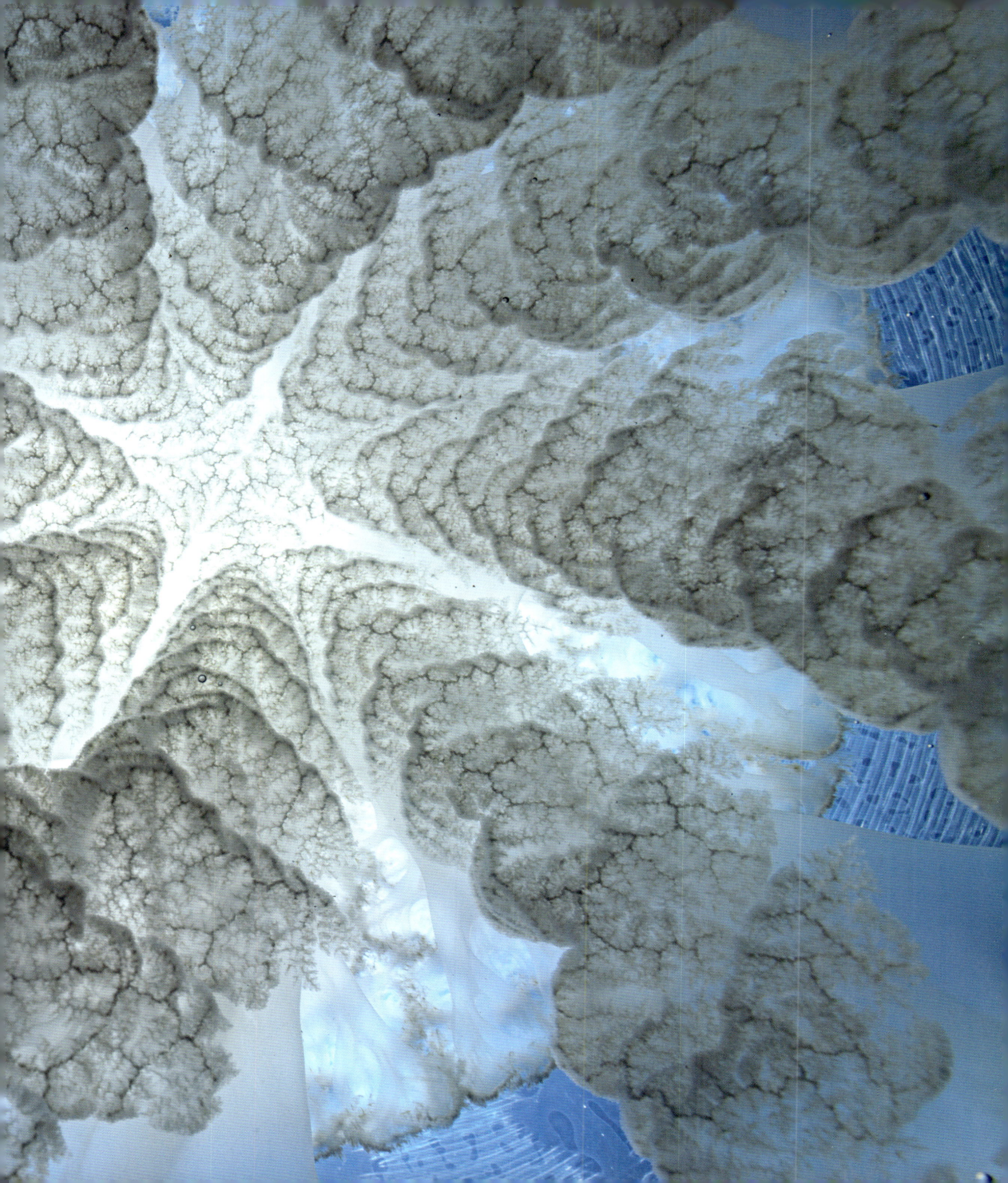

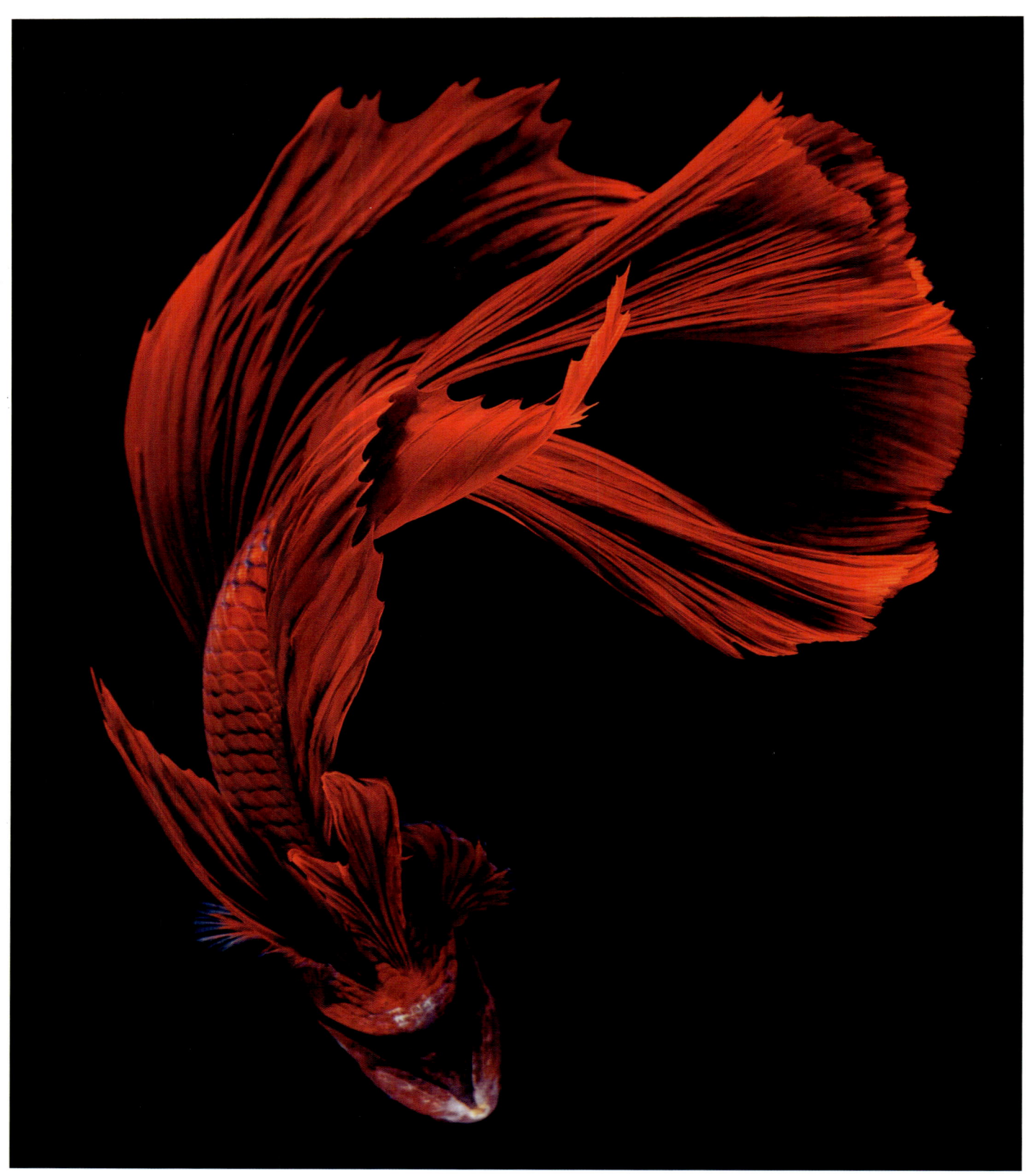

Visarute Angkatavanich
A bottom-up perspective of a betta, a popular aquarium fish prized for its vibrant color and delicately textured fins

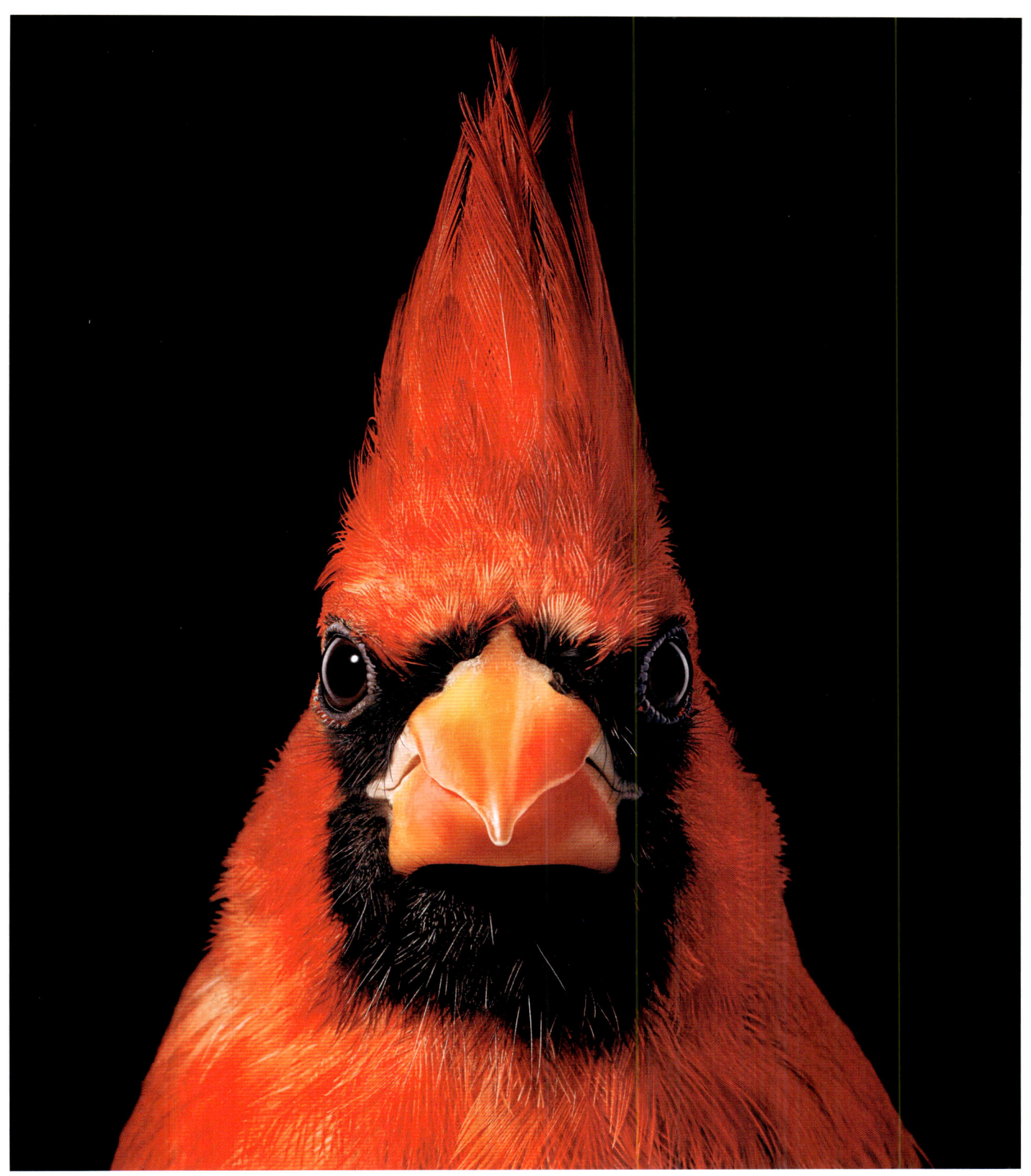

Tim Flach
A studio portrait of a male northern cardinal. Males use their vibrant red plumage to attract females.

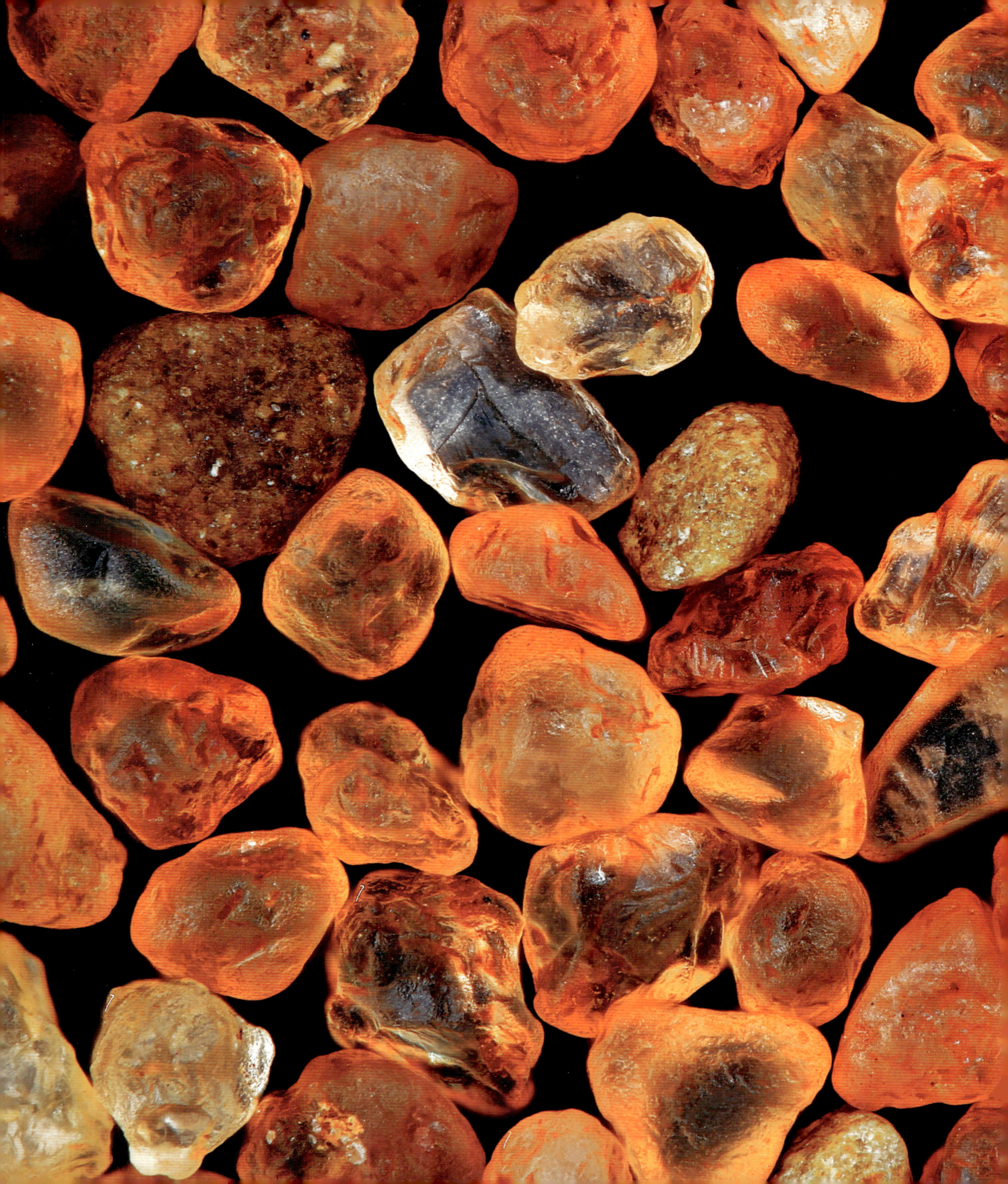

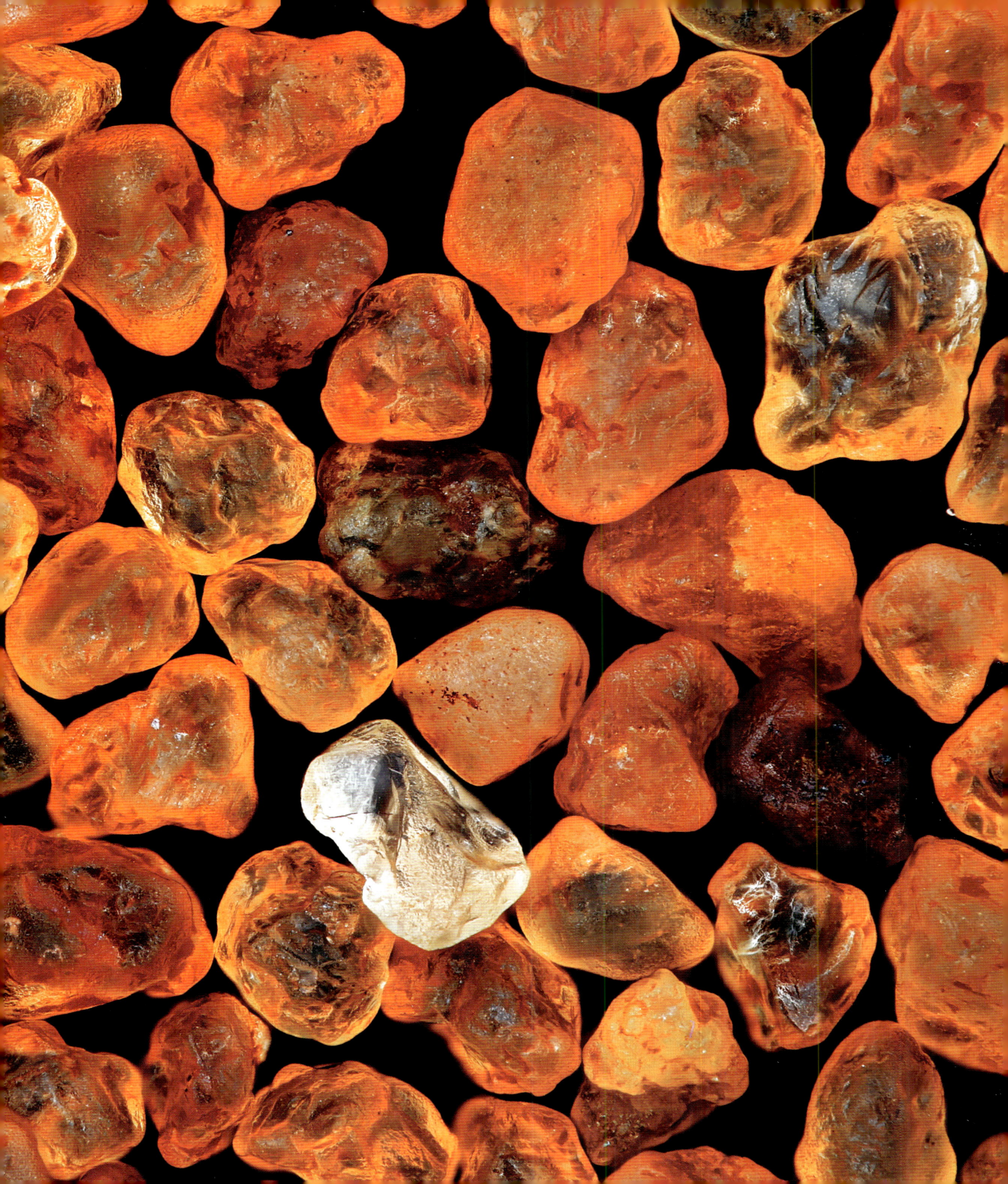

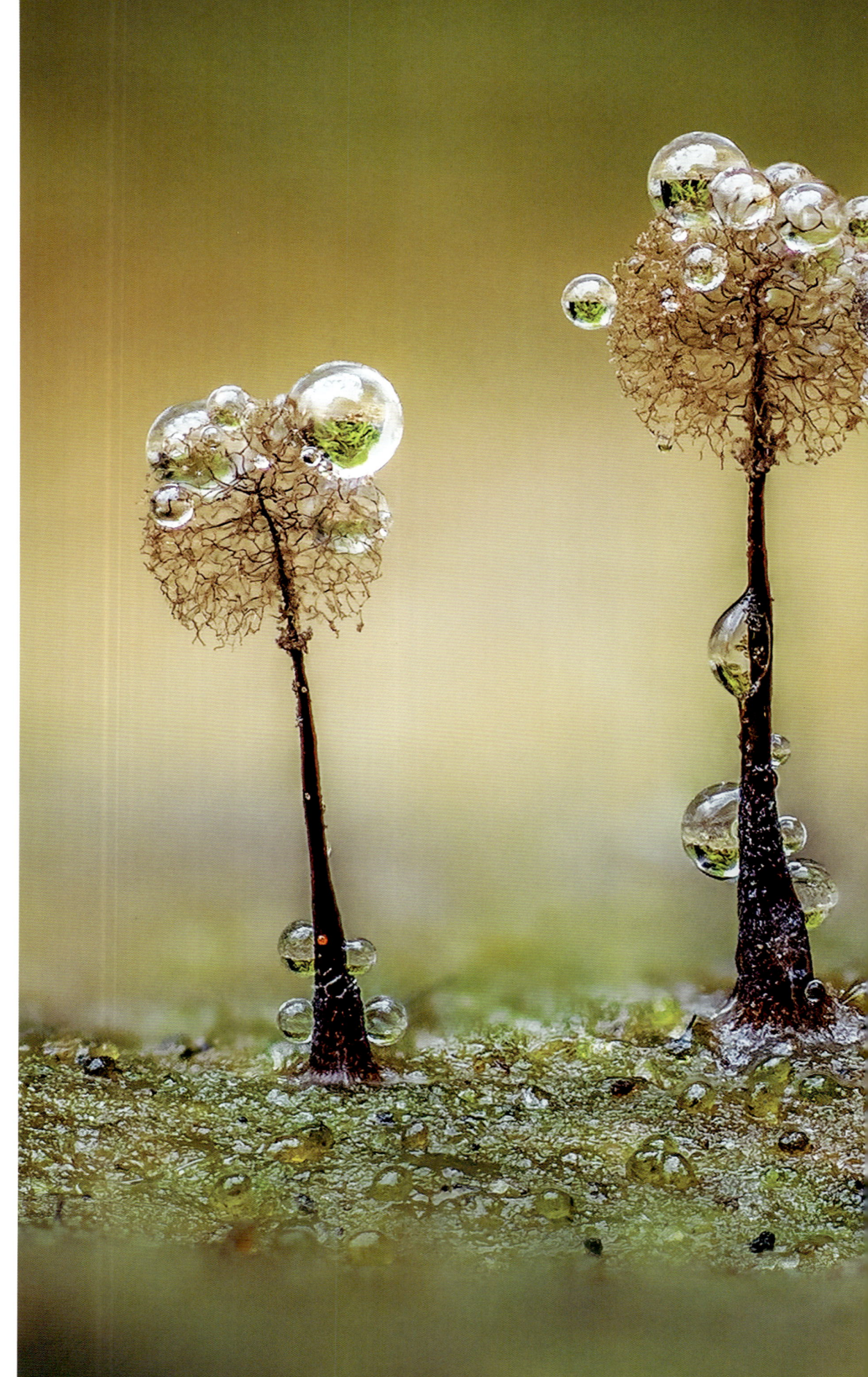

Barry Webb
The fruiting bodies of this slime mold extend just a few millimeters off the ground to release their spores.

PREVIOUS PAGES:
Xinpei Zhang
Focus stacking reveals the minute details of each grain of desert sand.

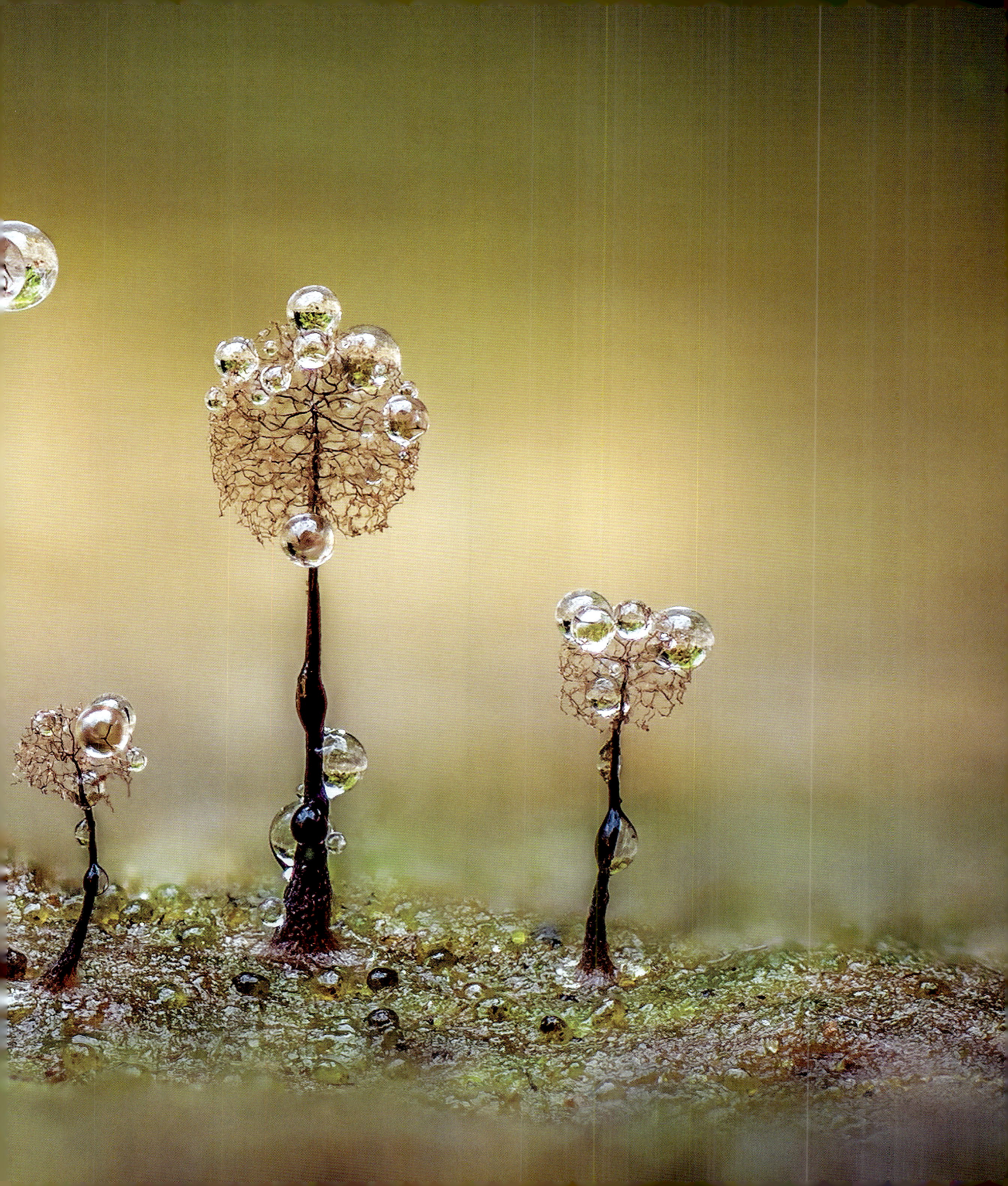

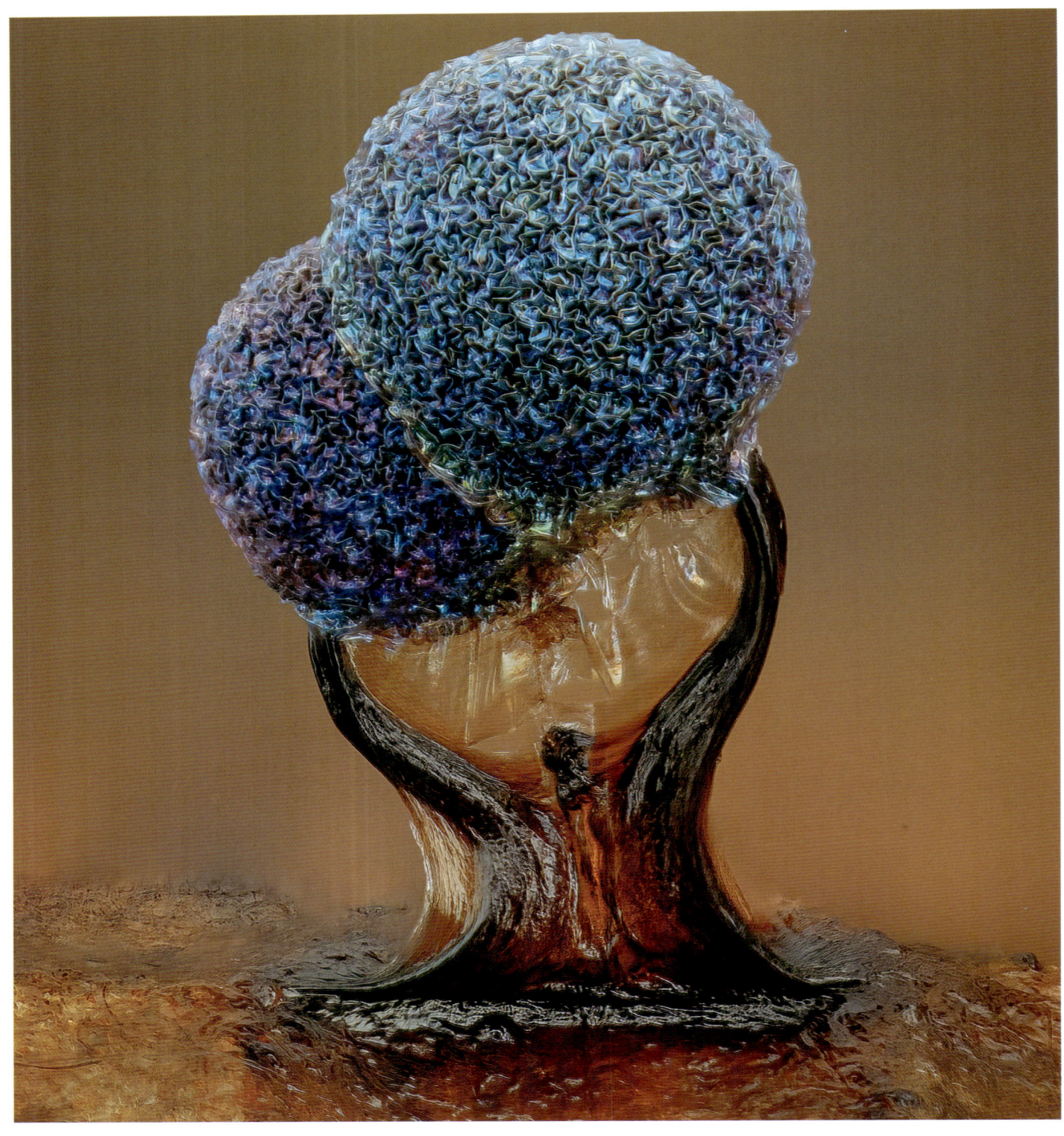

Alison Pollack
Two fruiting bodies of a slime mold stand less than a millimeter tall over a rotting leaf.

FOLLOWING PAGES:
Nathan Renfro
Geometric etching on the surface of a diamond

BY ENLARGING THESE ALIEN REALMS TO A SIZE
WE CAN COMPREHEND, THESE PHOTOGRAPHS
INVITE OUR EYES TO LINGER AND EXPLORE.
RICH IN TEXTURE AND DETAIL, THEY REWARD
THE CURIOSITY OF THE CAREFUL OBSERVER.

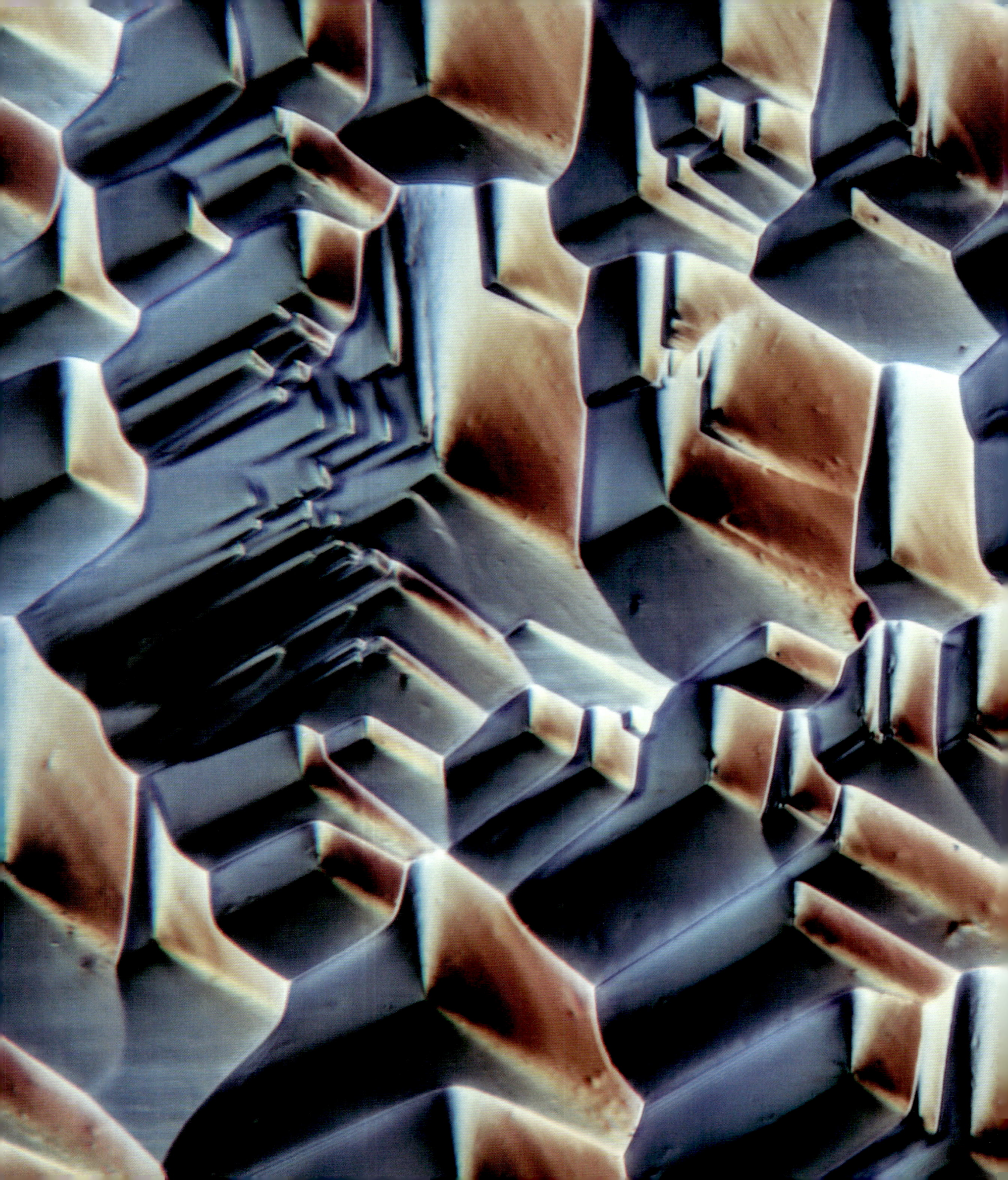

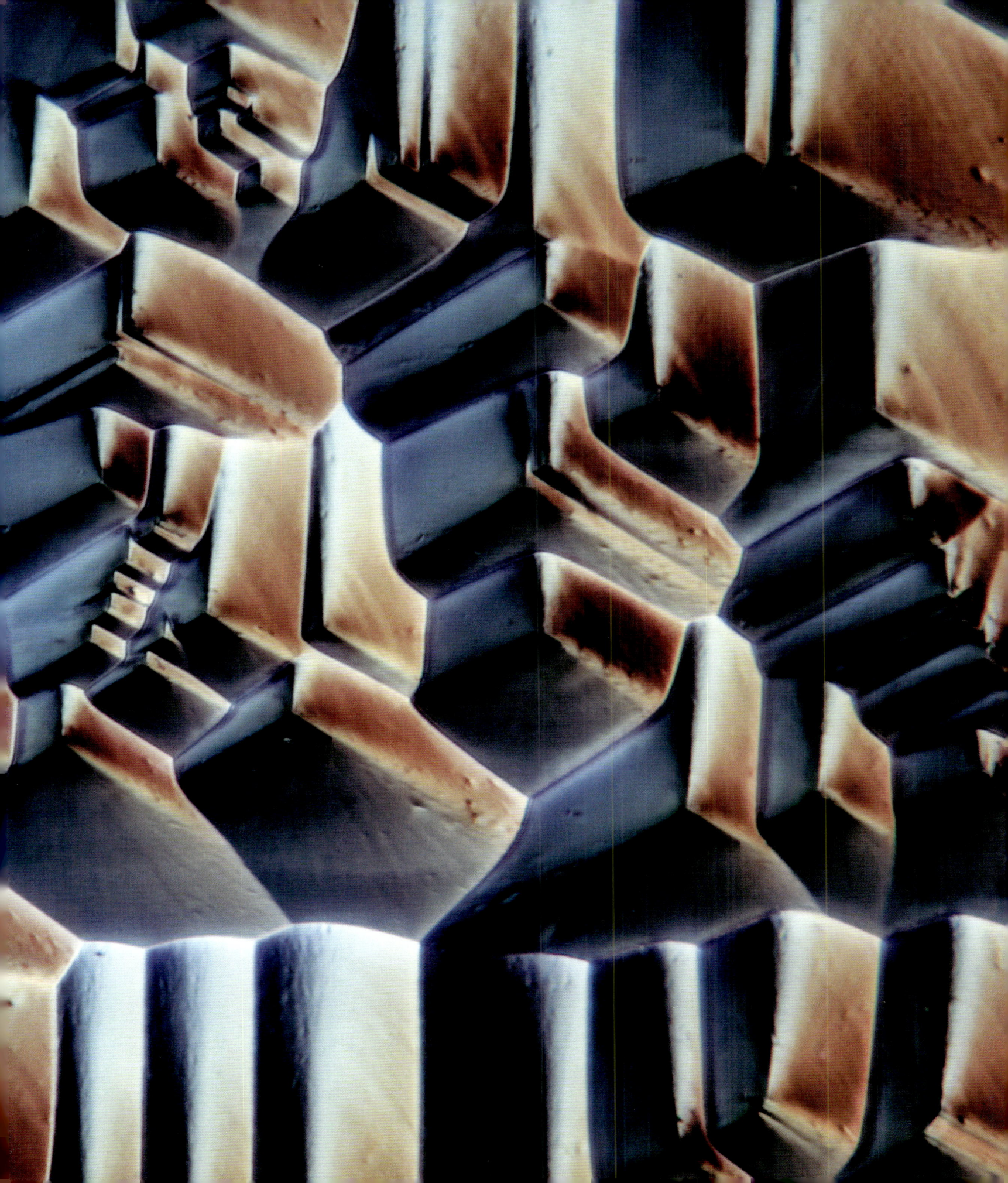

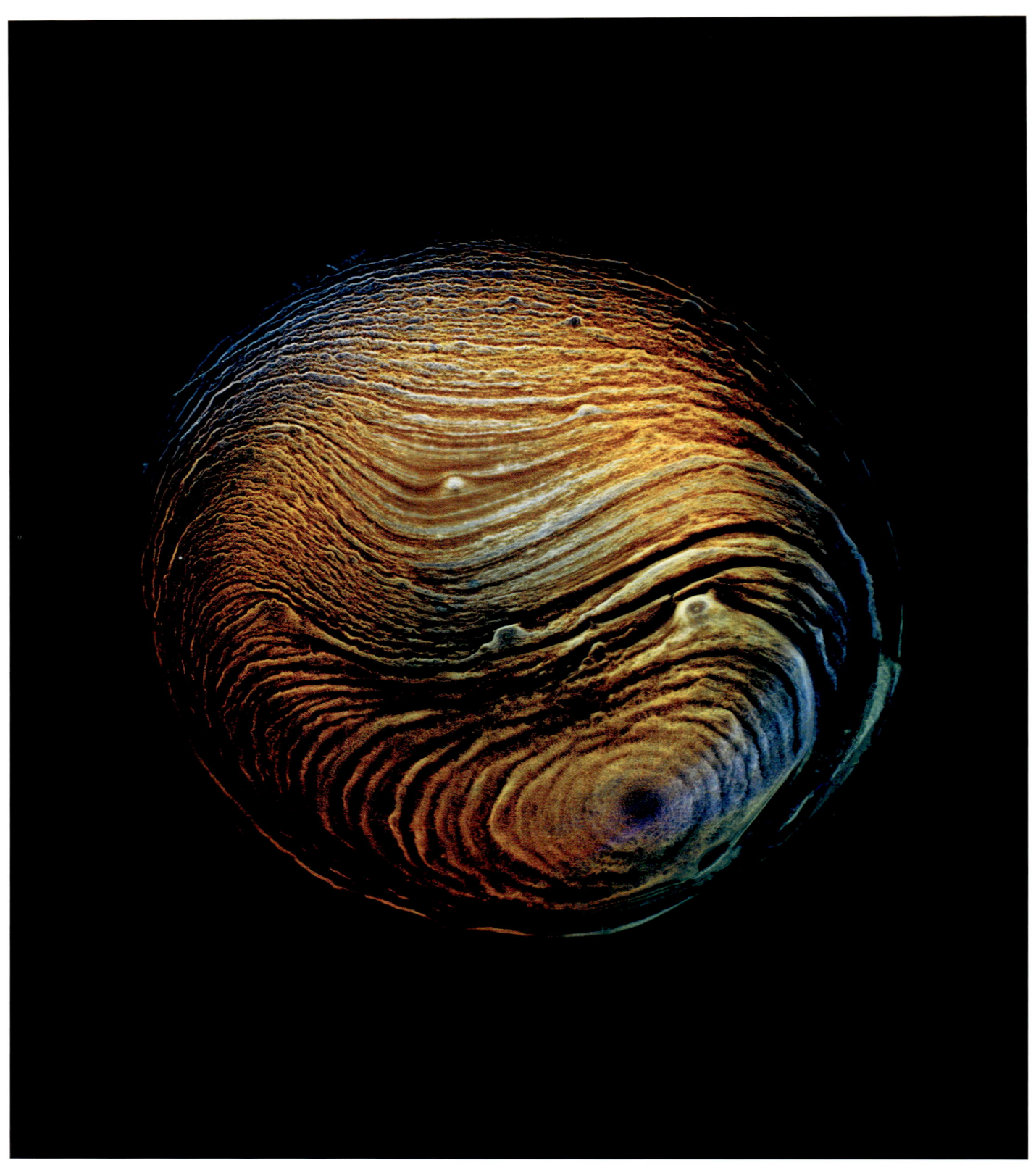

Ernie Button
Single malt Scotch leaves a layer of sediment behind when it dries in the bottom of a glass.
Each glass of whiskey produces a different pattern.

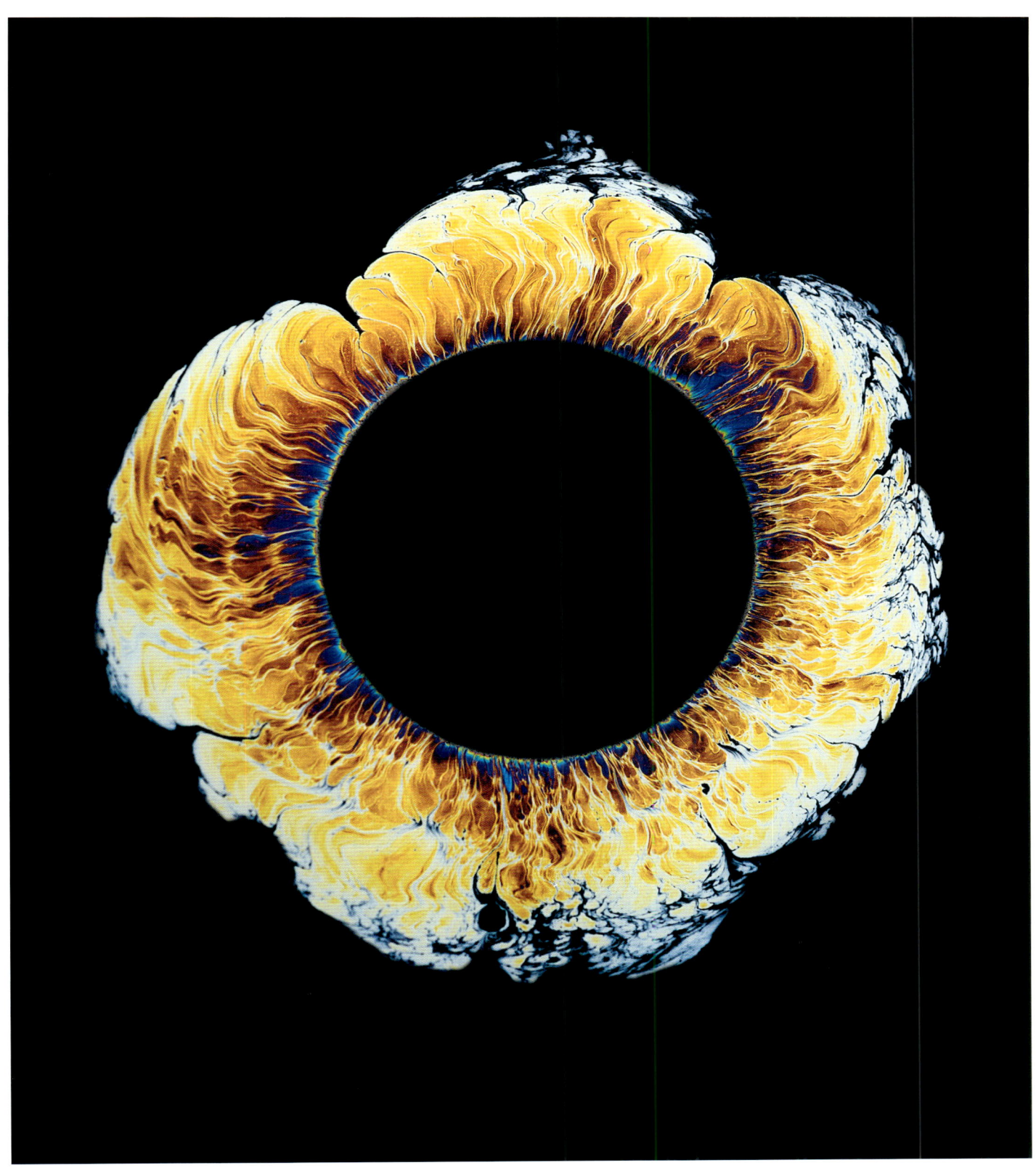

Fabian Oefner
A drop of oil expands across a pool of water.
The colors appear as light refracts through the thin film.

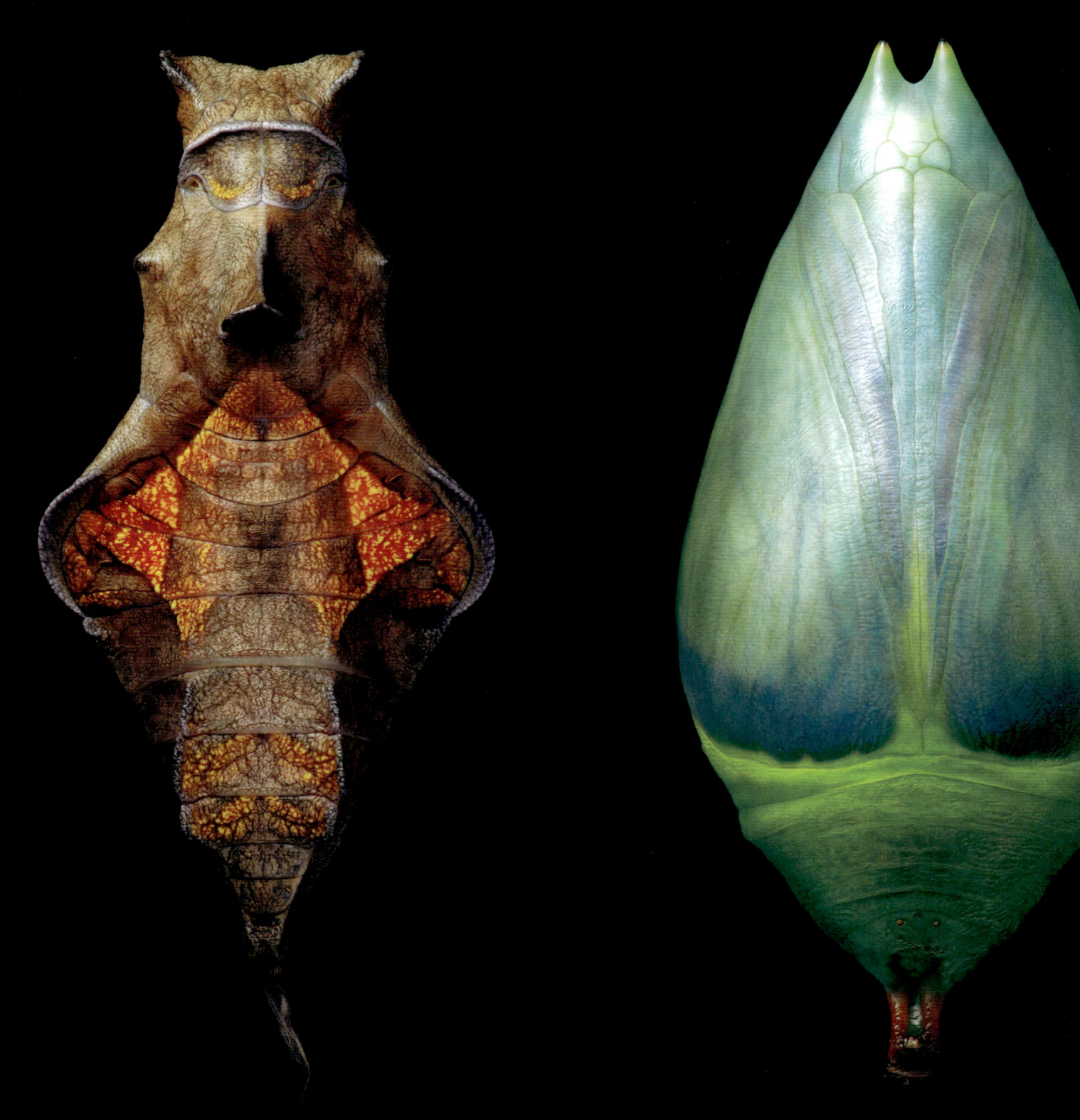

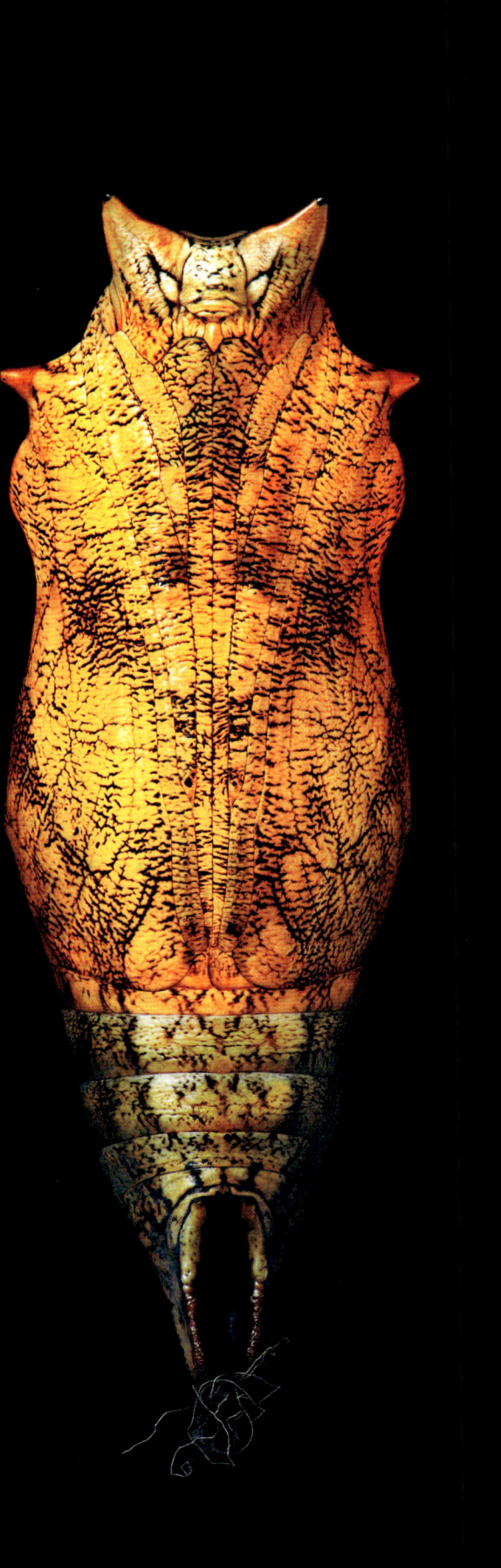

Tim Flach
From left to right: pupae
of a gold rim swallowtail,
a blue morpho, and a
mother-of-pearl butterfly

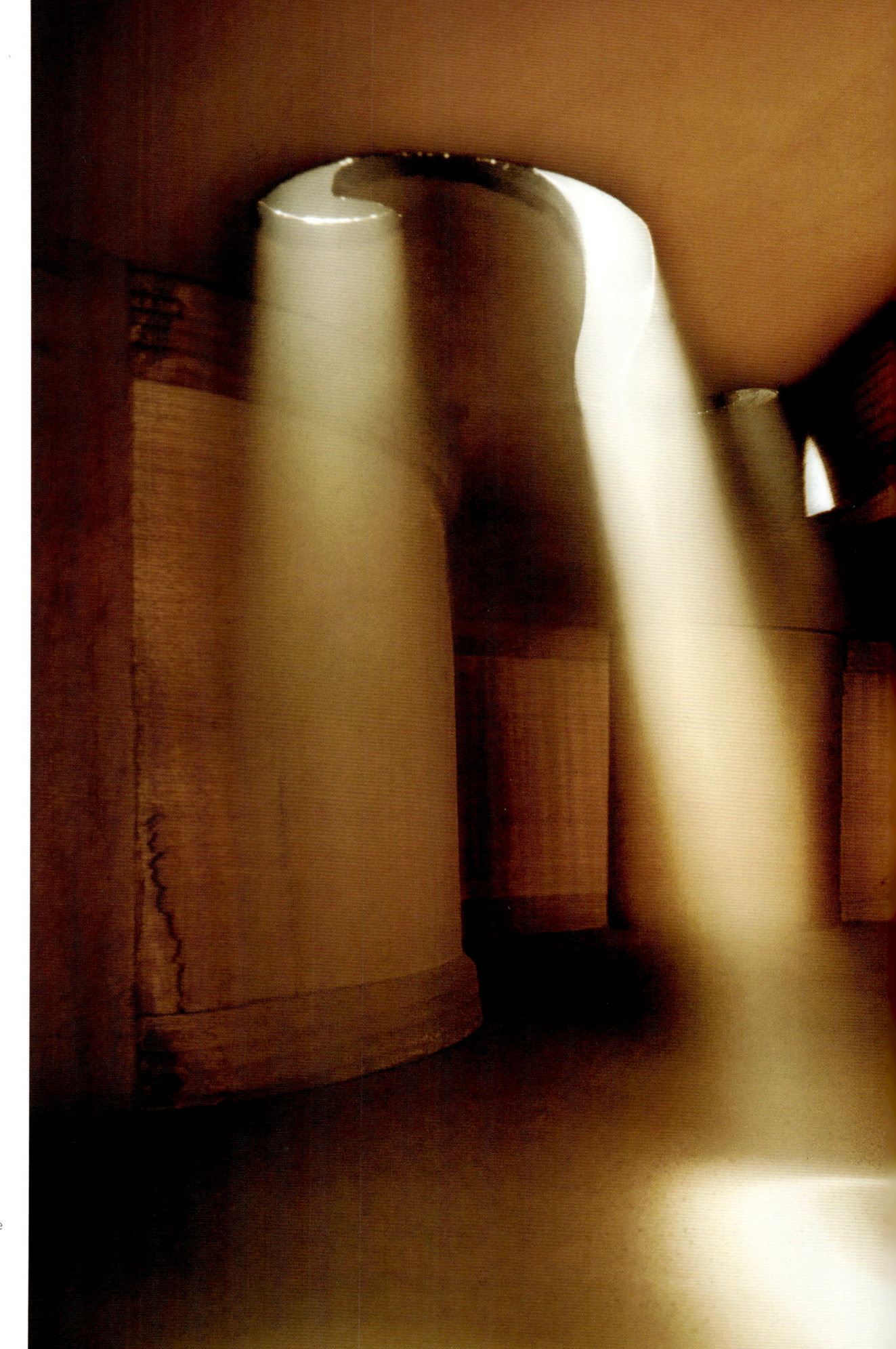

Andreas Mierswa and
Markus Kluska
Shafts of light illuminate
the inside of a bass viol.

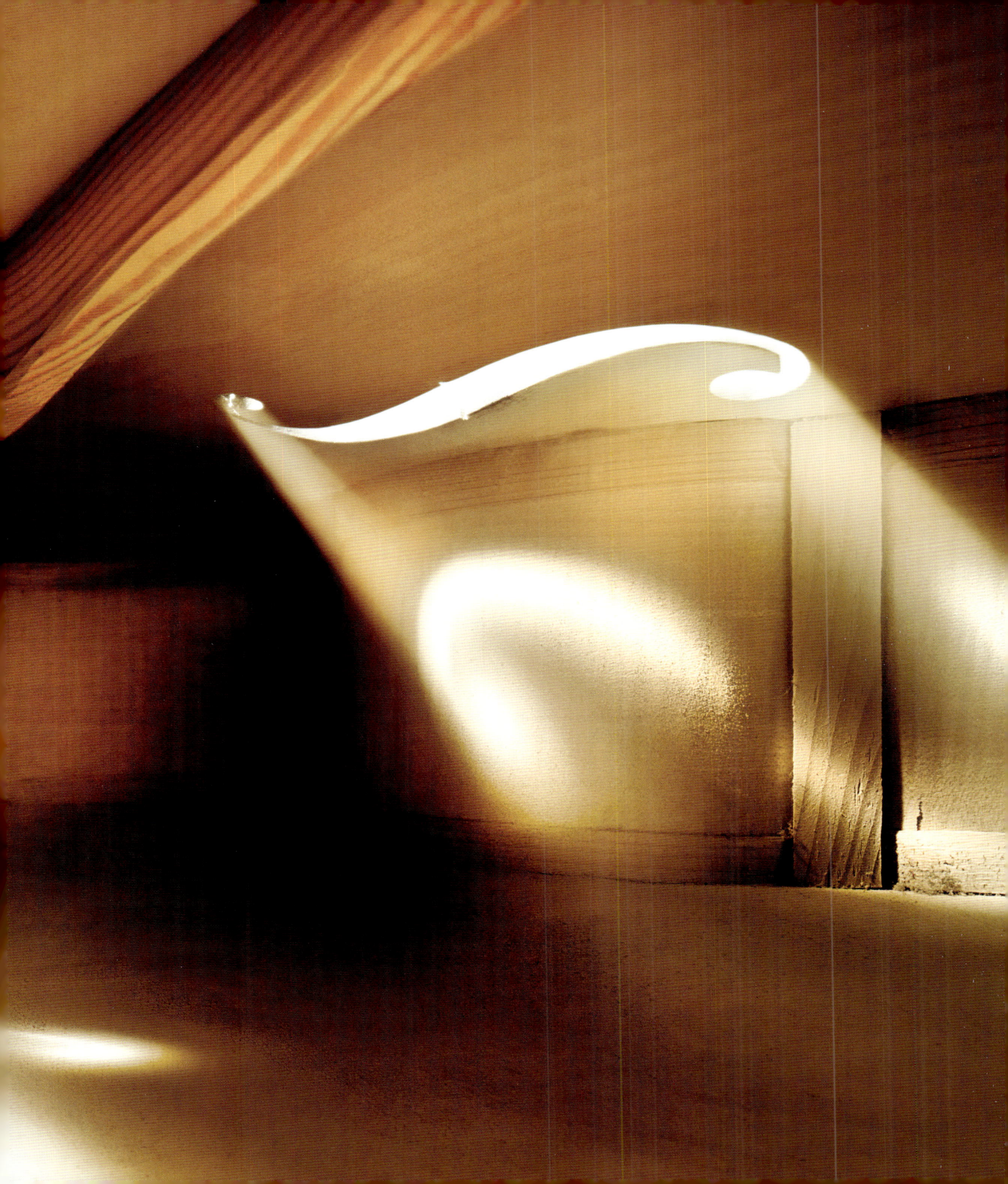

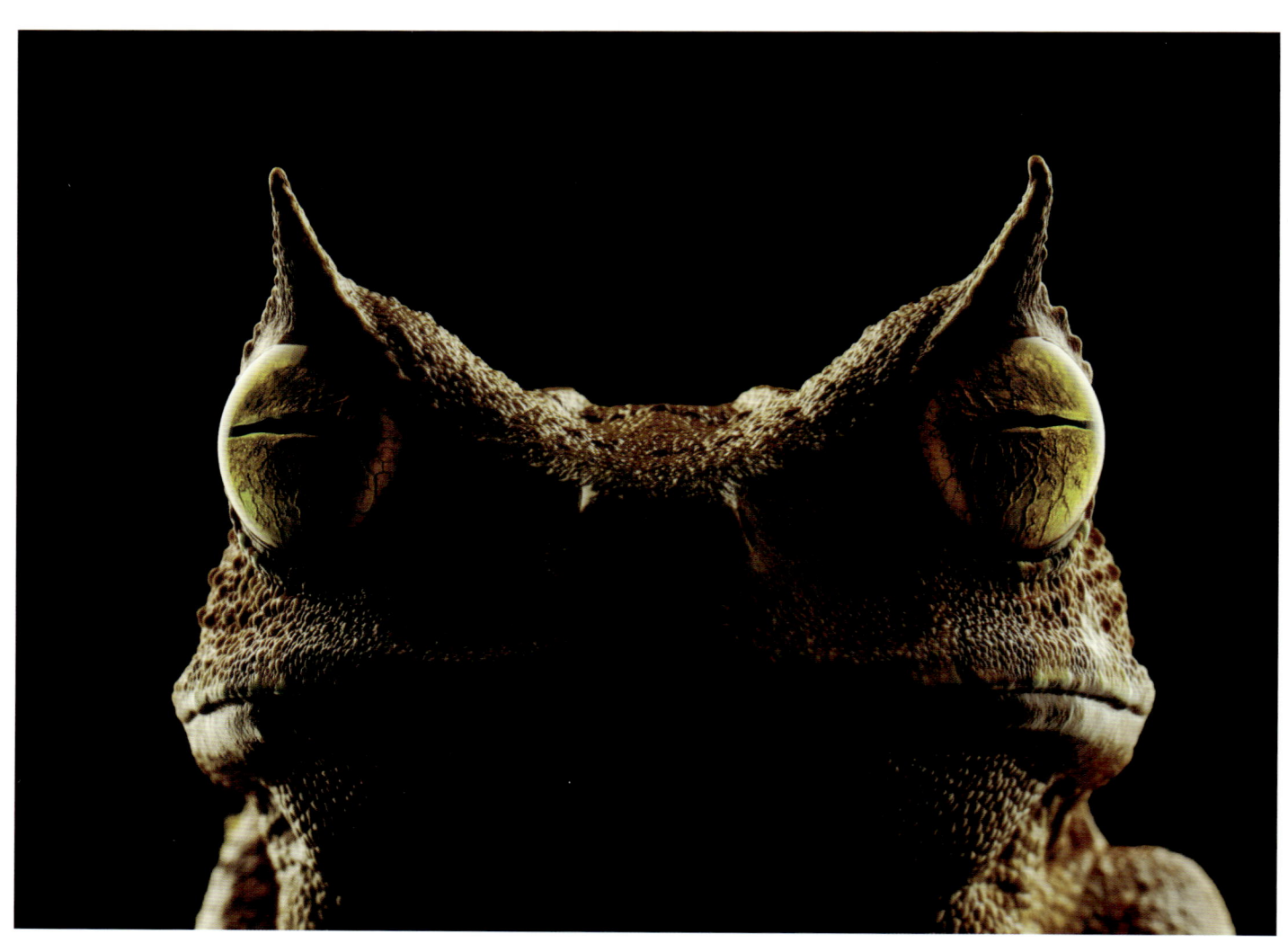

Javier Aznar
An endangered horned marsupial frog native to the Chocó rainforest of Ecuador

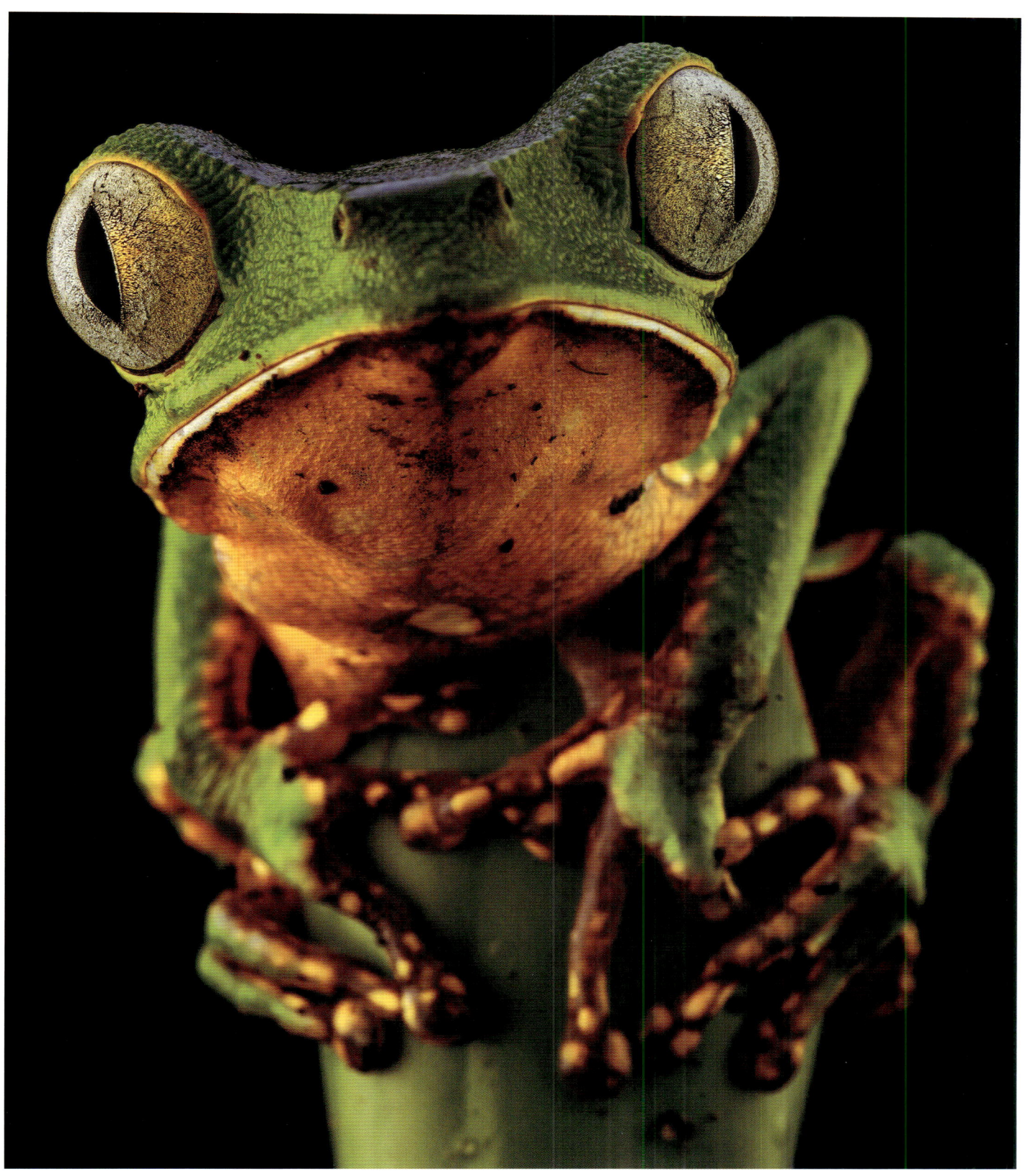

Javier Aznar
A monkey tree frog perched on a leaf in Yasuní National Park, Ecuador

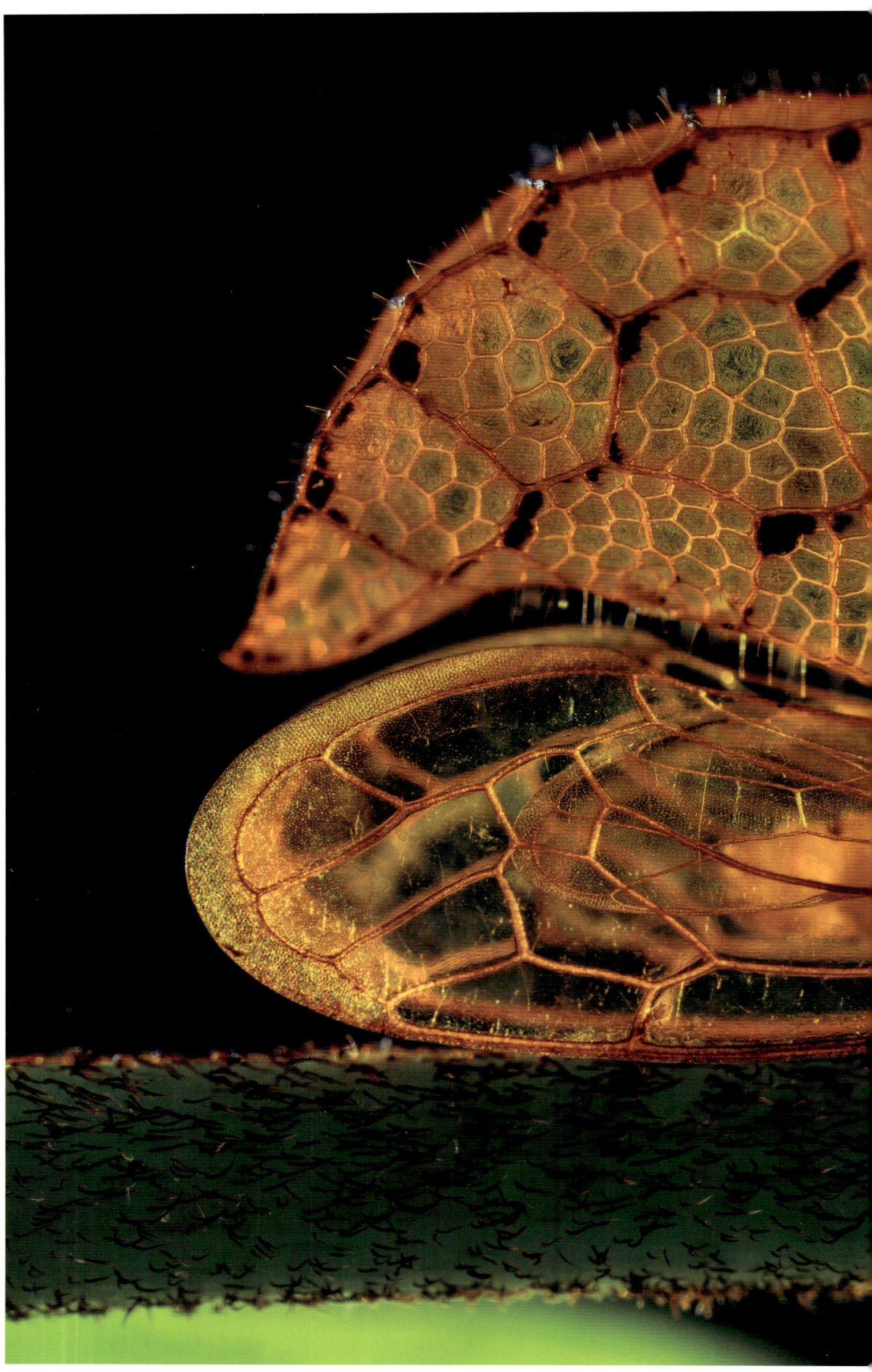

Javier Aznar
A seed-mimicking treehopper in Ecuador. This species has evolved an elaborate helmet to help it hide from predators.

FOLLOWING PAGES:
Dennis Breitsprecher
Magnified bundles of actin protein filaments illuminated by a technique called total internal reflection fluorescence microscopy

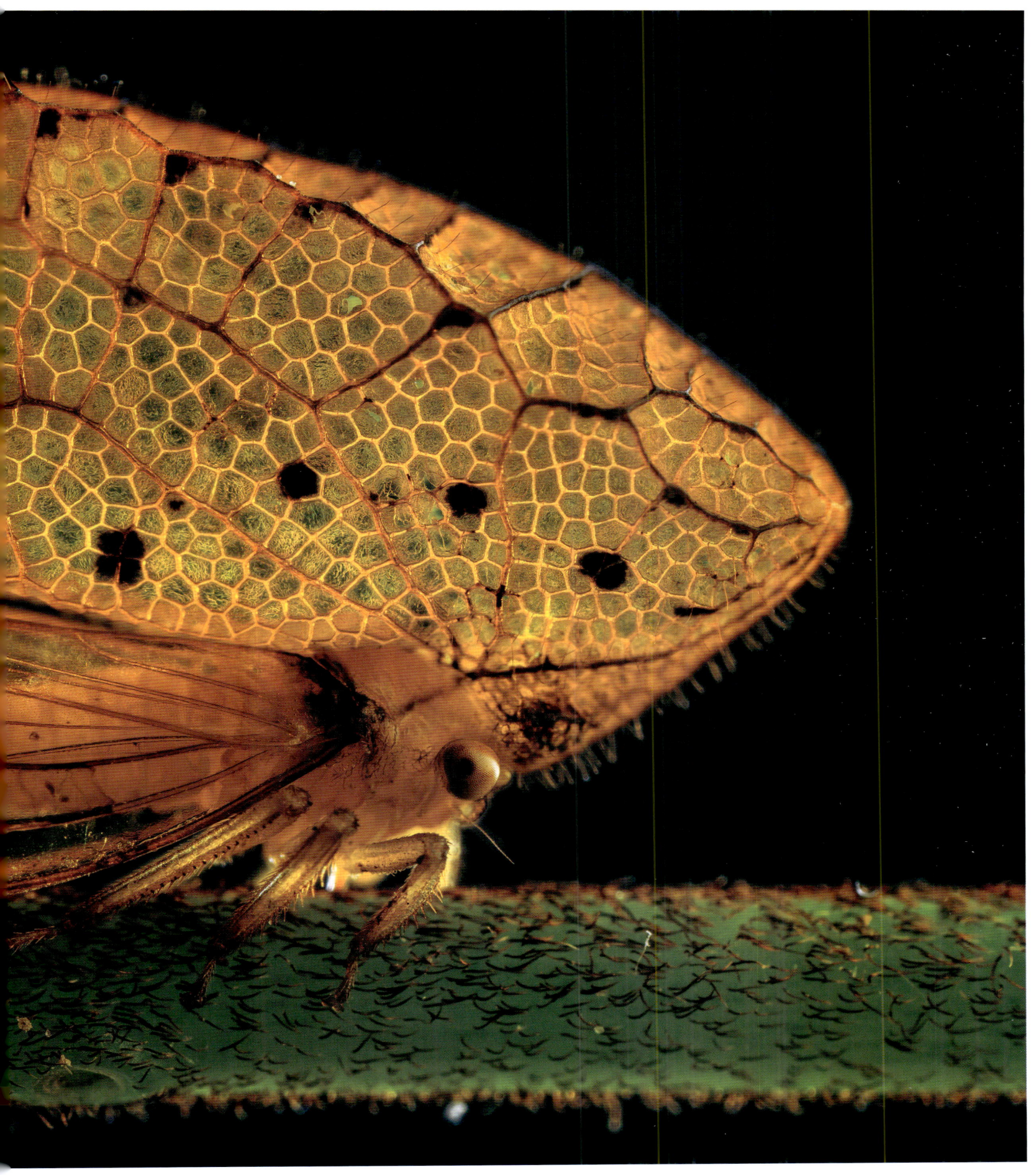

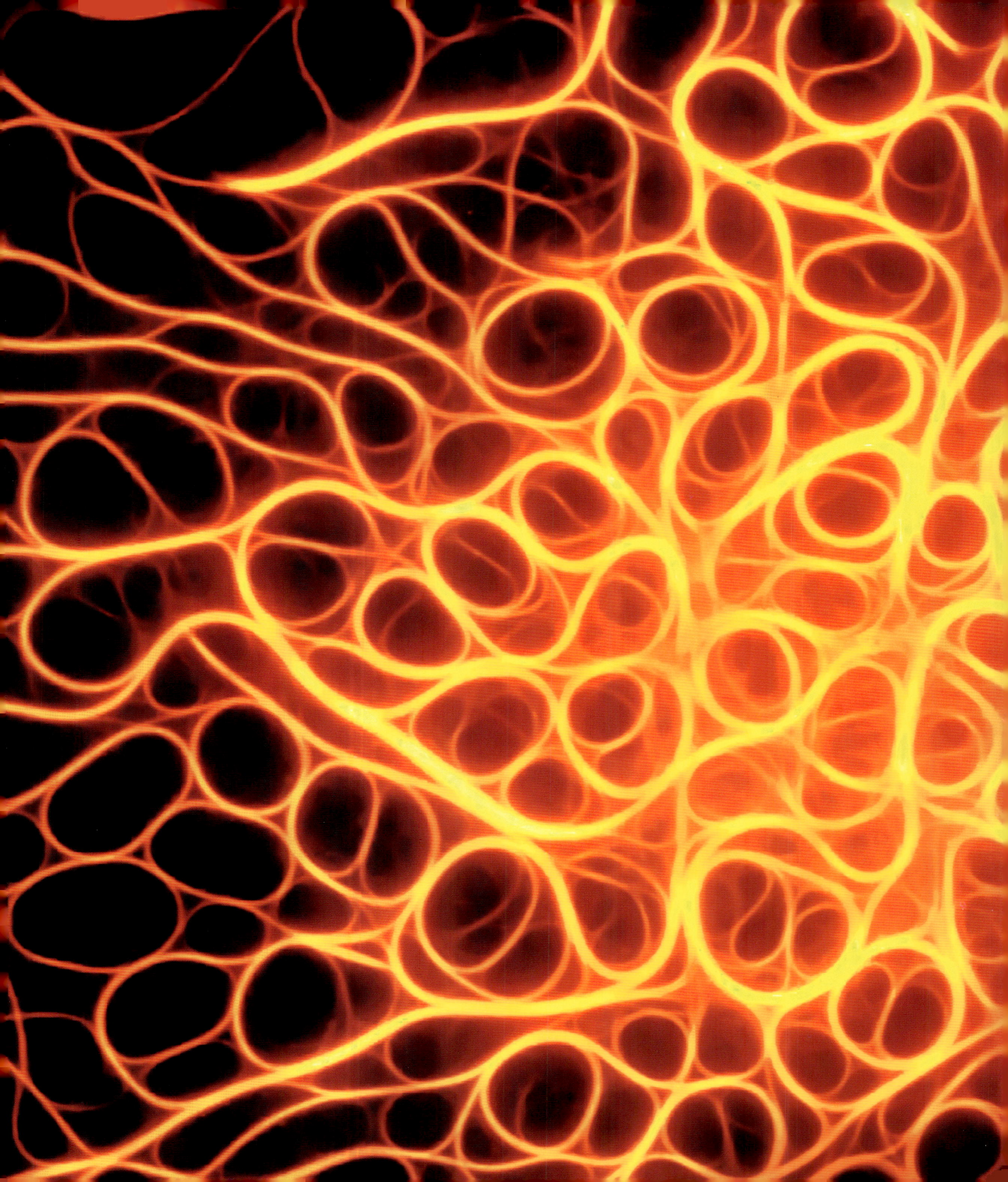

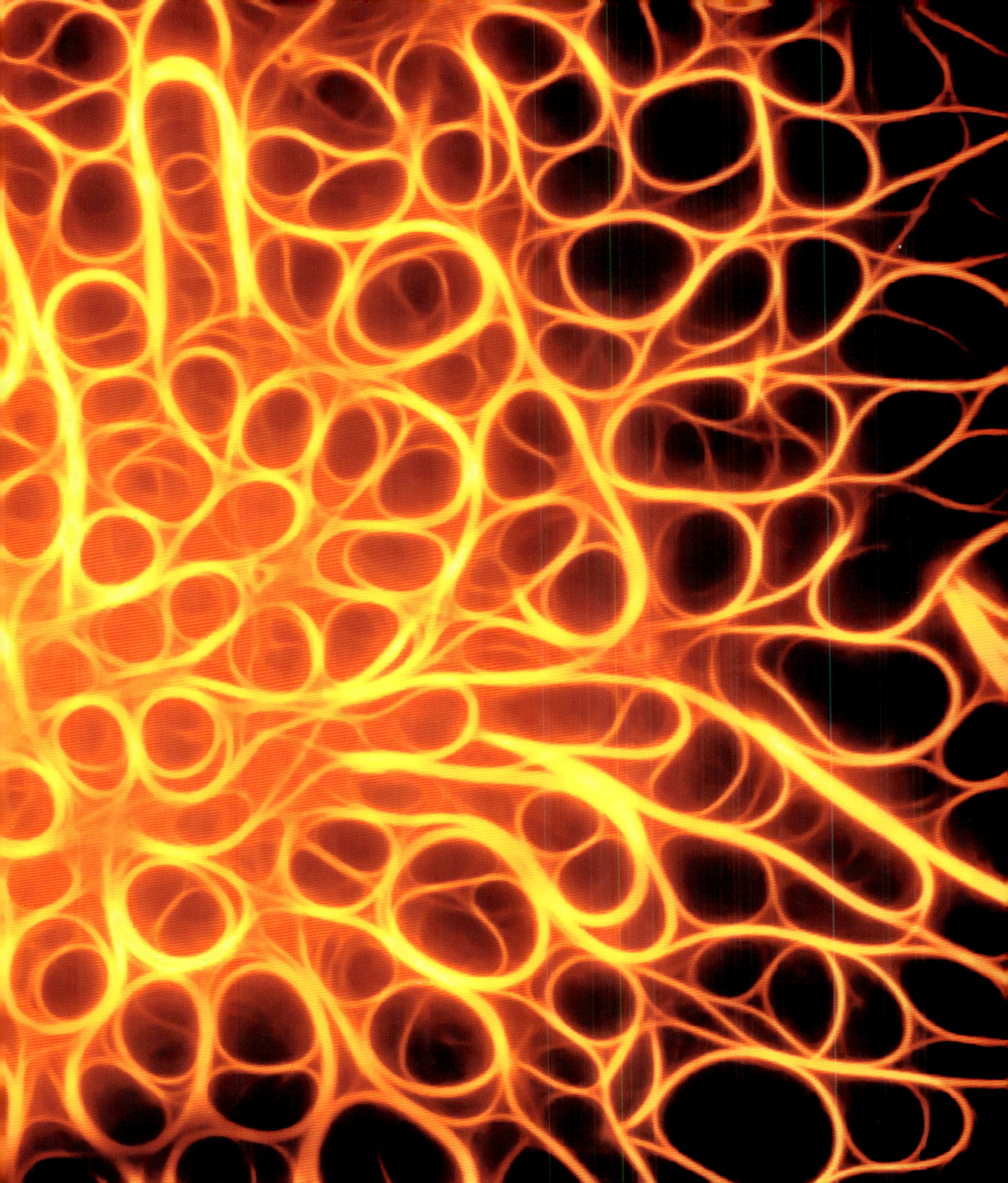

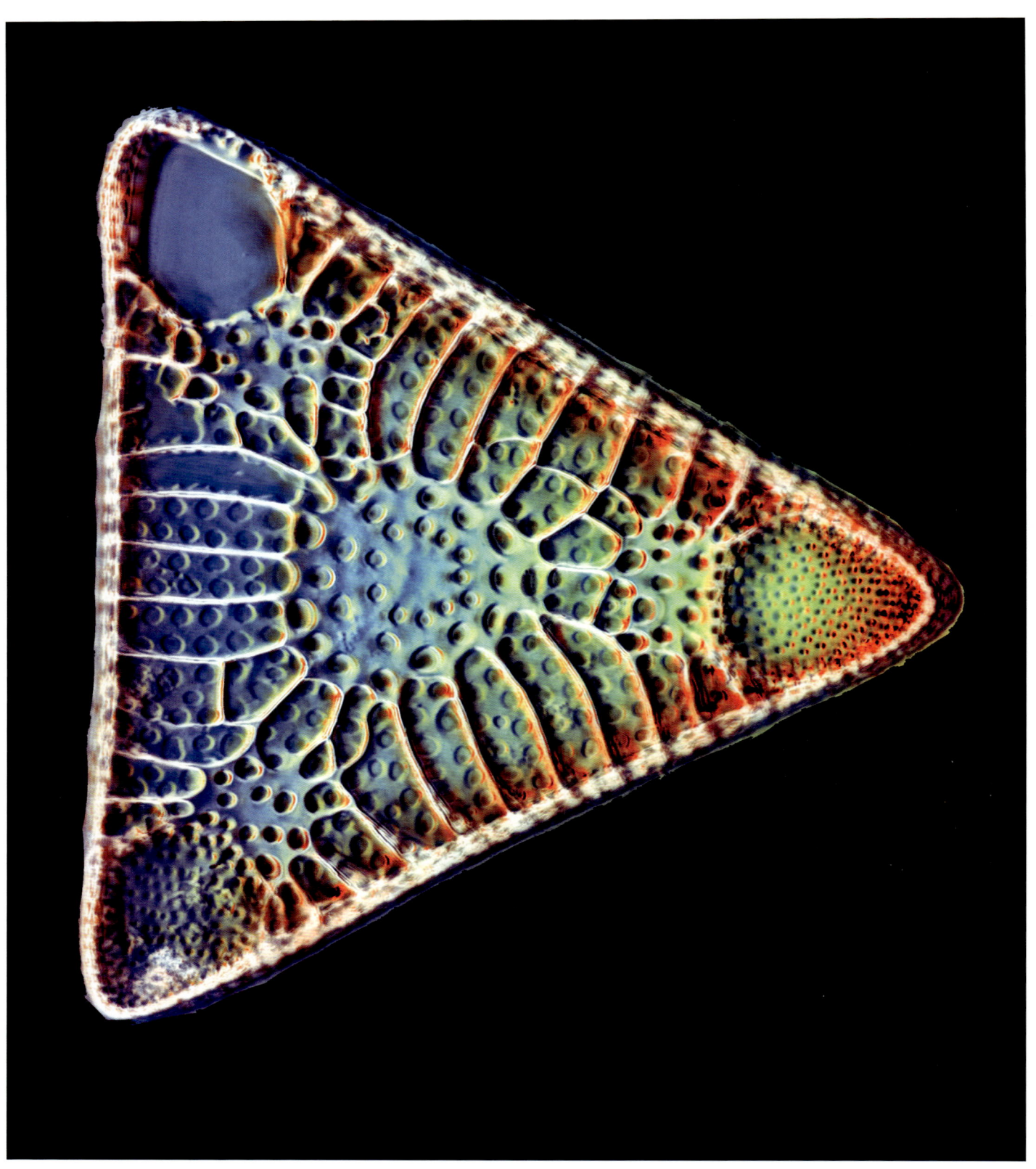

Jan Rosenboom
A diatom—a single-celled species of algae—turns rainbow colors
when photographed using polarized light.

THESE IMAGES REMIND US THAT OUR WORLD HOLDS WONDERFUL SURPRISES IF WE LEARN TO SLOW DOWN AND PAY CAREFUL ATTENTION.

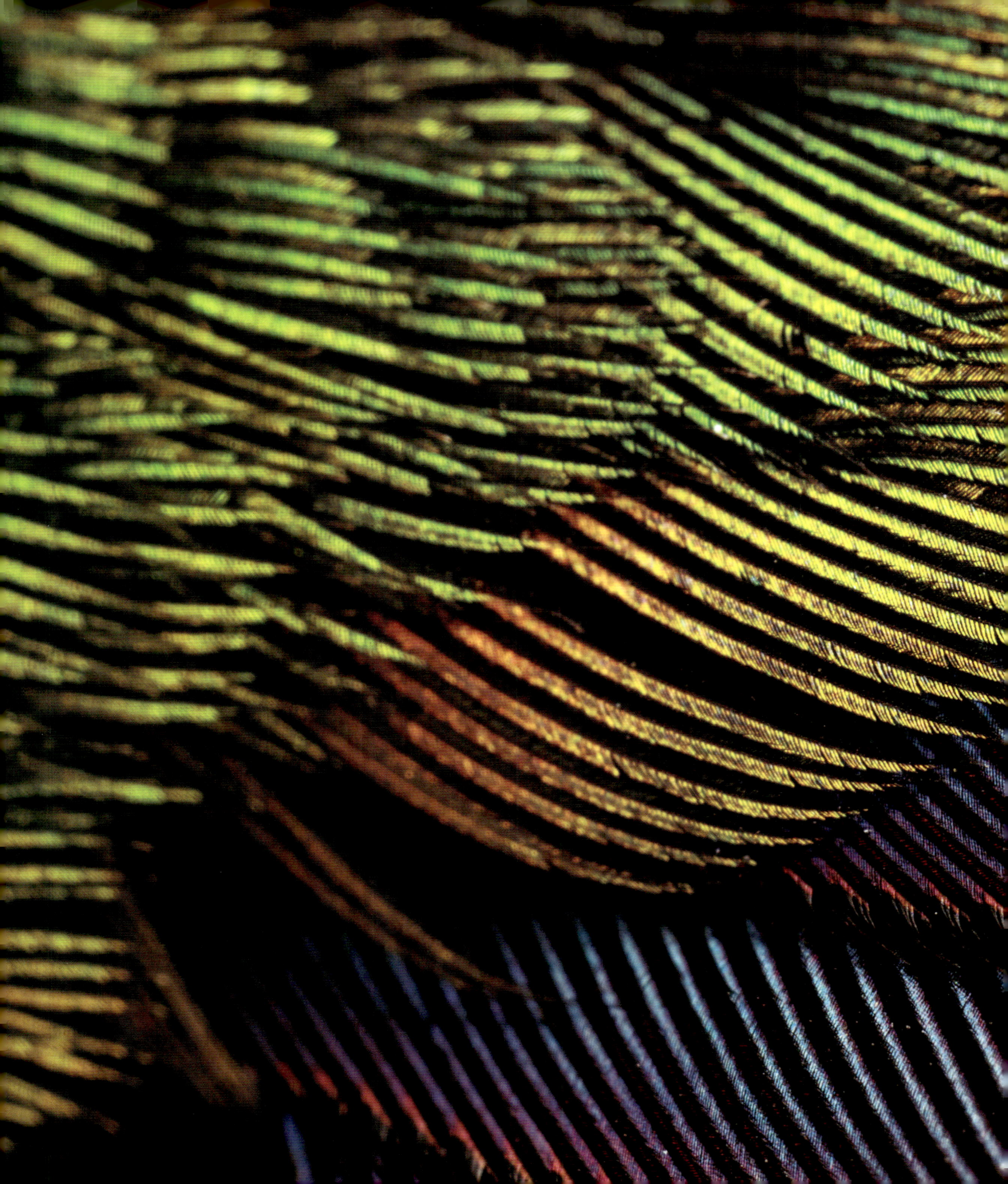

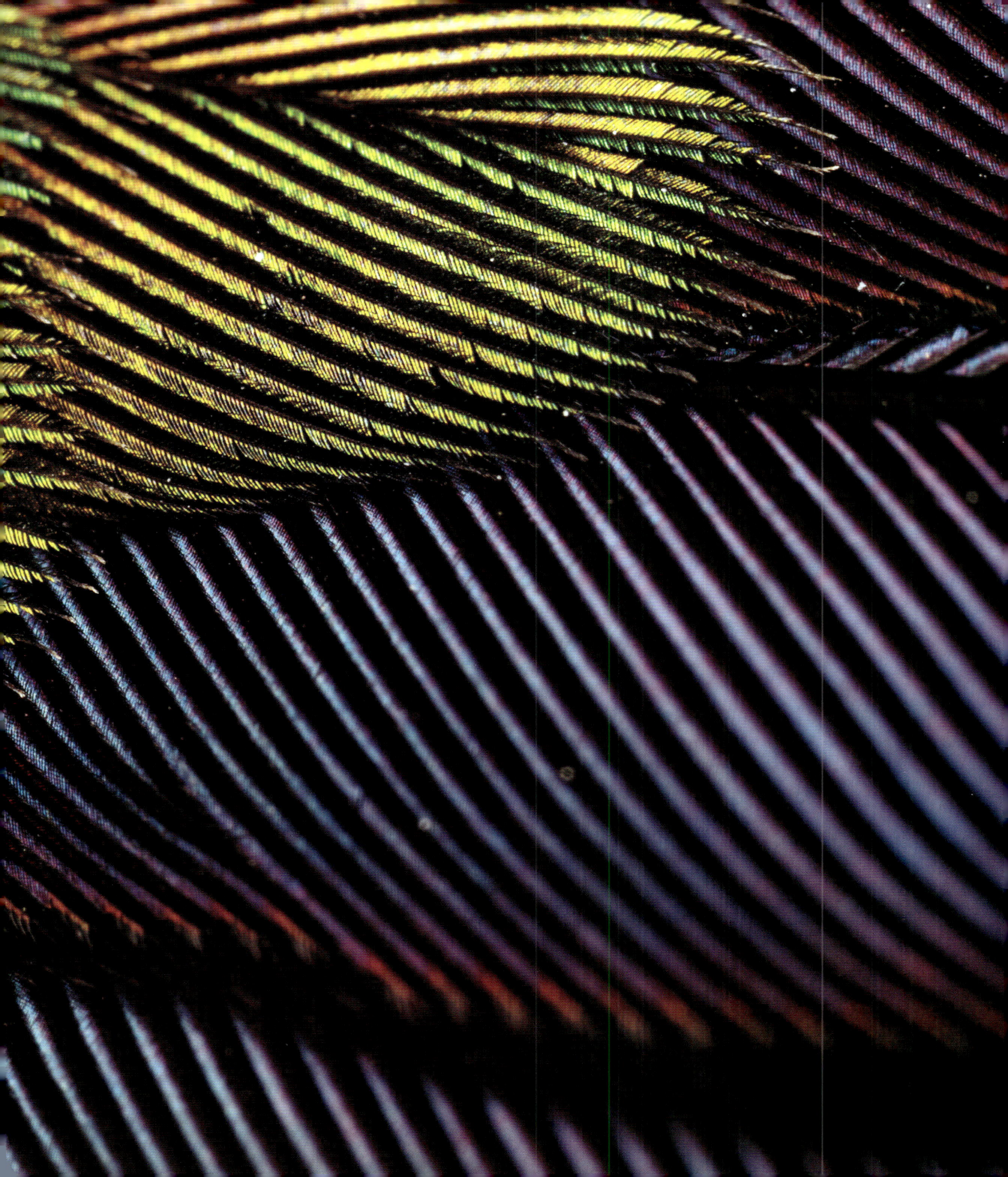

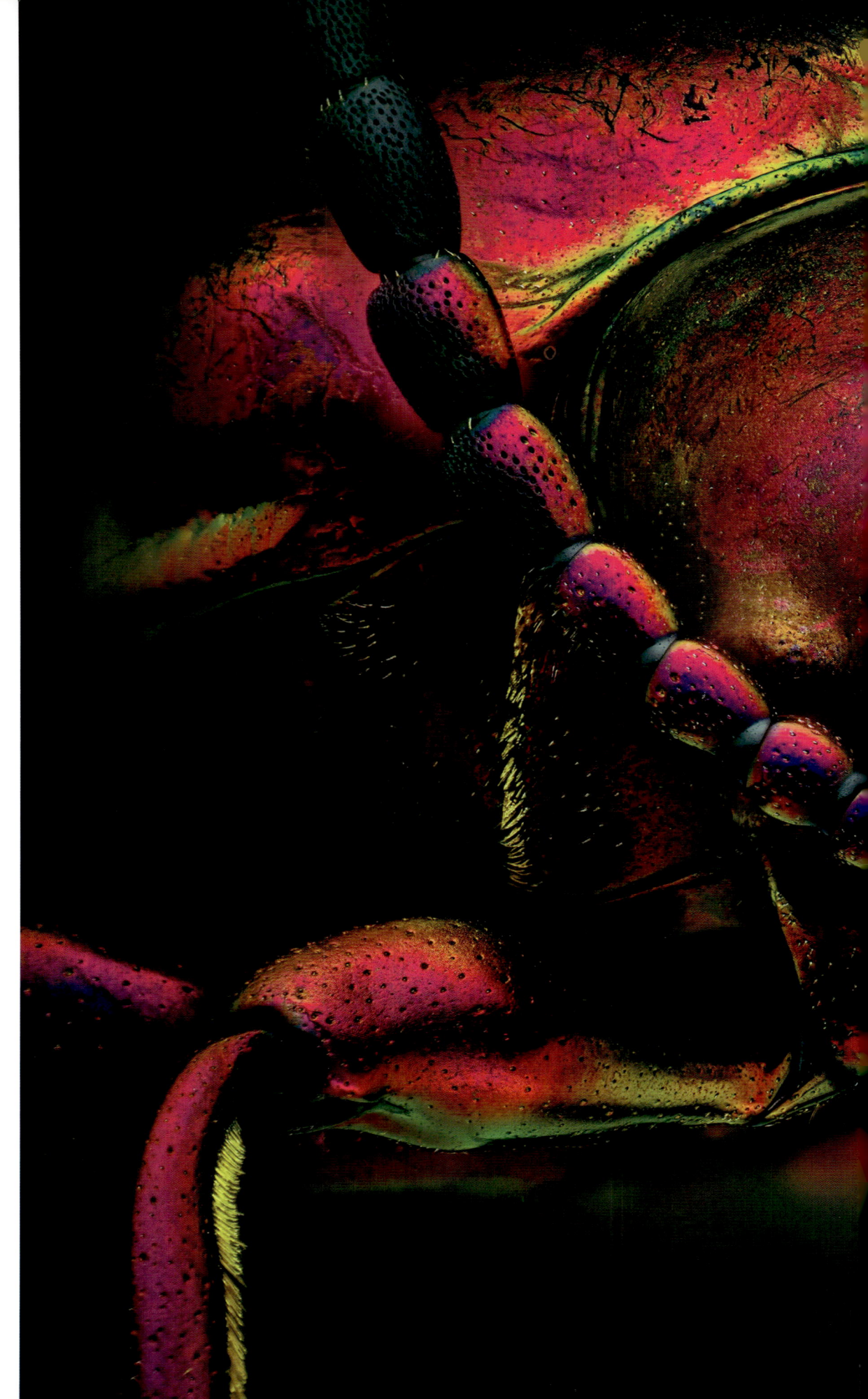

Yousef Al Habshi
A frog-legged leaf beetle native to Southeast Asia. This iridescent insect can grow to two inches in length.

PREVIOUS PAGES:
Anand Varma
Feathers of a Tyrian metaltail, a South American hummingbird

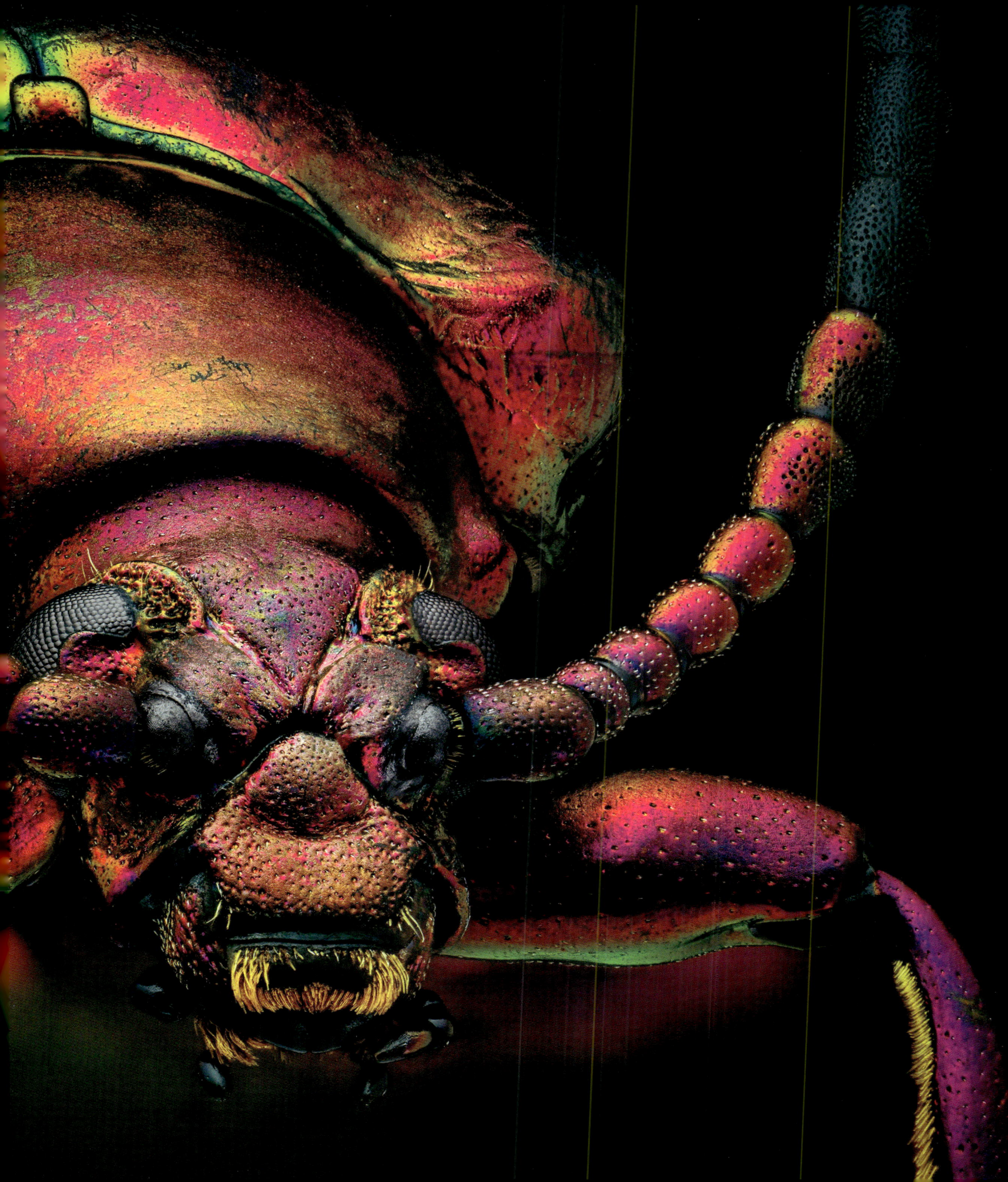

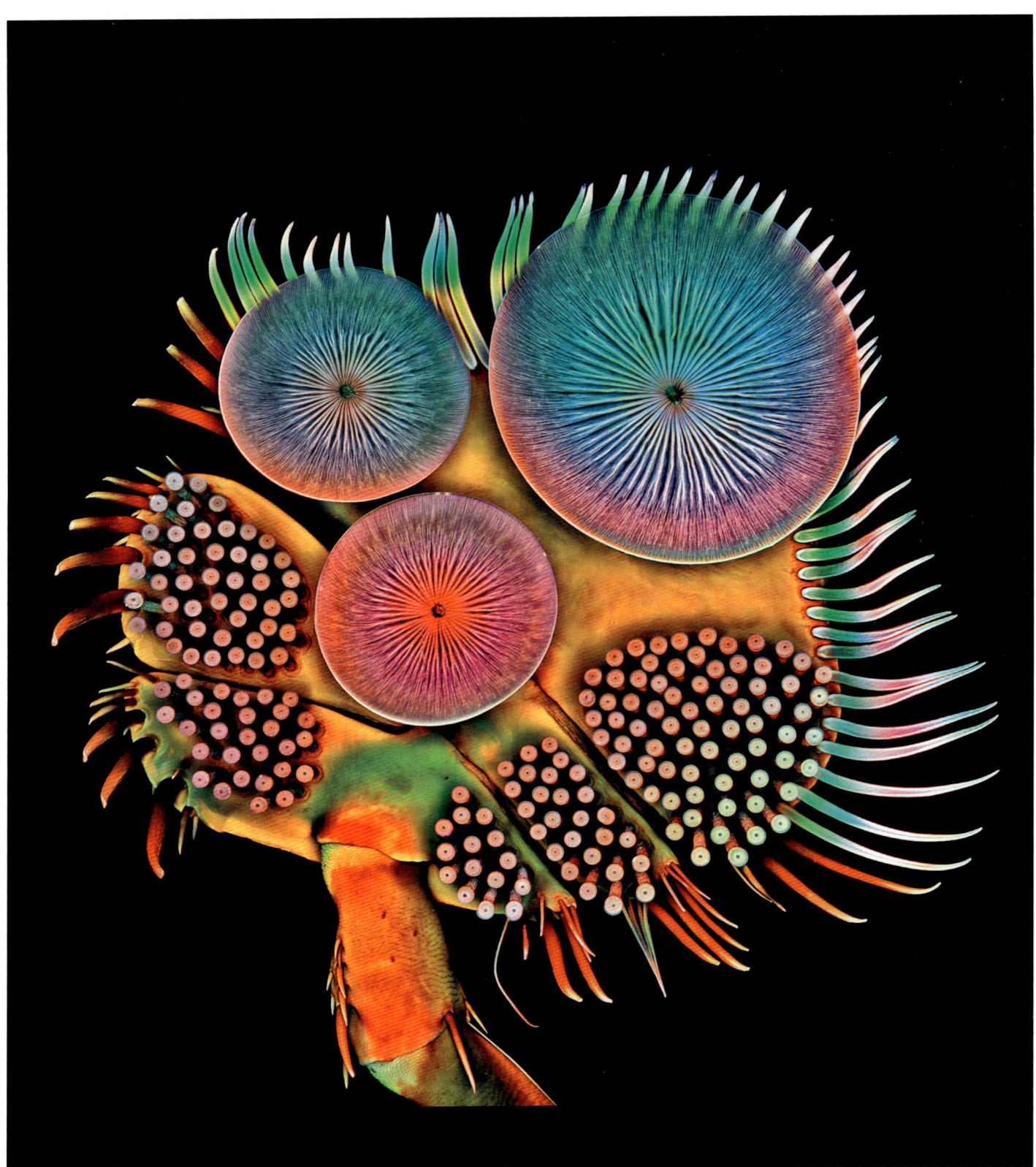

Igor Siwanowicz

The front foot of a male diving beetle, magnified by a confocal laser scanning microscope.
The three large disks are suction cups used by males to hold on to females during copulation.

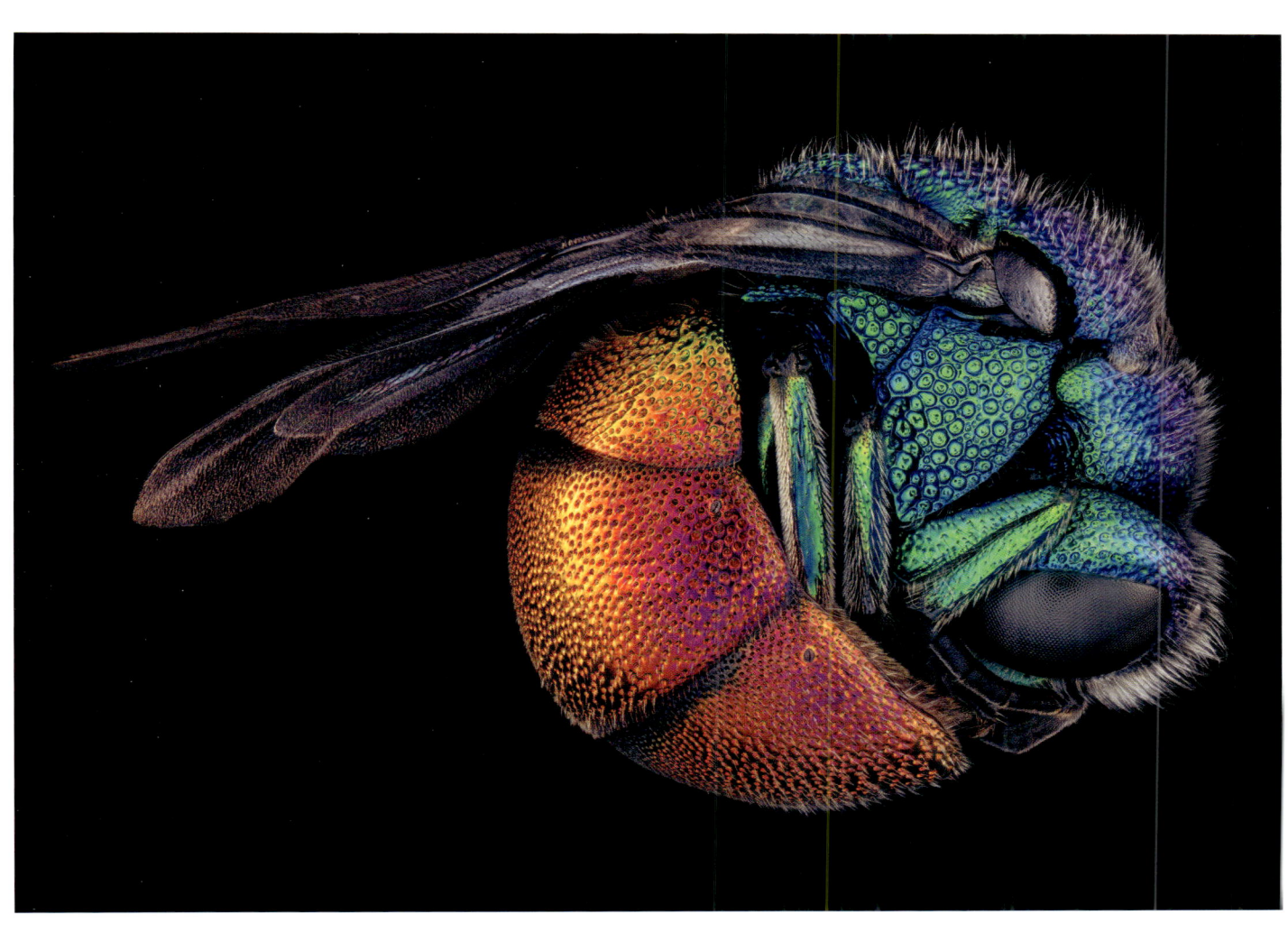

Thorben Danke
A cuckoo wasp. These wasps curl up defensively when threatened.

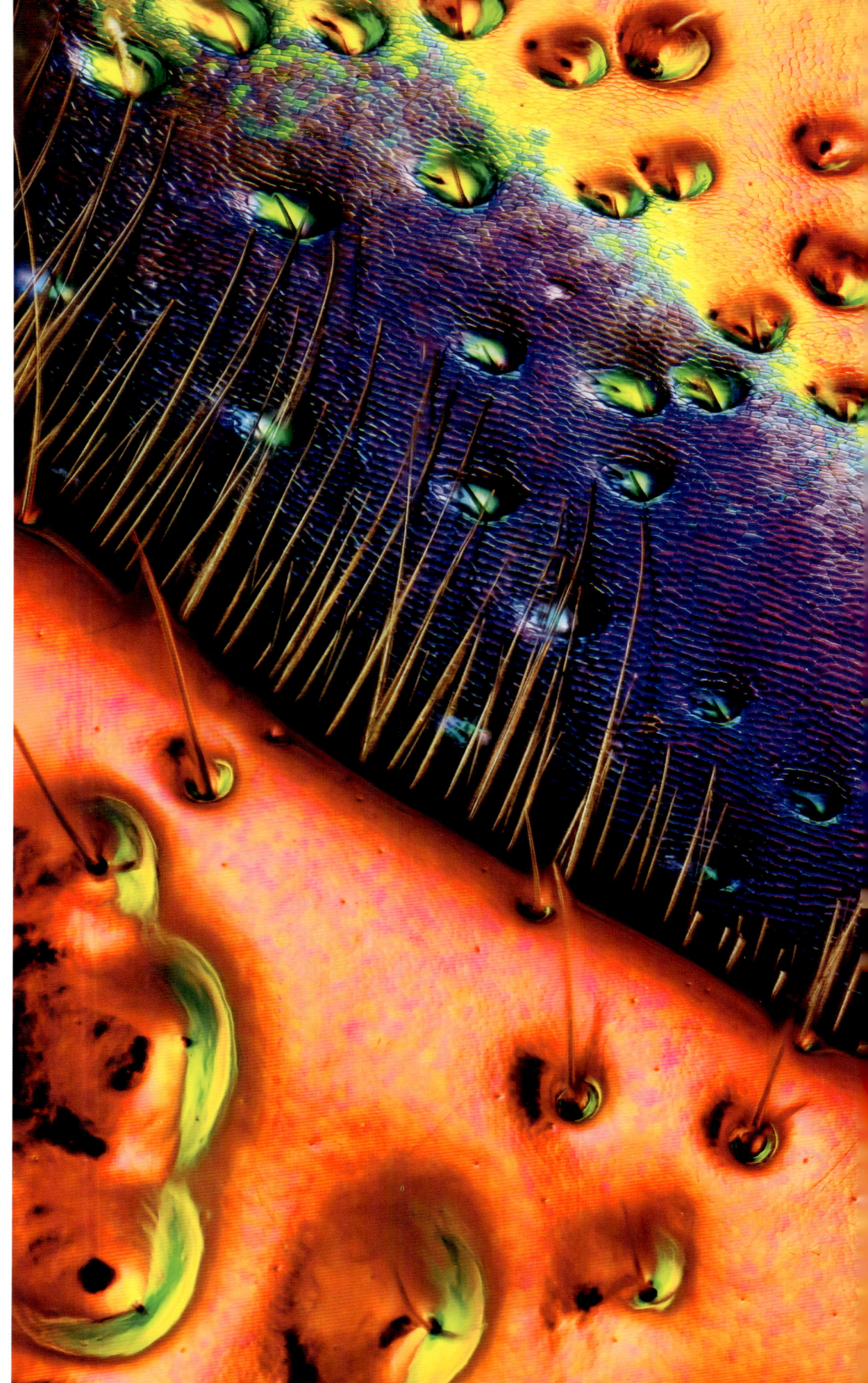

Charles Krebs
An extreme close-up of a jewel beetle's face. The dark crescent in the upper right corner is the edge of the insect's eye.

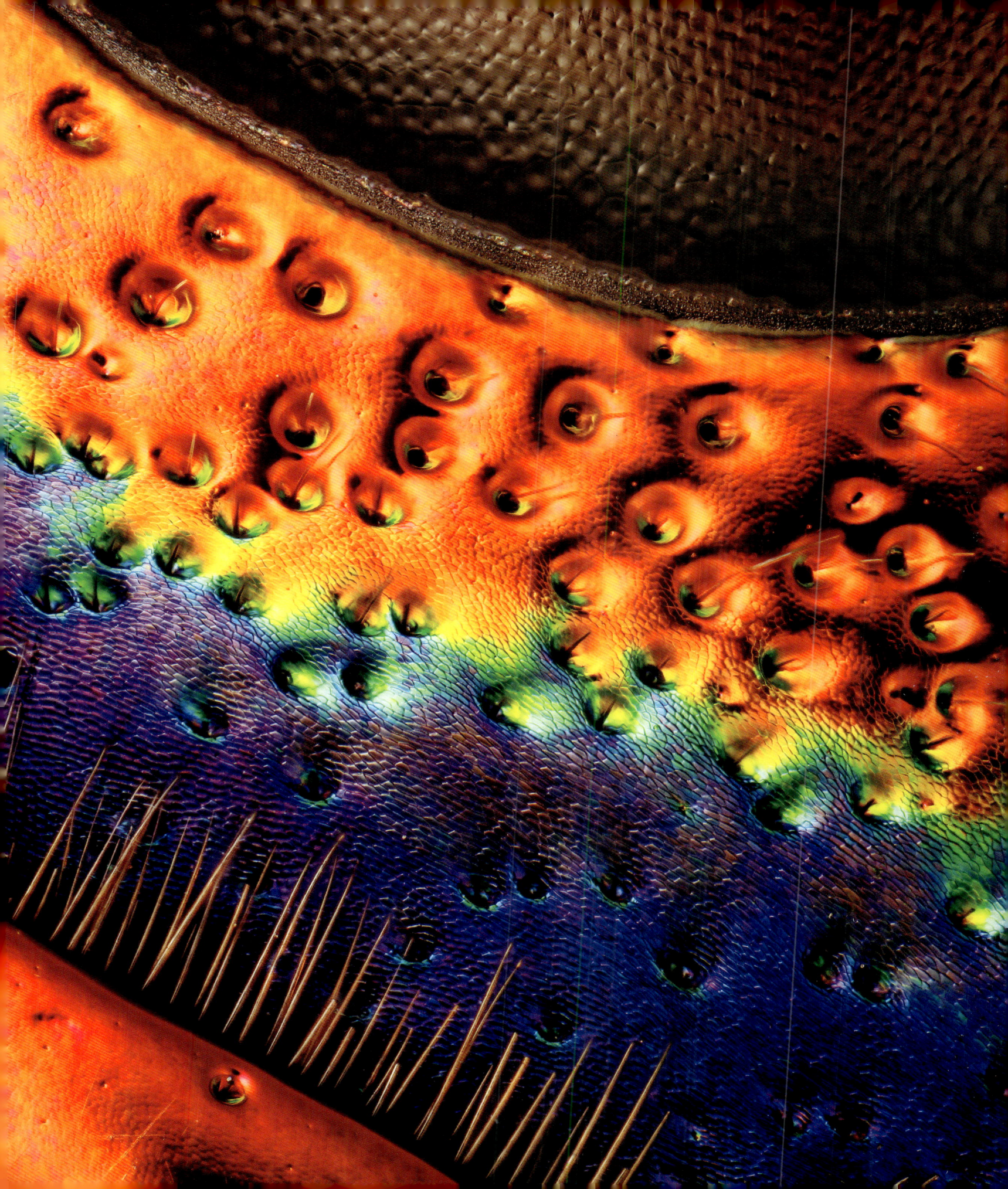

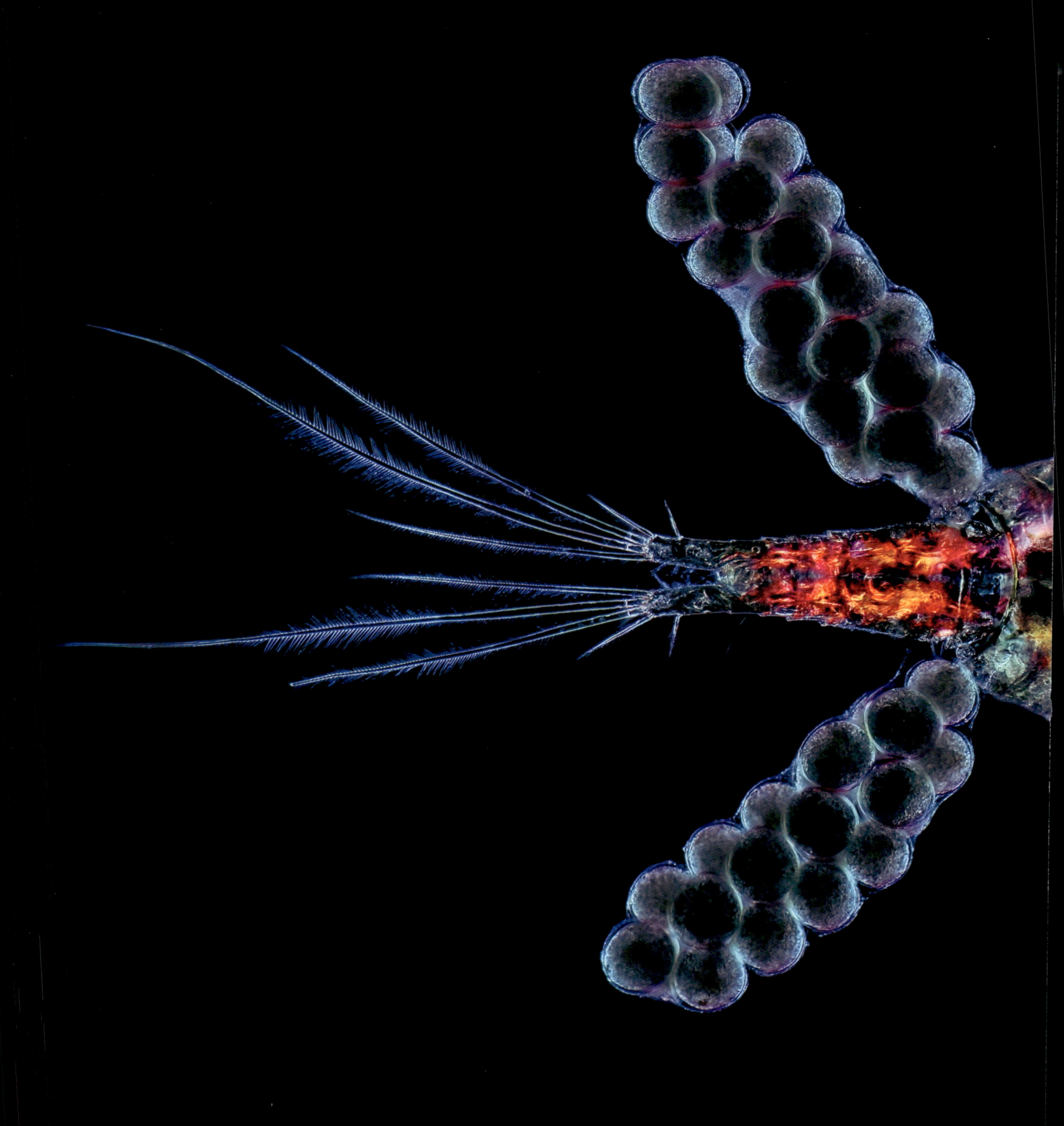

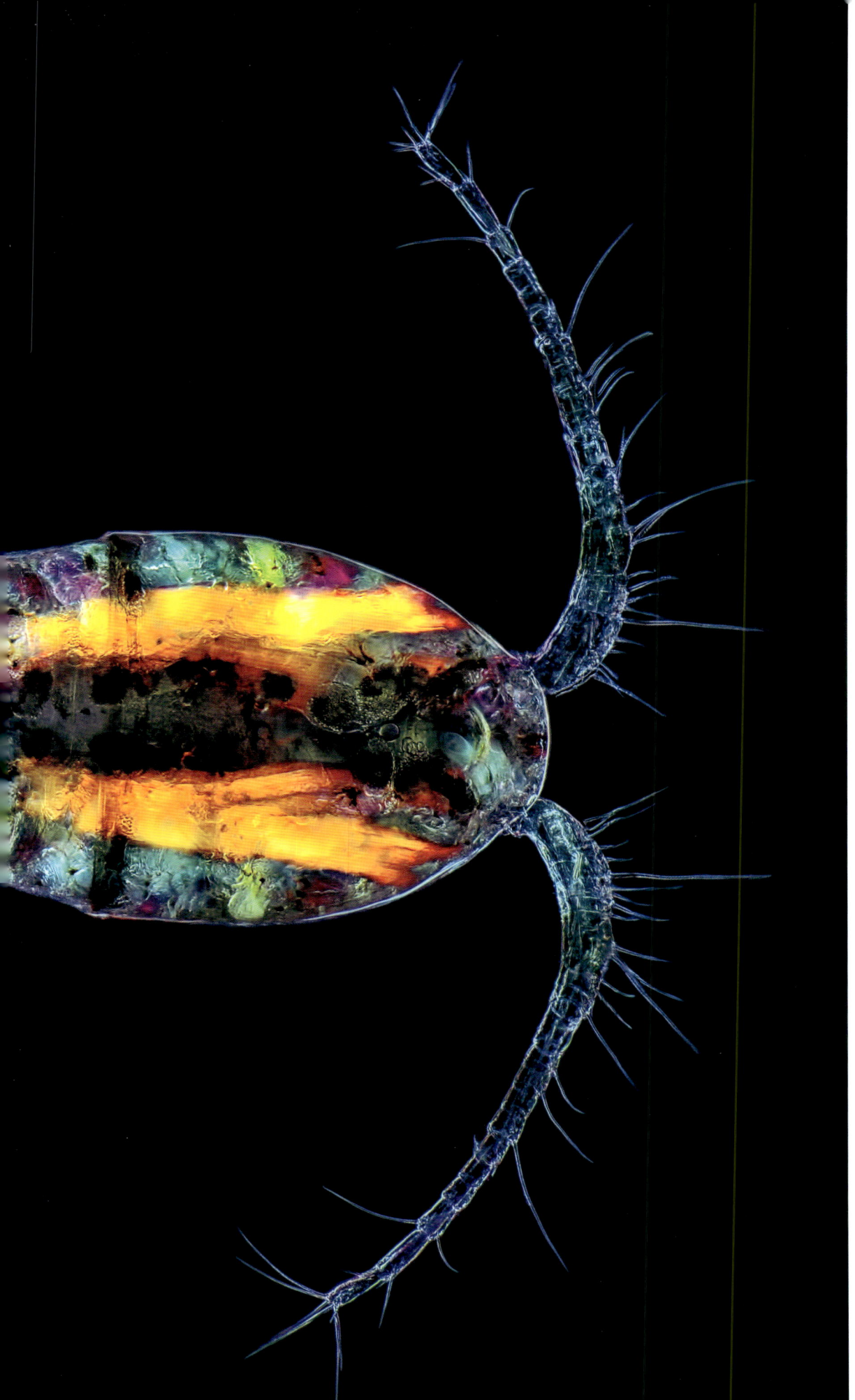

Rogelio Moreno
An oceanic creature called a copepod trails a clutch of eggs. Polarized light reveals intricate structures.

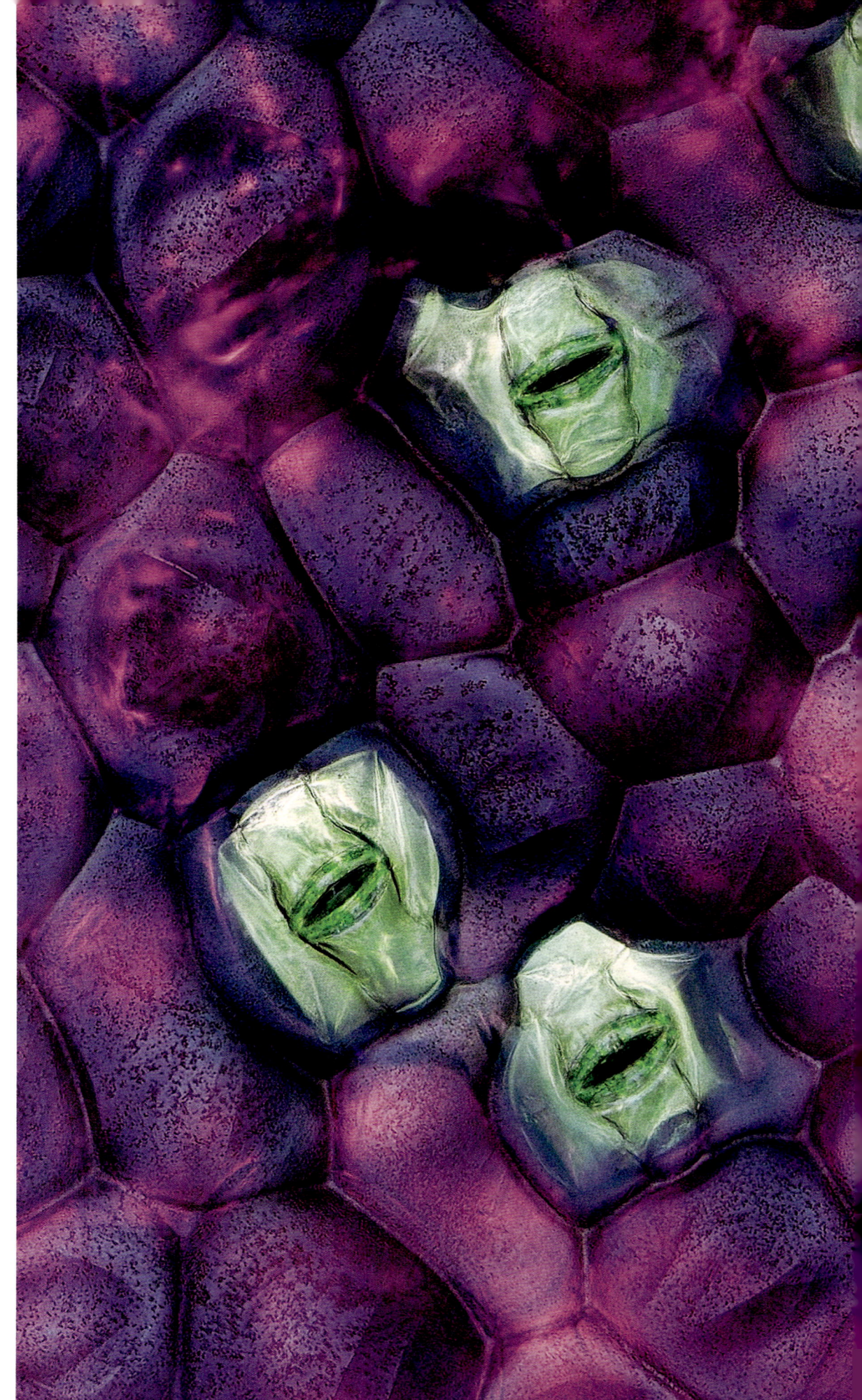

Jerzy Gubernator
This view displays less than a square millimeter of a zebra plant and reveals the leaf structures called stomata that plants use for respiration, bringing in carbon dioxide and releasing oxygen.

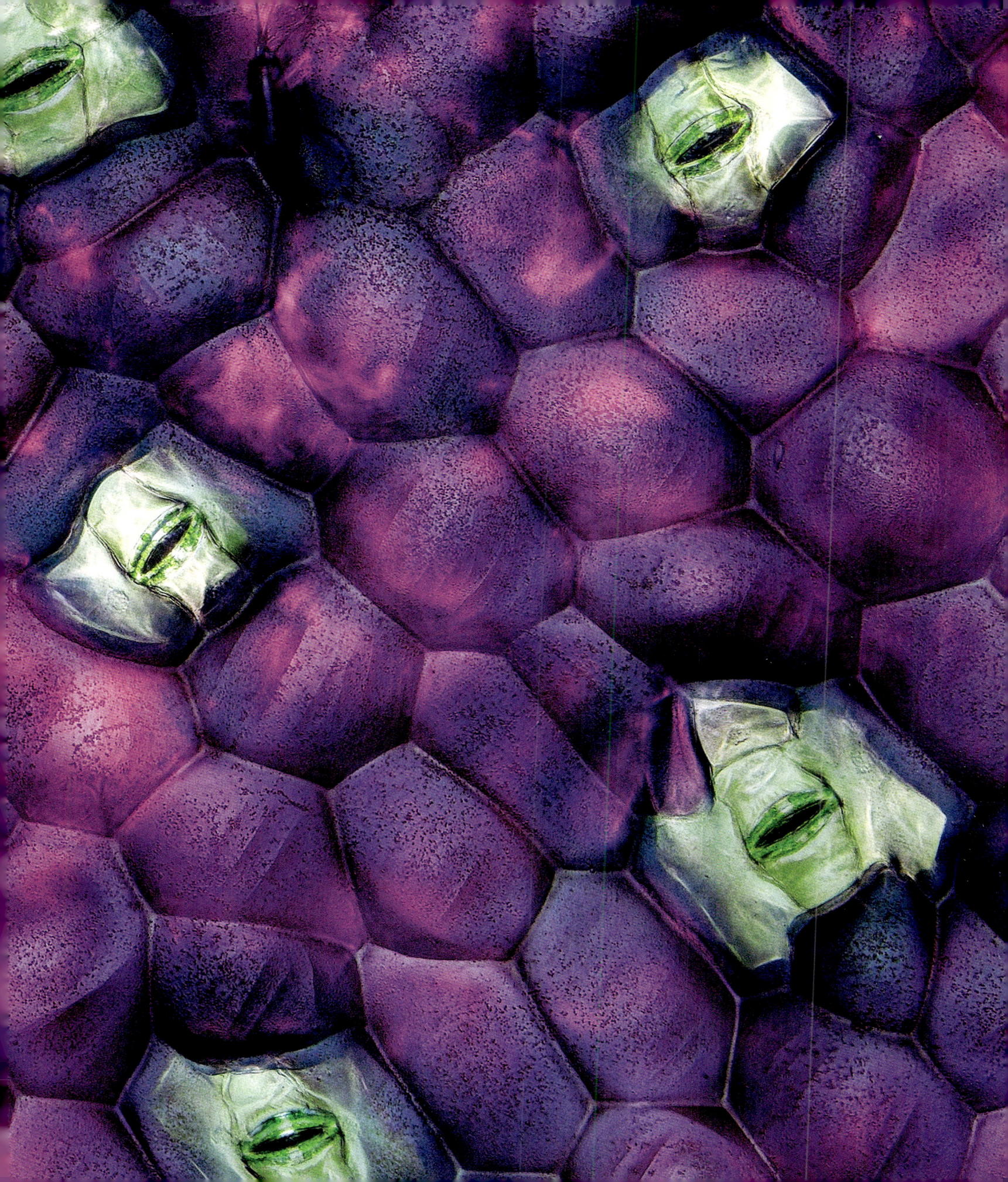

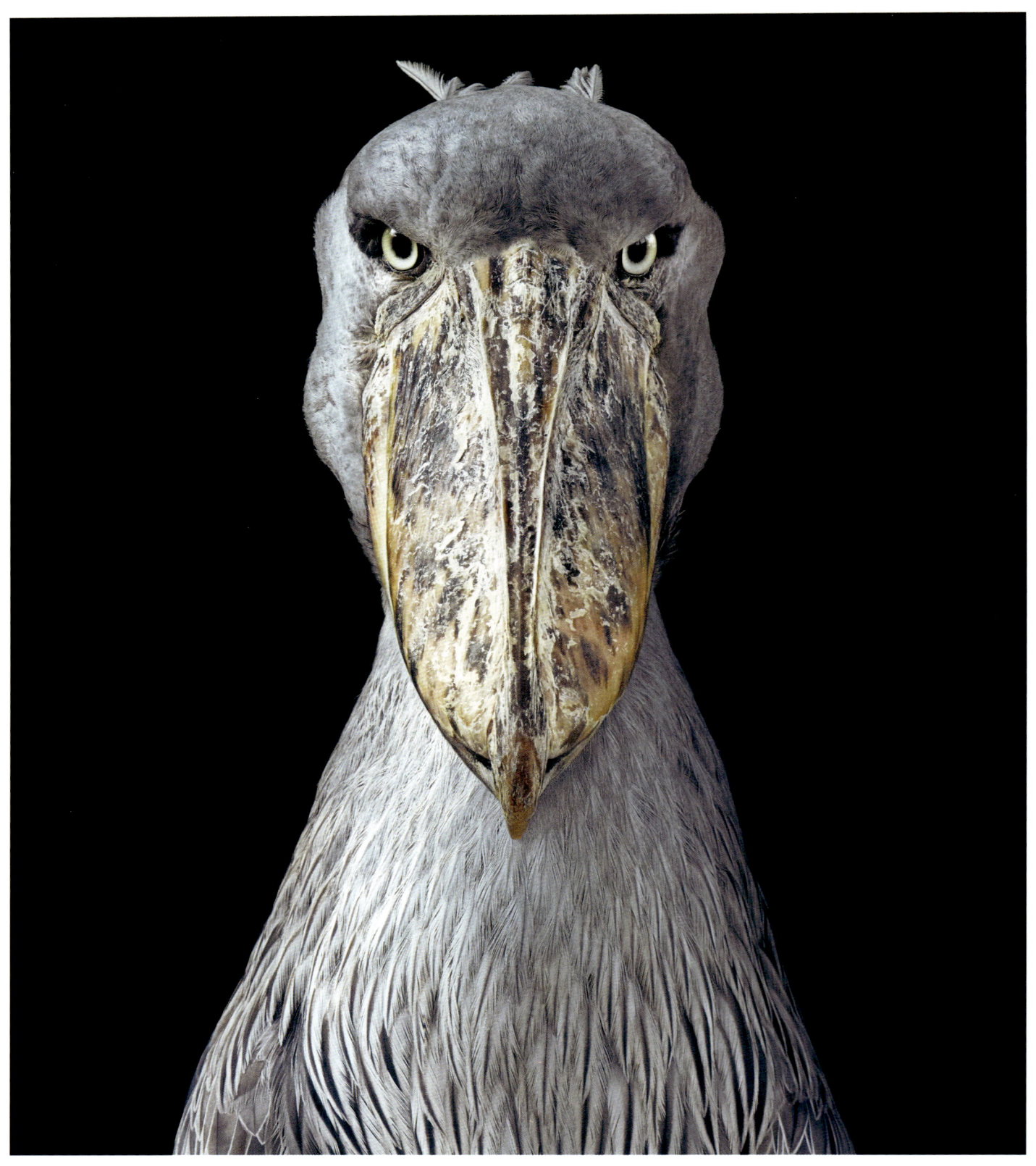

Tim Flach
An East African bird called a shoebill. These predators stand up to five feet tall
and use their impressive bills to capture prey such as snakes and eels.

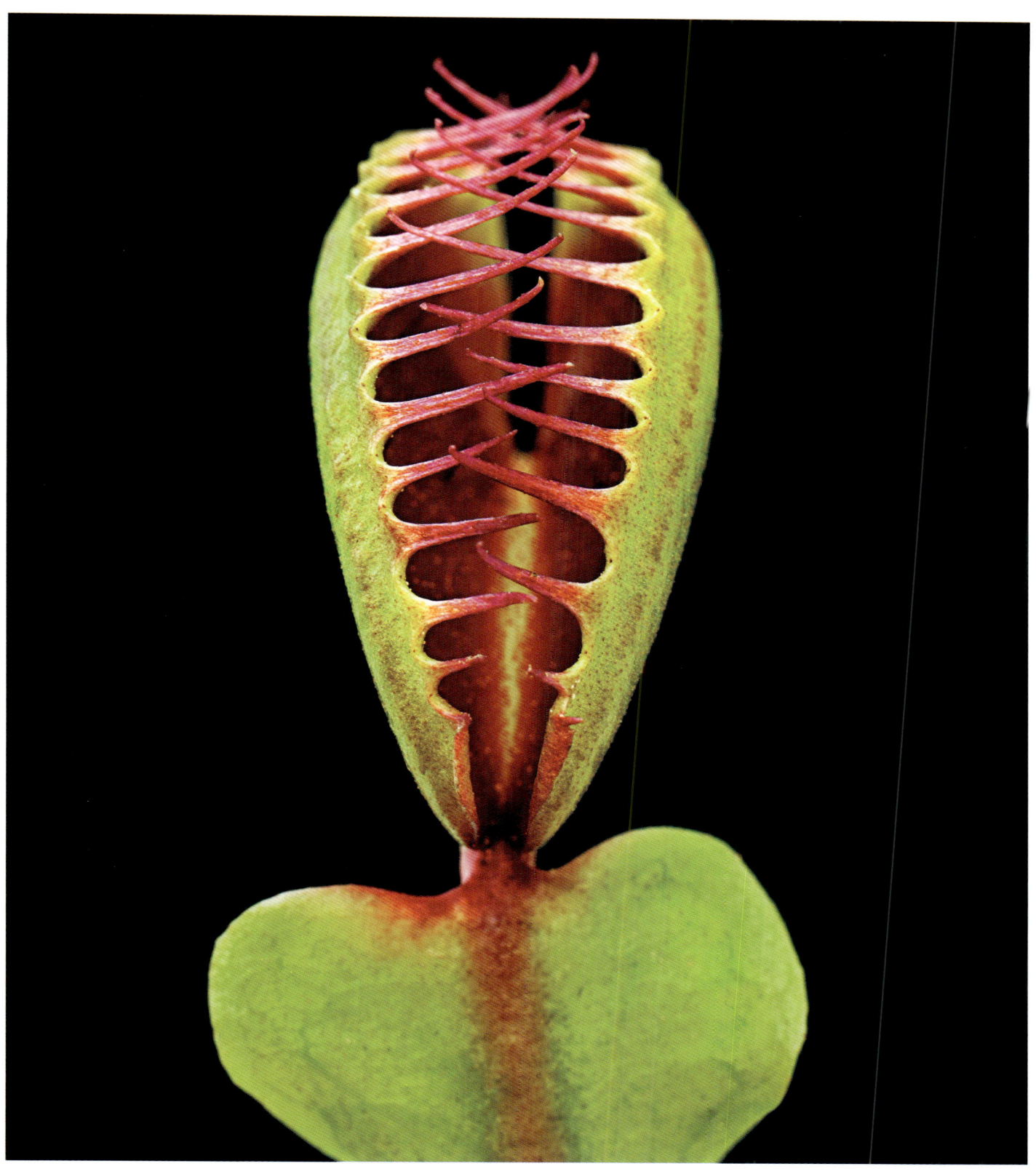

Helene Schmitz
The jaws of a carnivorous Venus flytrap plant open to allure spiders and insects
and then shut, the bristles interlocking to trap the plant's prey.

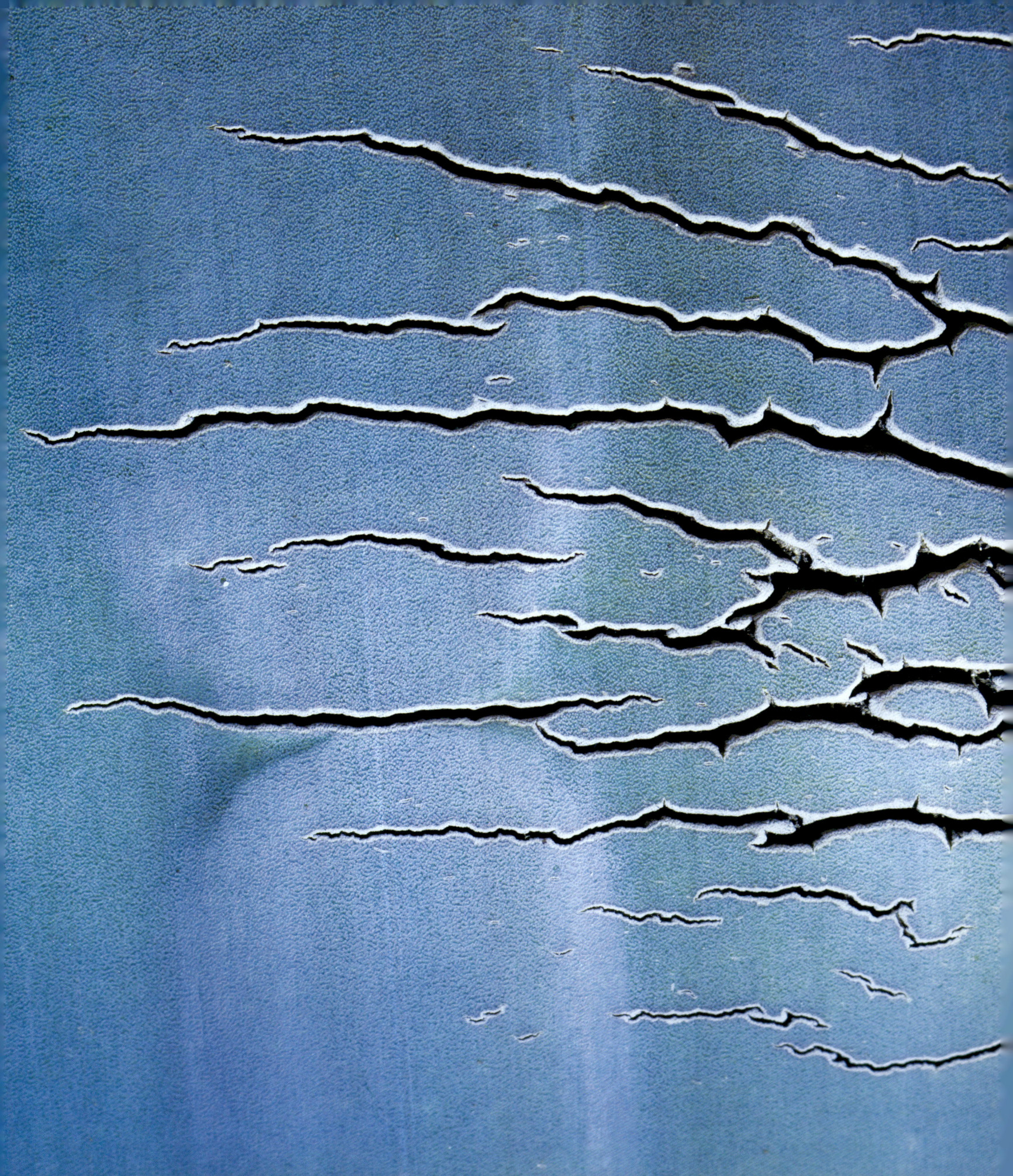

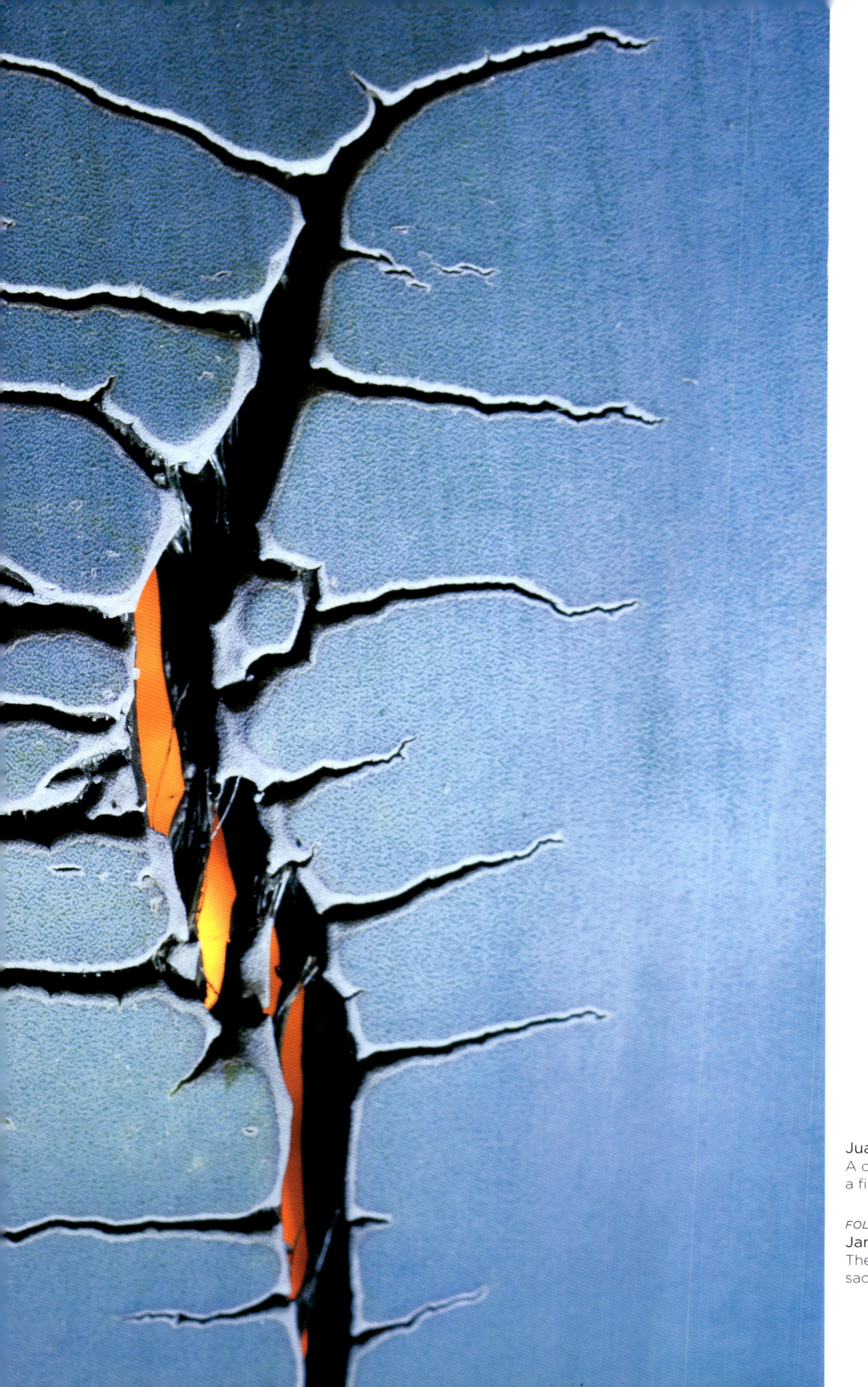

Juan Jesús González Ahumada
A crack in an agave leaf reveals
a fire burning behind it.

FOLLOWING PAGES:
Jan Vermeer
The sporangia, or spore-holding
sacs, of a common slime mold

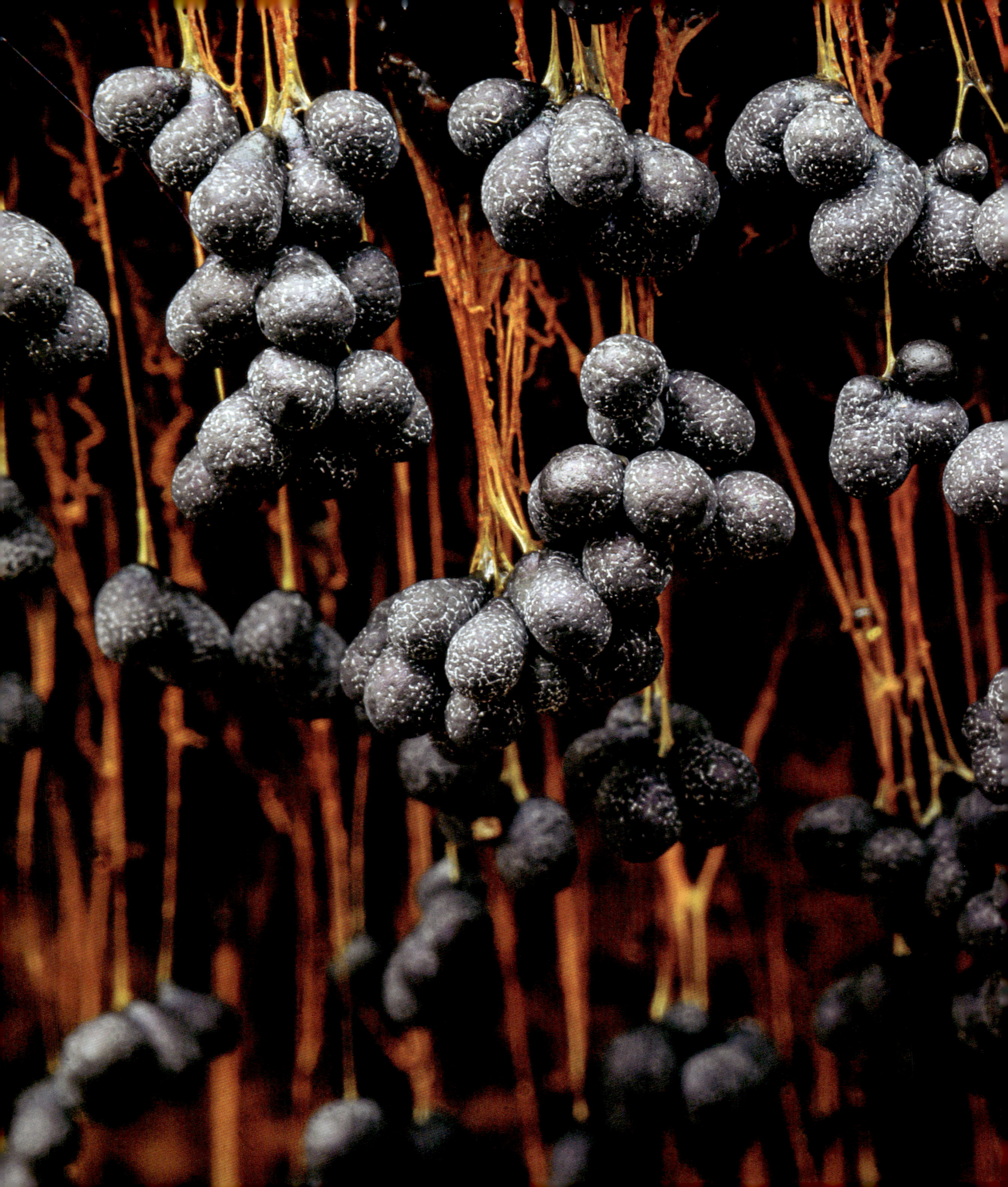

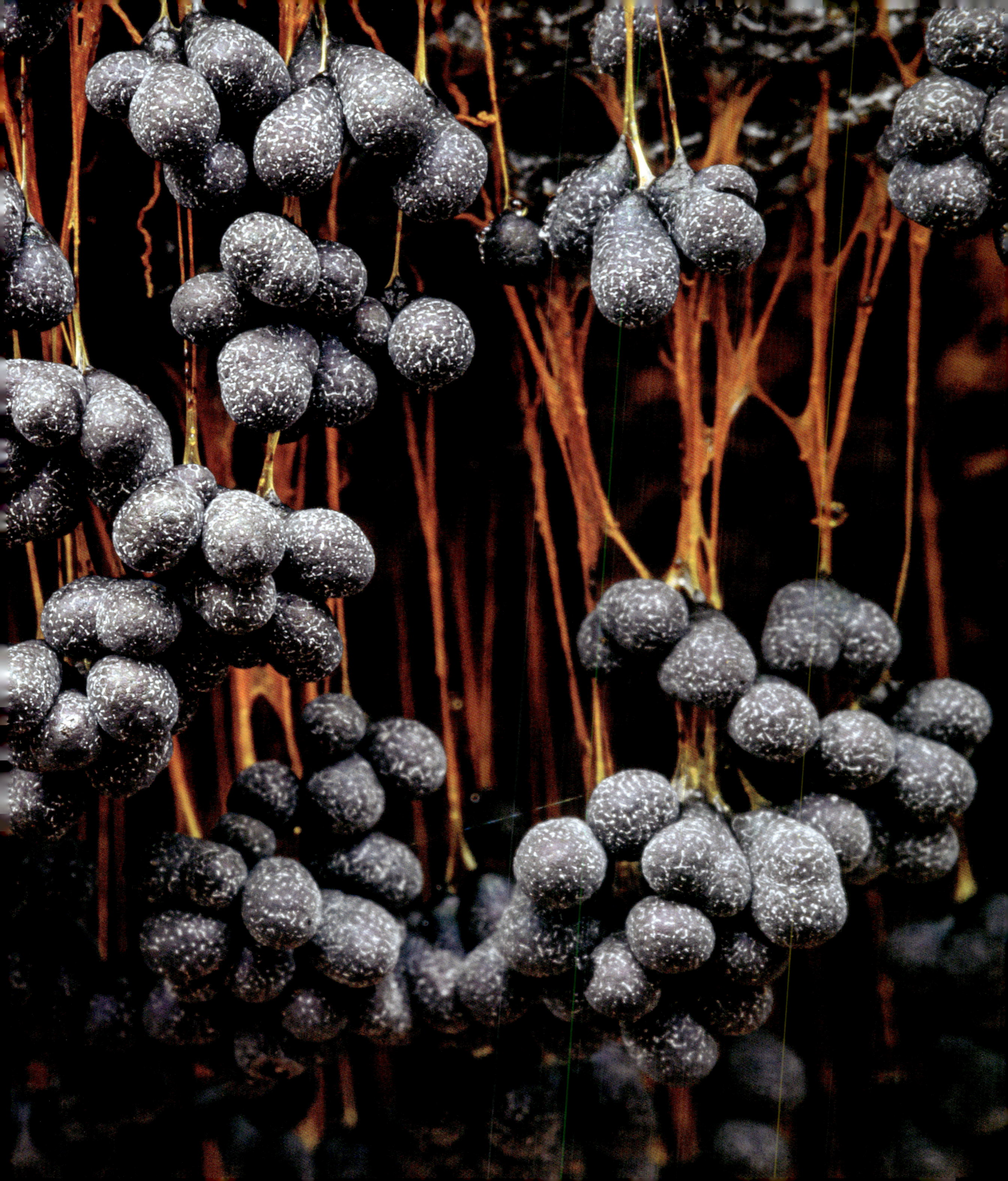

EVEN FAMILIAR OBJECTS
ARE GIVEN NEW LIFE
WITH A CLOSE-UP LENS.

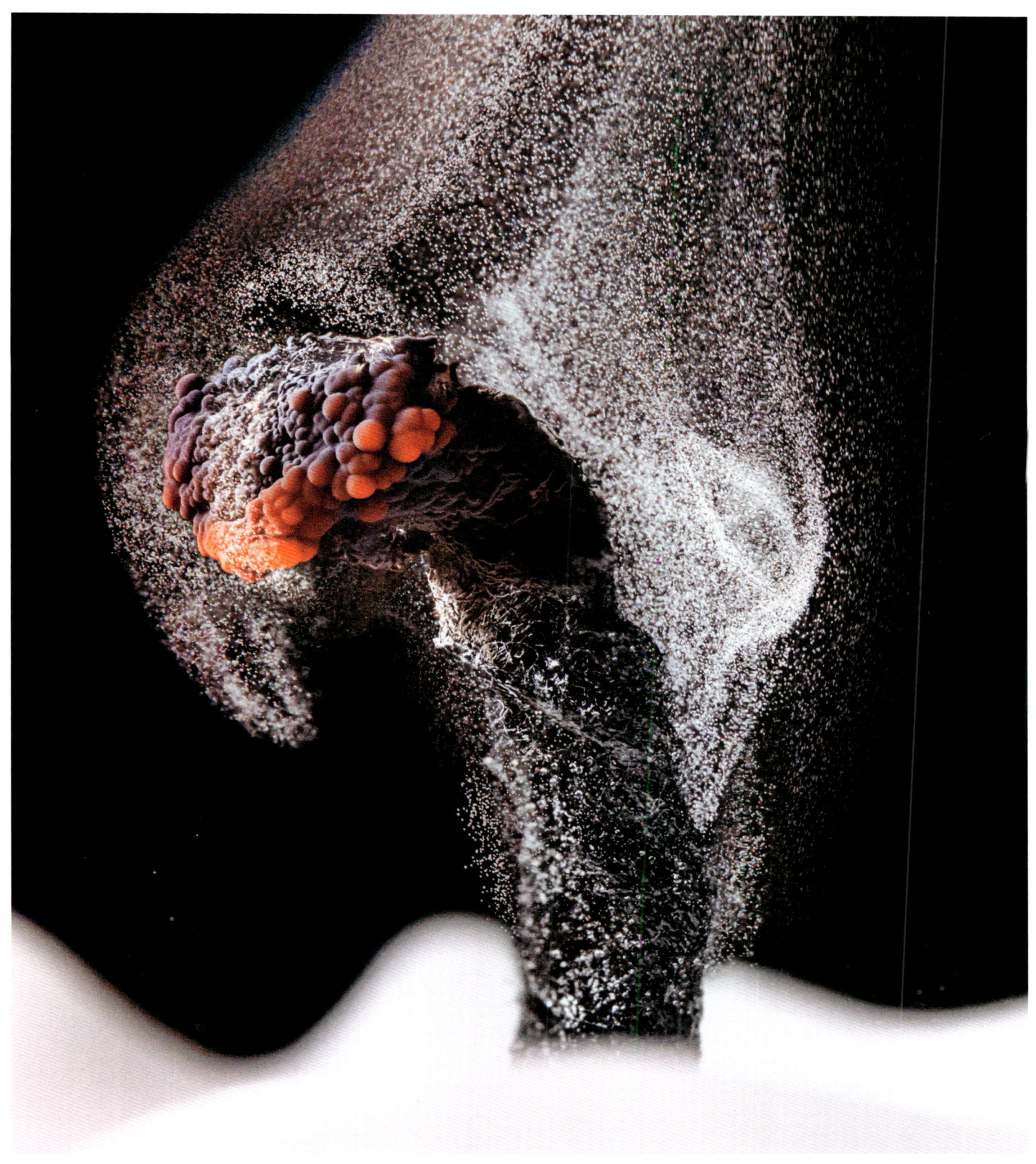

Ole Bielfeldt
Smoke from a recently extinguished candle. The residual heat from the glowing wick
breaks down wax to release a stream of unburned carbon particles.

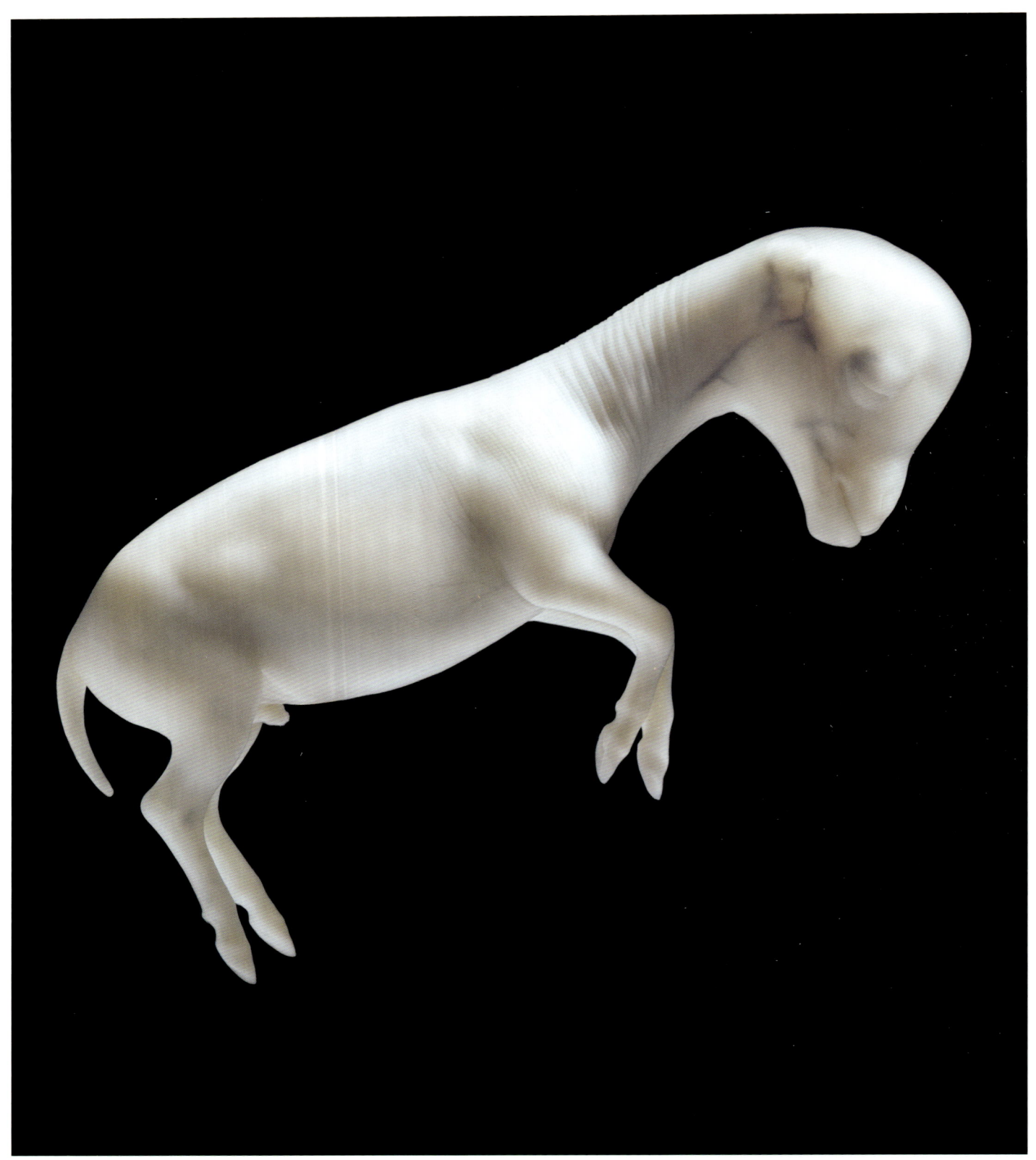

Tim Flach
A 10-week-old embryo of a Thoroughbred horse
measures only three inches long.

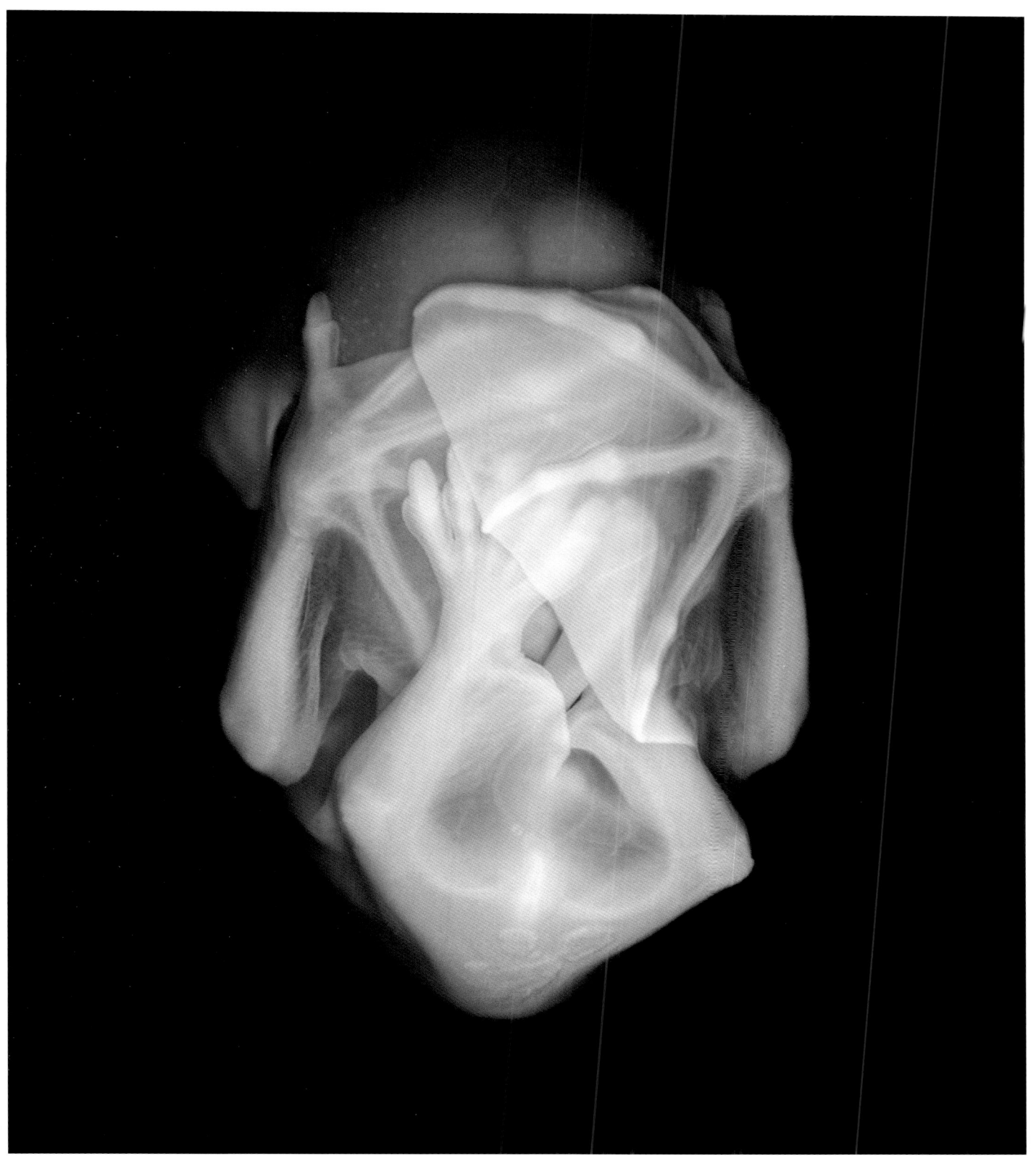

Zuzana Vavrušová and Richard Behringer
A 10-week-old embryo of a short-tailed fruit bat measures just over half an inch long
at a point in its development that scientists call the peekaboo stage.

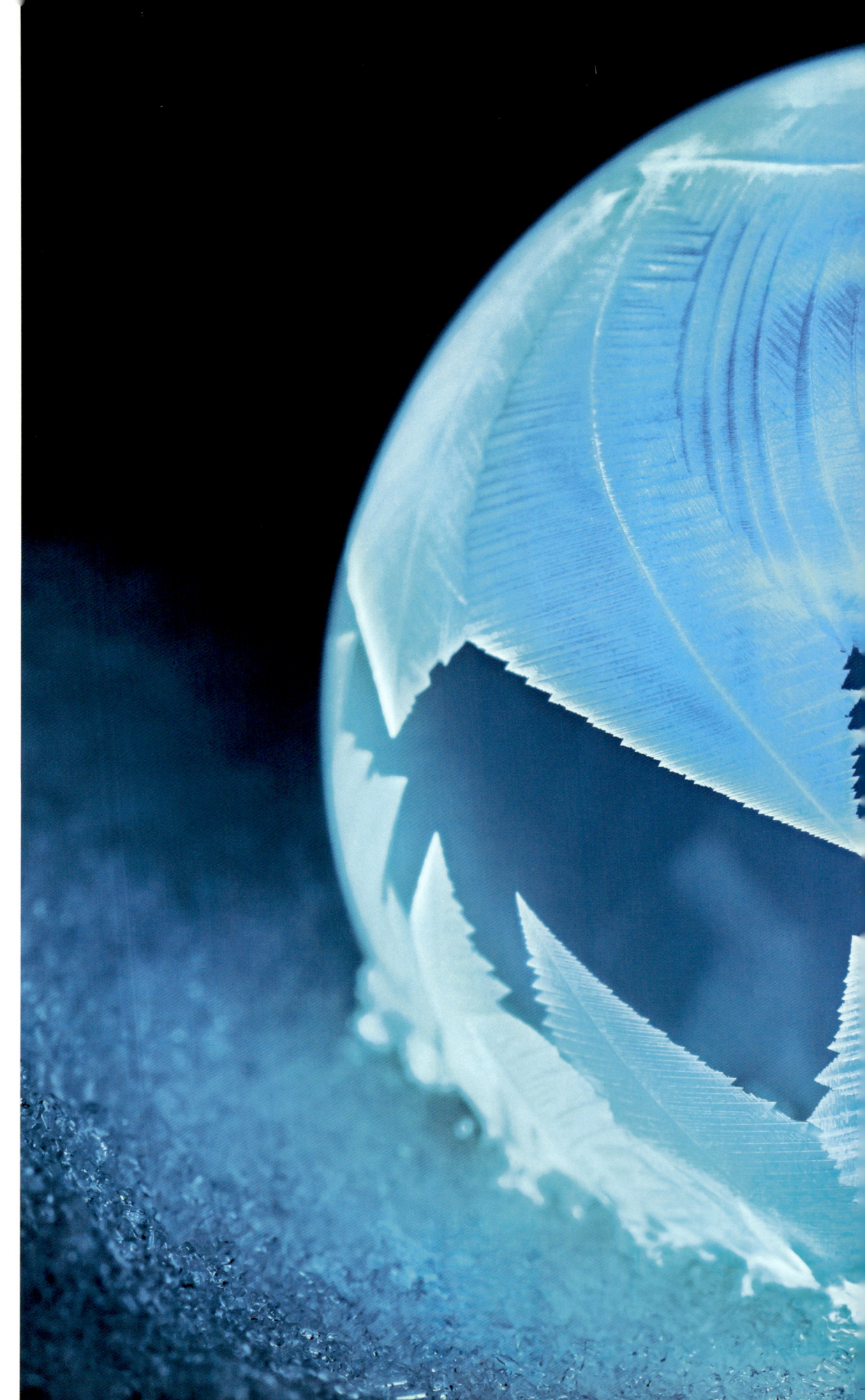

Don Komarechka
Ice crystals spread across
the surface of a soap bubble
as it freezes on a bed of snow.

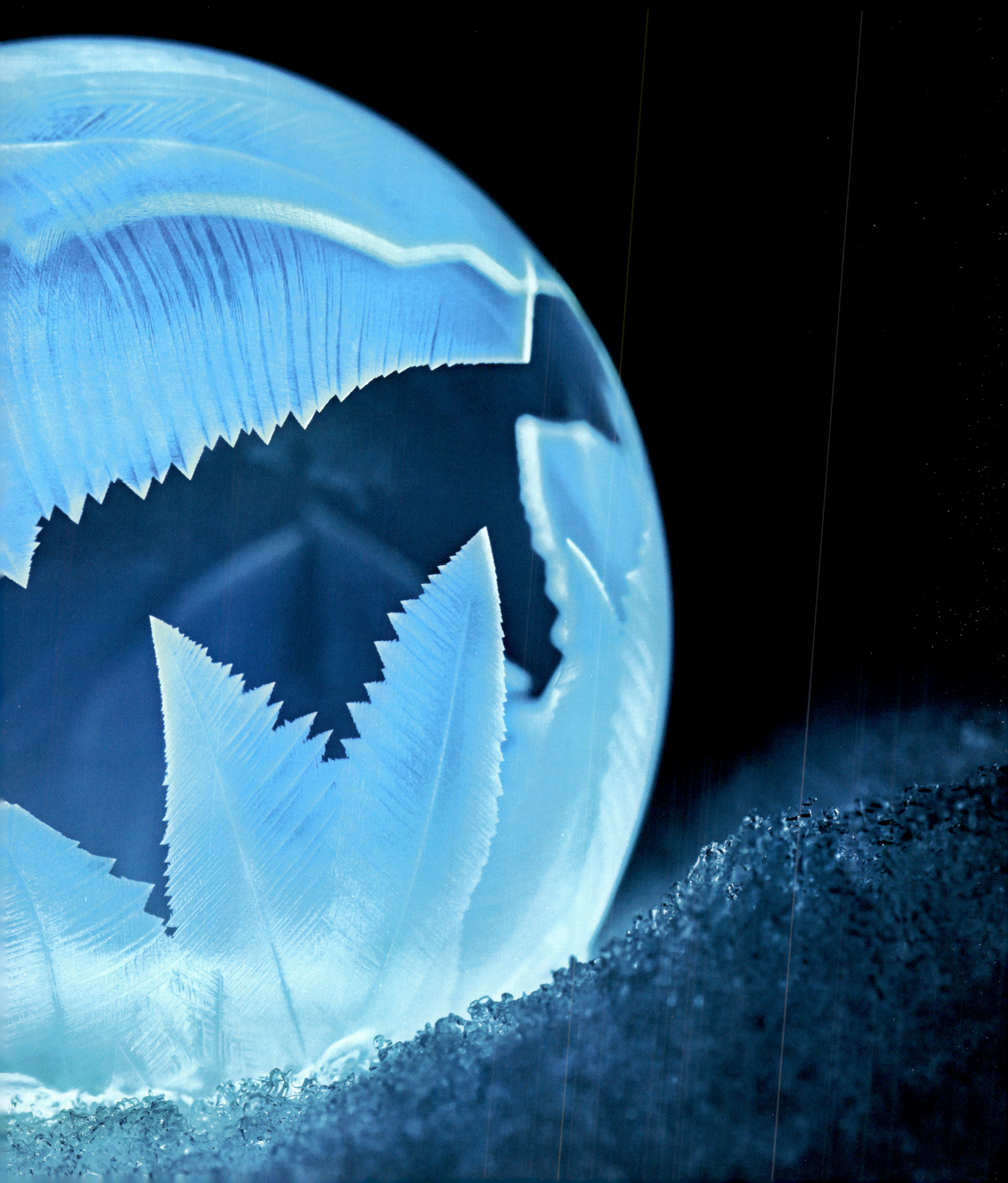

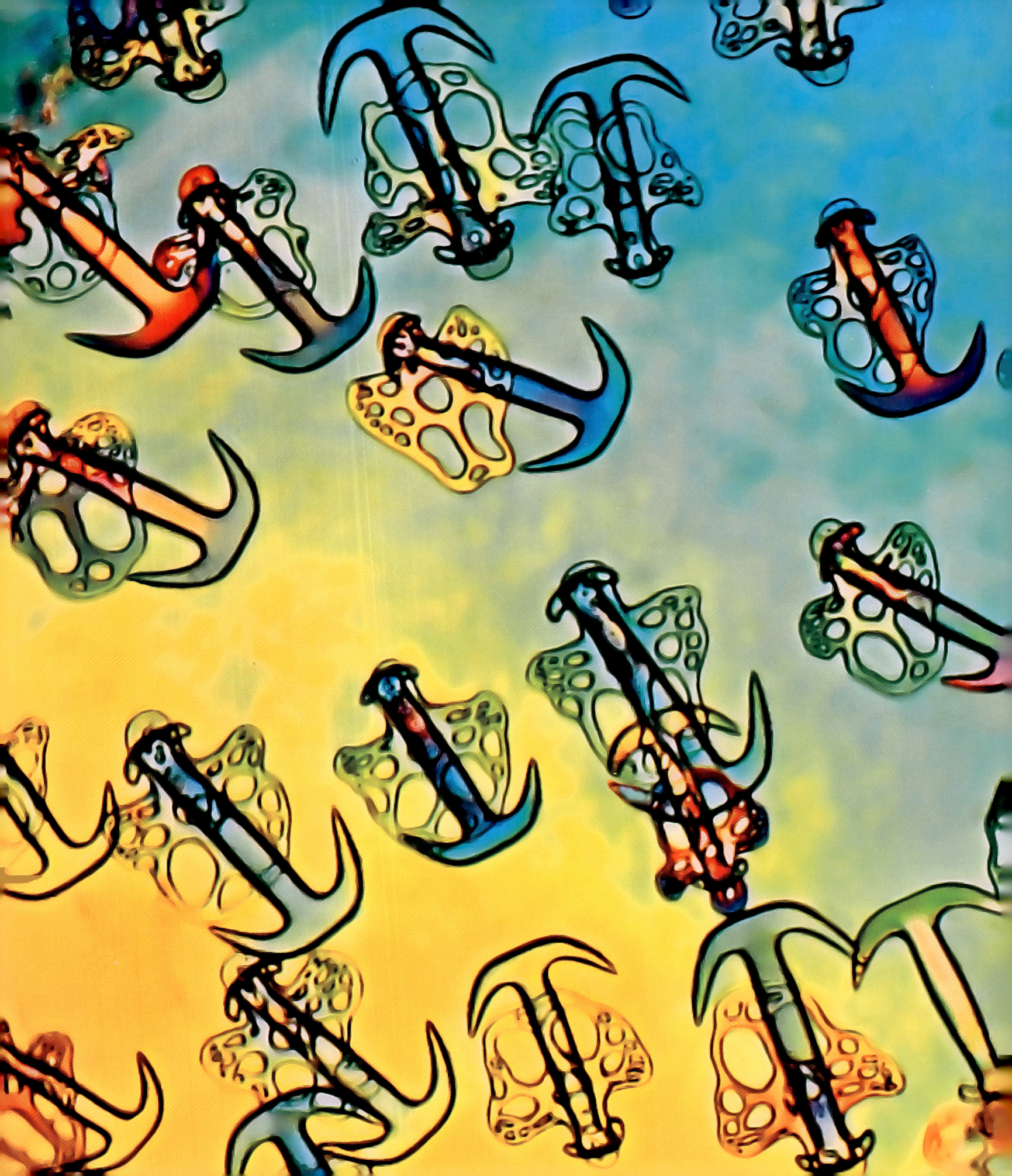

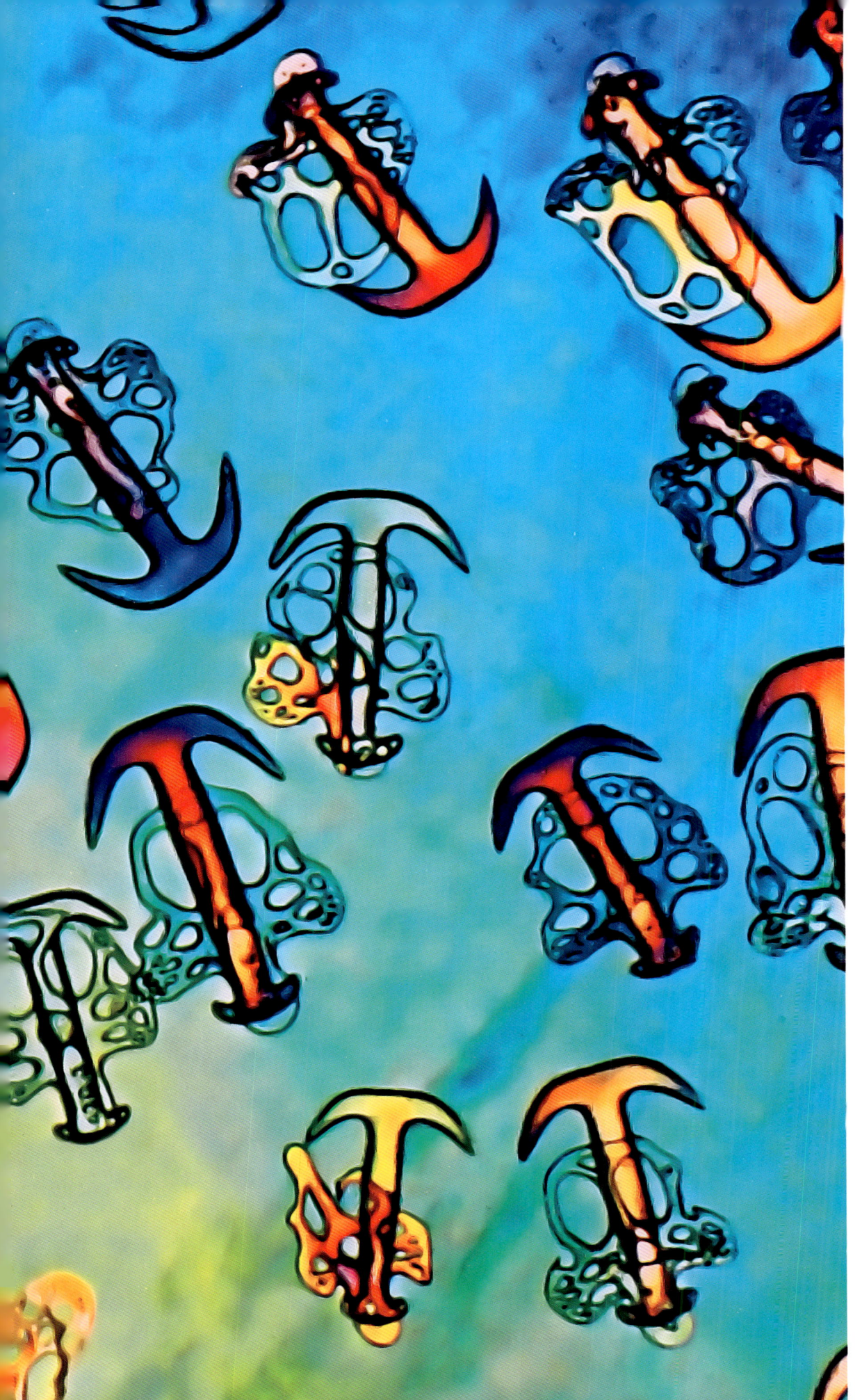

Christian Gautier
A magnified view of
a sea cucumber's skin.
The anchor-like structures,
called ossicles, provide
protection and may he p
the animal burrow.

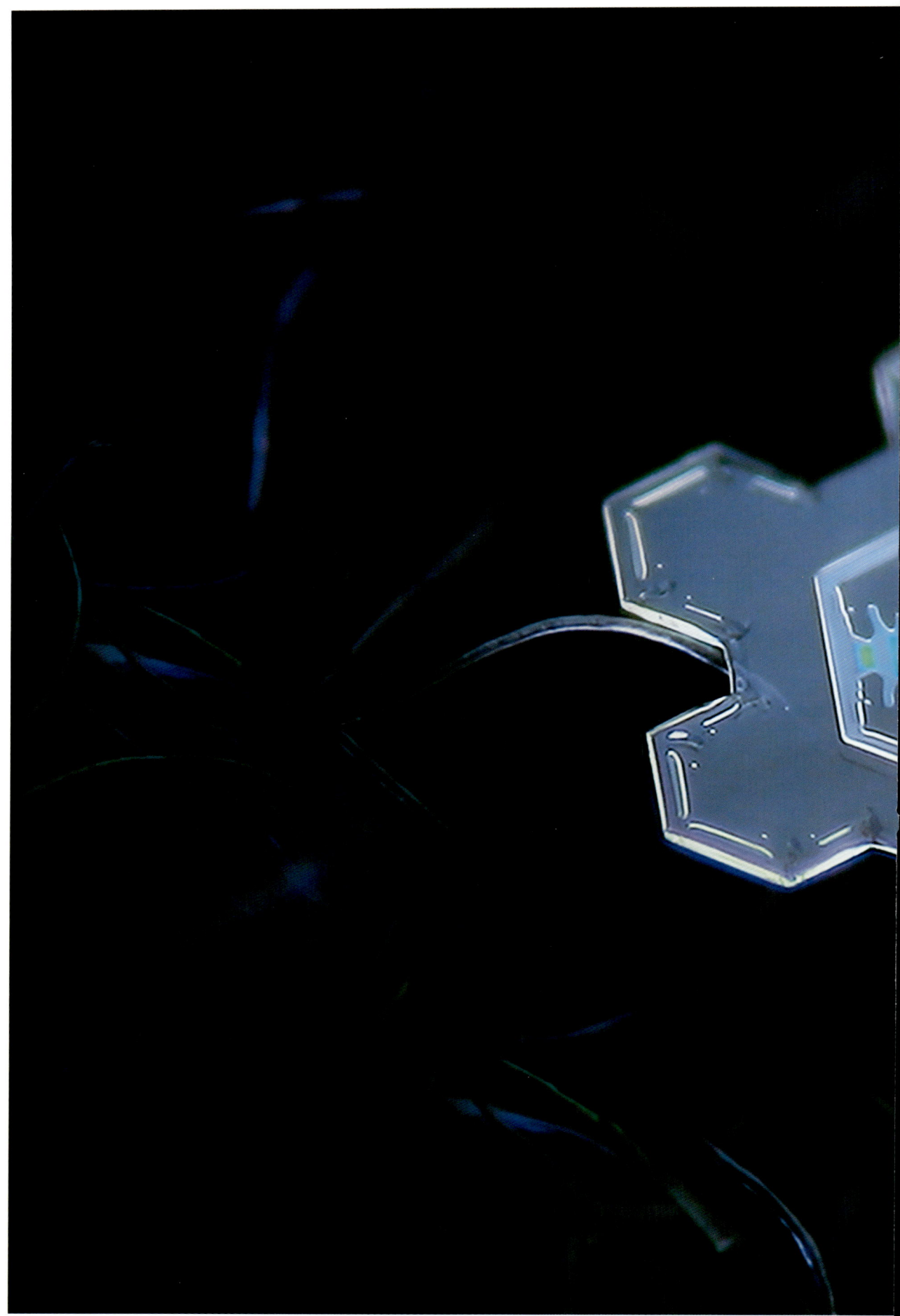

Alexey Kljatov
A snowflake rests on wool, captured on the photographer's porch in Moscow.

FOLLOWING PAGES:
Nick Selway
A vein of hot lava from Hawaii's Kilauea volcano glows through a crack in the rock.

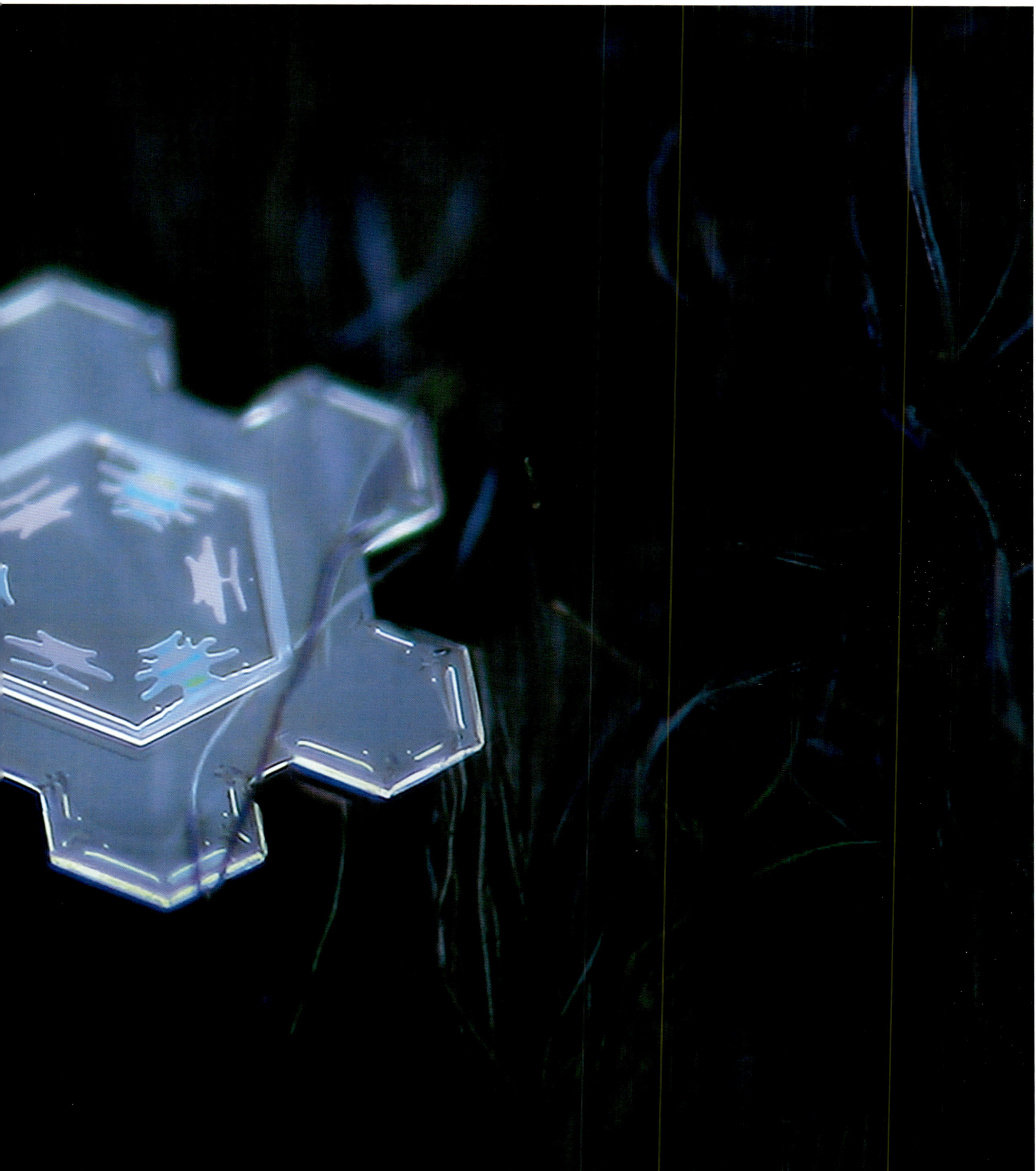

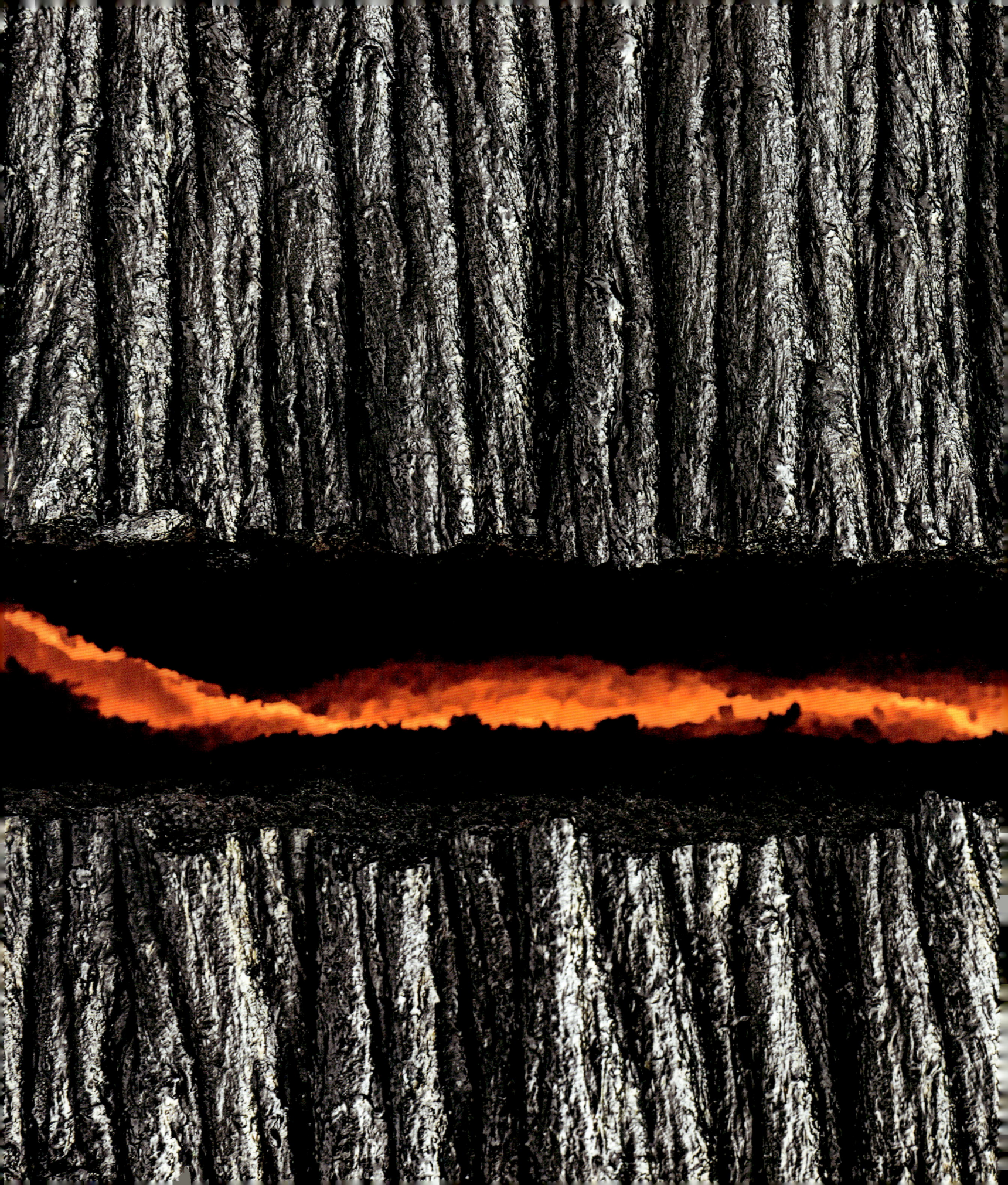

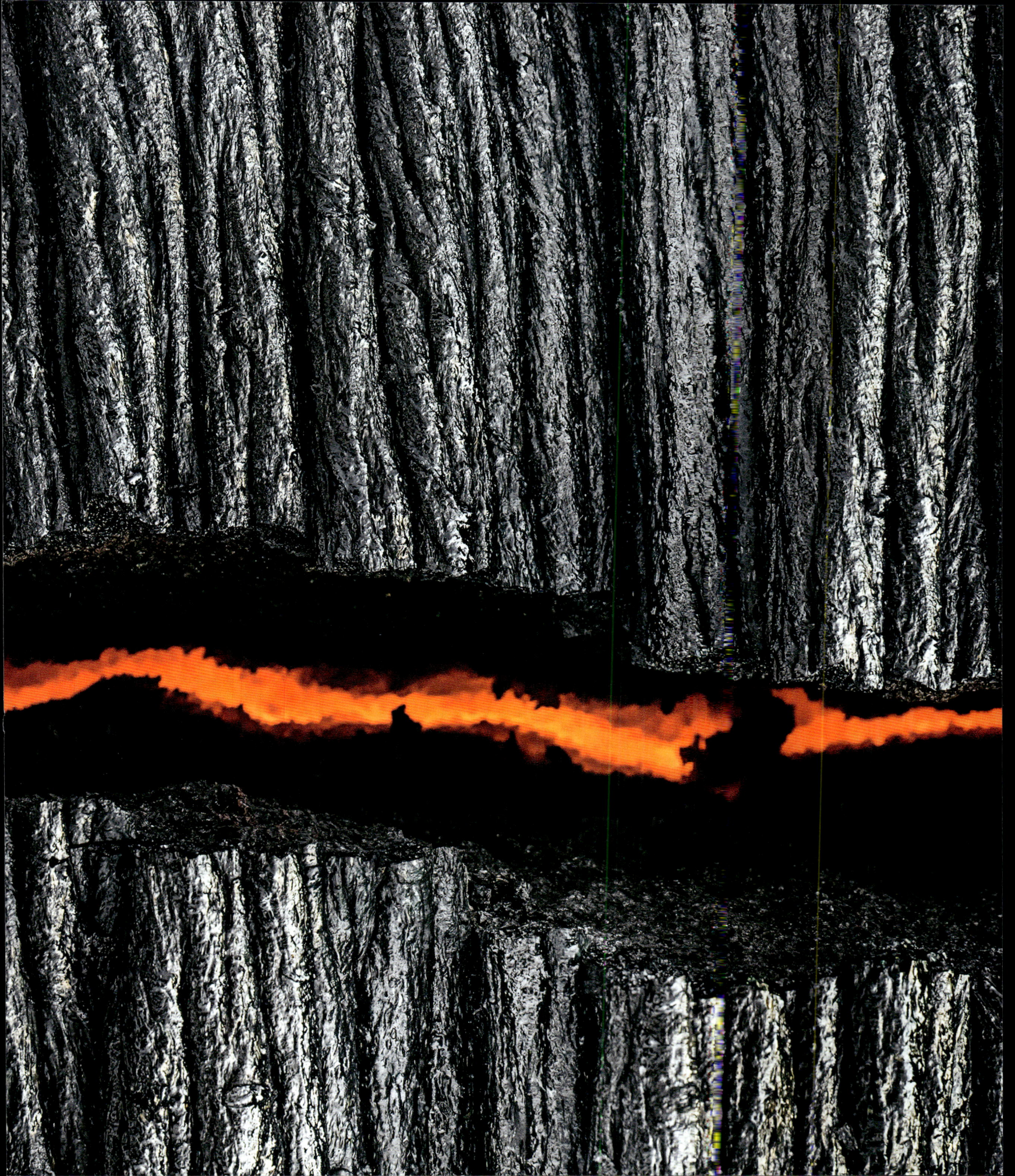

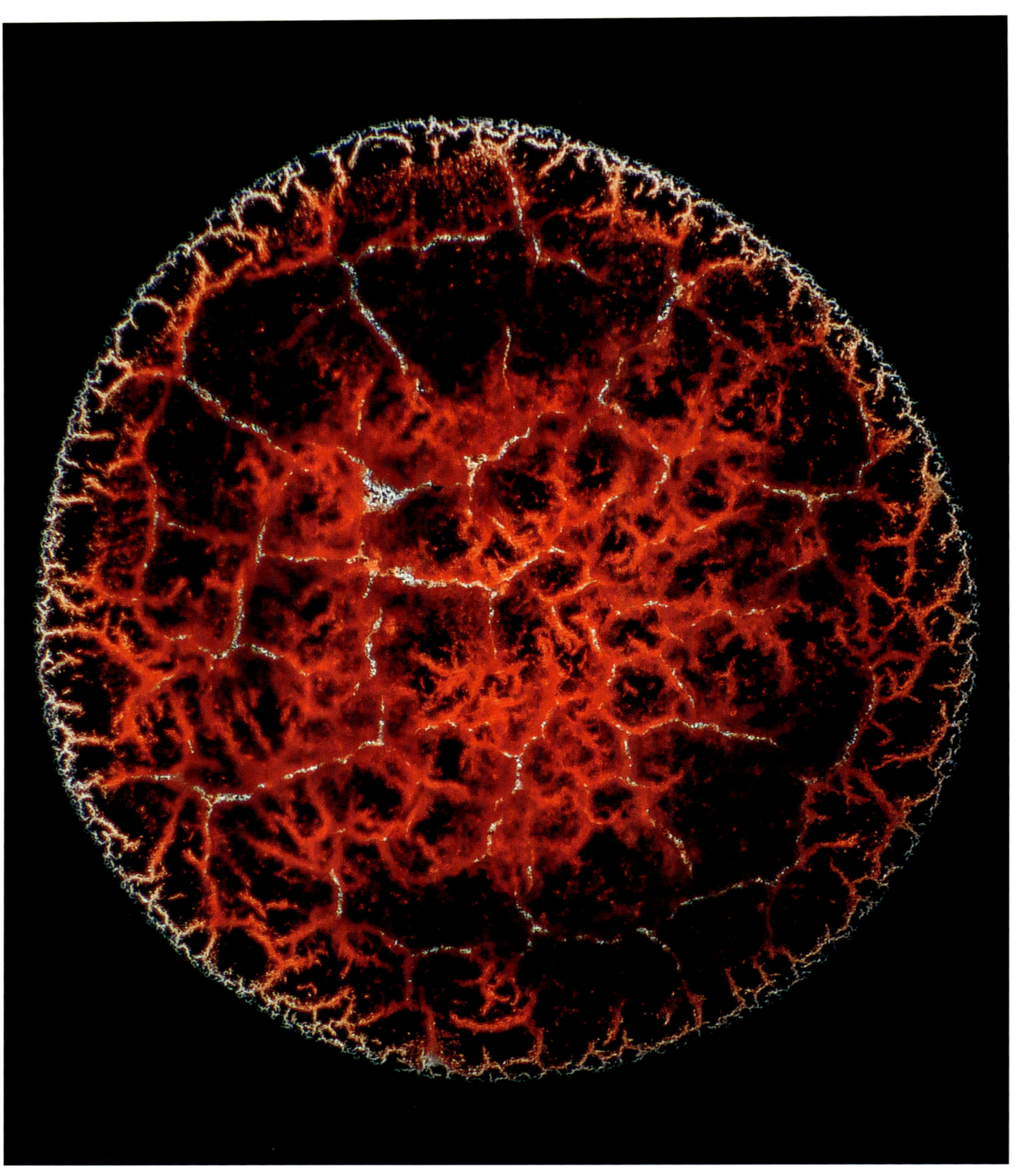

Maurice Mikkers
A dried drop of blood

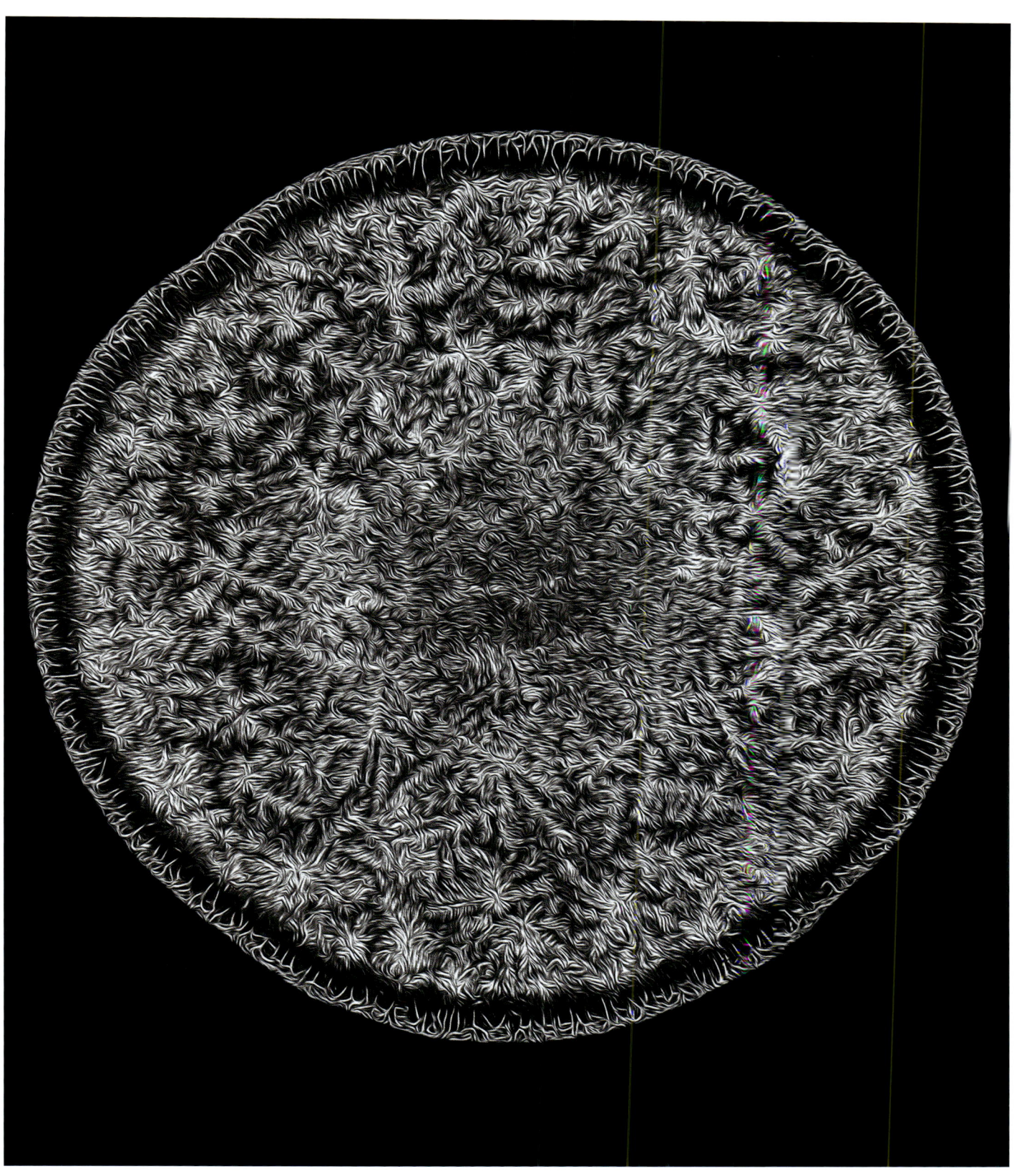

Norm Baker
A dried human teardrop

INFLUENCES
FRANS LANTING

Frans Lanting figured out how to capture the personality of a snail. That's what I thought when I first came across this image in his beautiful book *Life: A Journey Through Time.*

By bringing the camera down to ground level, Frans invites us to look at this little creature with a generosity of attention that we would never have brought to the subject on our own. We are not looking down on this snail, as we would if we had come across it on a sidewalk. Instead, we see it eye to eye, as we would a fellow being our own size. The wrinkled texture of its skin and its tentative posture add a sense of intimacy to a scene that would have been dismissed as mundane without Frans's careful consideration of perspective and detail. By magnifying this snail to a size I could relate to, this image gave me a fresh appreciation for its quiet beauty.

When I started photographing honeybees, I hoped to create the same impact. I wanted to establish a connection between people and bees that would not have otherwise existed between beings of such different sizes. But how does someone relate to a bee?

I had read that Chris Mullin's lab at Penn State University was using a special contraption to study honeybee memory, and I thought this could be interesting to photograph. My original plan was to

Frans Lanting
A close-up of a land snail

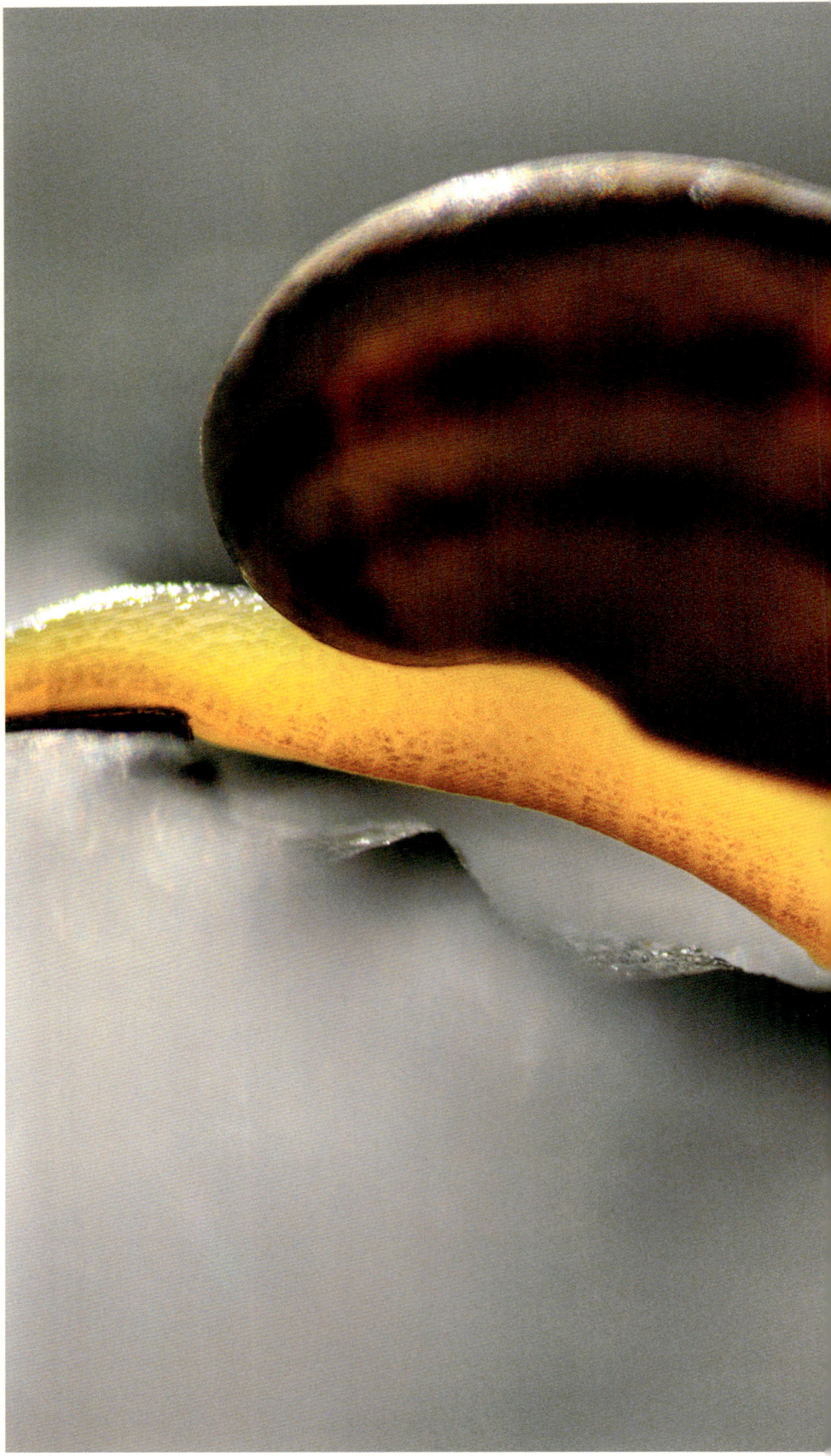

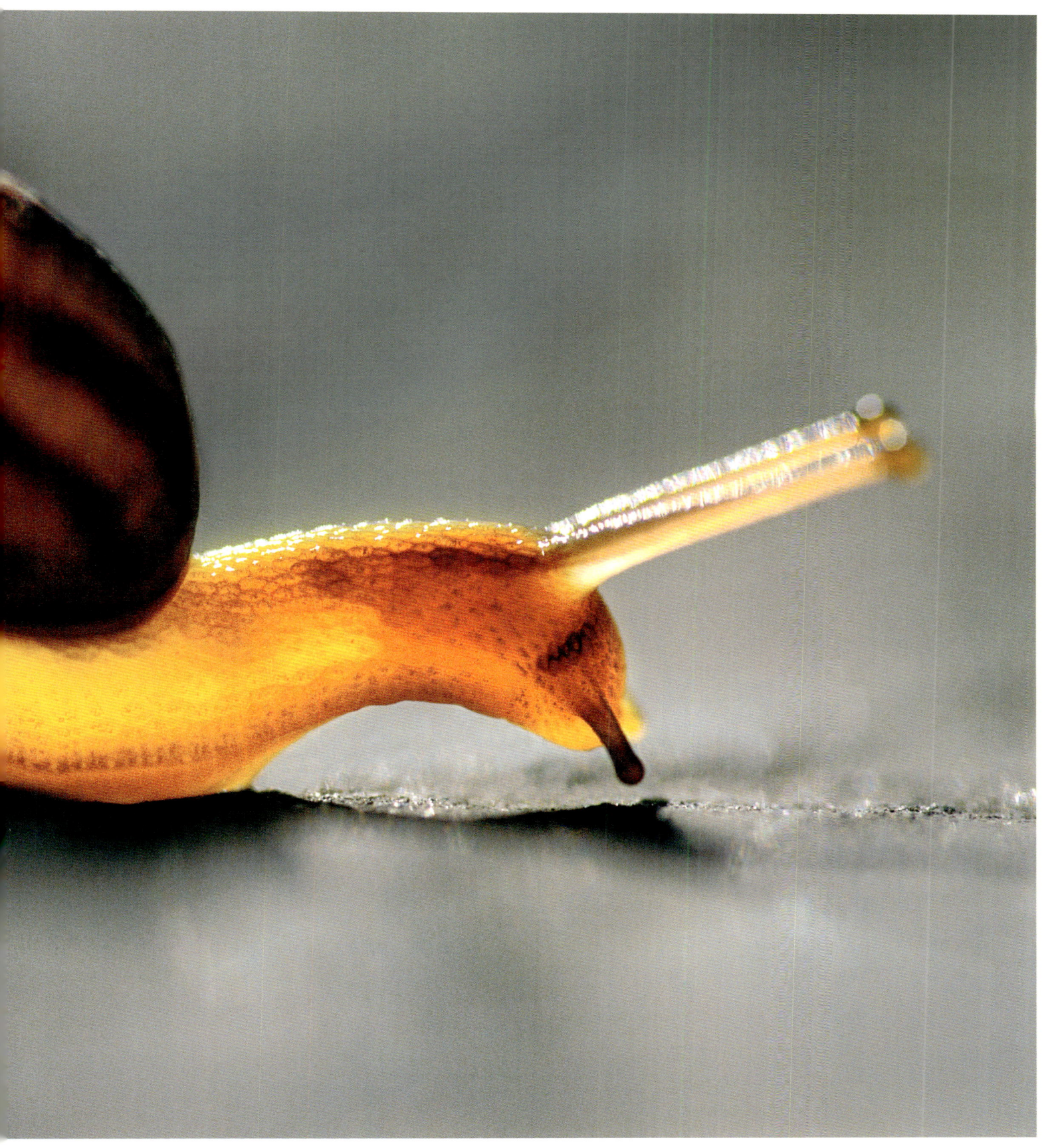

include all the tubes, wires, and machinery that went into the experiment. But it was too hard to make sense of all the components, and the bee appeared as just a tiny speck in the middle of the apparatus. I gave up and decided to cut out most of the scientific equipment and zoom in on just the bee. When I did, I realized that I could see the bee's tongue sticking out to lick the cotton swab as if it were an ice-cream cone.

Just as Frans Lanting has given us a new way of seeing snails, the resulting image expanded the size of this tiny scene to give us an intimate look into this bee's world.

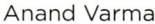

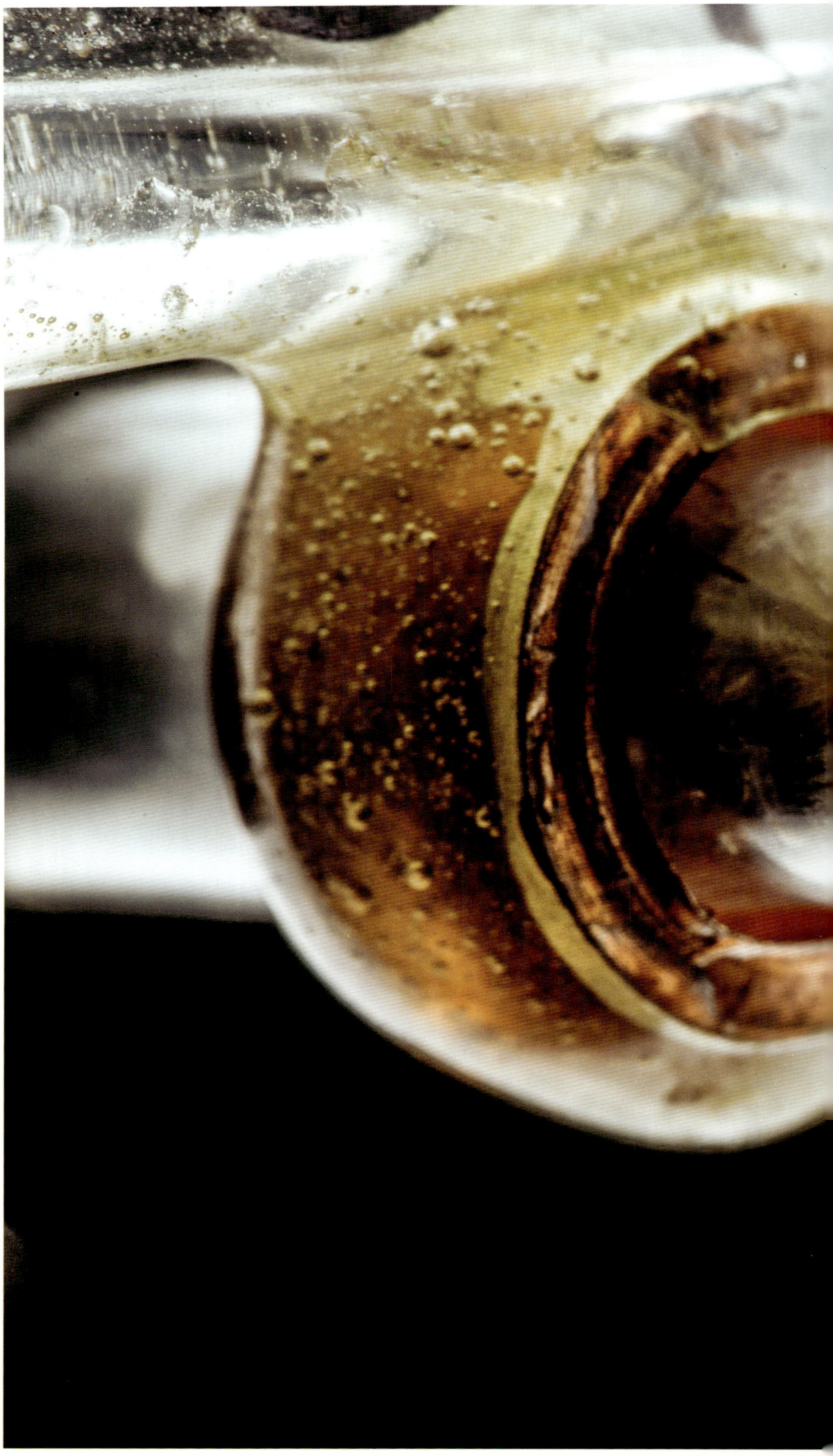

Anand Varma
A close-up of a bee sipping sugar water from a cotton swab, part of an experiment to measure the effect of agricultural chemicals on honeybee memory. Researchers compared the rate at which two groups of bees—one group that had been exposed to farm sprays and one that had not—learned that puffs of scented air will be followed by sugary rewards.

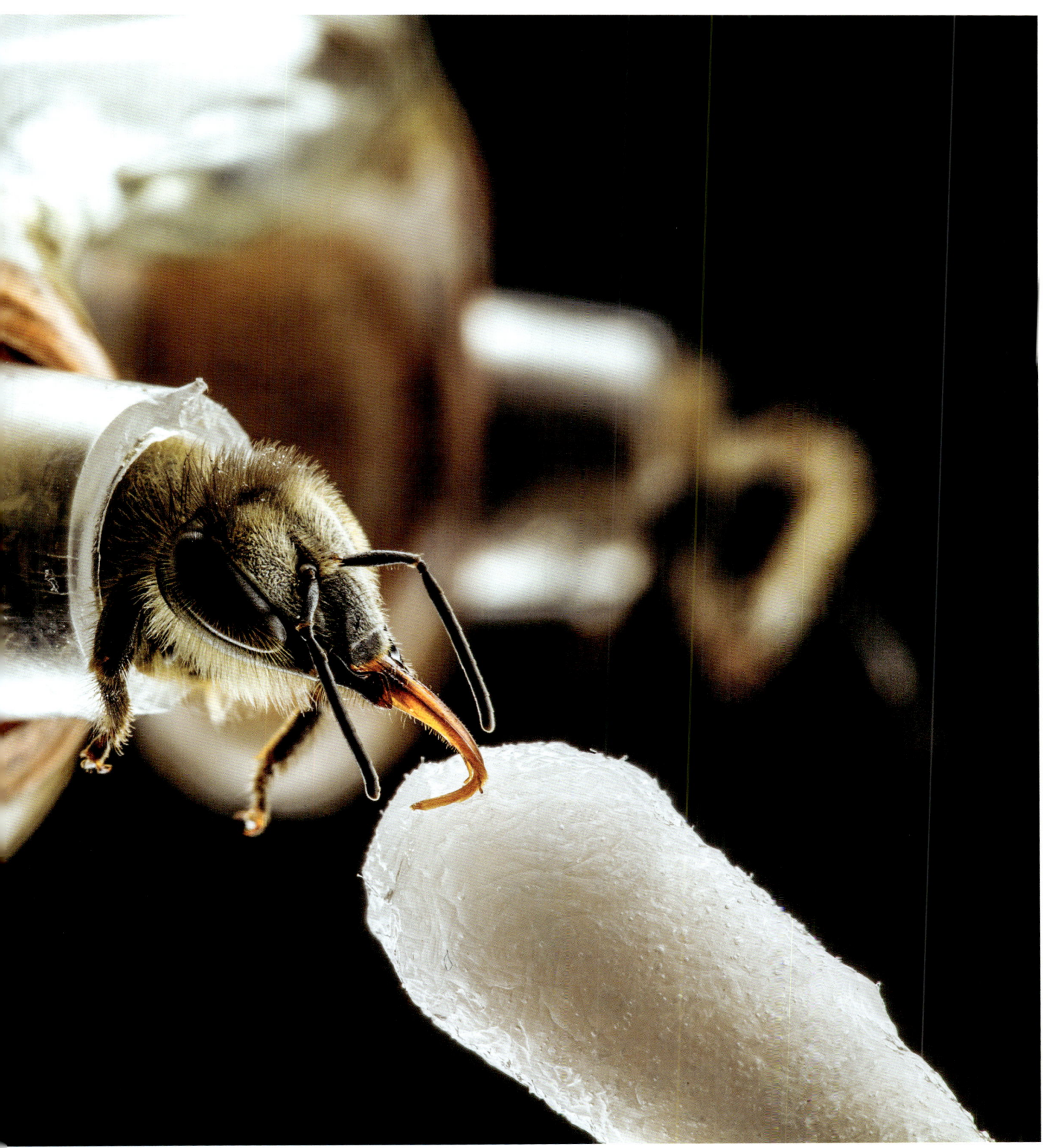

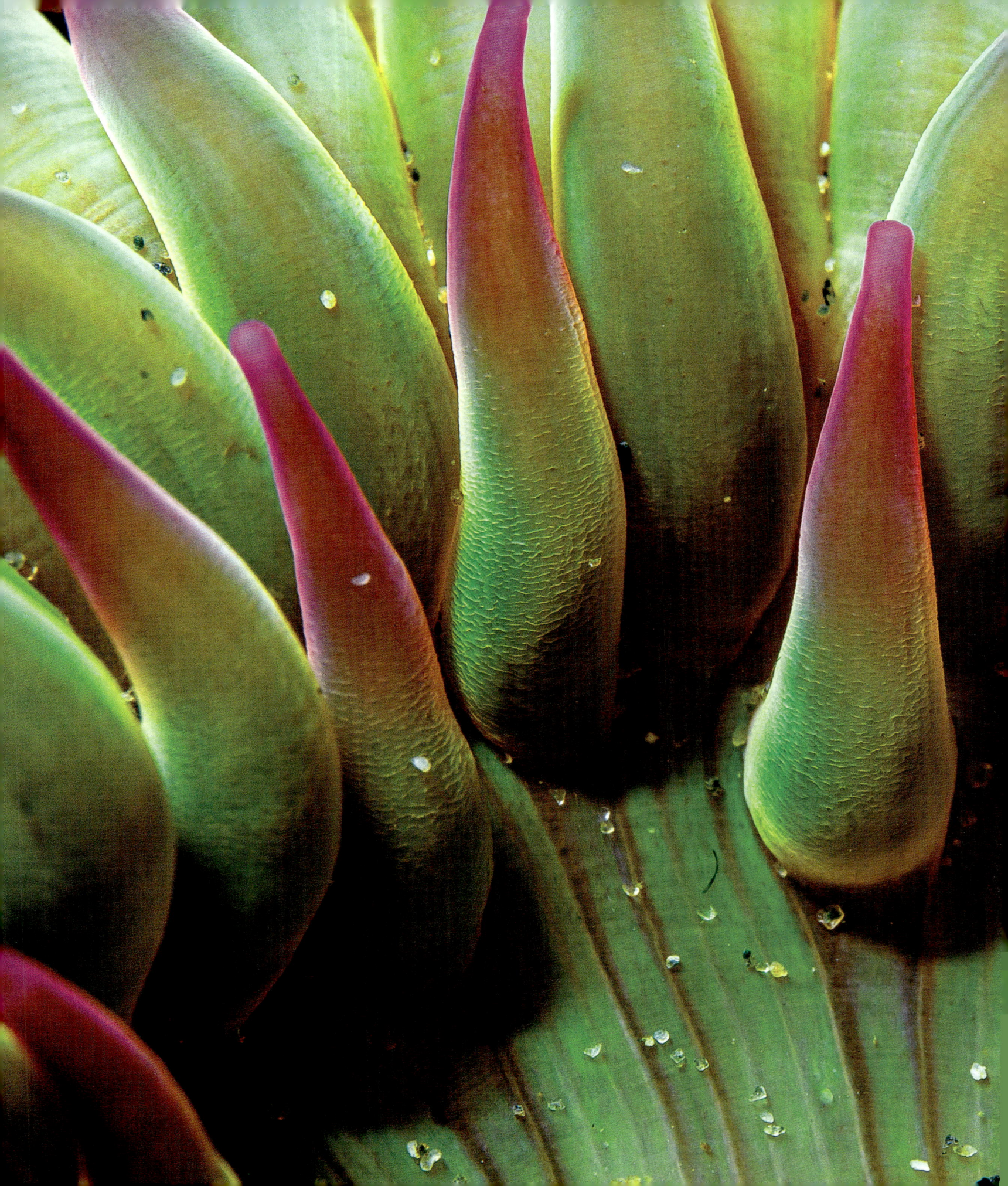

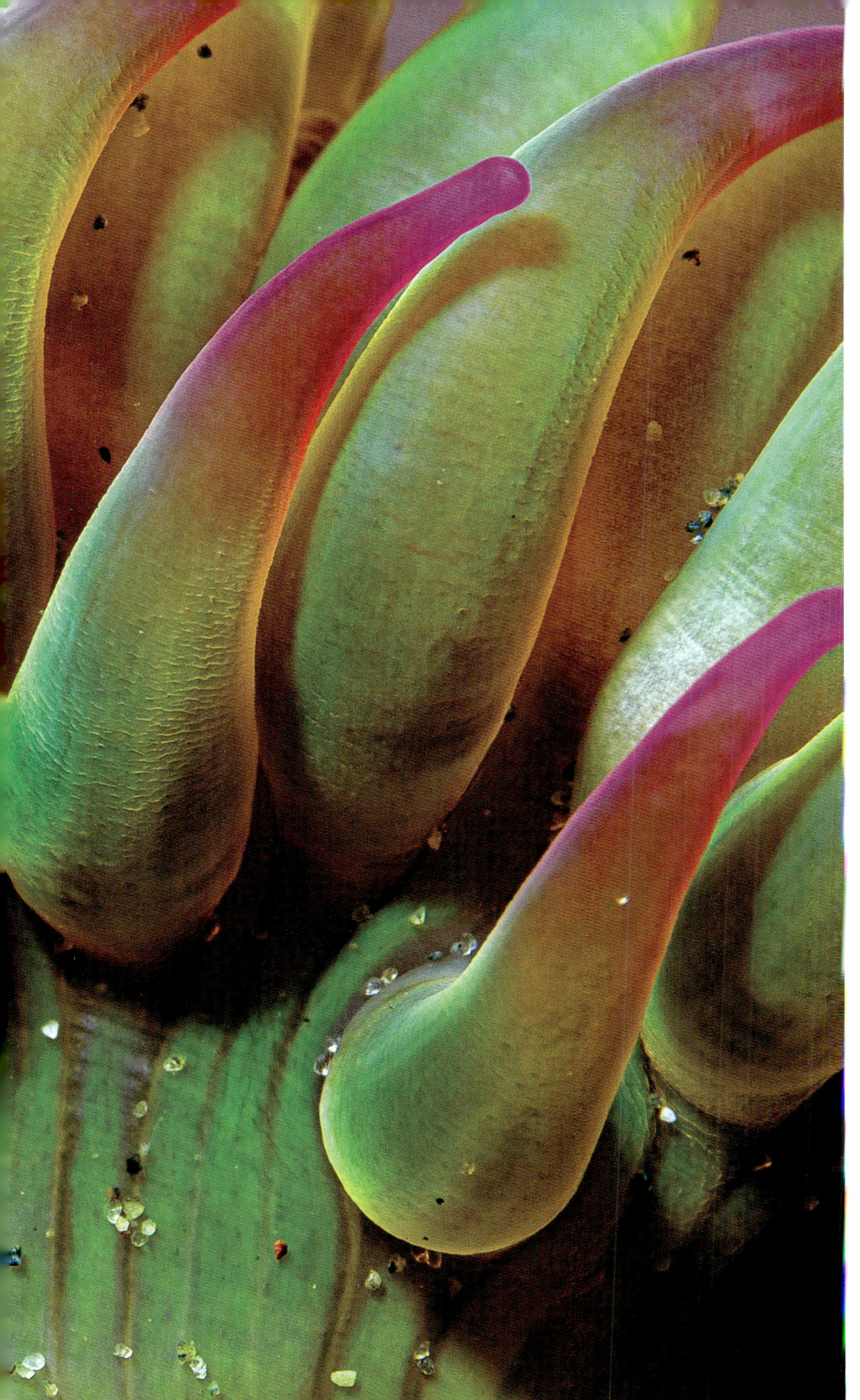

Anand Varma
The tentacles of an
aggregating anemone,
a species found along
the Pacific coast of North
America that uses special zed
tentacles to wage war
on neighboring colonies
of anemones

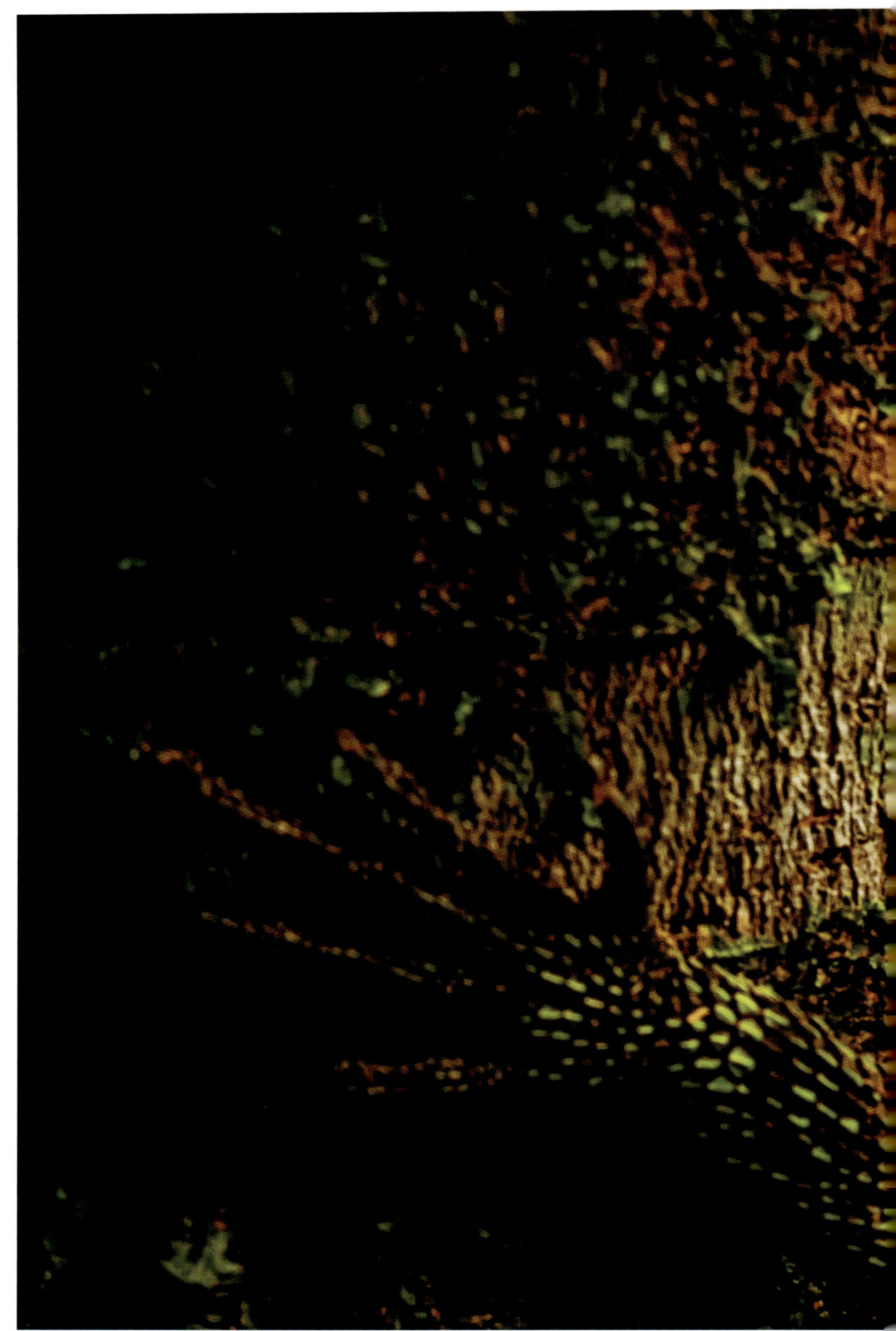

Javier Aznar
A red-eyed dwarf iguana
rests on a tree in the Chocó
rainforest of Ecuador.

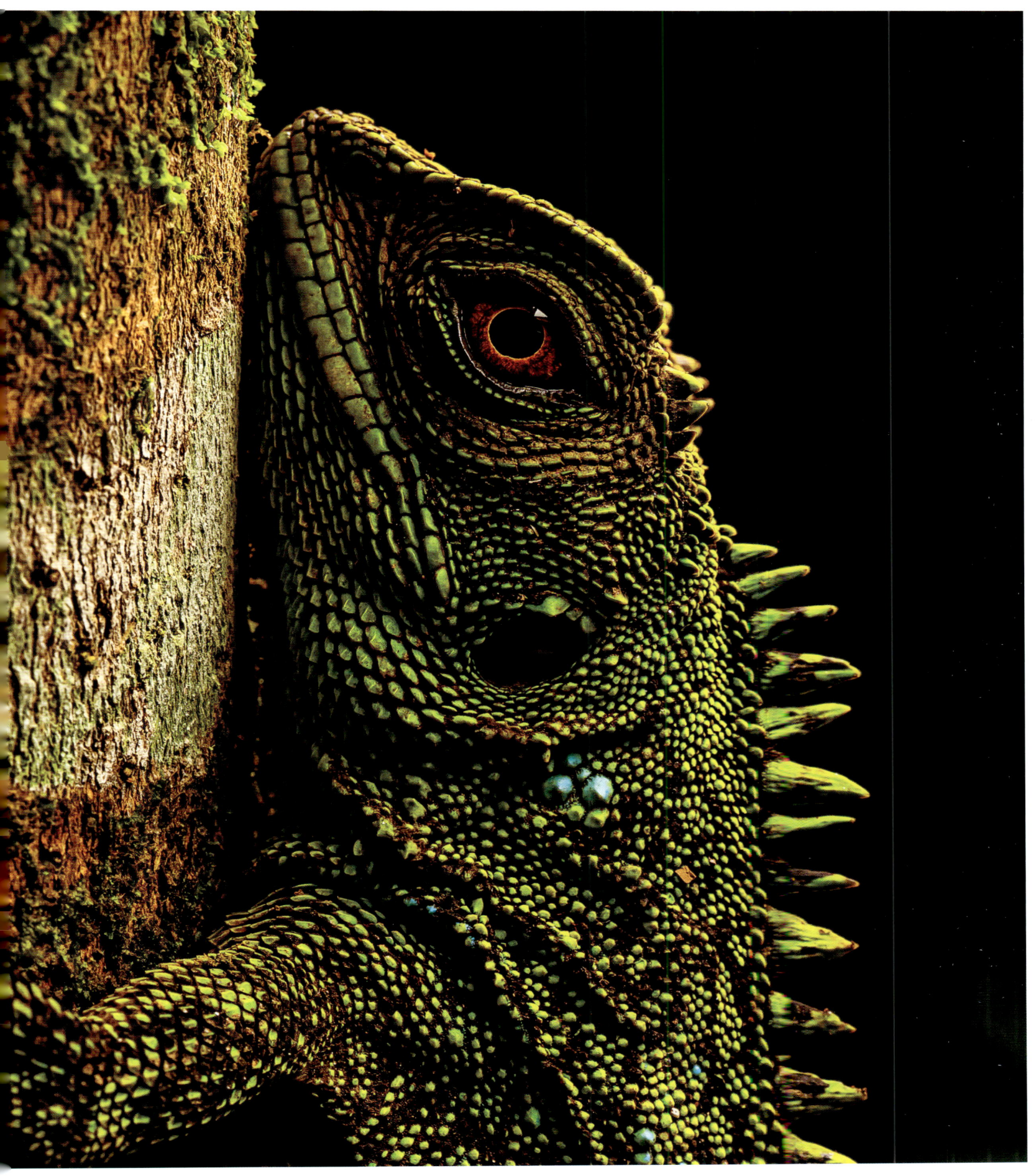

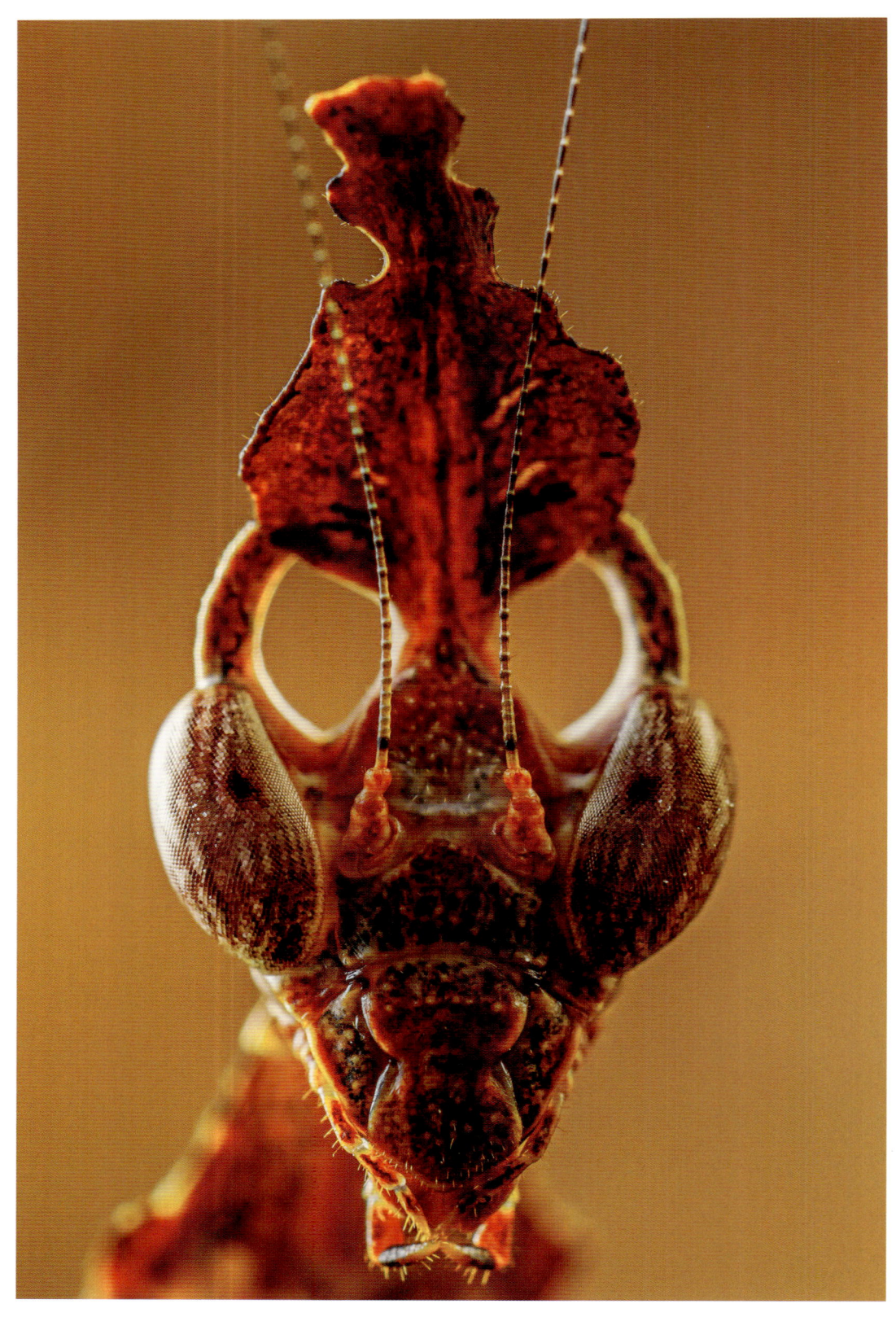

Javier Aznar
A dragon mantis camouflages itself by mimicking
leaves in Yasuní National Park, Ecuador.

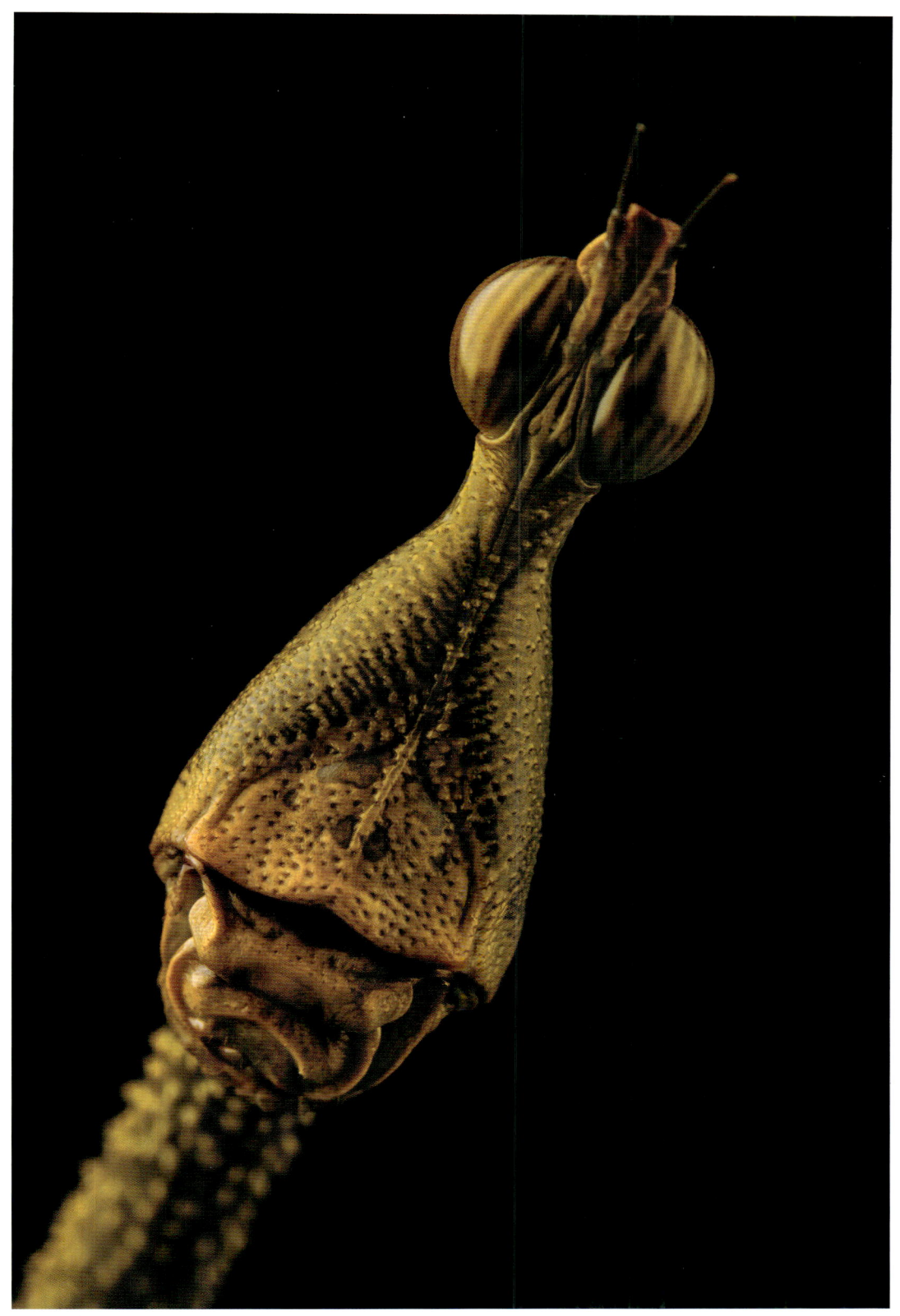

Javier Aznar

This flightless stick grasshopper, native to the tropics of the Western Hemisphere
relies on camouflage to hide from predators.

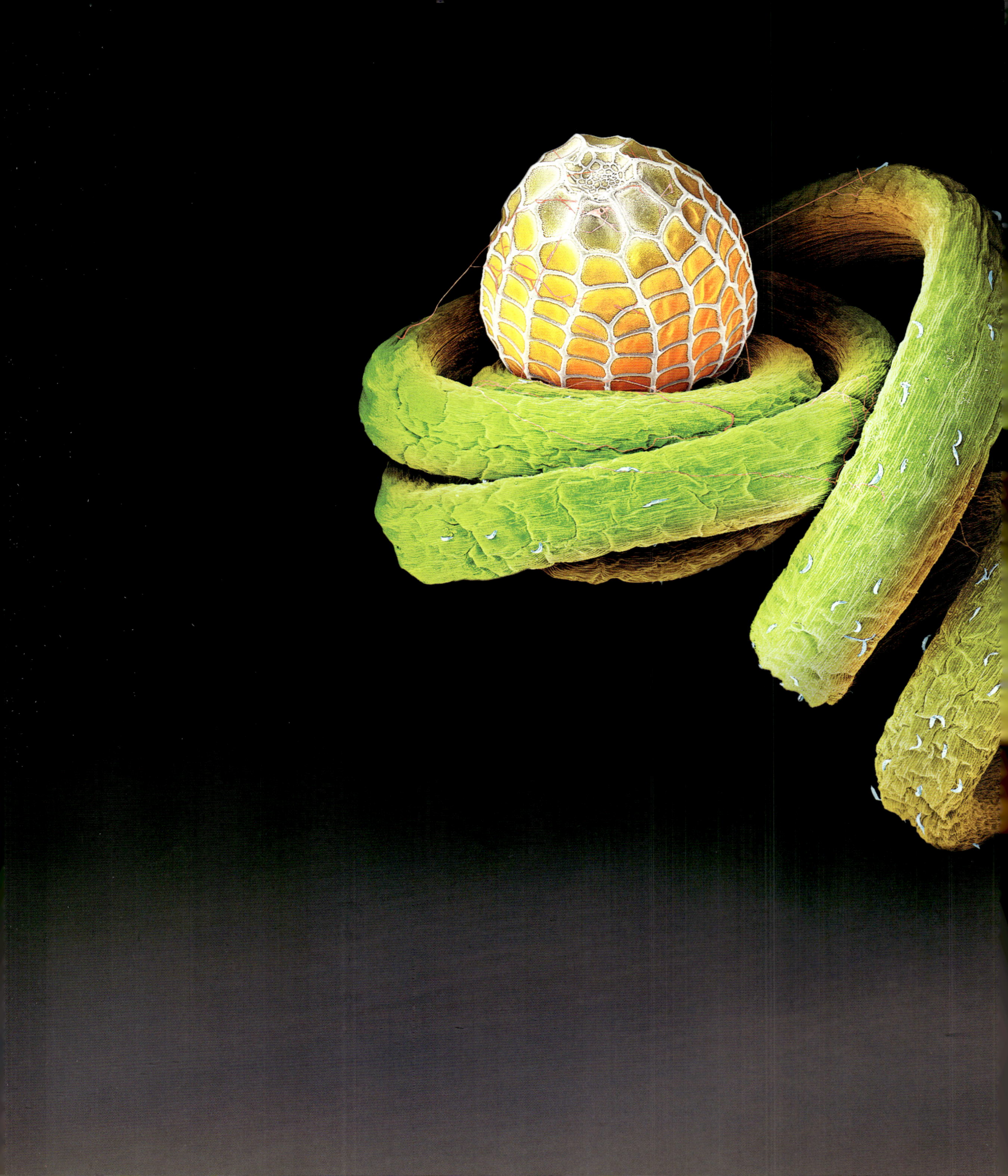

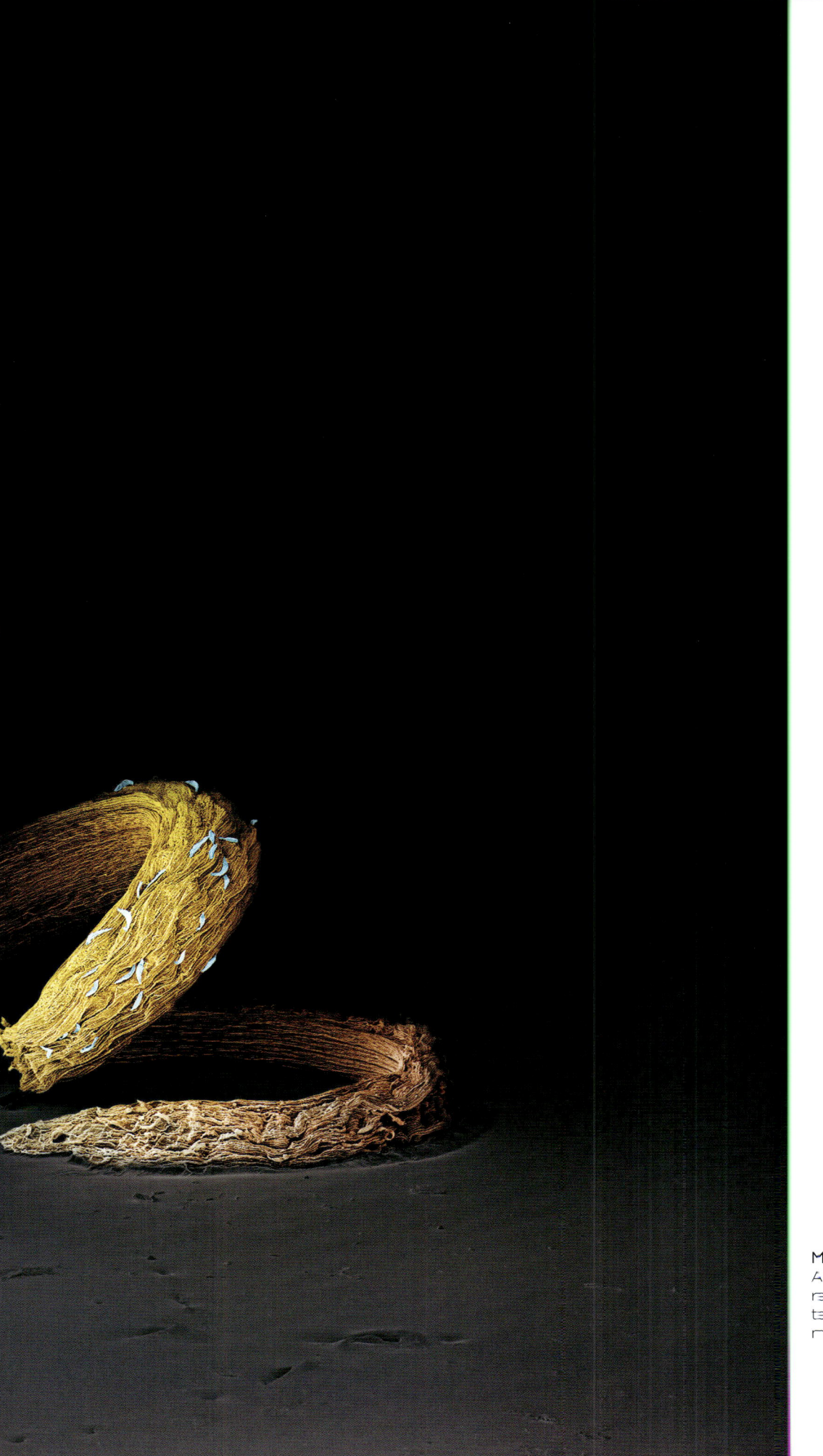

Martin Oeggerli

An egg of a Julia butterfly rests on a coiled passionflower tendril in this scanning electron microscope image.

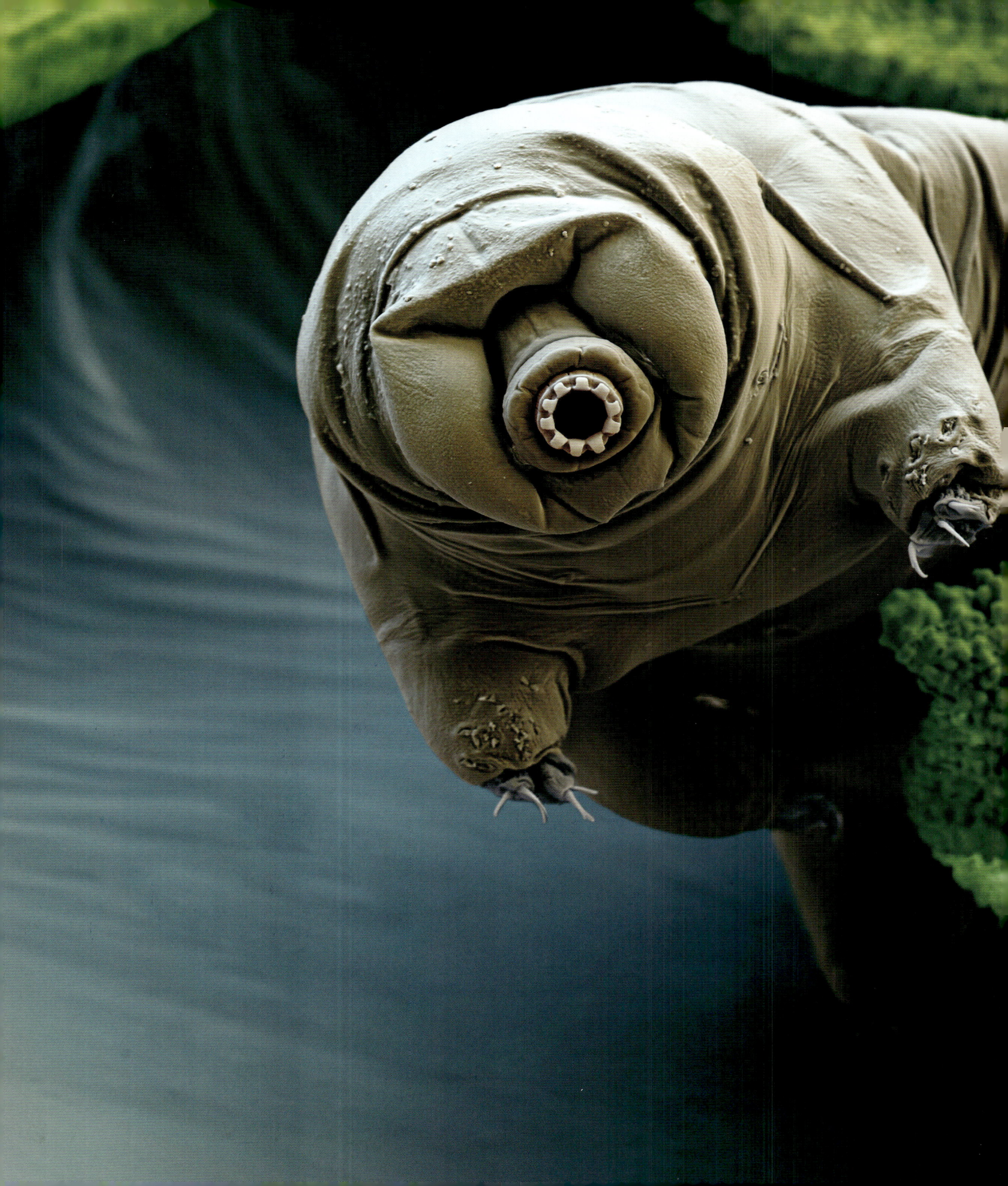

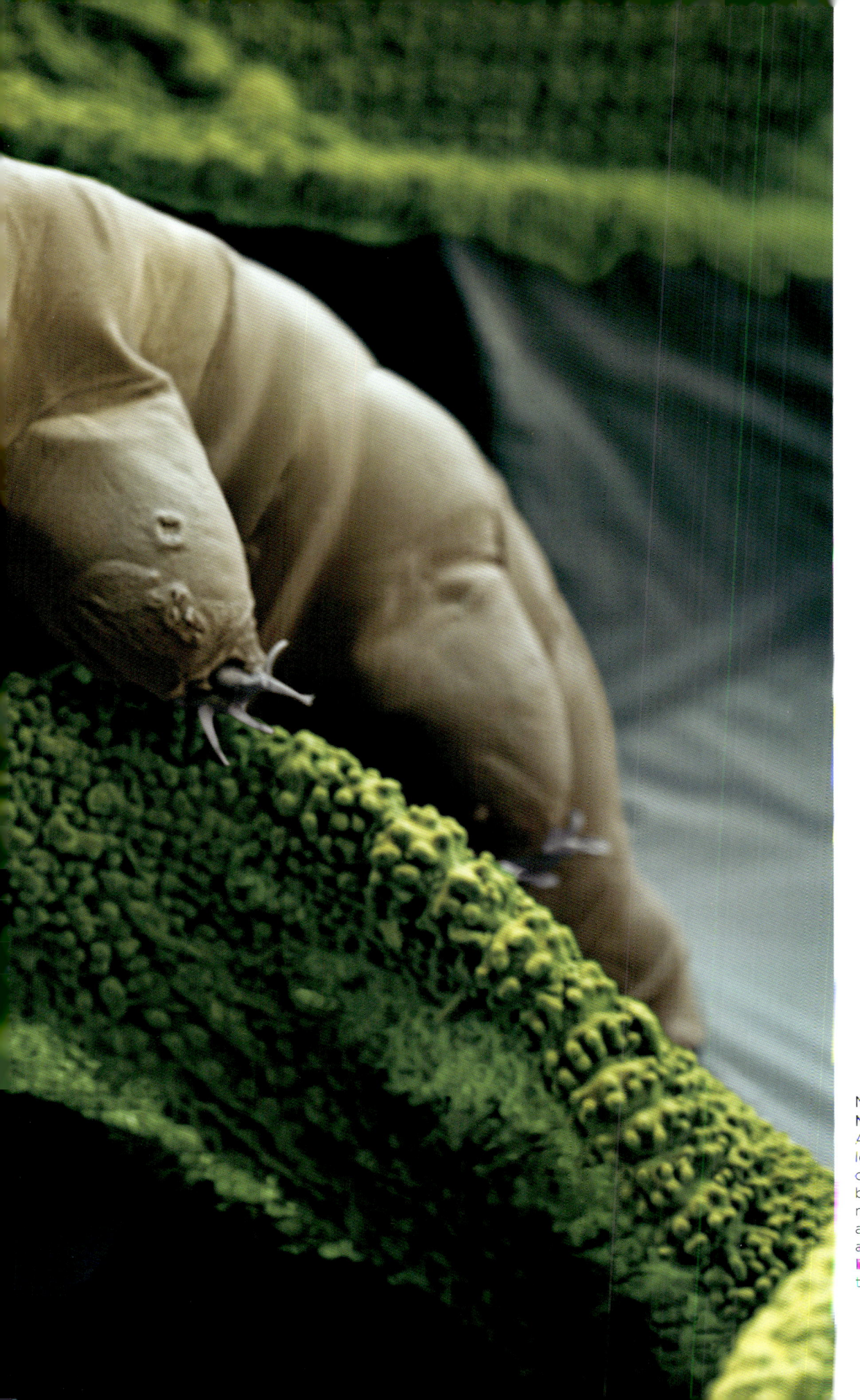

Nicole Ottawa and Oliver Meckes
A tardigrade, measuring less than a millimeter long, captured on a bit of moss by a scanning electron microscope. These creatures, also known as water bears are one of the most resilient life-forms, even surviving in the vacuum of space.

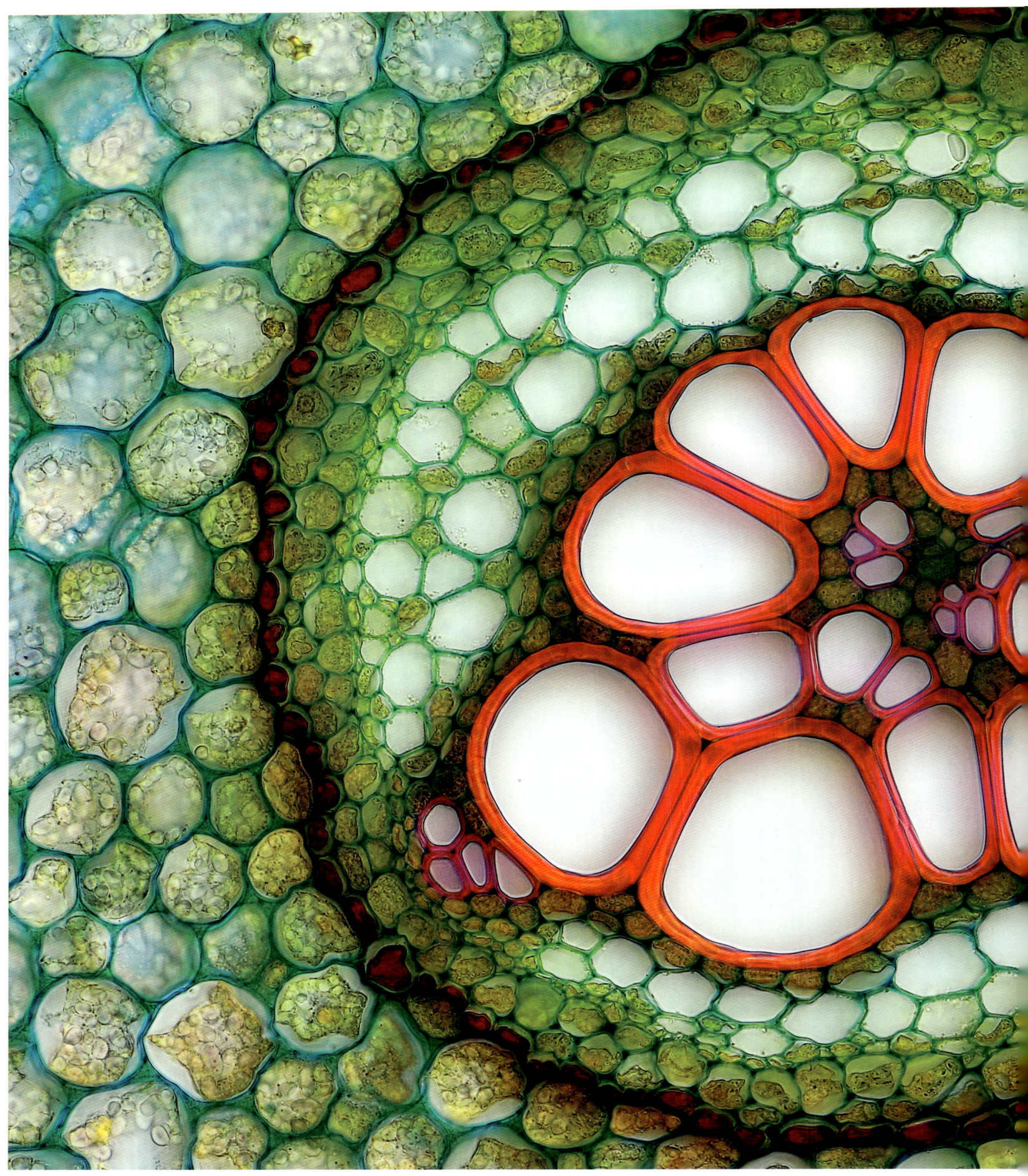

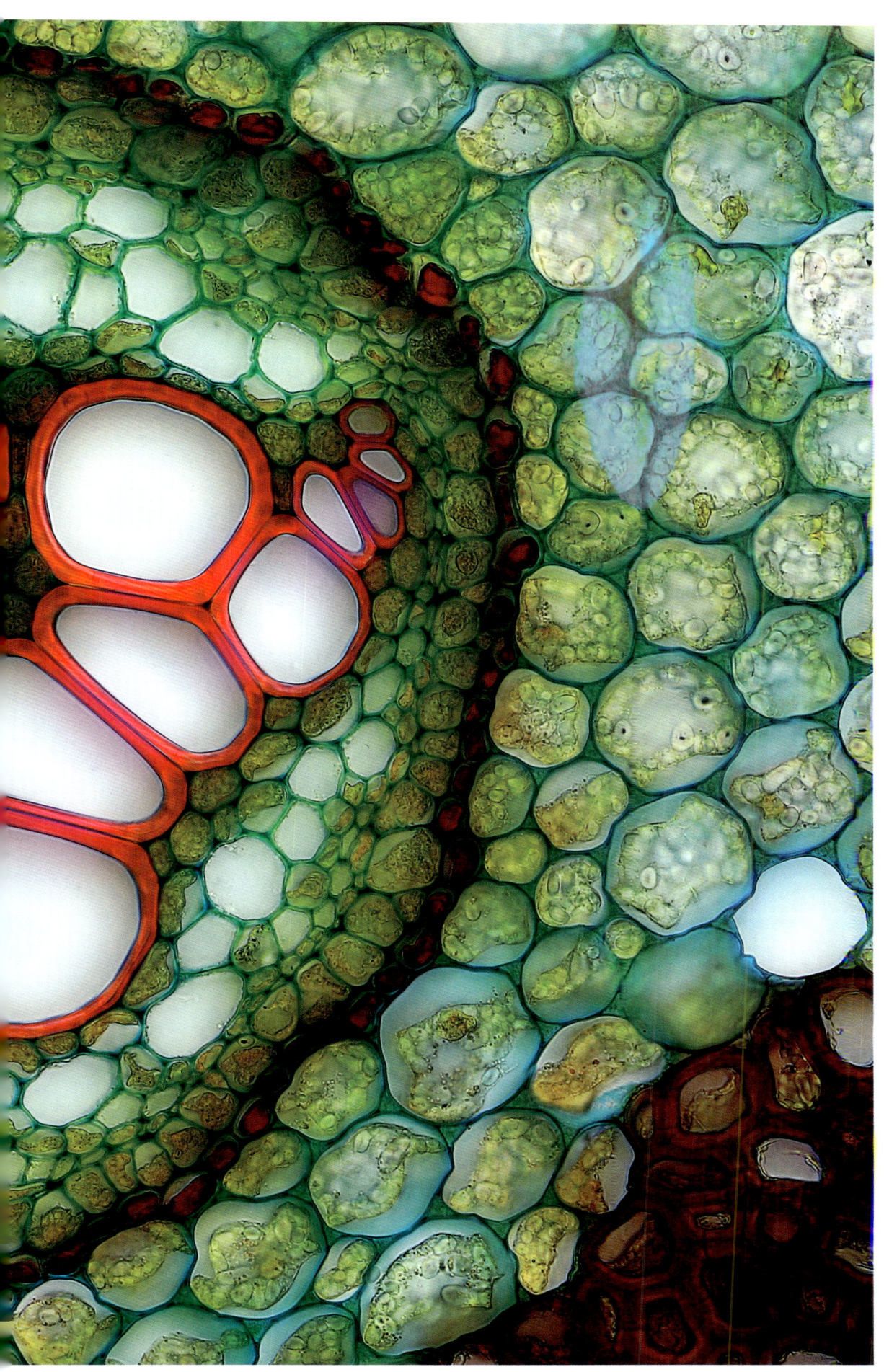

Anatoly Mikhaltsov
A cross section of an ostrich fern stem shows the internal plumbing the plant uses to transport water and nutrients.

FOLLOWING PAGES:
Martin Oeggerli
Specialized microrobots called MagMites embedded on a silicon wafer. These half-millimeter-wide robots could one day be used for biomedical applications like drug delivery or surgery.

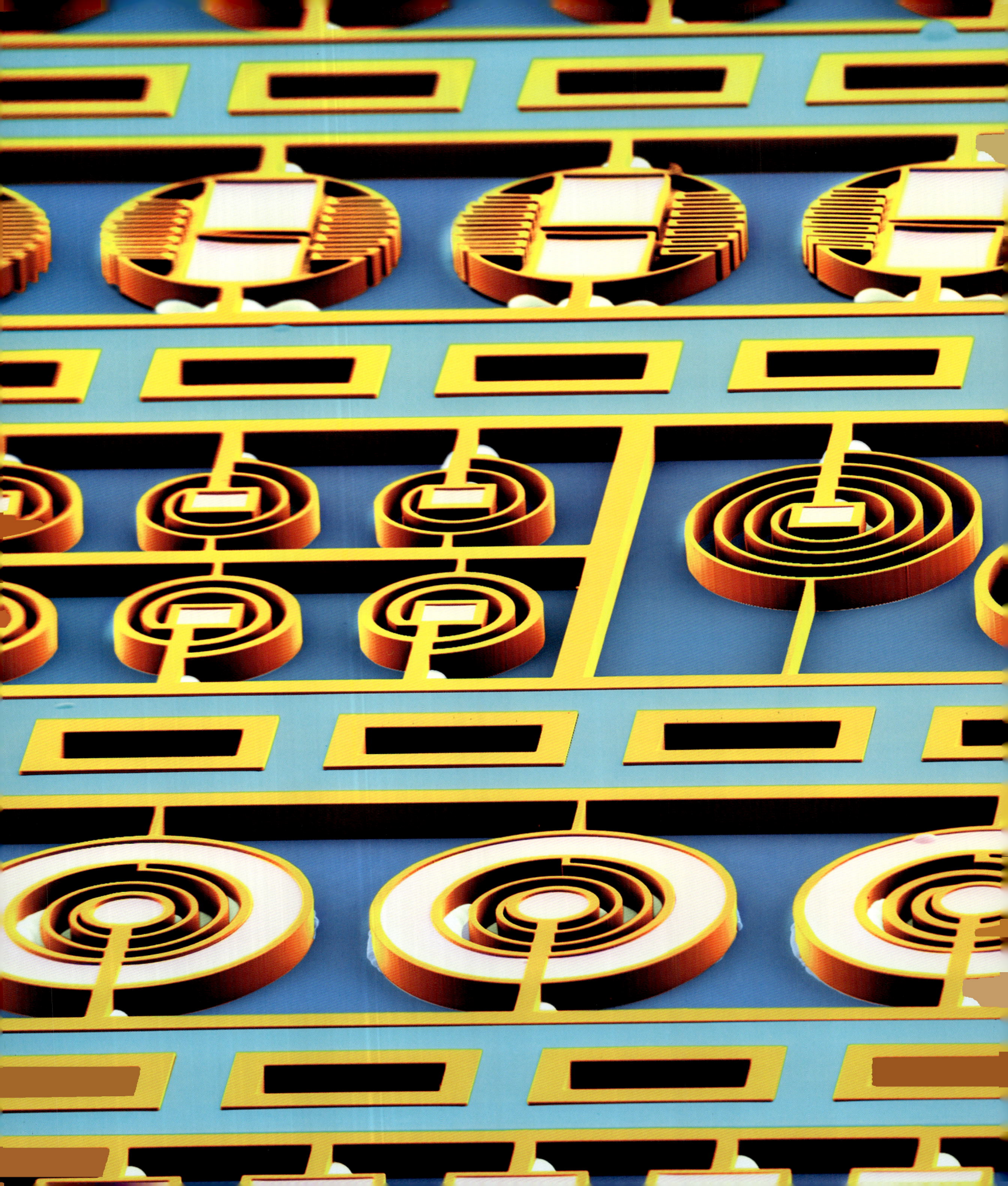

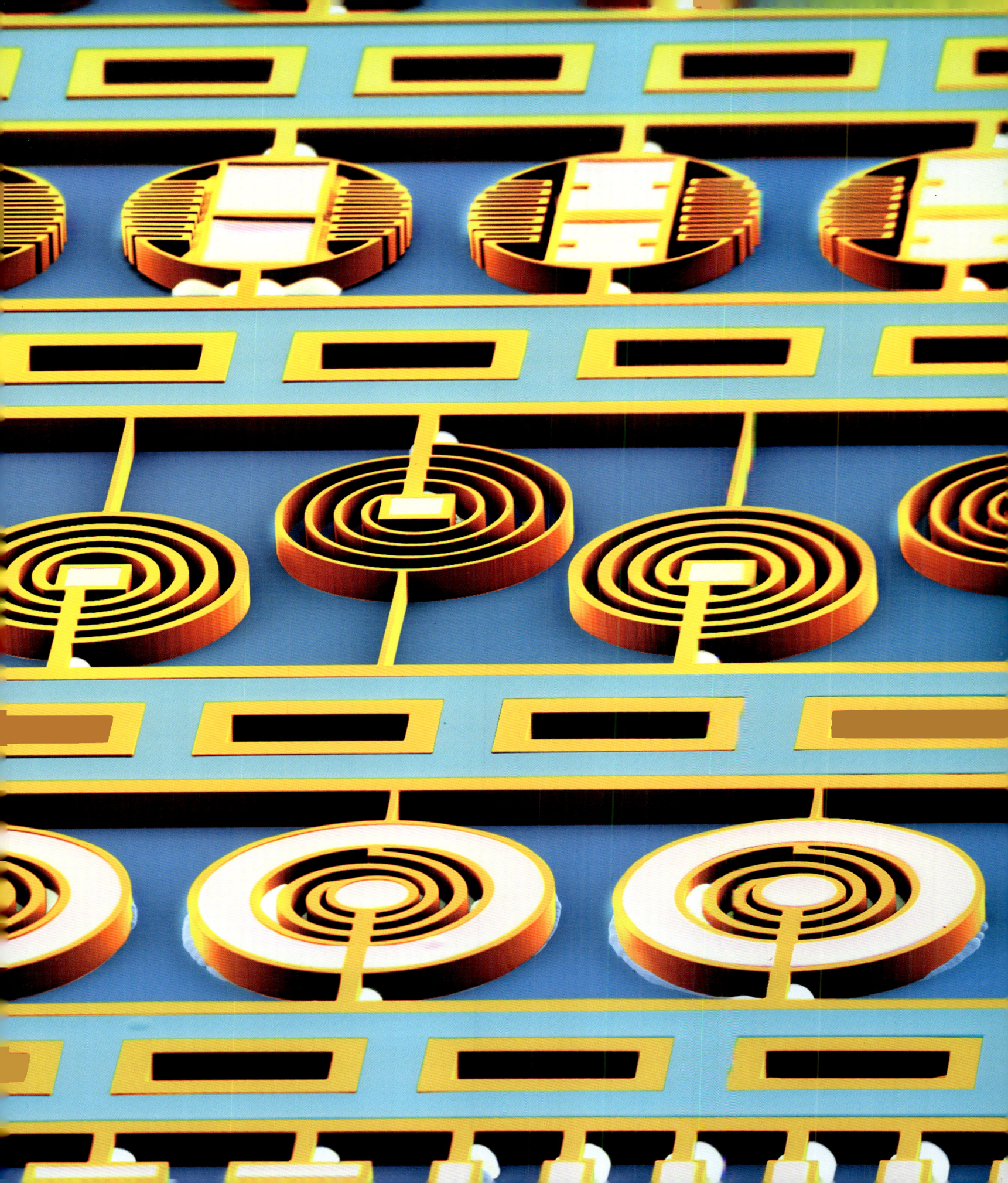

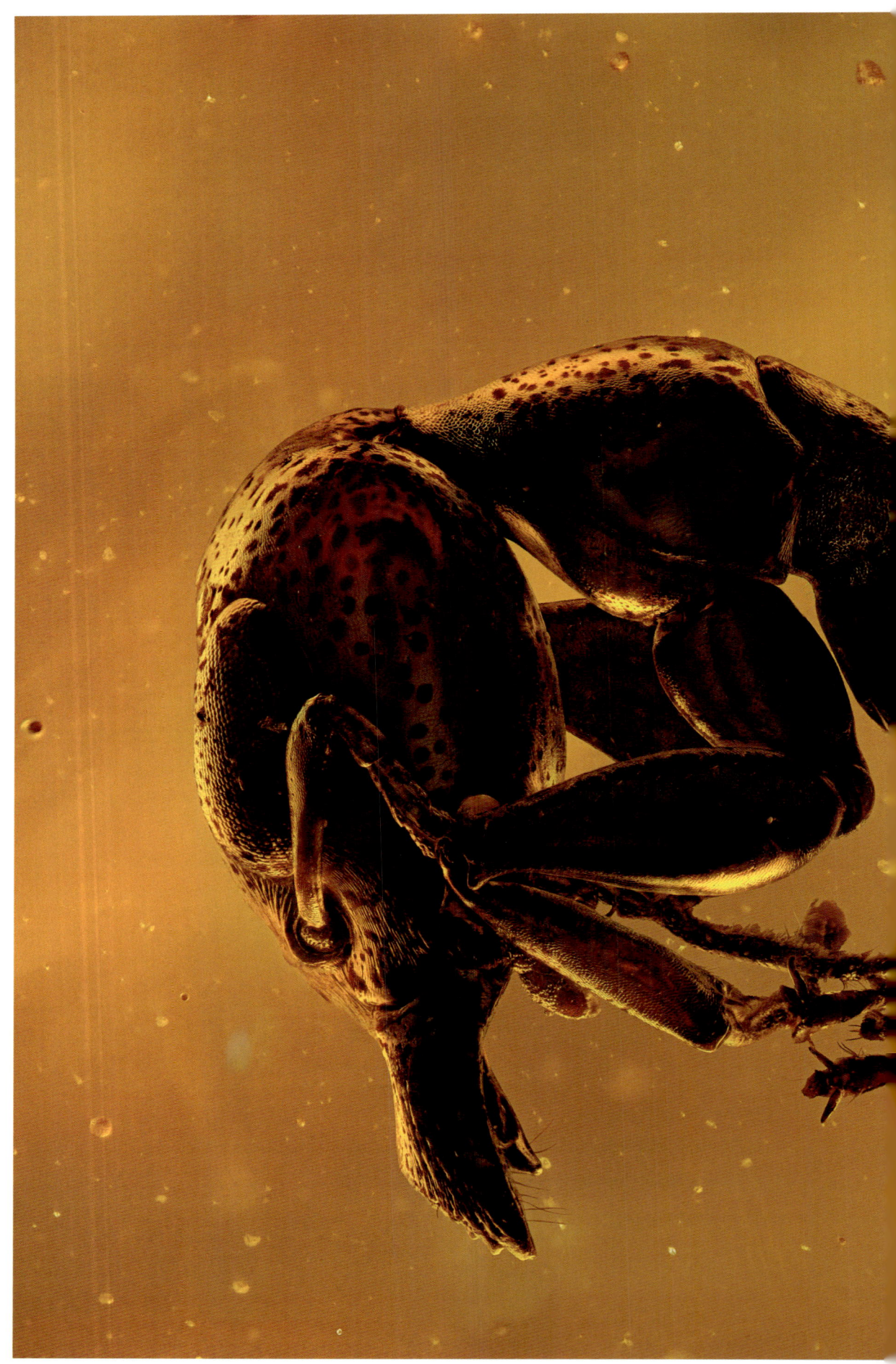

Levon Biss
An ant encapsulated in amber more than 40 million years ago. Amber—fossilized tree resin—preserves the structure and DNA of the organisms trapped within it.

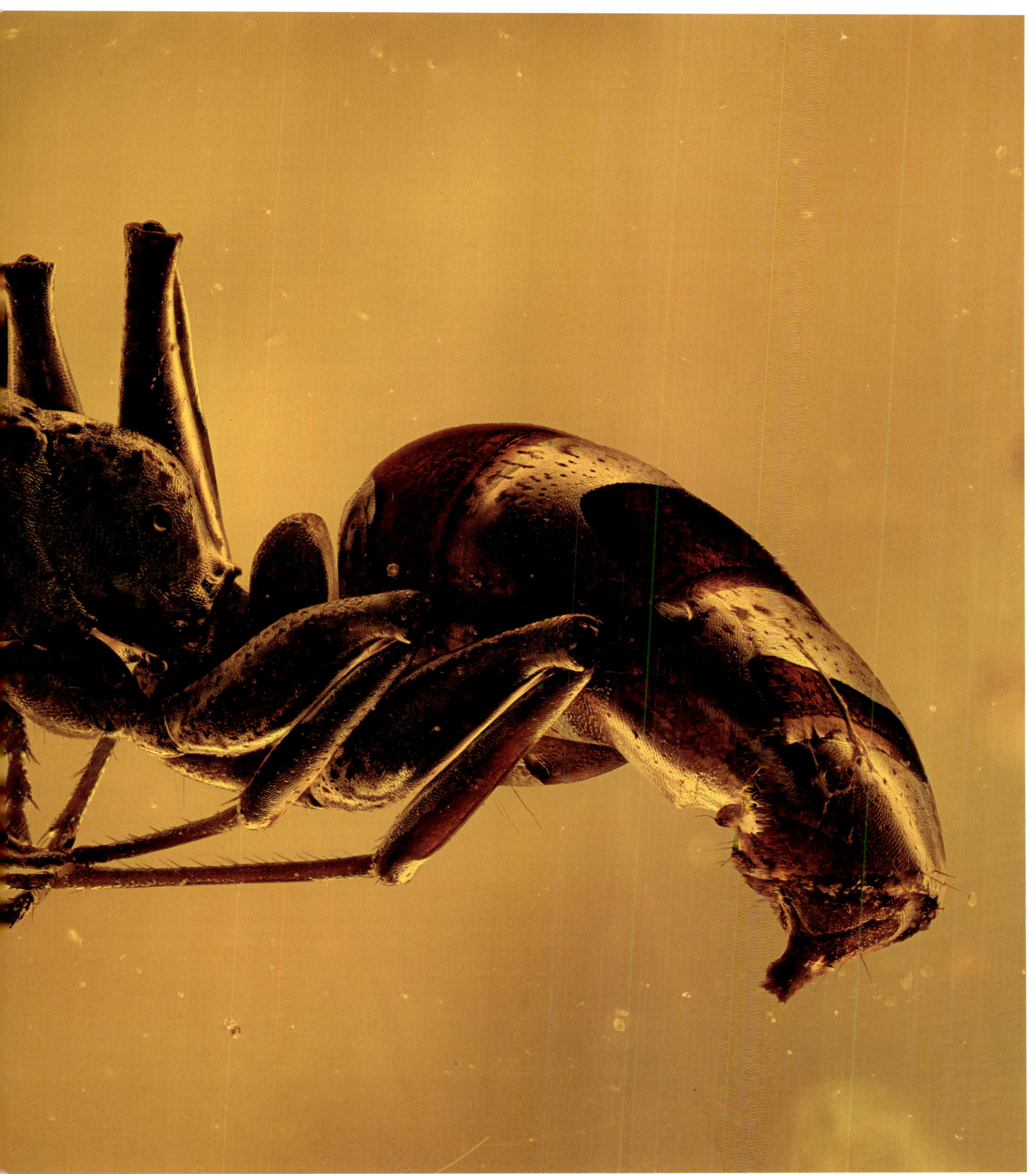

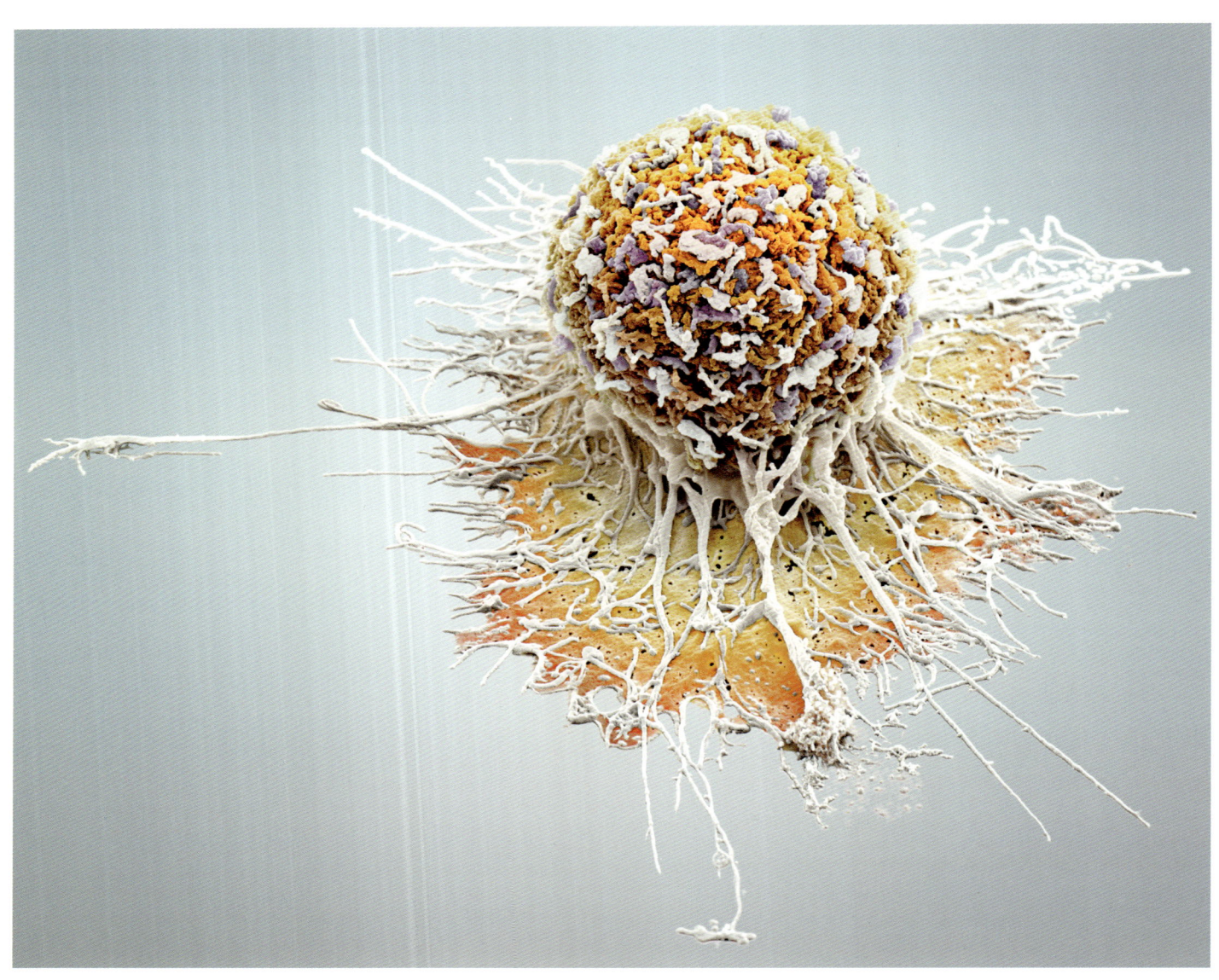

Martin Oeggerli
An invasive cancer cell, recorded with a scanning electron microscope,
propels itself by extending appendages from its base.

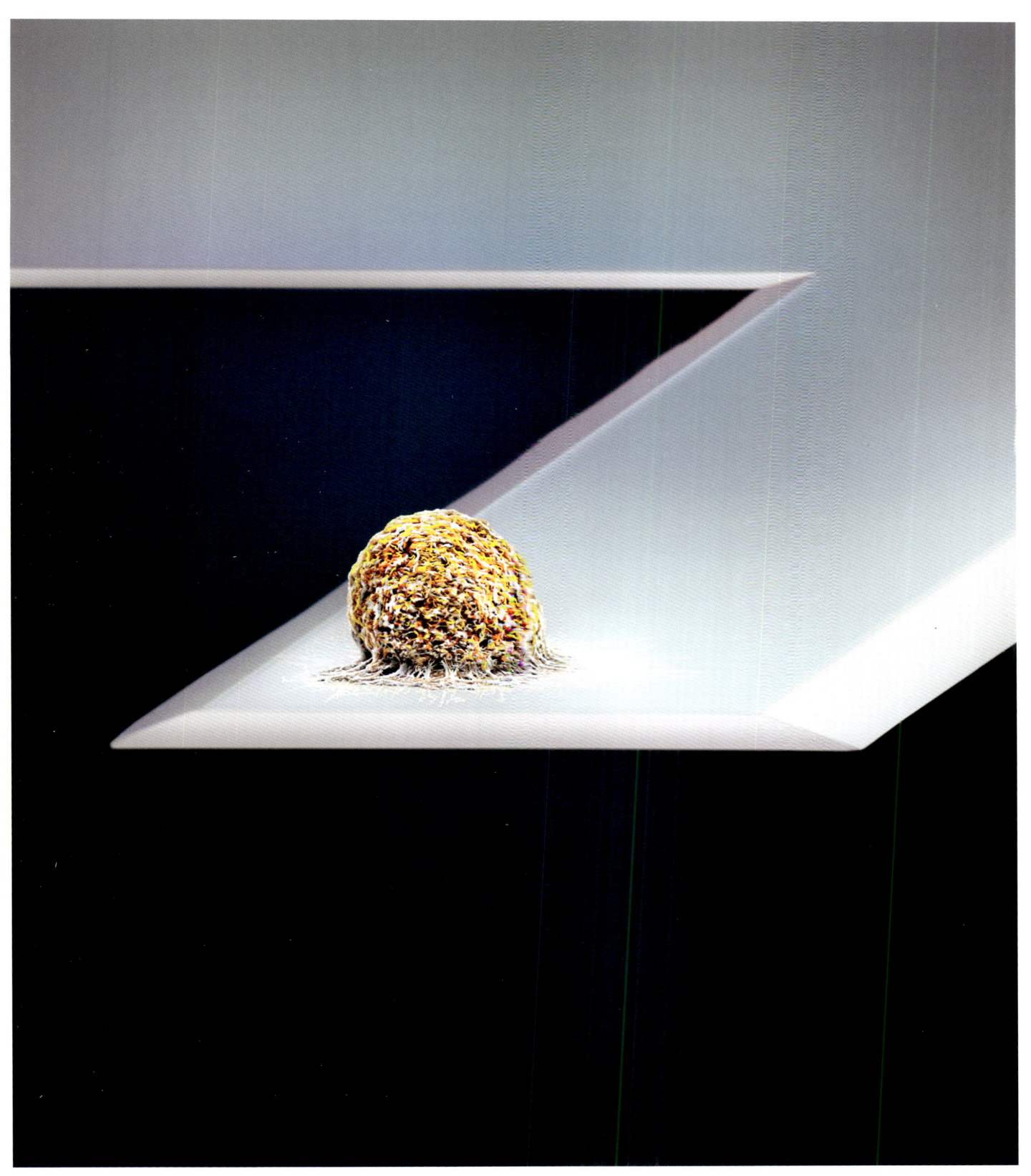

Martin Oeggerli
A human cell rests on a tiny cantilever designed to measure its weight

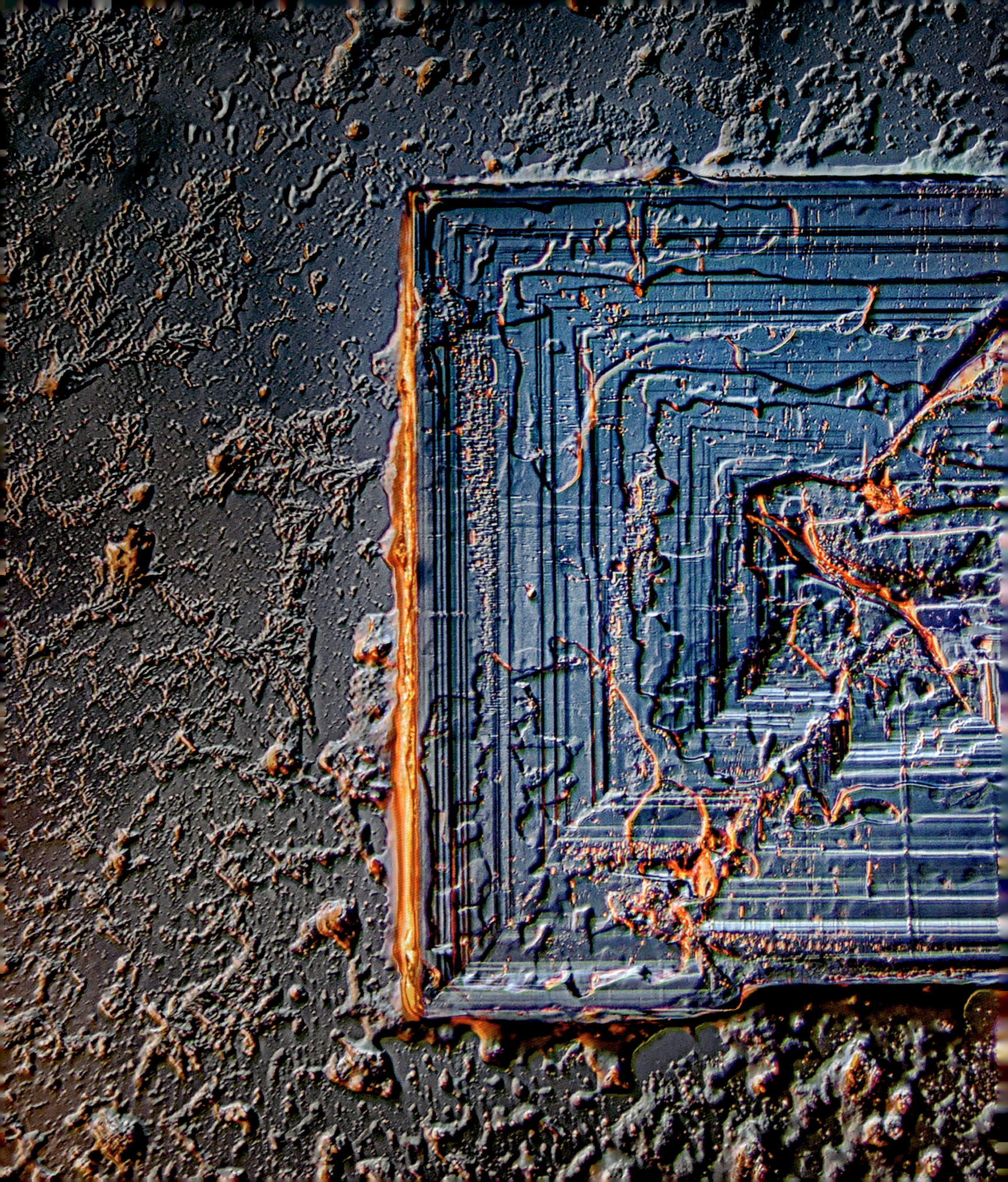

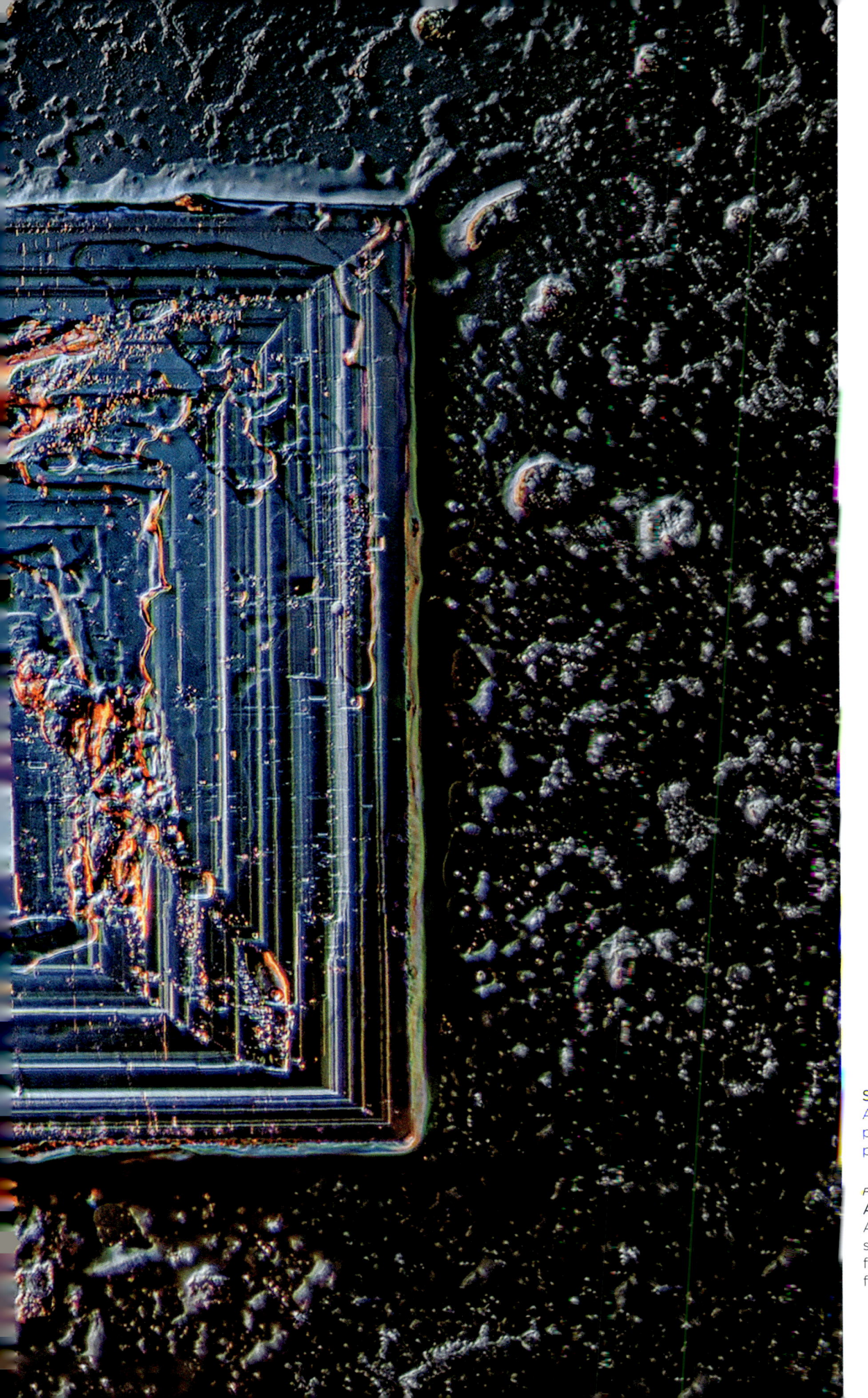

Saulius Gugis
A crystal of table salt
photographed with
polarized light

FOLLOWING PAGES:
Alexander Semenov
A magnified view of a sea star
shows the spiky clubs it uses
for protection and the fleshy
fingers it uses to breathe.

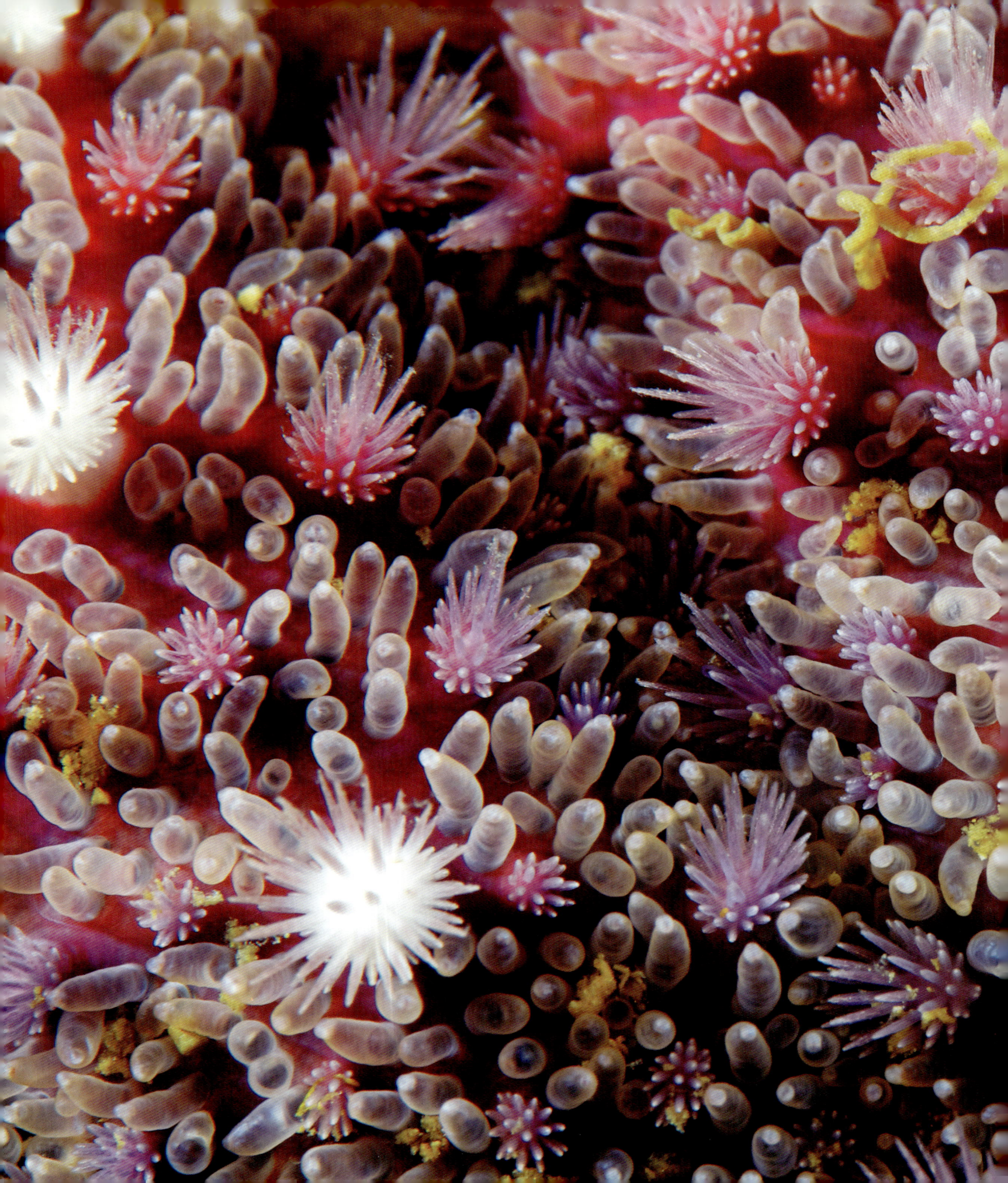

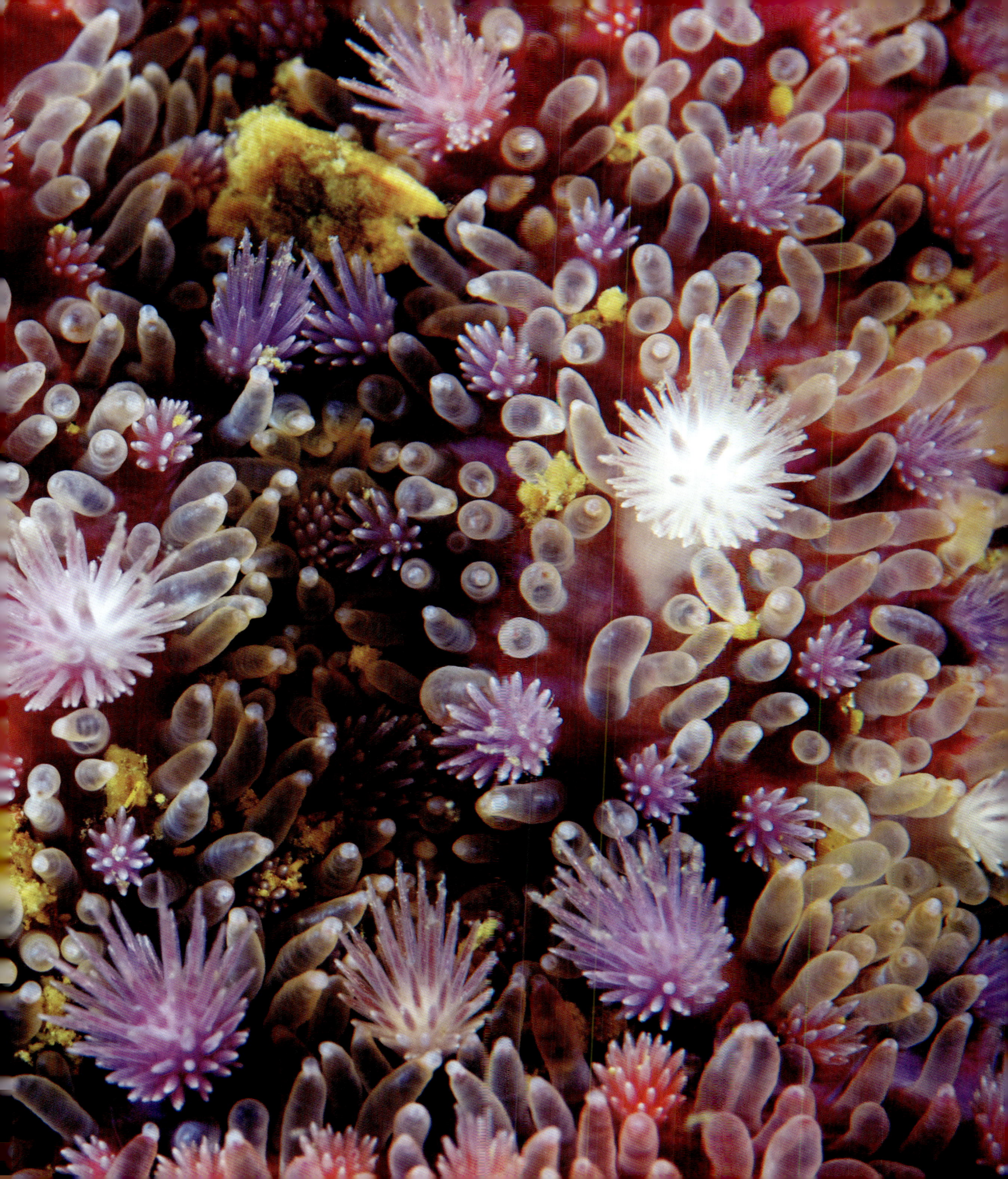

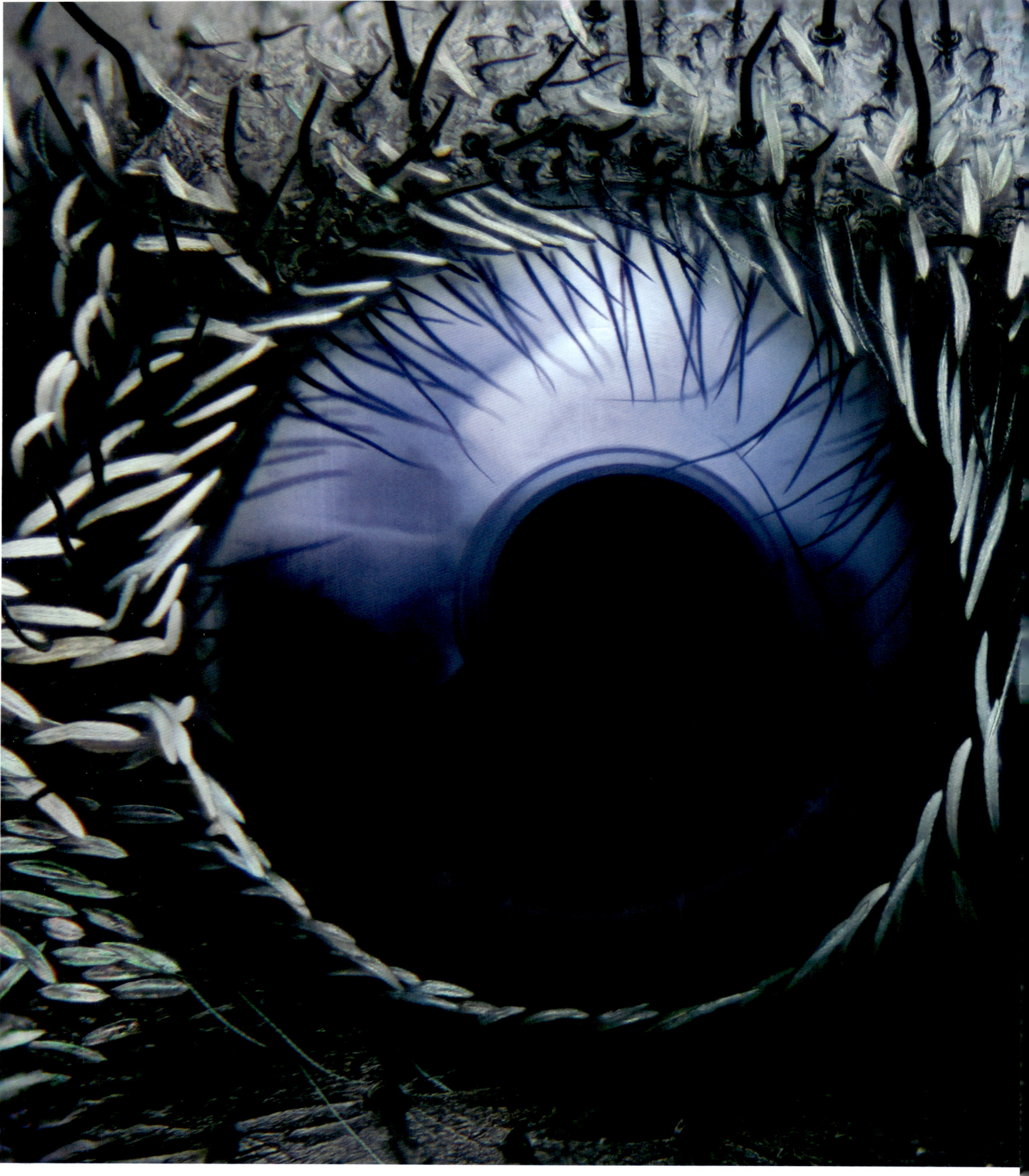

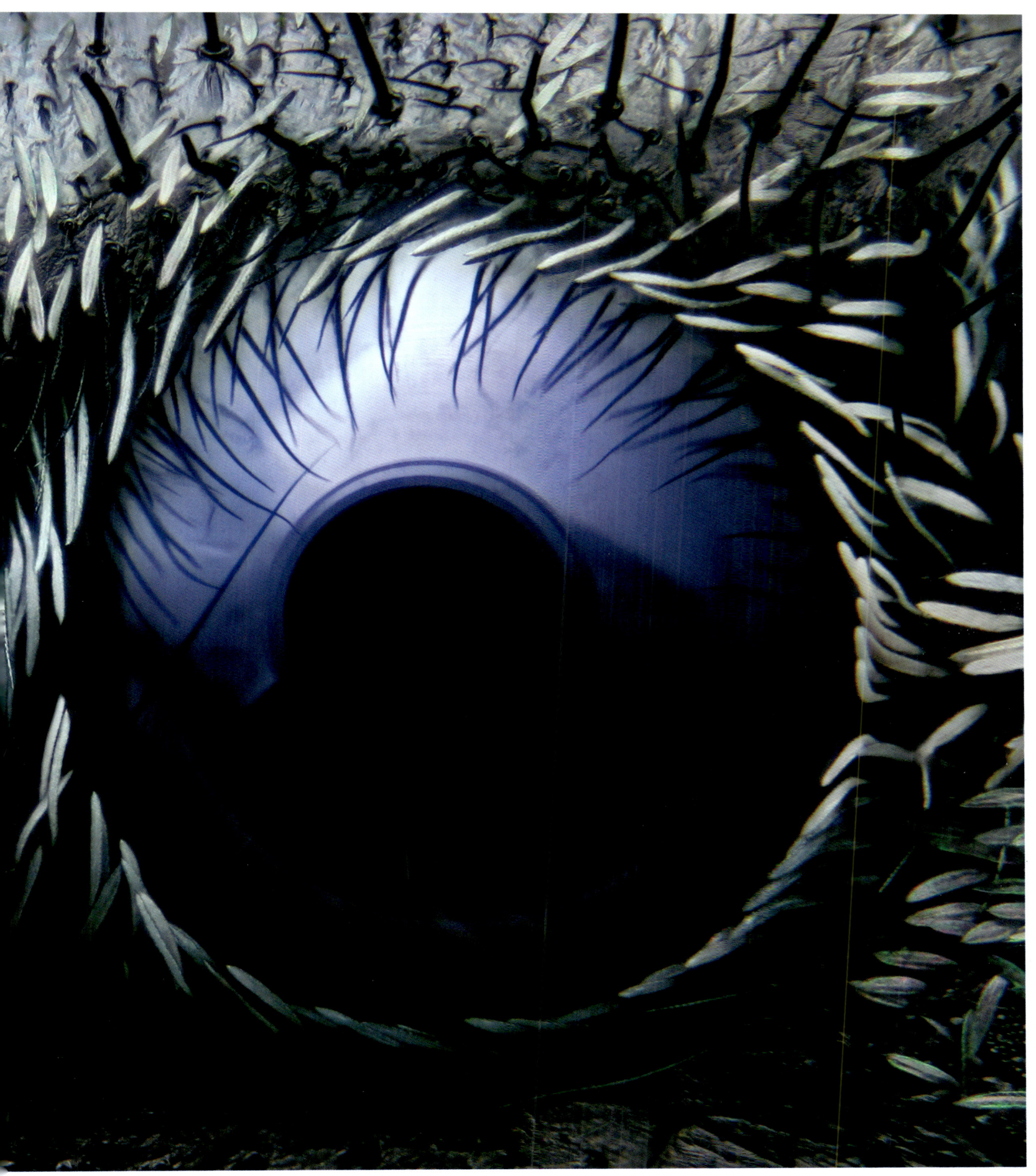

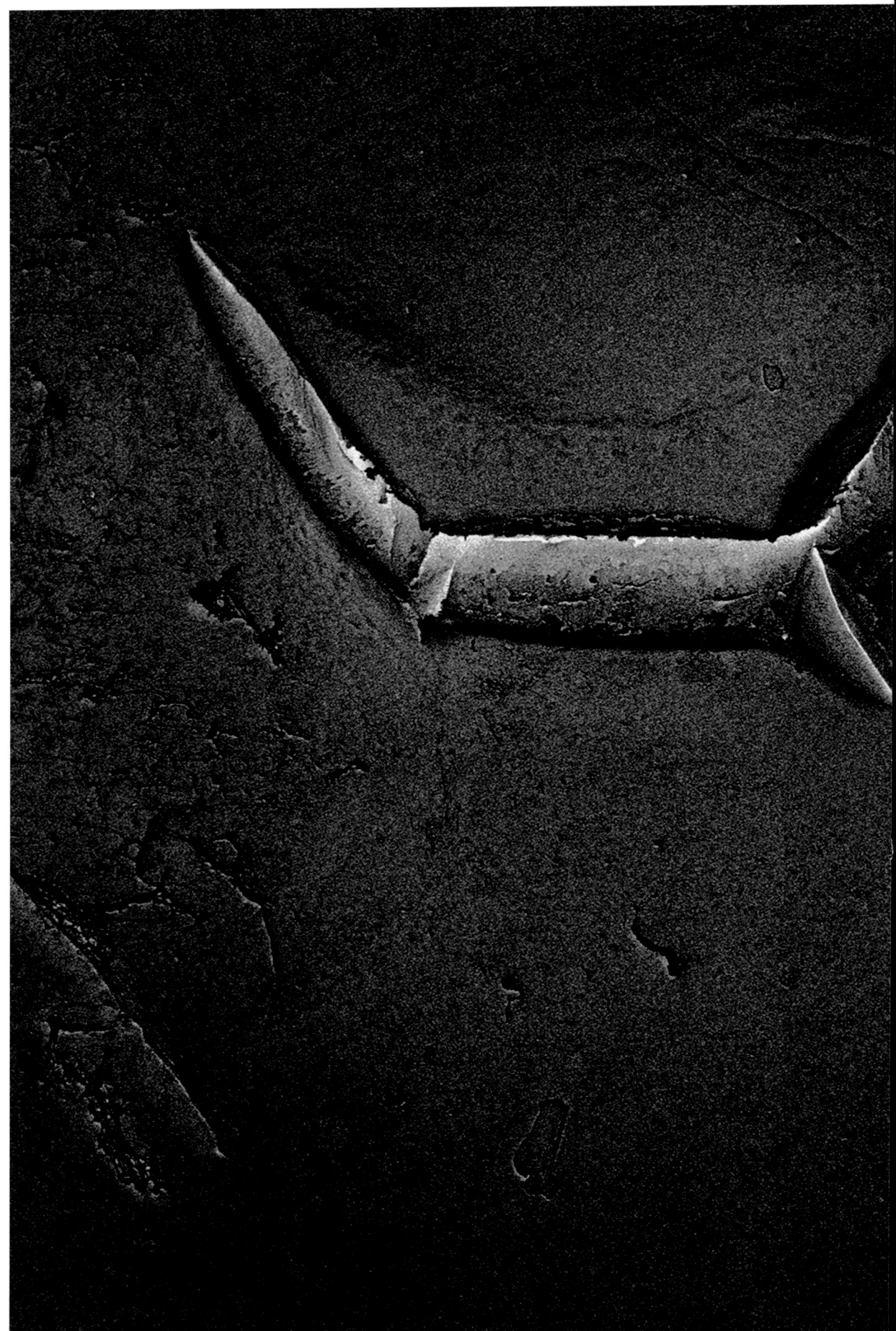

Caren Alpert
The seed of a kiwifruit,
magnified by a scanning
electron microscope

PREVIOUS PAGES:
Noah Fram-Schwartz
The front-facing eyes of
a jumping spider give it
excellent depth perception,
allowing it to move and
hunt with precision.

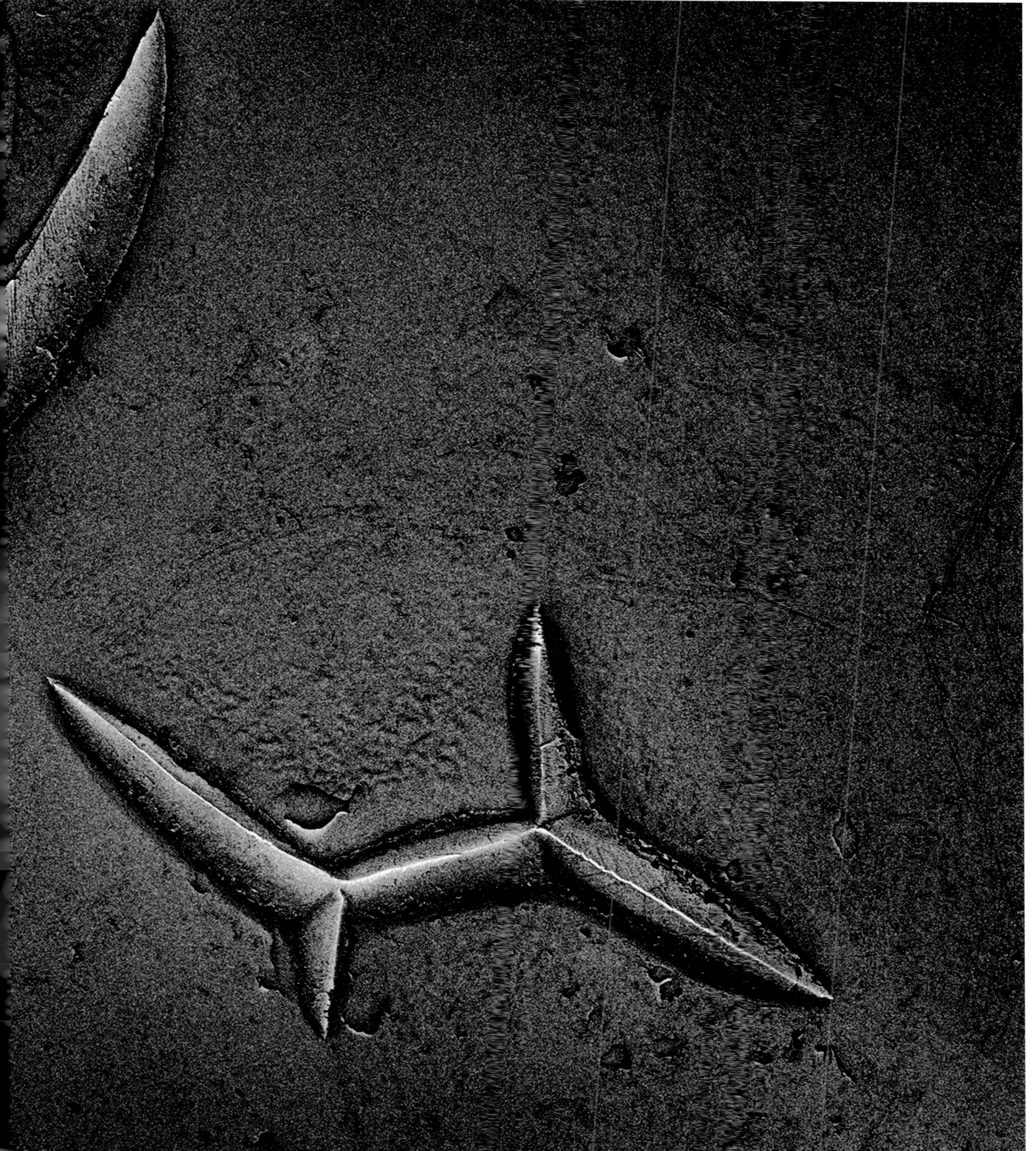

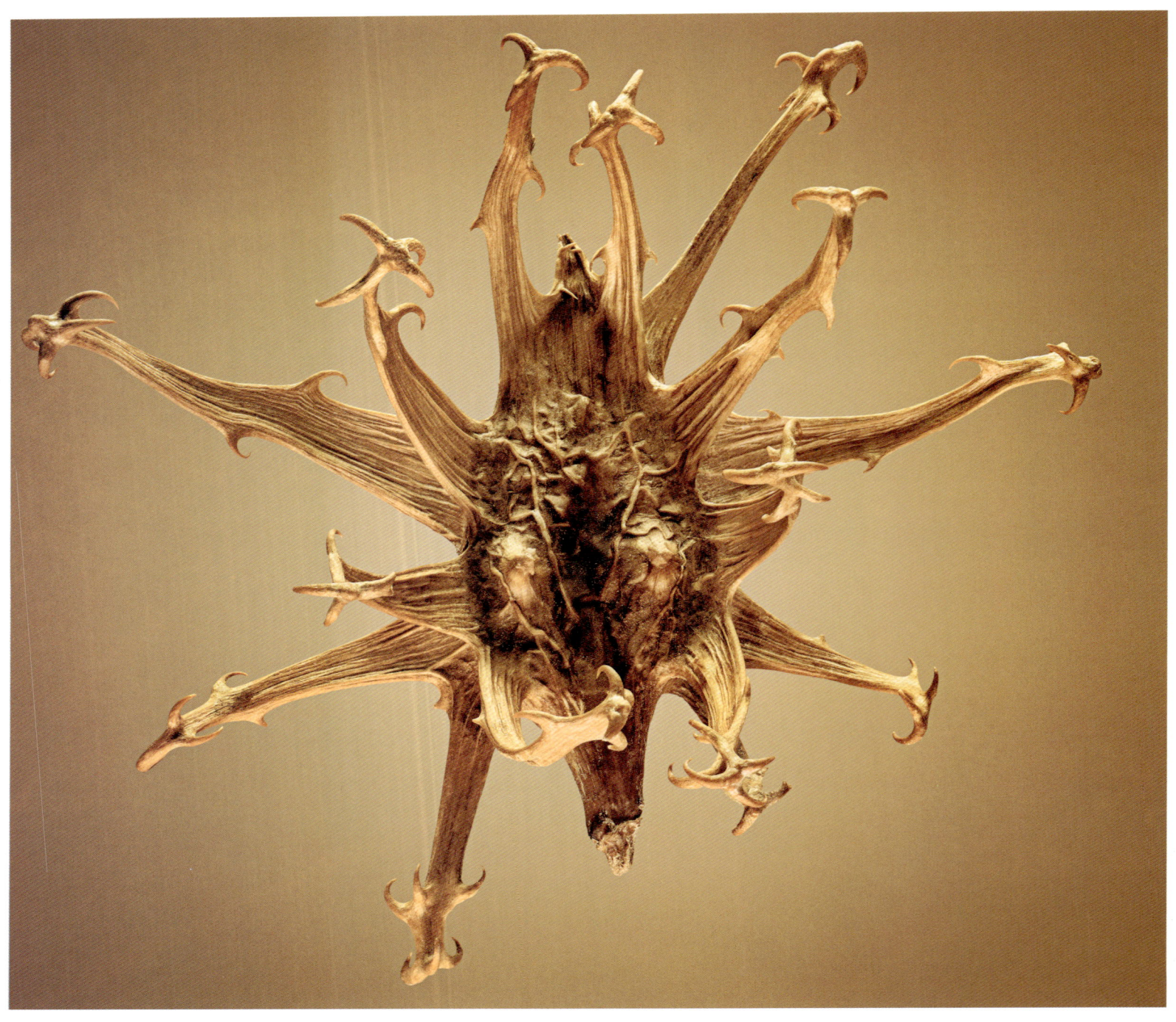

Dillon Marsh
The seeds of this South African grass evolved hooks to help it latch on to
the fur of passing animals, earning it the nickname "hitchhiker plant."

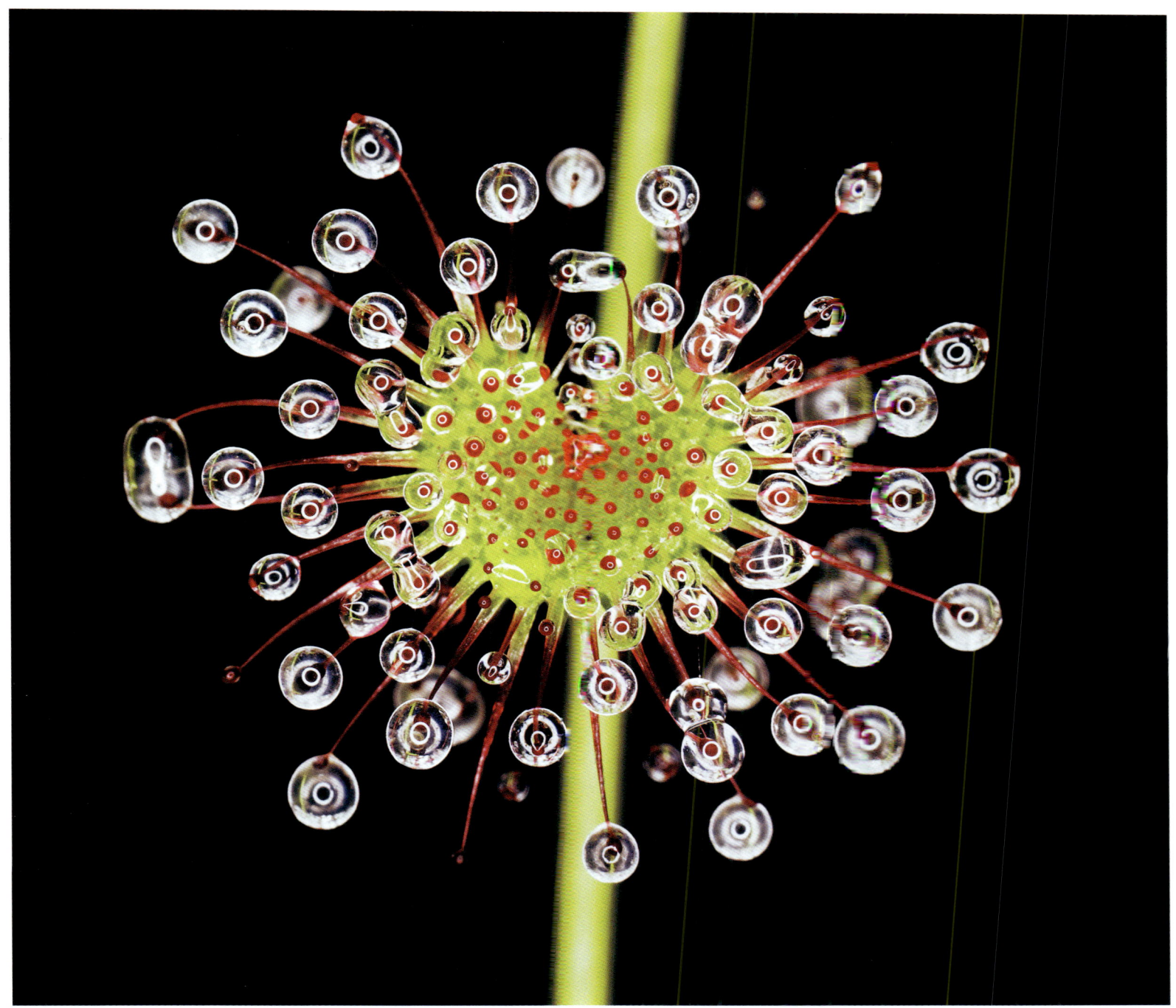

Helene Schmitz
The tiny tentacles of leafy sundew, a carnivorous plant native to Western Australia,
offer sticky droplets to allure insect prey.

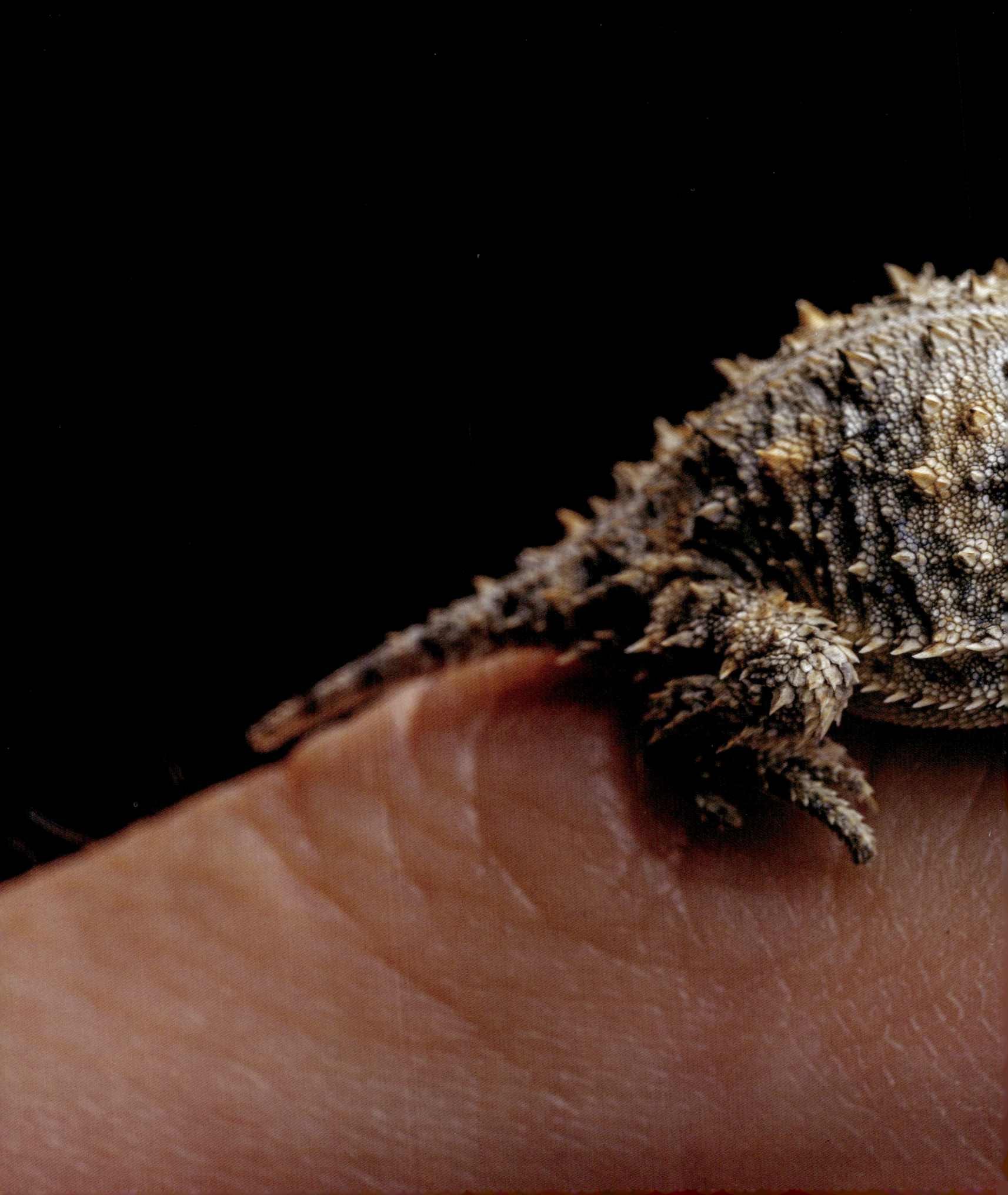

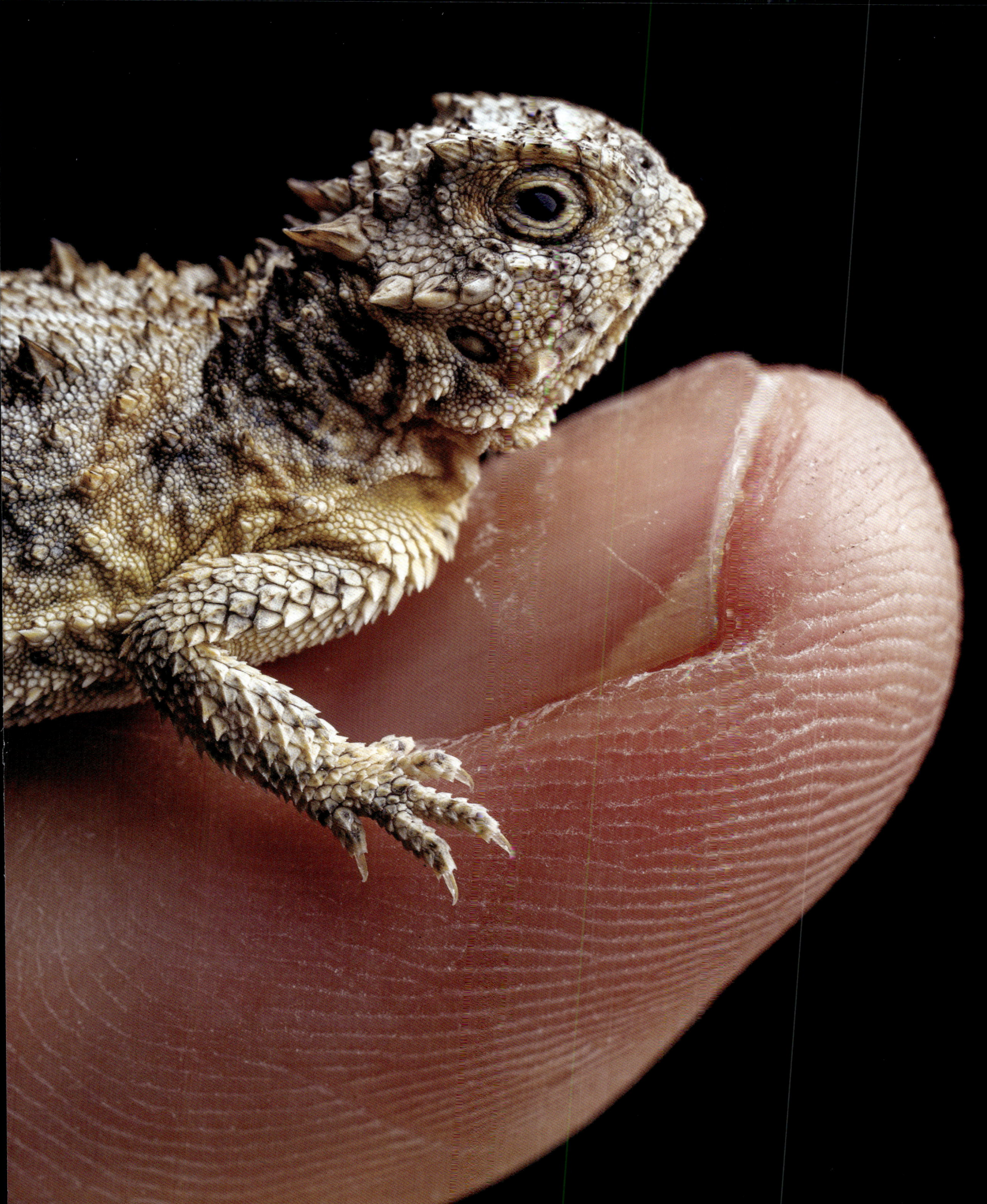

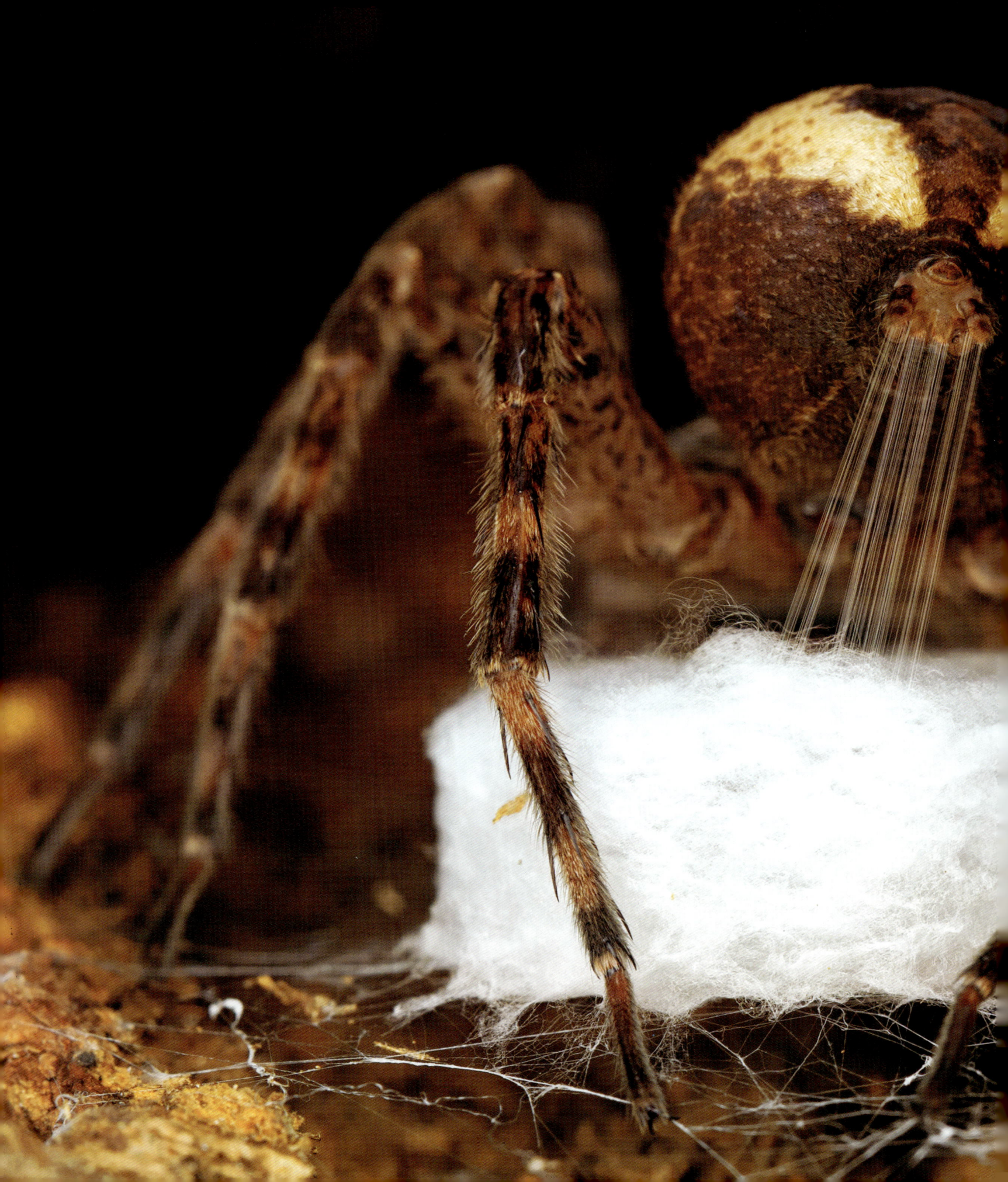

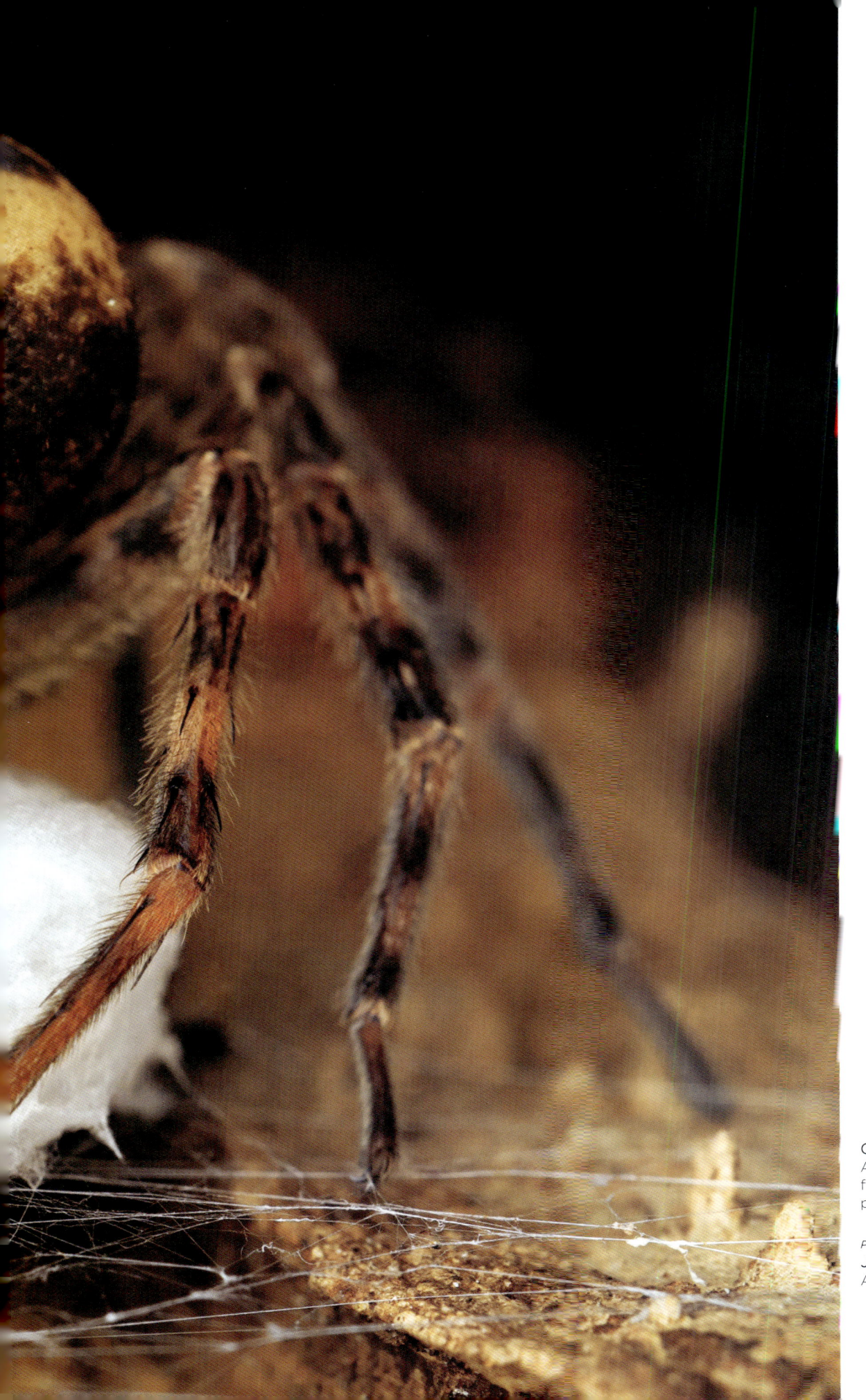

Gil Wizen
A fishing spider extrudes silk
from its spinnerets to weave a
protective cover over its eggs.

PREVIOUS PAGES:
Javier Aznar
A baby Texas horned lizard

119

MARTIN OEGGERLI

Martin Oeggerli is a freelance science photographer who holds a doctorate degree in medical molecular biology from the University of Basel, Switzerland. He specializes in scanning electron microscopy, which allows him to create intricately detailed images of subjects far too small for our naked eyes to see. His work has brought him significant acclaim and awards, including the German Prize for Scientific Photography (first prize, 2011), the International Photography Awards (first prize, 2011), and the Lennart Nilsson Award, recognizing photography that advances scientific inquiry (2022).

ANAND VARMA: Can you describe how you take these photographs?

MARTIN OEGGERLI: Of course. I specialize in using a scanning electron microscope. I'm not using a camera that records light. I'm actually using a technique that scans the surface of the subject. The advantage of this technology is that you can magnify incredibly small things. There are also some disadvantages. It's quite complicated and difficult to prepare the samples. You cannot shoot straight at something, like with a camera. You're very limited with the angles and perspectives. This is a complex thing, and the technology is also quite expensive, so it's not available for everyone.

AV: Can you take me through the steps required to create one of your photographs?

MO: Most of the things that I'm interested in are biologically or medically relevant samples, and these often contain a lot of water—but I can only work with dry samples. If you want to have a look at the surface of a tomato leaf, for example, it's 98 percent water. So each sample goes through a dehydration process in a special

machine. It basically heats the sample in a pressure chamber, and that minimizes artifacts that otherwise would be prominent. If you magnify something, it really needs to be conserved as naturally as possible.

Also, I need to cover my samples with a thin layer of gold to make them electronically conductive. And then you can put the sample into the high-vacuum chamber of the microscope and scan the sample, literally pixel by pixel, with a very thin electron beam from an electromagnetic lens. This electron beam scans the surface, and slowly, slowly, the image builds up. I can scan a nice image after 10 to 20 minutes, then i can start to color.

AV: How do you add color to your images?

MO: It's the most time-consuming part but also the more creative part of my job. I will use colors that are close to reality to produce an image that is photorealistic, but I can also use a clever selection of colors to make someone look at something that I find interesting. And then I can tell a story with my image.

At first you're overwhelmed with small details, as you would be when

walking through a garden. A typical electron scan has so many small things everywhere. Sometimes there's almost too much. They can take away the attractivity and the main message you want to tell.

AV: How long does this colorization step take?

MO: It depends on the final size of the image. If it's fine-art museum quality, that takes much longer than a small picture. It also depends on how many small details you see and how sharp everything is. If you have a picture at a very low depth of field and not so many things are sharp, that's a little bit faster. But if everything is sharp, you have to be very precise in every part of the picture. My longest picture took 187 days. That was absolutely crazy, but I just had to do that.

AV: It took you 187 days to color a single image? What image was that?

MO: It was a retina of a human being. That's a sample, of course, that is so precious. A human being saw the world through this structure, and I have the honor to look

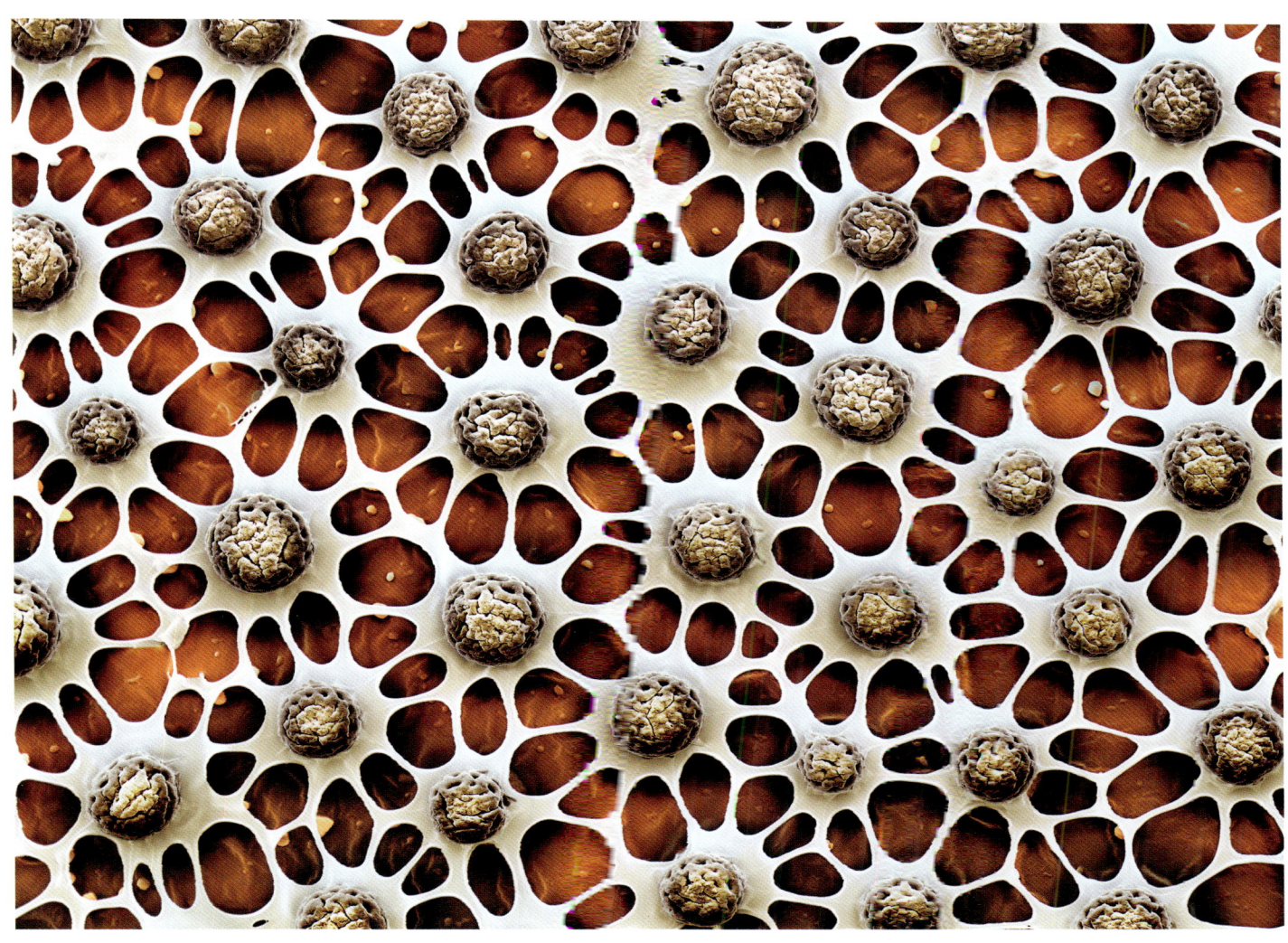

Martin Oeggerli
The water-repellent surface of a mosquito egg, which traps a thin layer of air,
enabling the egg to float

FOLLOWING PAGES:
Martin Oeggerli
The book scorpion grows to only a few millimeters long and hunts lice
that feed on the bindings of old books.

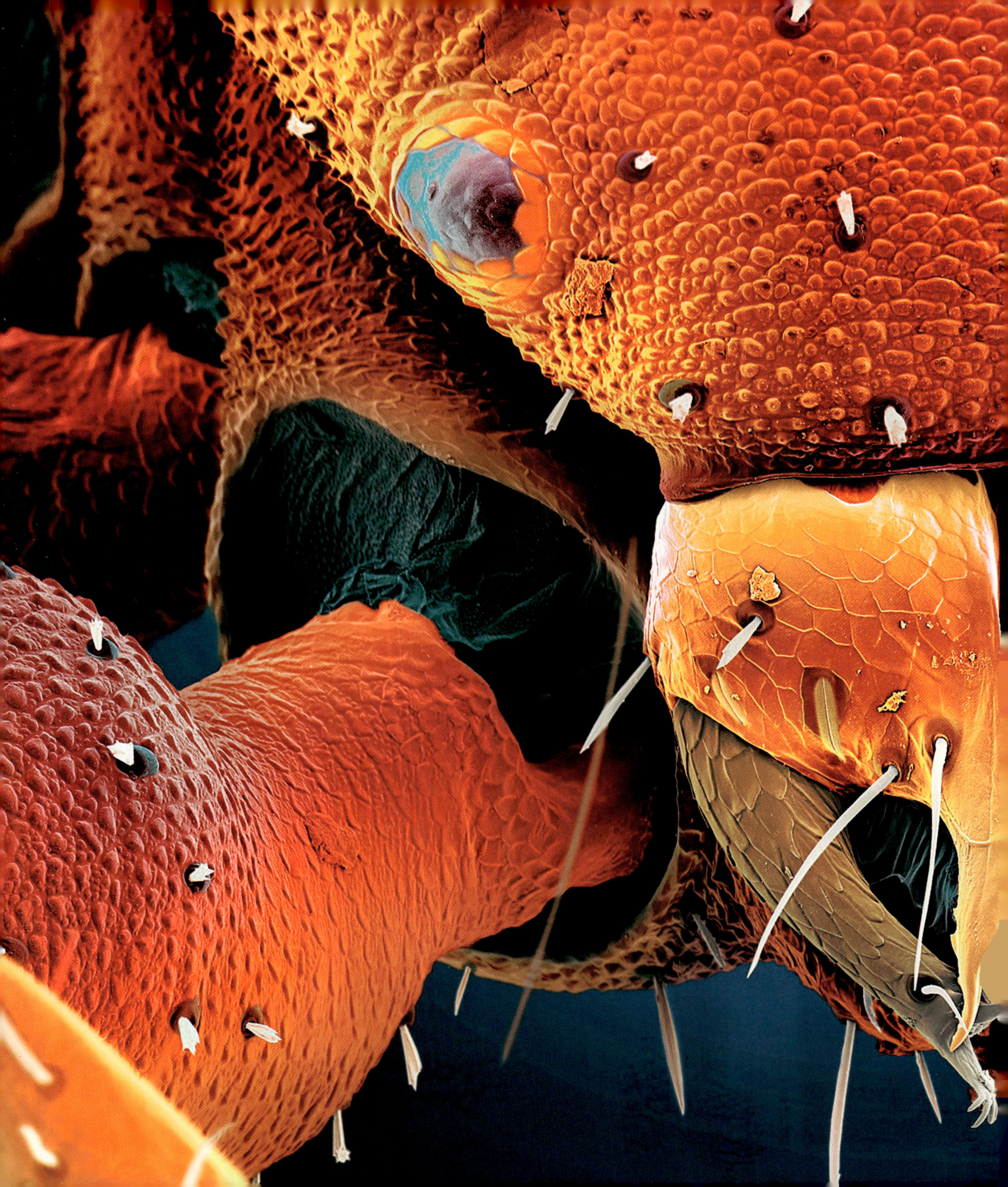

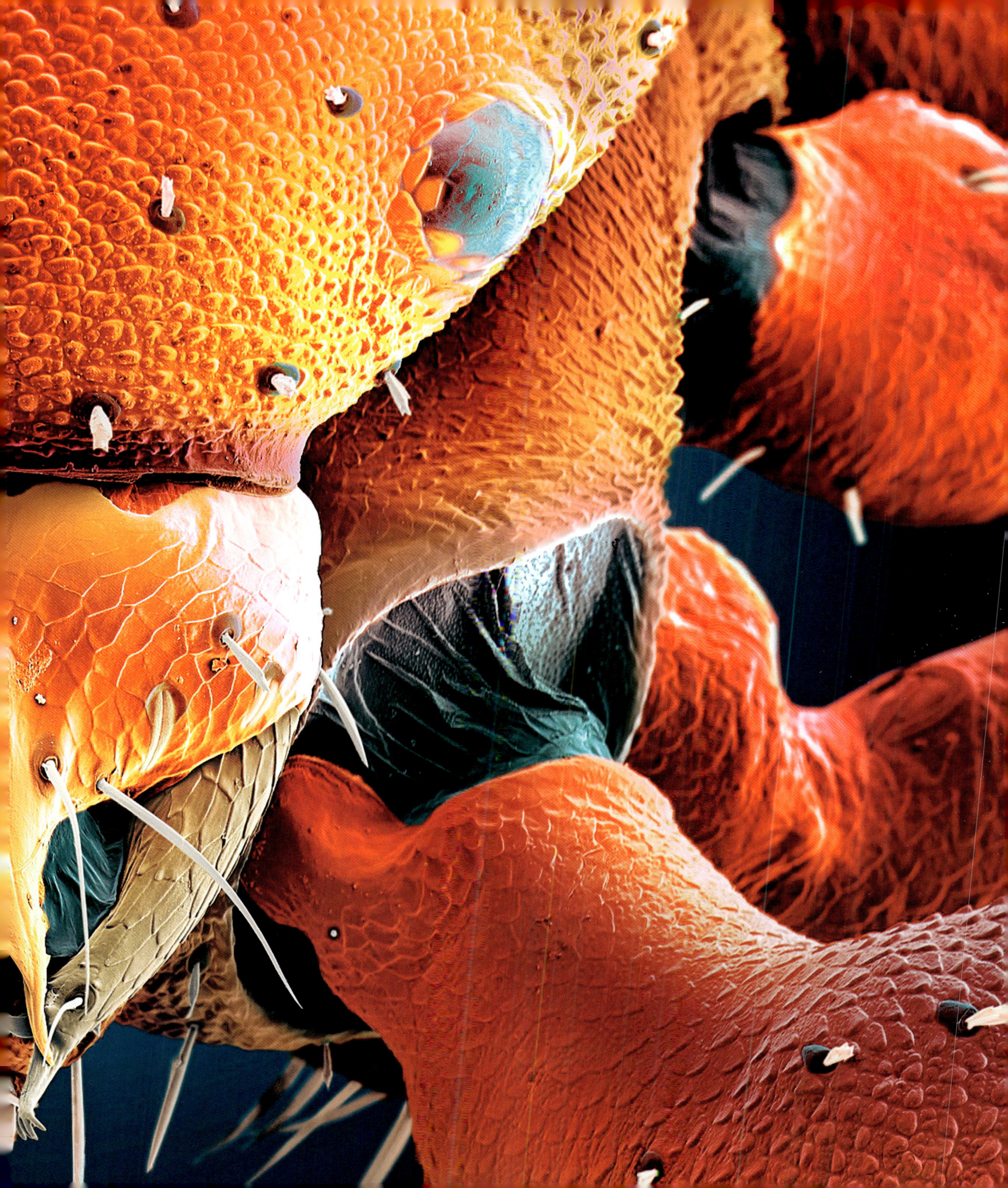

at this under the microscope and show it to other people. I just had to go that extra bit. Our retina is built up by 120 million rods and six million cones. There are so many different receptors. You have receptors for different colors, for movement, for dark, for daylight. It's absolutely fascinating. How do you say in English "when some cold water runs down your back"?

AV: "Spine-tingling"?

MO: Yes, exactly that—when you think about how complex this is.

AV: That's incredible. How do you choose what subject you want to work on?

MO: Some projects start from pure curiosity. For example, when I was a little boy, I played in the garden of my grandfather. He had a big barrel of rainwater to water the garden, and there were tiny structures swimming on the water's surface. And I would push them, and they wouldn't sink—they'd float right back up. My grandfather told me they were rafts of mosquito eggs. Twenty years later, when I was having lunch at a university hospital garden, I saw one of these rafts floating by, and I thought, "I am a scientist. Why am I not exploring these things under the microscope? There must be something special that makes it float." And it turns out there is an extremely beautiful network that spans the mosquito's egg that keeps it floating in a vertical orientation. If it didn't, the mosquito would hatch on the surface of the water and would be trapped there by the surface tension. It's fascinating how evolution designs the diversity on Earth. I hope others find it fascinating that an animal they don't like could make this.

AV: You eventually made an entire series about insect eggs.

MO: Yes. I found a clutch of eggs in the garden of my parents. I thought they were butterfly eggs—they looked so nice, and I colored them in fine-art quality. Obviously, people like butterflies, so I thought I could sell them. But after some research I found that these were the eggs of a stinkbug [*laughs*]. So it was not very commercially successful. But in the meantime, I did have some butterfly eggs from a breeder in Switzerland. I sent *National Geographic* those images, and their response was three words: "Beautiful, beautiful, beautiful." That was my second article in the magazine.

AV: It seems like your most memorable images come from unlikely places.

MO: My heart always beats a little bit for these animals, the ones that go below the radar. The mites, the mosquitoes, the nasty things [*laughs*]. That's why I like showing mites on a white background. I think it's a second chance to convince someone that it's important. I want to teach my children, and maybe the viewers, that they have to look at something for themselves, like a scientist, and come to their own conclusion. Can you imagine something more beautiful than someone falling in love with a tiny animal, invisibly small, and finding out it's a mite?

AV: What made you fall in love with photography?

MO: I was drawing when I was a kid. I always liked nature, and I was a member of *National Geographic* since I was very young because I liked the images so much. But I didn't take any pictures myself. My father was a hobby photographer, and one day he decided to send me a Nikon Coolpix. I got this big box in front of the door—I was living as a student in a very cheap part of the city, and it's miraculous the box wasn't stolen—and I thought it was sent to the wrong address, it was so big. It was not my birthday; it was not Christmas. But I took it upstairs and unpacked it. That was a fantastic moment, because with a digital camera I could see if my pictures looked good right away. I think in my first months I made 20,000 pictures.

It was my hobby. I didn't understand that I should show them to somebody else. But people at my lab saw them and said I should enter competitions. I finally understood that other people like the images that I make. It's strange. Some things in life, we push hard and it doesn't work, but the images that I made, they just worked. When I finally got into *National Geographic,* the artist Ai Weiwei sent me a letter and said he greatly admires my innovative works. And that blew my mind [*laughs*].

AV: I have a very similar origin story. My father was also an amateur photographer who let me borrow his Nikon Coolpix. That's amazing. I wanted to ask about a specific image. What's the story behind the book scorpion?

MO: I have a passion for parasites and these crazy little critters. I was very ambitiously working on a project about the eye, and I was looking for animals with highly developed eyes and others with very primitive eyes. The scorpion was almost too big for the microscope, but the eye was such a primitive construction that I decided to take a headshot and show that there were animals with

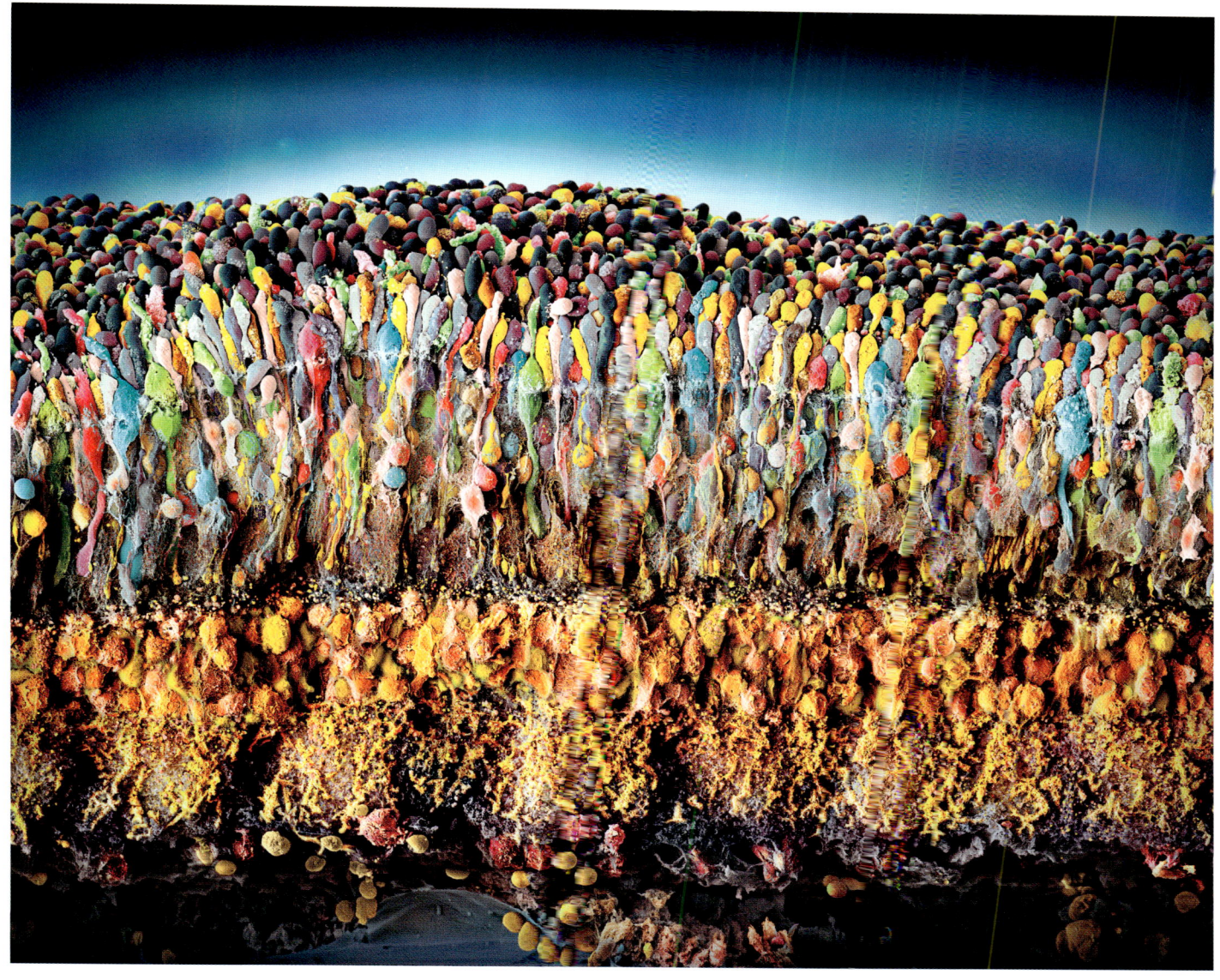

Martin Oeggerli
A cross section of a human retina.
The photographer worked for 187 days to colorize this image.

almost no eyes at all. It can probably detect black and white, but aside from that, it is not highly developed.

AV: Are there new frontiers in your work you're excited to explore?

MO: I just started a project about the start of life. When a sperm meets an egg cell and begins to develop into an embryo in utero, you cannot see that—so it's a very interesting thing. As a father of two kids, that interests me very much.

I've also been working since 2015 on a project about the microbiome. We just did a small movie with a company in Switzerland featuring my five-year-old daughter, because I collected from her excrements very small portions one month after birth, six months after birth, and 10 or 12 months after birth I wanted to document the increase in her microbiome as she grew.

AV: What motivates you to show people such small subjects?

MO: I like to show them the fascination I have when I see these structures. I have an opportunity to use high-tech machines that not so many people have access to, so I feel an obligation to show what I can see.

125

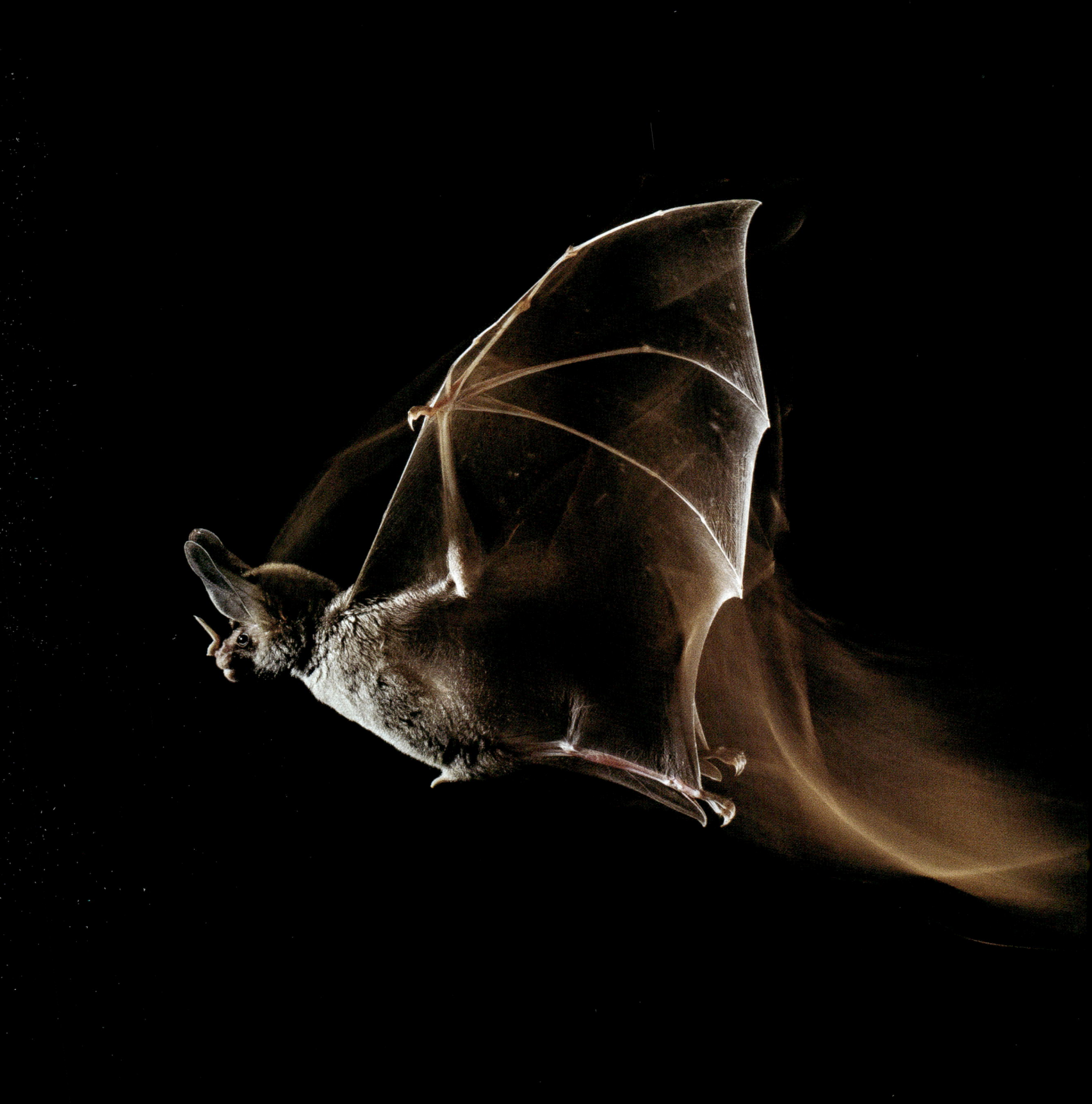

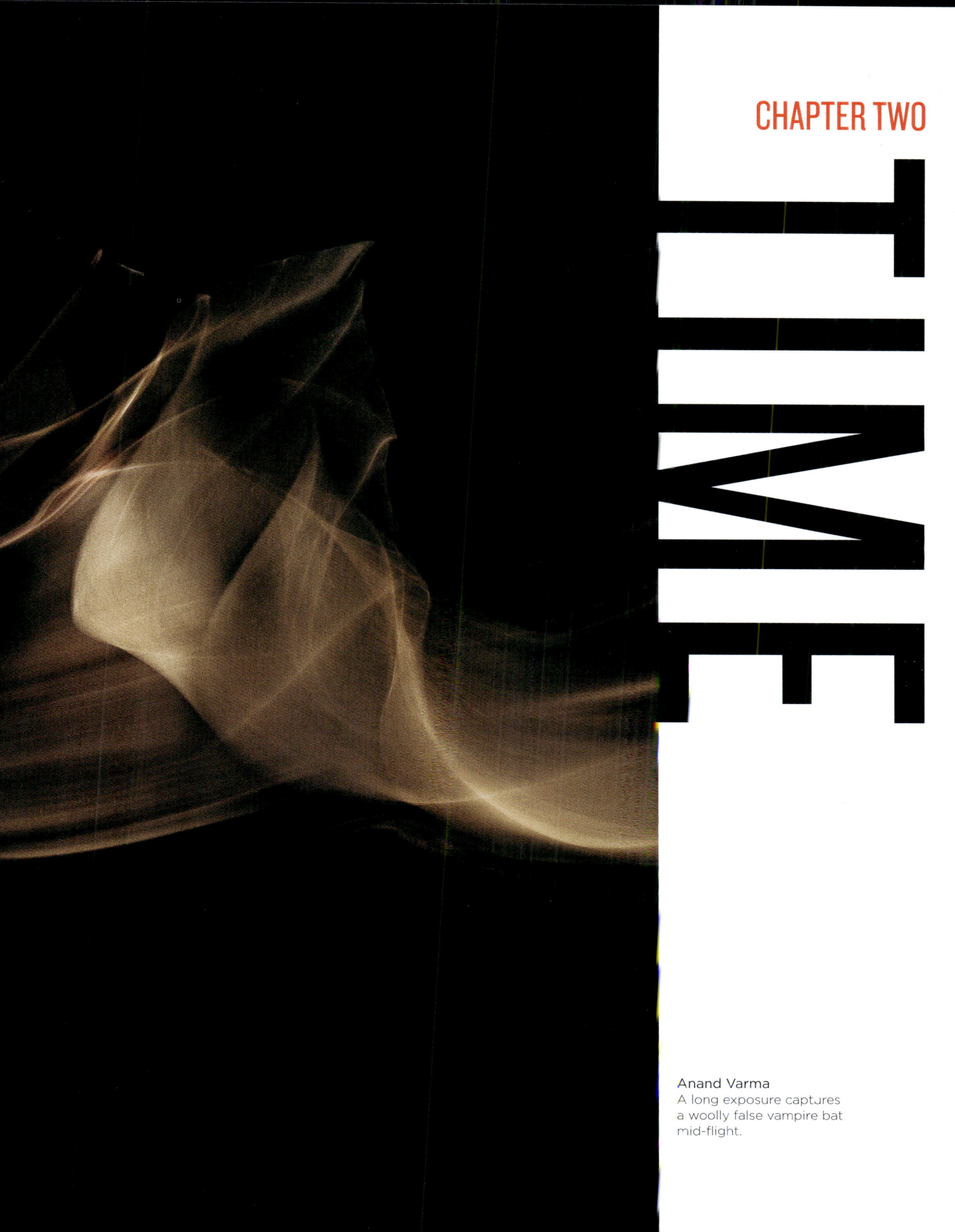

CHAPTER TWO

TIME

Anand Varma
A long exposure captures
a woolly false vampire bat
mid-flight.

IN 2015

National Geographic magazine asked me to photograph the research of world-renowned bat expert Rodrigo Medellín. I met him in the Yucatán Peninsula of Mexico, where his team had located a family of woolly false vampire bats, one of the largest and rarest species in the Americas. He had asked me to bring an enclosure so that we could work with the bats in captivity, so I brought a collapsible batting cage used for baseball practice. I set up the cage at the foot of the bed in my hotel room, and we released a captured bat inside.

Normally, bats are frustrating subjects to photograph. Masters of the nocturnal world, their speed and stealth make them nearly impossible to track with a camera. But fortunately, with a little patience and raw chicken, you can convince carnivorous species like false vampire bats to cooperate. Using bits of meat as a reward, Rodrigo developed a clever technique of enticing the wild bat to fly across the cage on command. That allowed me to predict where it would be ahead of time, so I could create a more complex shot. I set the camera to take the image over a third of a second. This alone would have recorded only a blurry trace of the bat's swift movement. But before the camera completed its exposure, I triggered a brief flash of light to create a crisply defined image of the bat superimposed onto its impressionistic form.

We often assume a photograph can only capture a single moment, but this photograph simultaneously illustrates two different time intervals. One is too fast for us to see, the other too slow. Our natural perception lies somewhere in between, as a series of moments strung together into a movie whose pace we cannot change. Thus, photography gives us a unique and flexible means to explore and portray the passage of time. With a little trickery, a photograph can portray an experience of time otherwise unseen.

When pioneering 19th-century photographers Eadweard Muybridge and Étienne-Jules Marey captured the first high-speed photographs, they revolutionized how we perceive movement. Their images of a horse trotting or a man jumping, for example, transformed mundane gestures into intricate visualizations. Their work unlocked our collective imagination, leading to a century and a half of innovative motion studies.

PHOTOGRAPHY GIVES US A UNIQUE AND FLEXIBLE MEANS TO EXPLORE AND PORTRAY THE PASSAGE OF TIME.

Markus Reugels's high-speed image, for instance, captures the mind-boggling interplay between colliding water droplets (page 132). Lorenzo Montezemolo uses a 30-second-long exposure to show the fog waves rolling through San Francisco Bay (page 152). Stephen Wilkes's composites compress the activity of a frenetic day into a single frame (pages 130 and 176). Shinichi Maruyama combines 10,000 individual photographs of a dancer to give us a unique view of the human body in motion (page 181).

Each of these images stretches or compresses time beyond what our naked perception can detect. Unlike movies, photographs let us spend as much time as we choose to look, probe, and wonder at the mysteries they reveal—a luxury amid the chaotic bustle of our everyday lives. ■

FOLLOWING PAGES:
Stephen Wilkes
A composite of many images shows life in New York's Times Square over a 12-hour period.

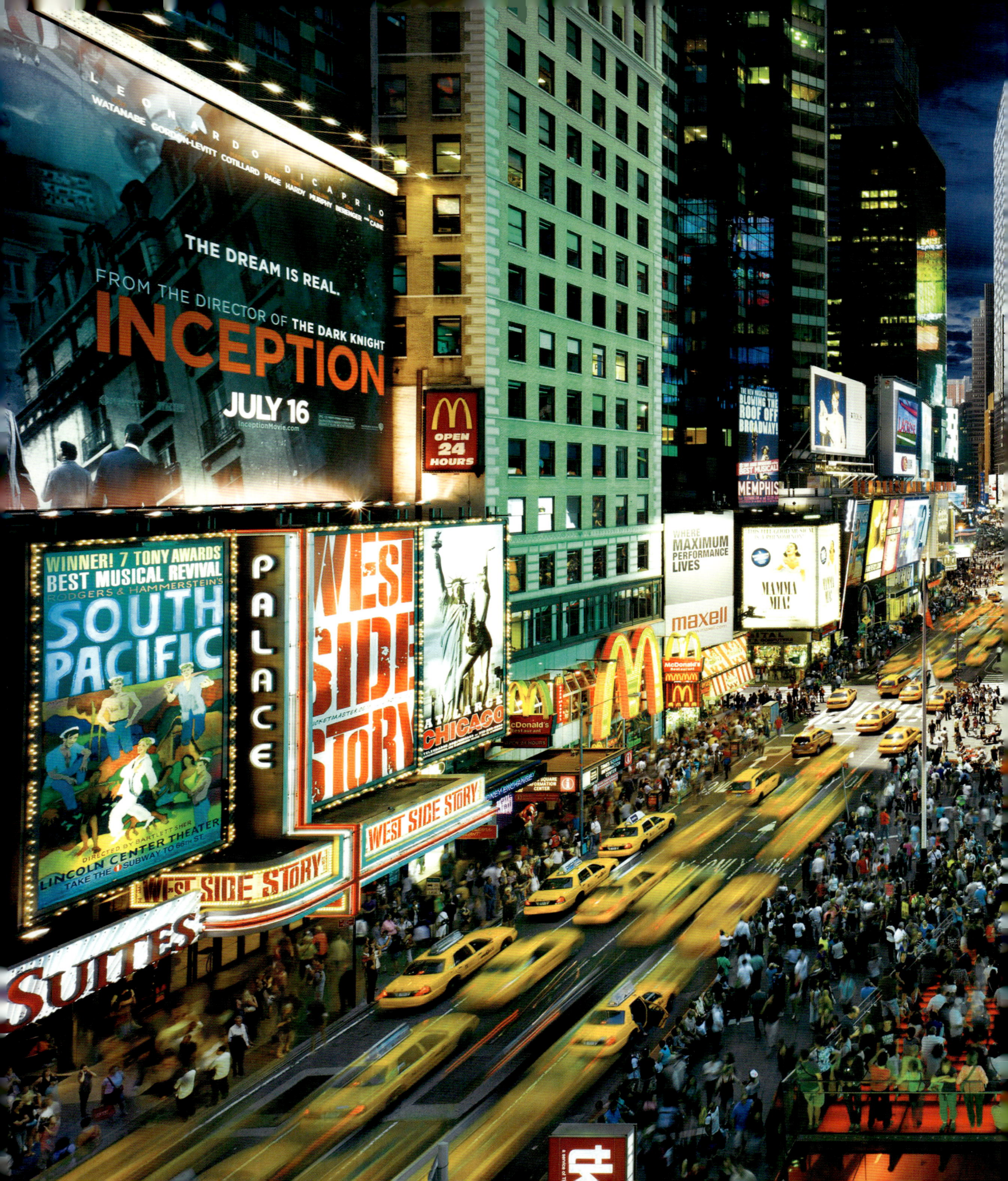

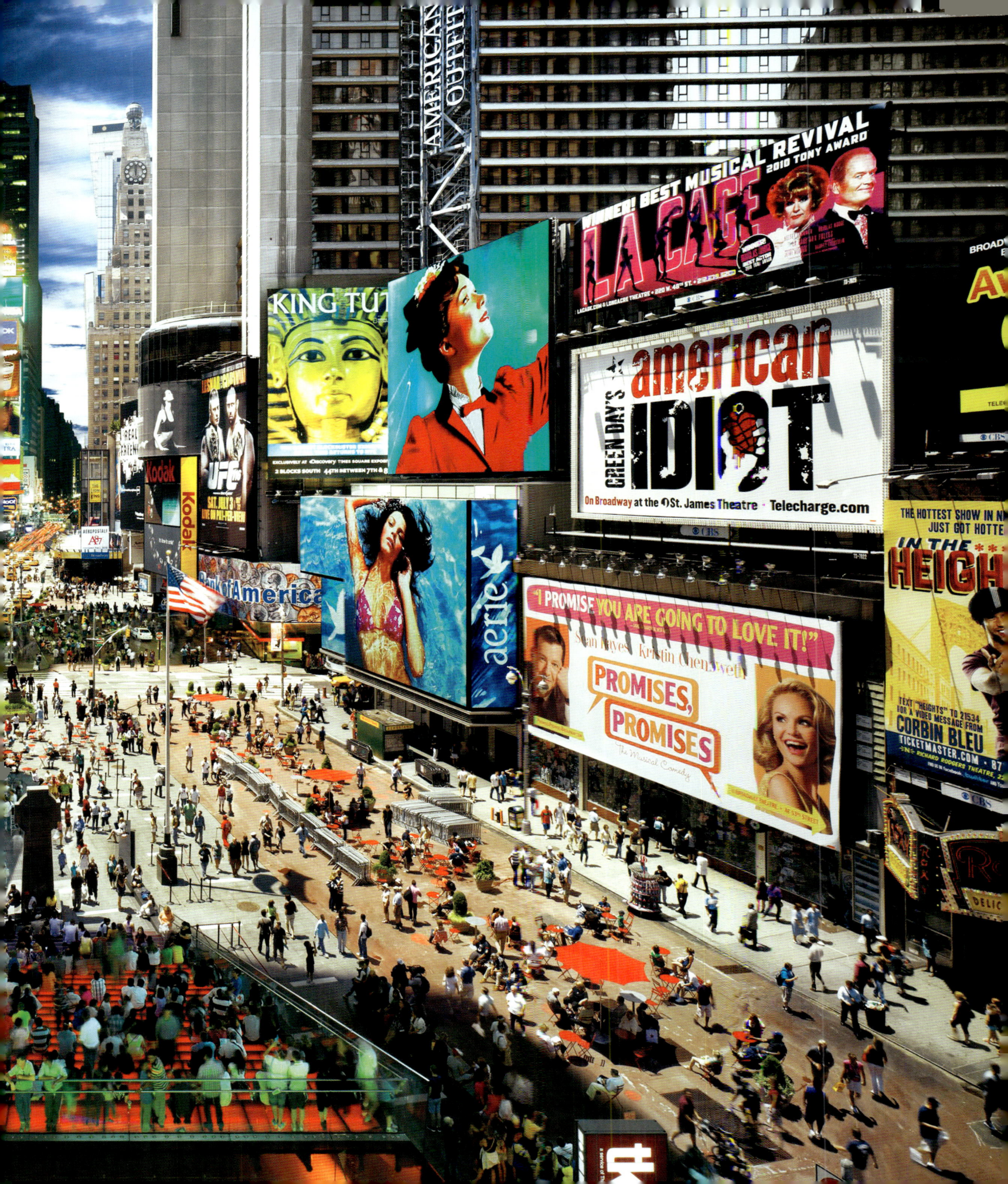

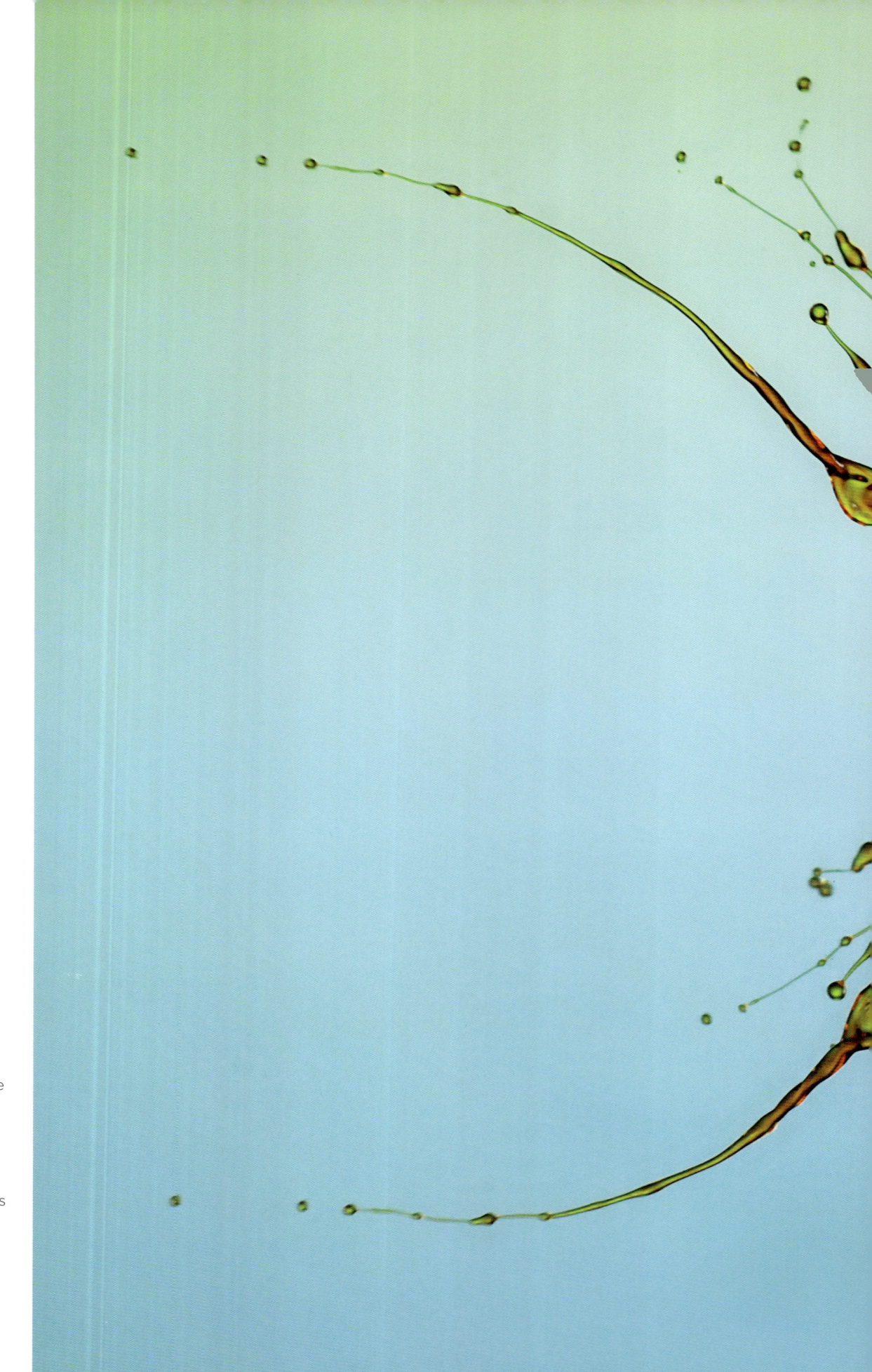

Markus Reugels
Colored water droplets bounce off a reflective surface and collide with one another.

FOLLOWING PAGES:
Kurt Lawson and Sean Goebel
An hour-long exposure reveals the movement of stars above Half Dome in Yosemite National Park. Hikers' head-lamps trace the path to the summit.

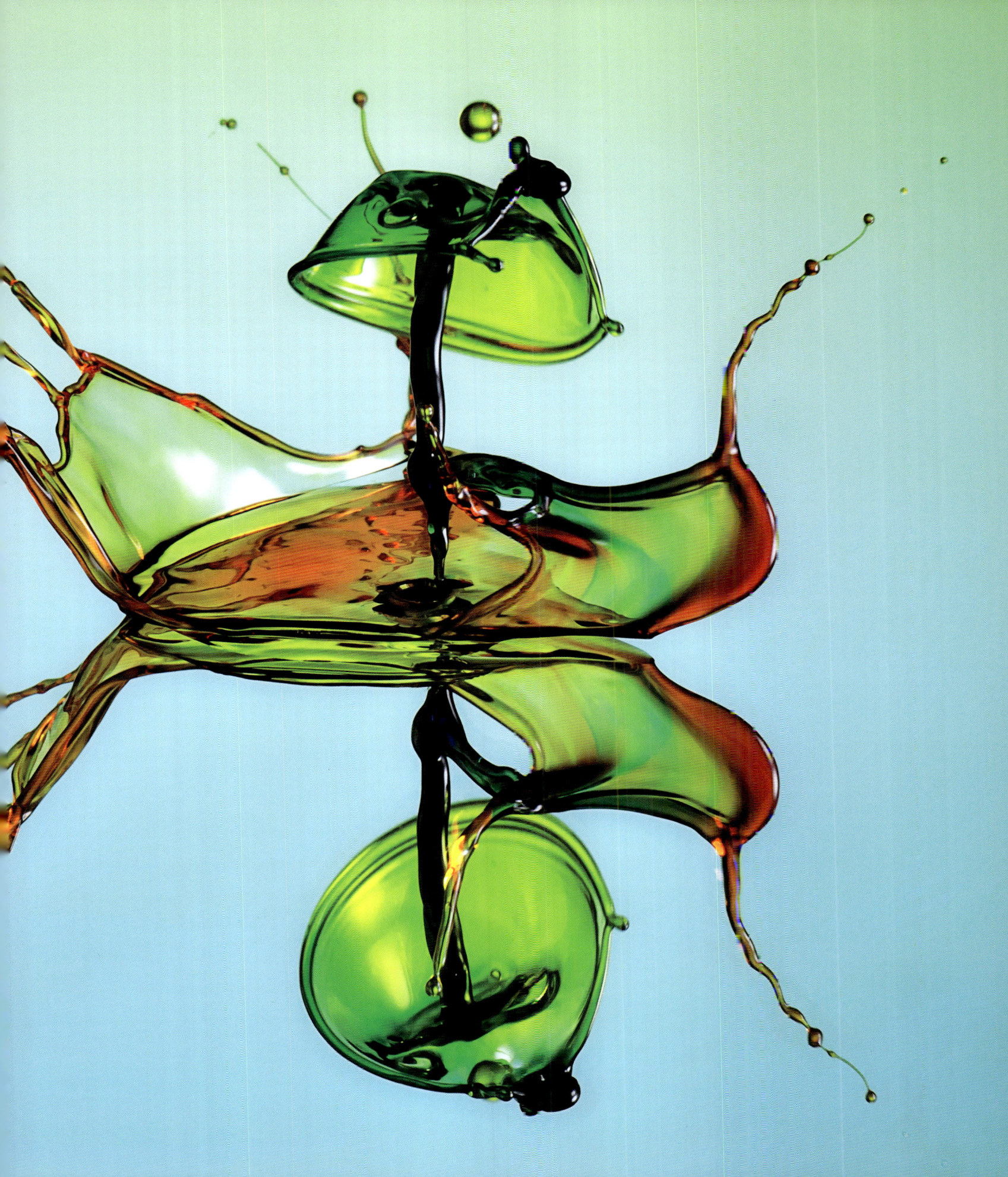

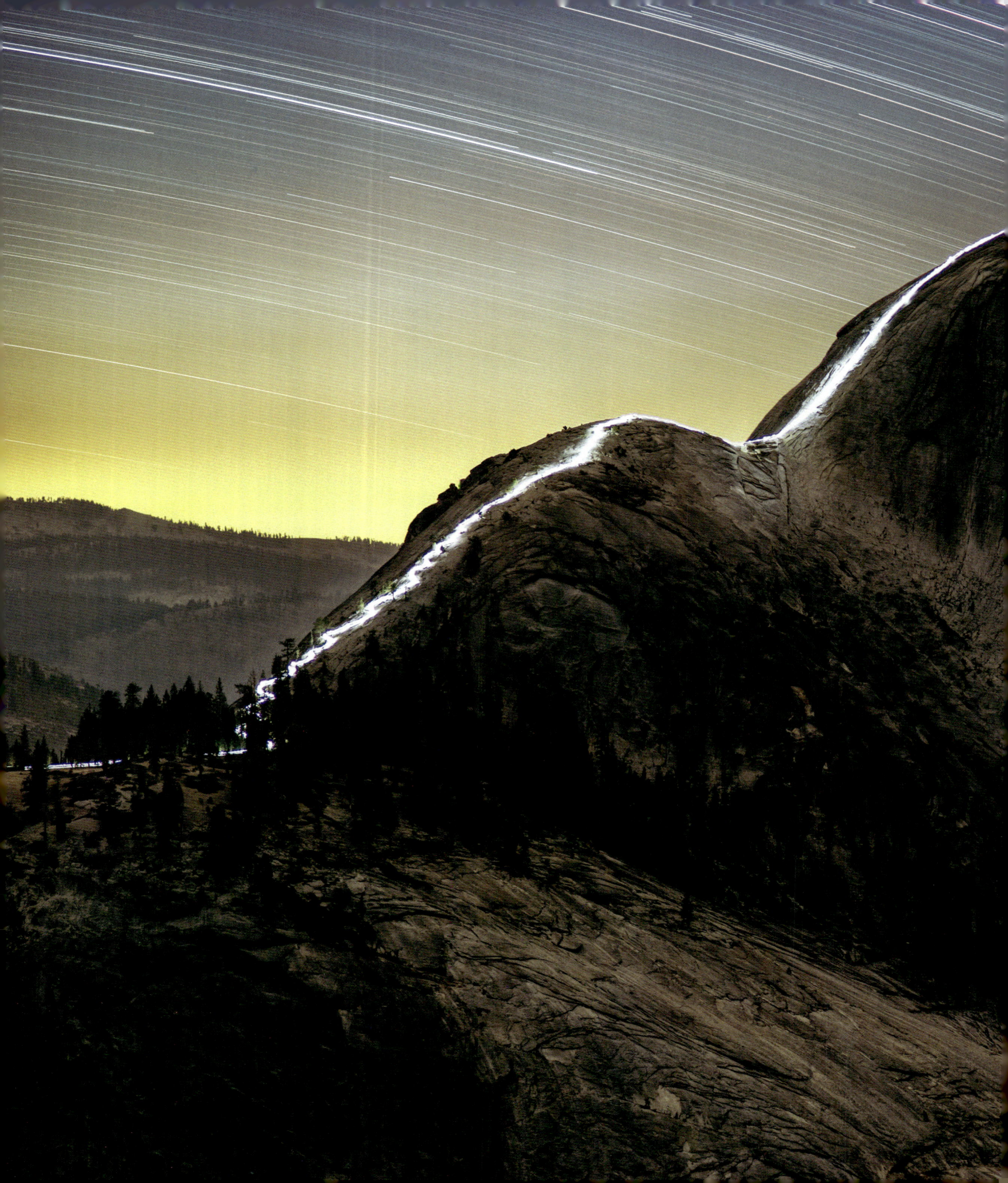

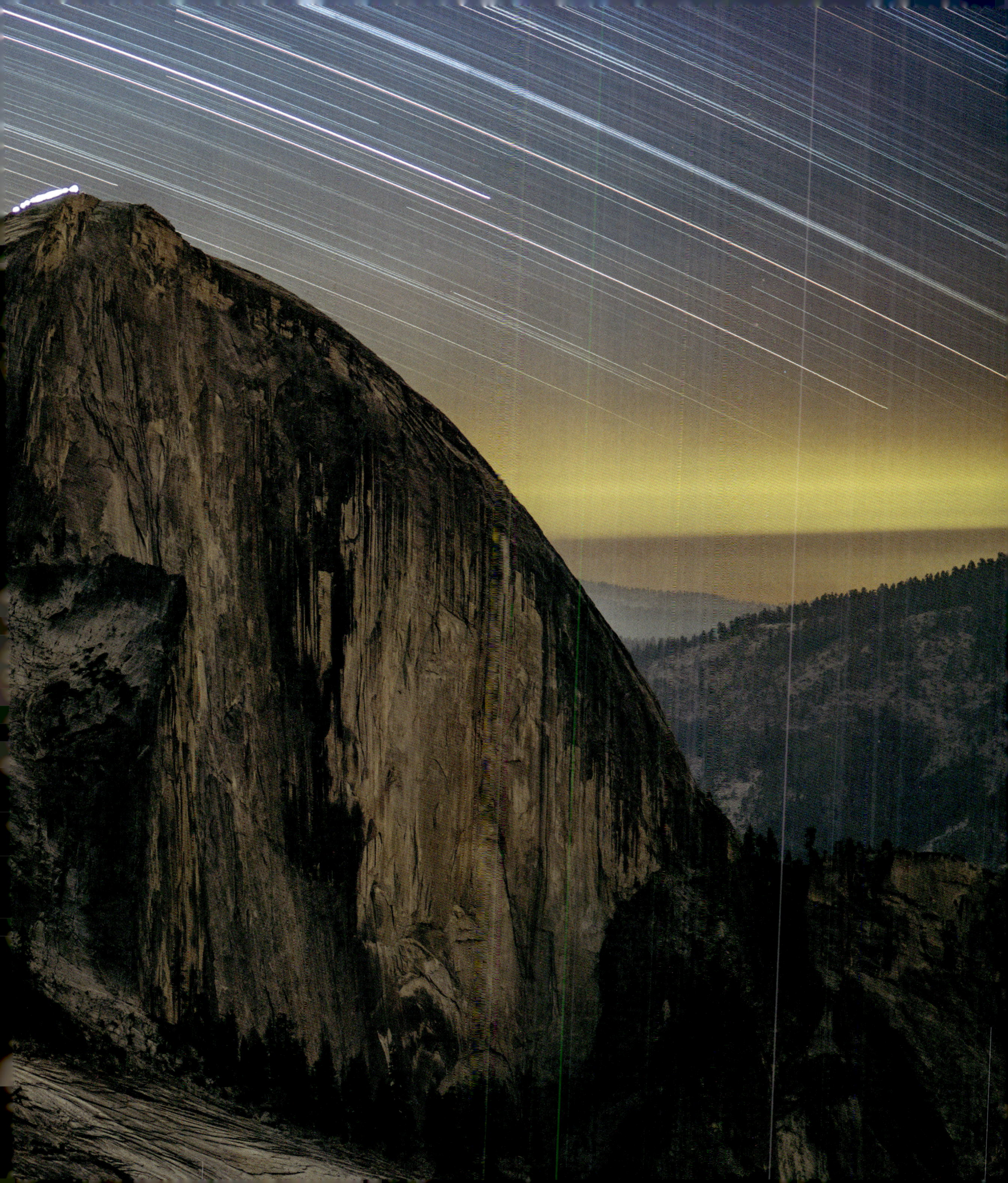

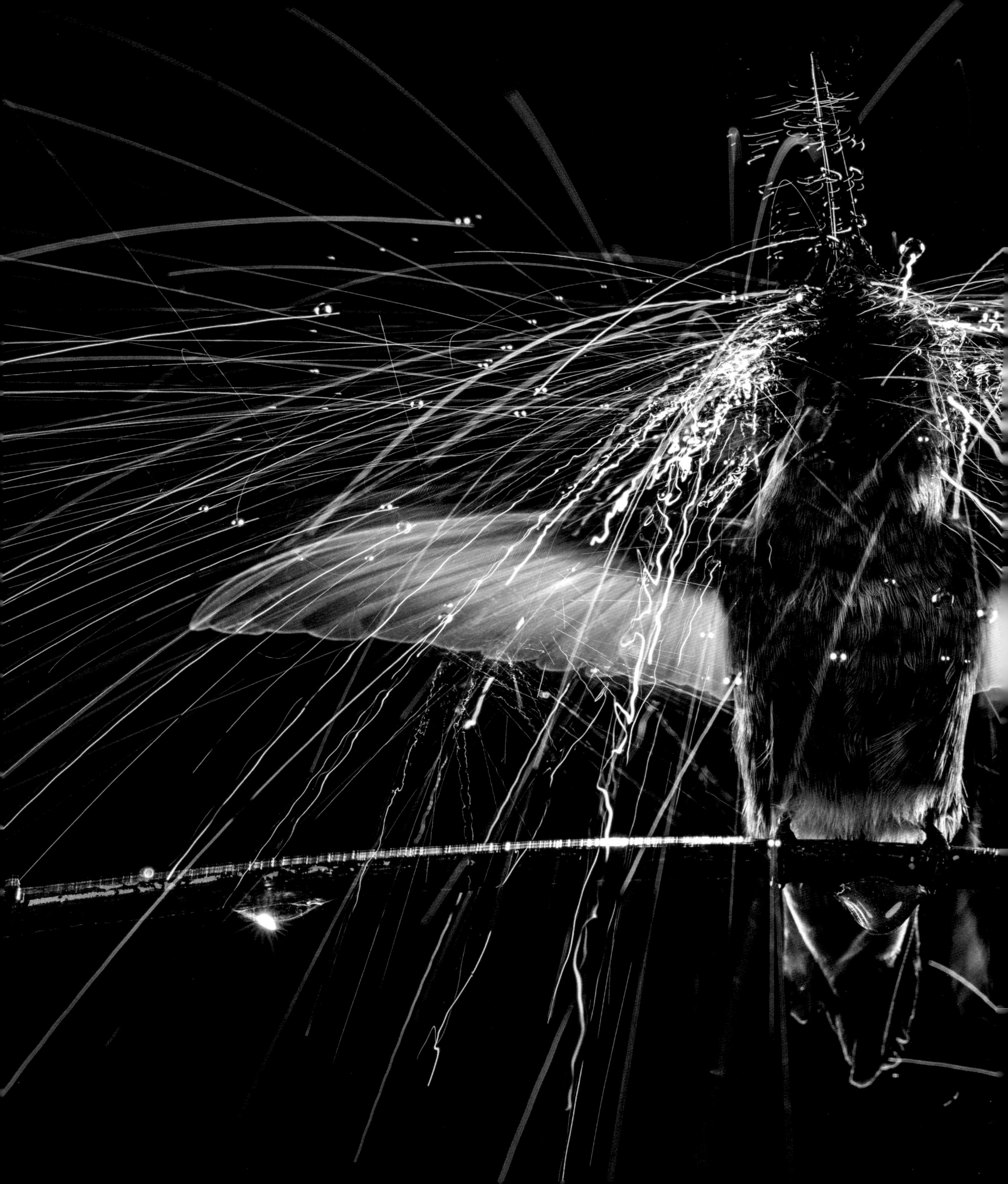

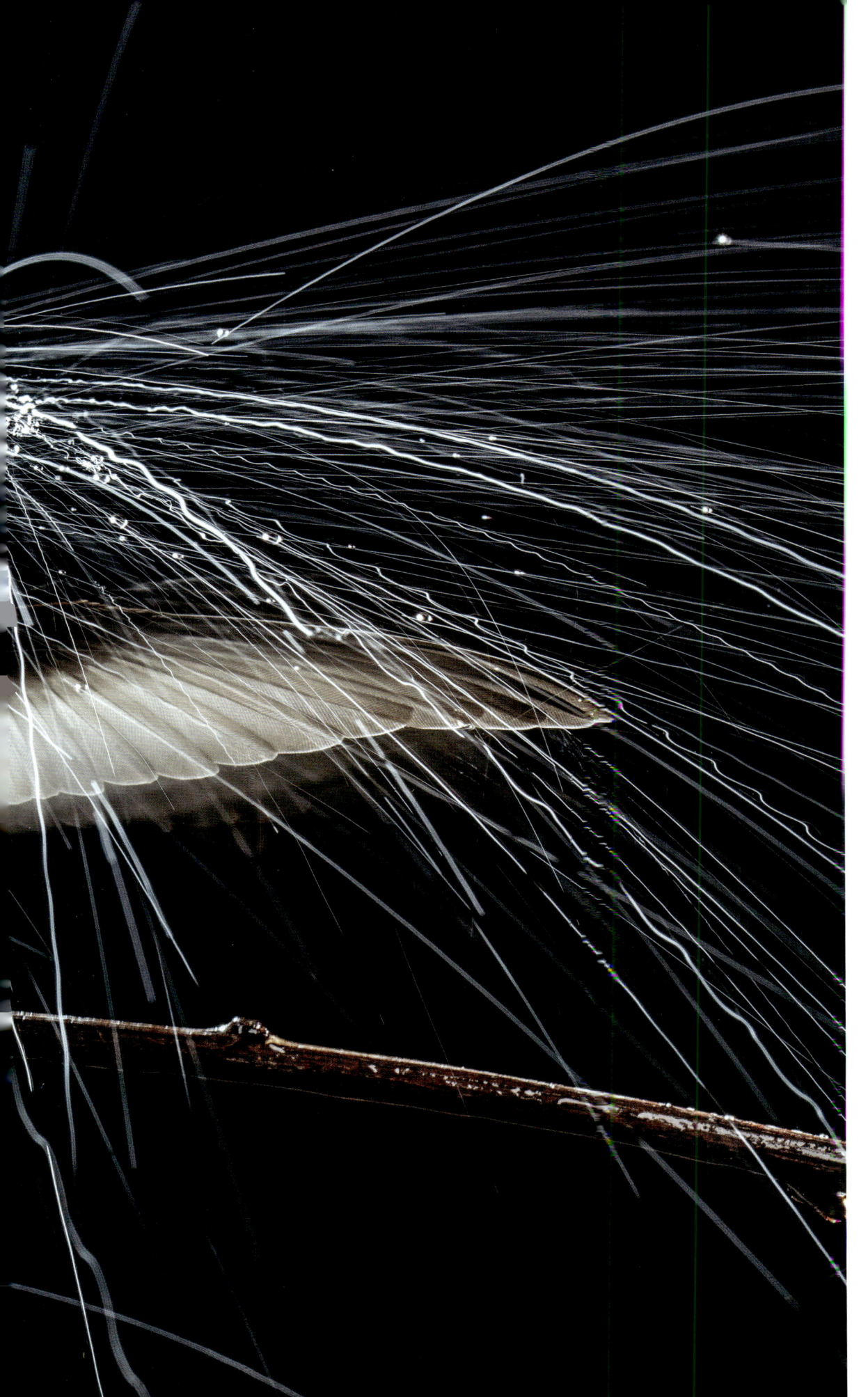

Anand Varma

An Anna's hummingbird shakes itself dry during an experiment conducted at the Dudley Lab at the University of California, Berkeley. A long exposure captures the streams of water while a single burst of light freezes the moment and defines the bird's body.

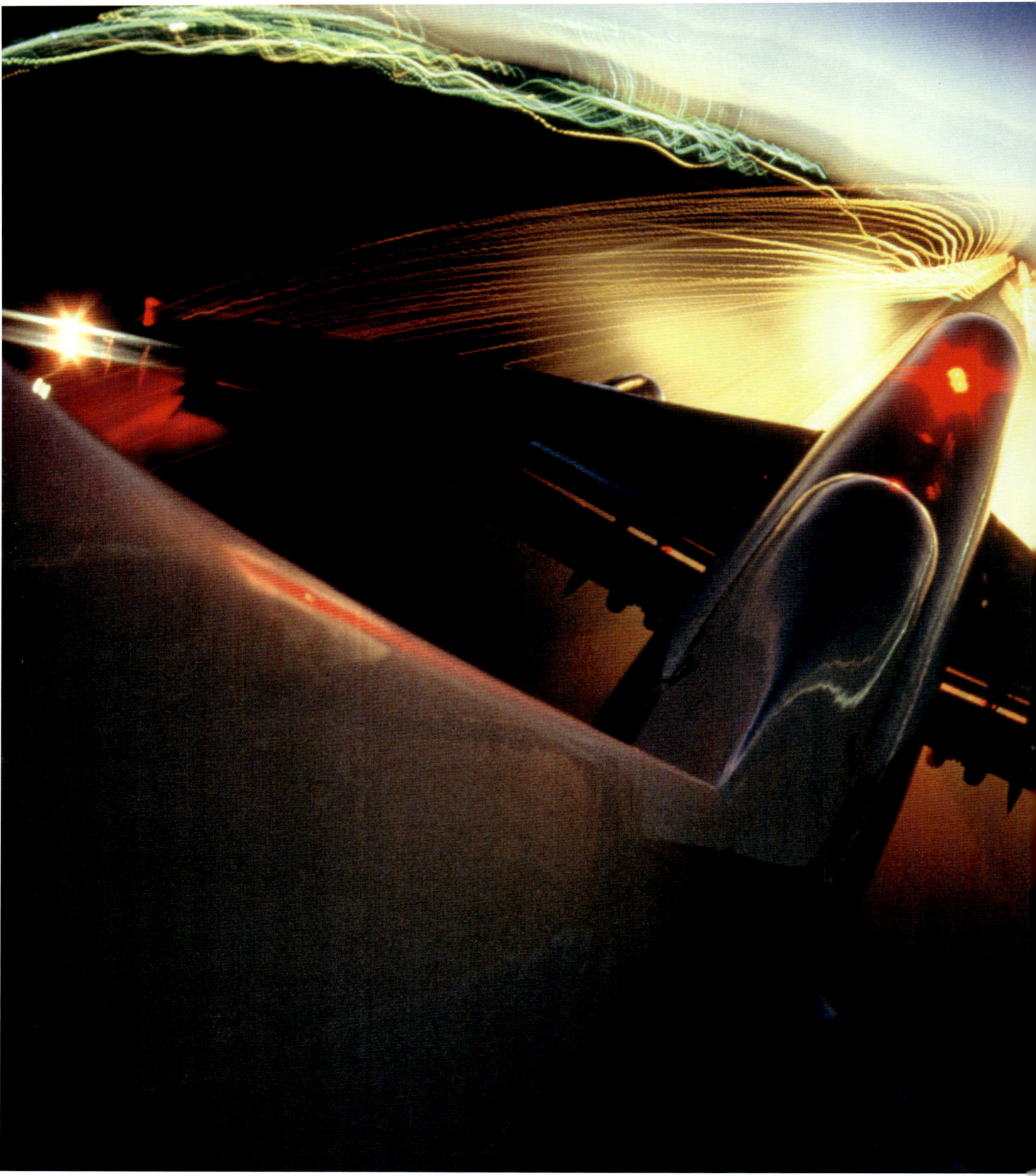

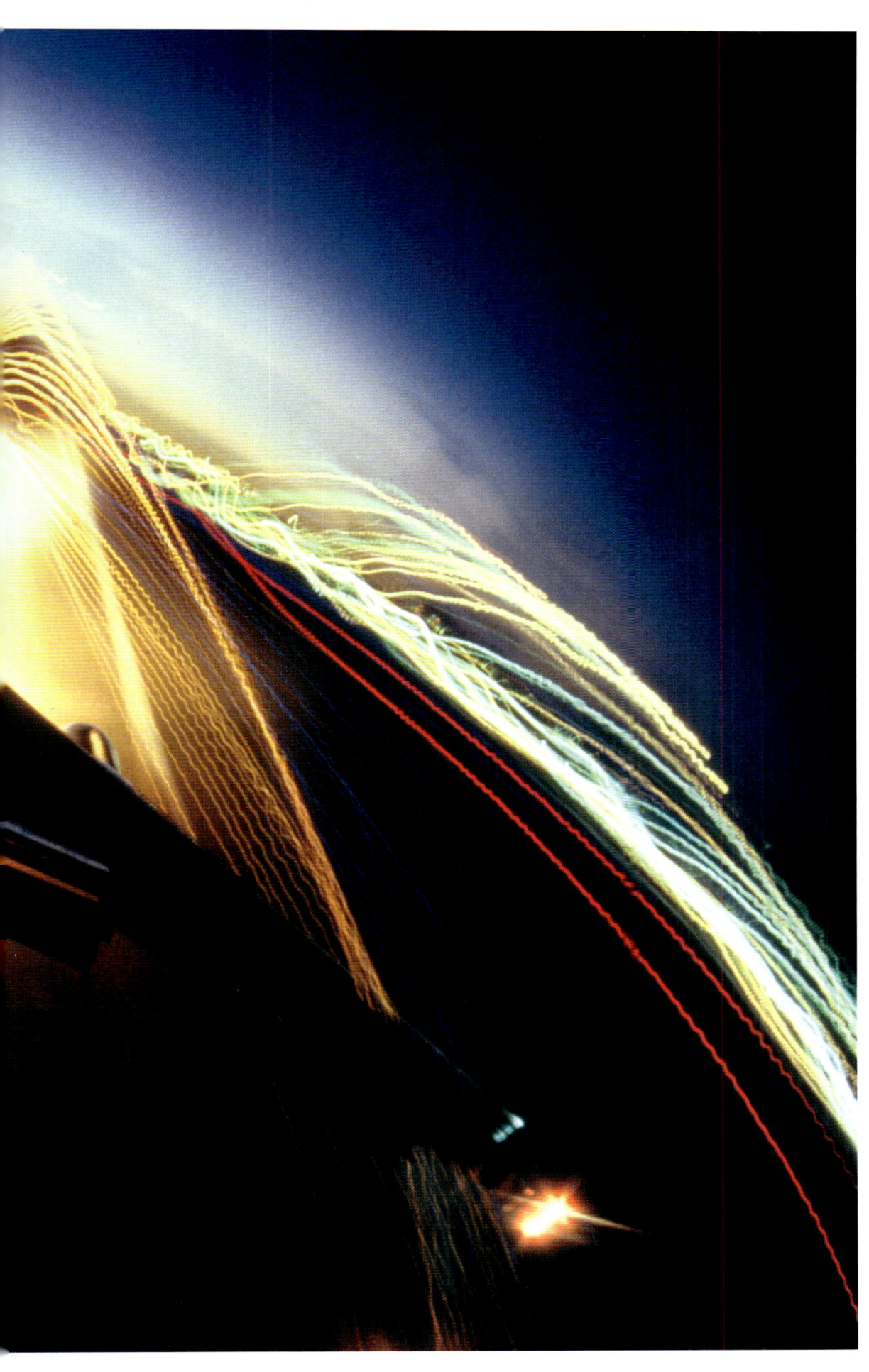

Bruce Dale
A jumbo jet lands in Palmdale, California. A camera affixed to the tail of the plane captures the blur of city and runway lights

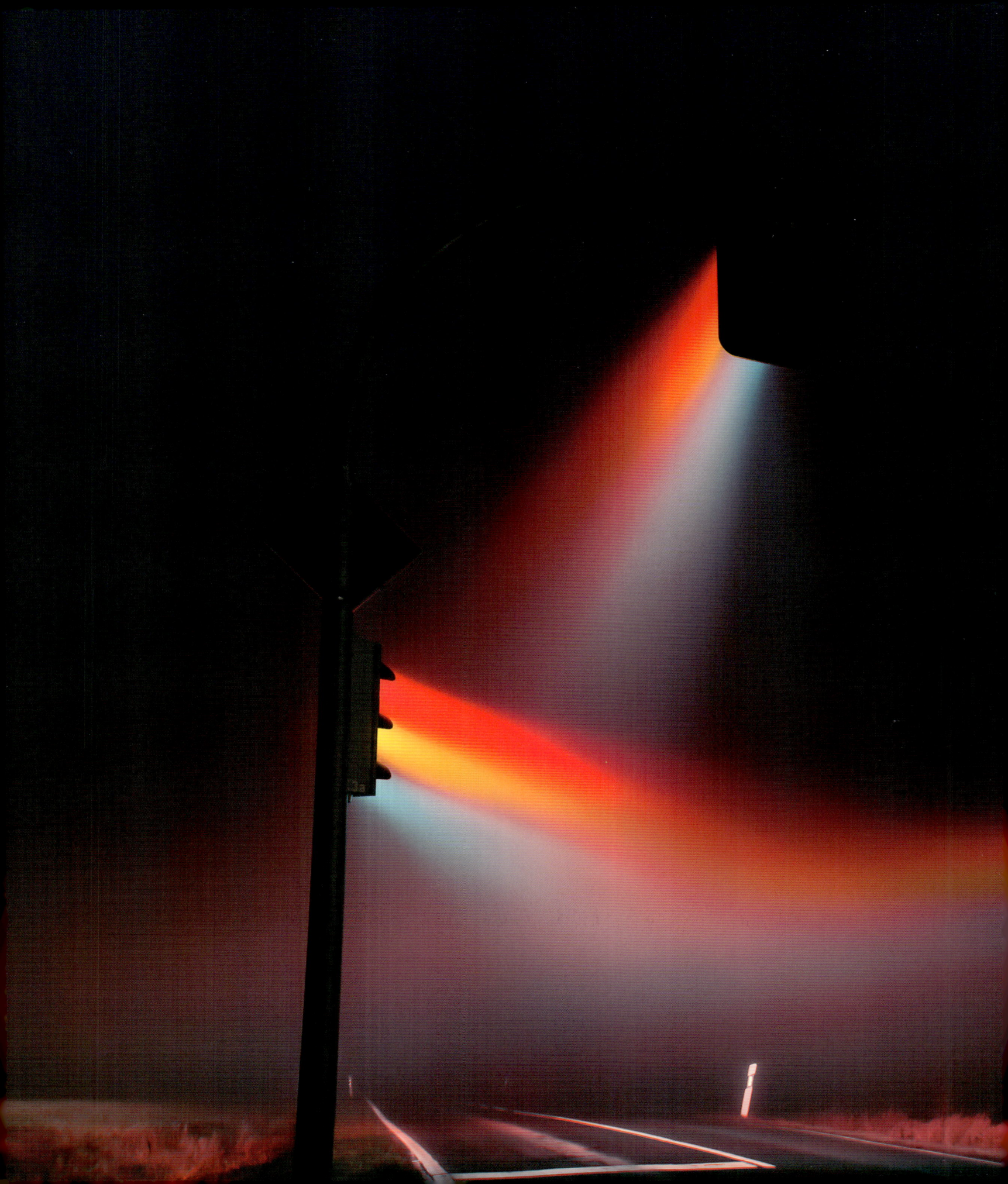

Lucas Zimmermann
Traffic lights photographed
with a long exposure on a
foggy night near Weimar
Germany

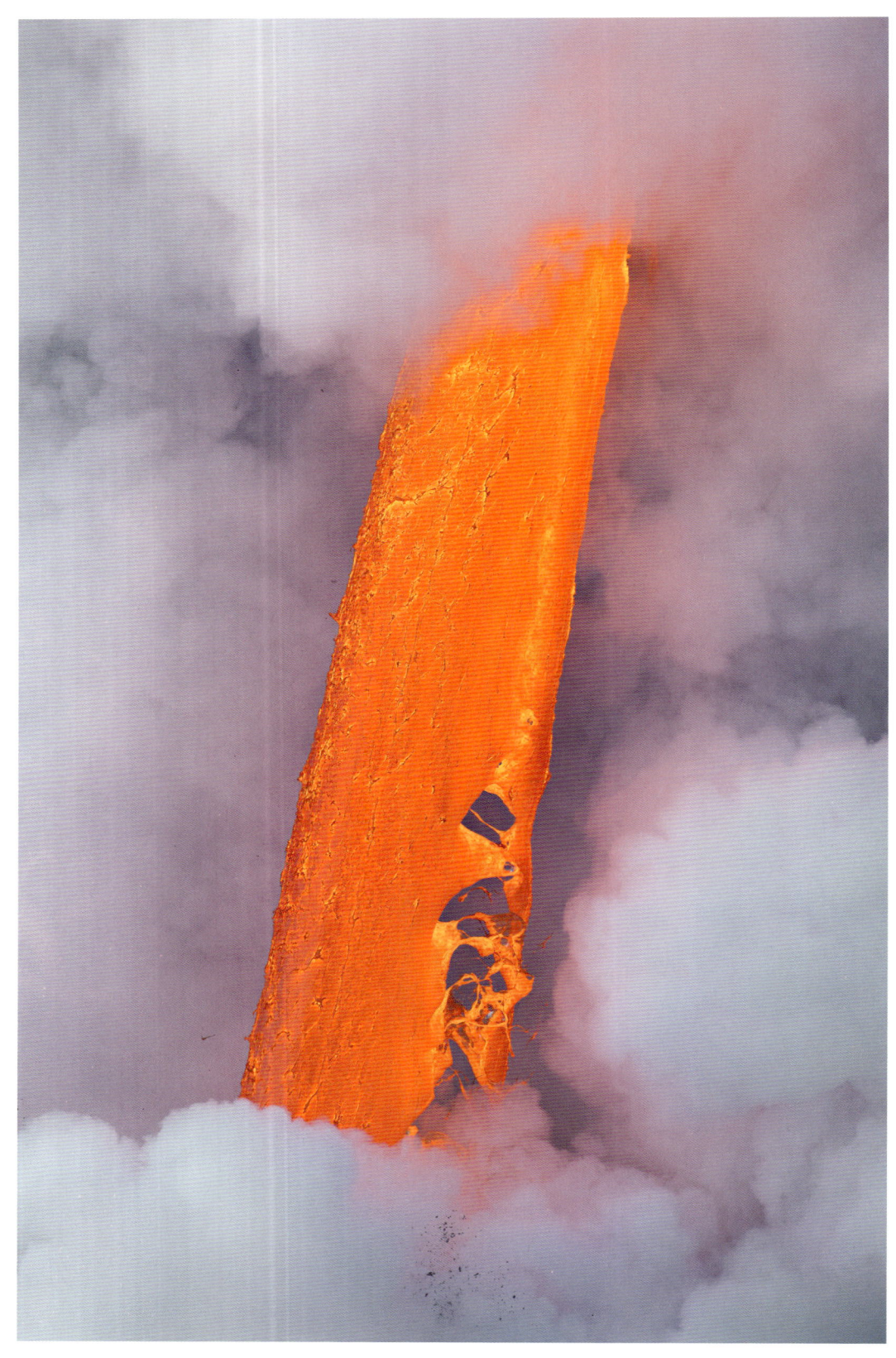

Andrew Dunn
Lava from the Kilauea volcano in Hawaii creates billows of steam as it hits the ocean.

PHOTOGRAPHS LET US SPEND AS MUCH TIME AS WE CHOOSE TO LOOK, PROBE, AND WONDER AT THE MYSTERIES THEY REVEAL—A LUXURY AMID THE CHAOTIC BUSTLE OF OUR EVERYDAY LIVES.

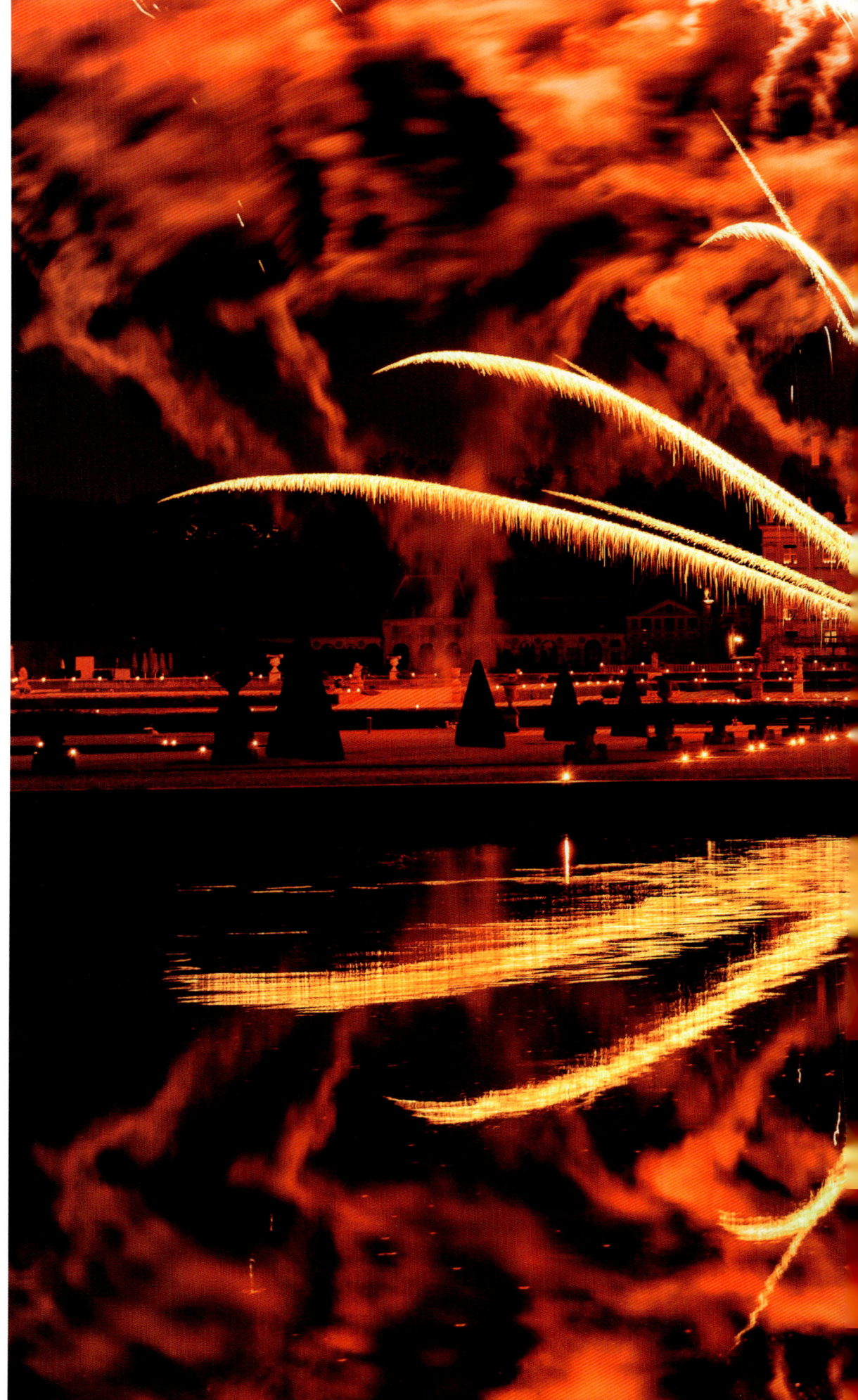

Diane Cook and Len Jenshel
Fireworks illuminate the
gardens of Vaux-le-Vicomte,
Maincy, France.

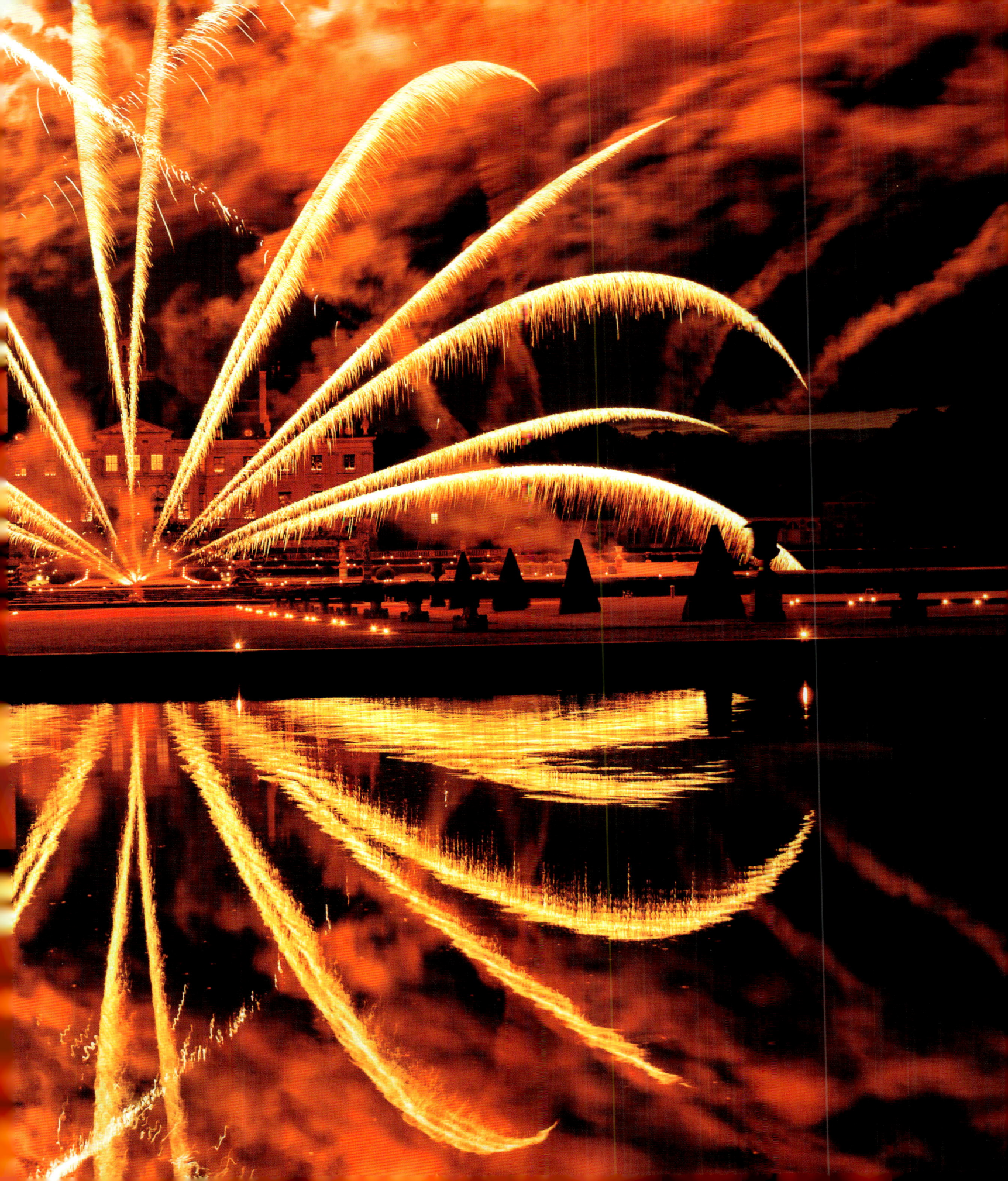

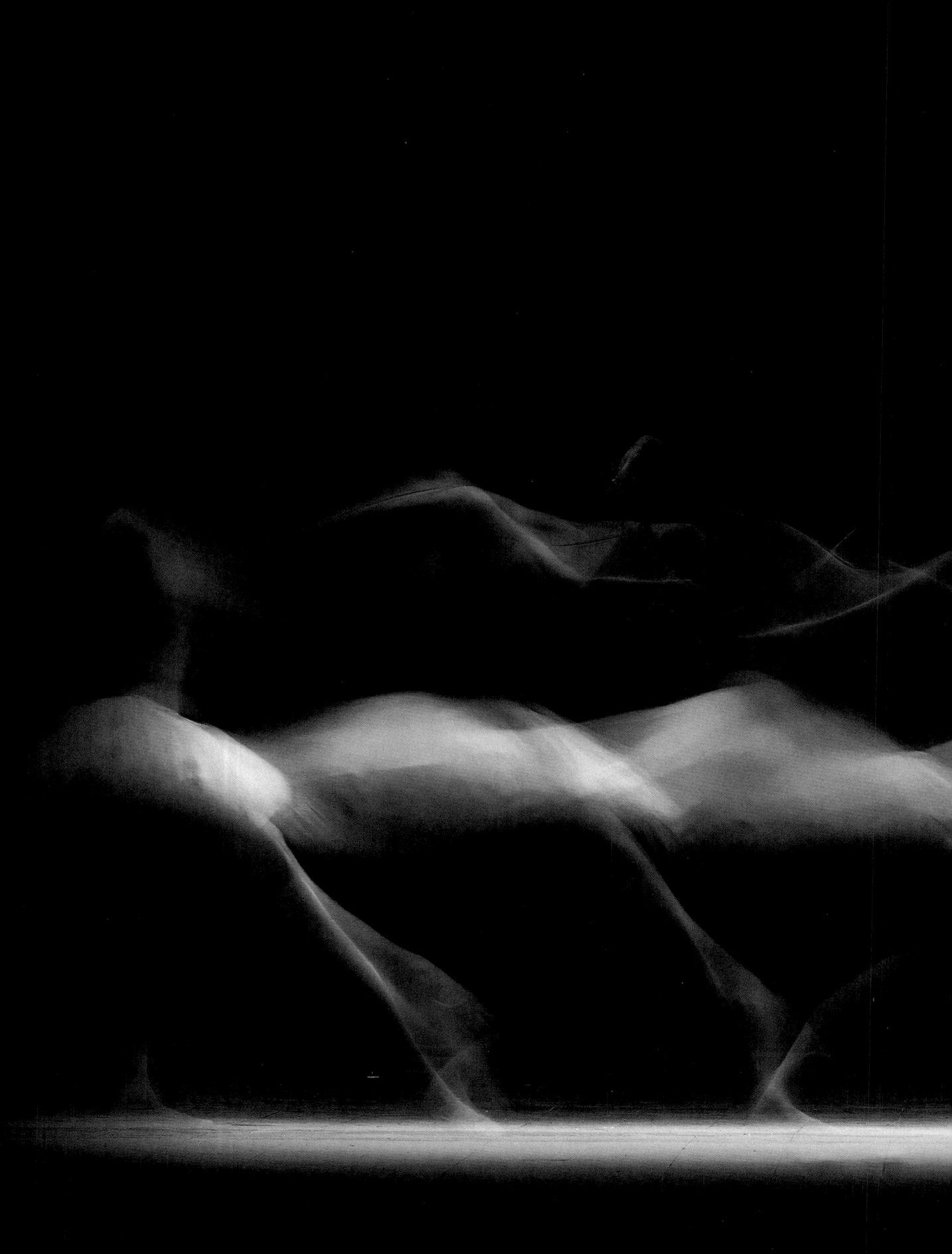

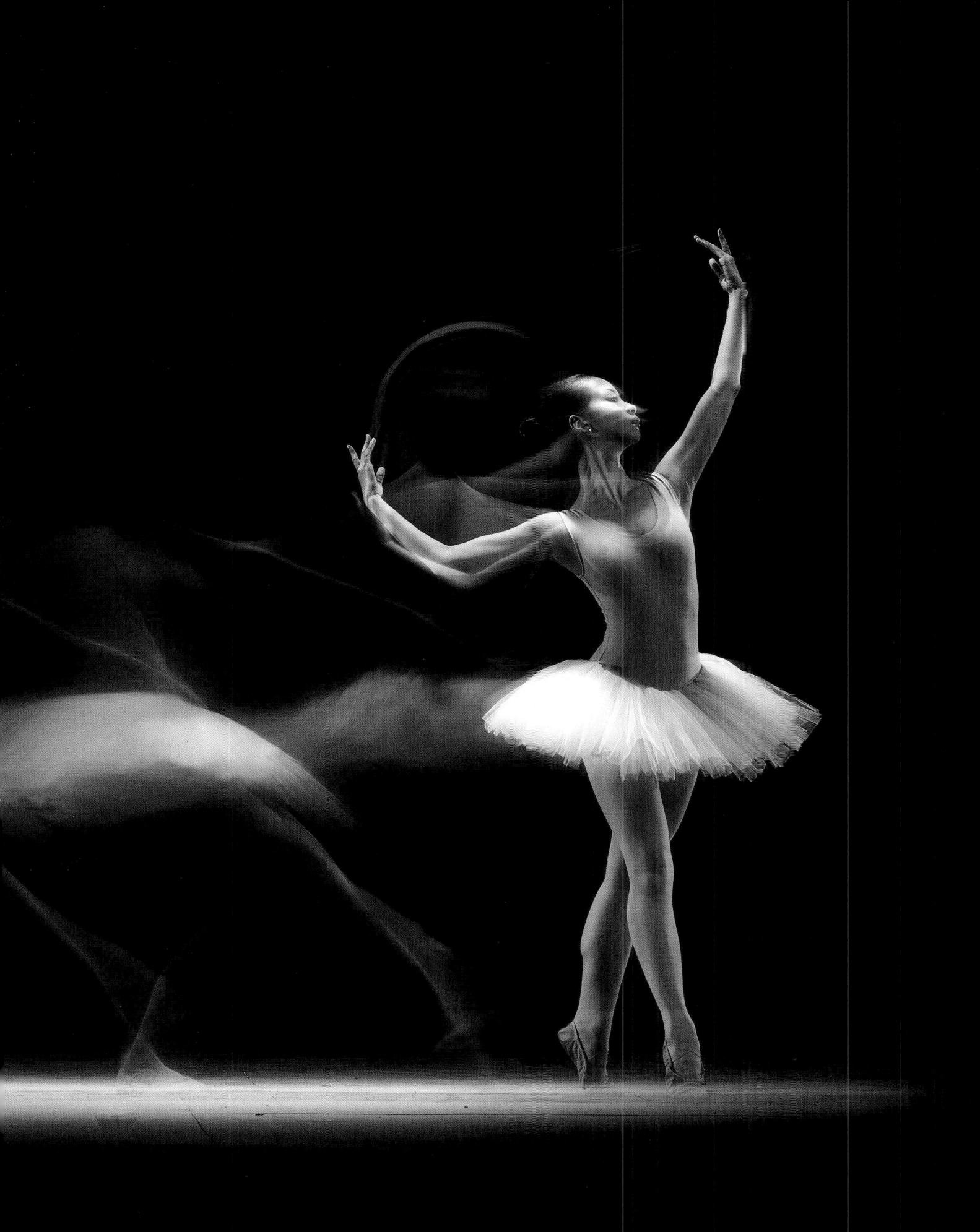

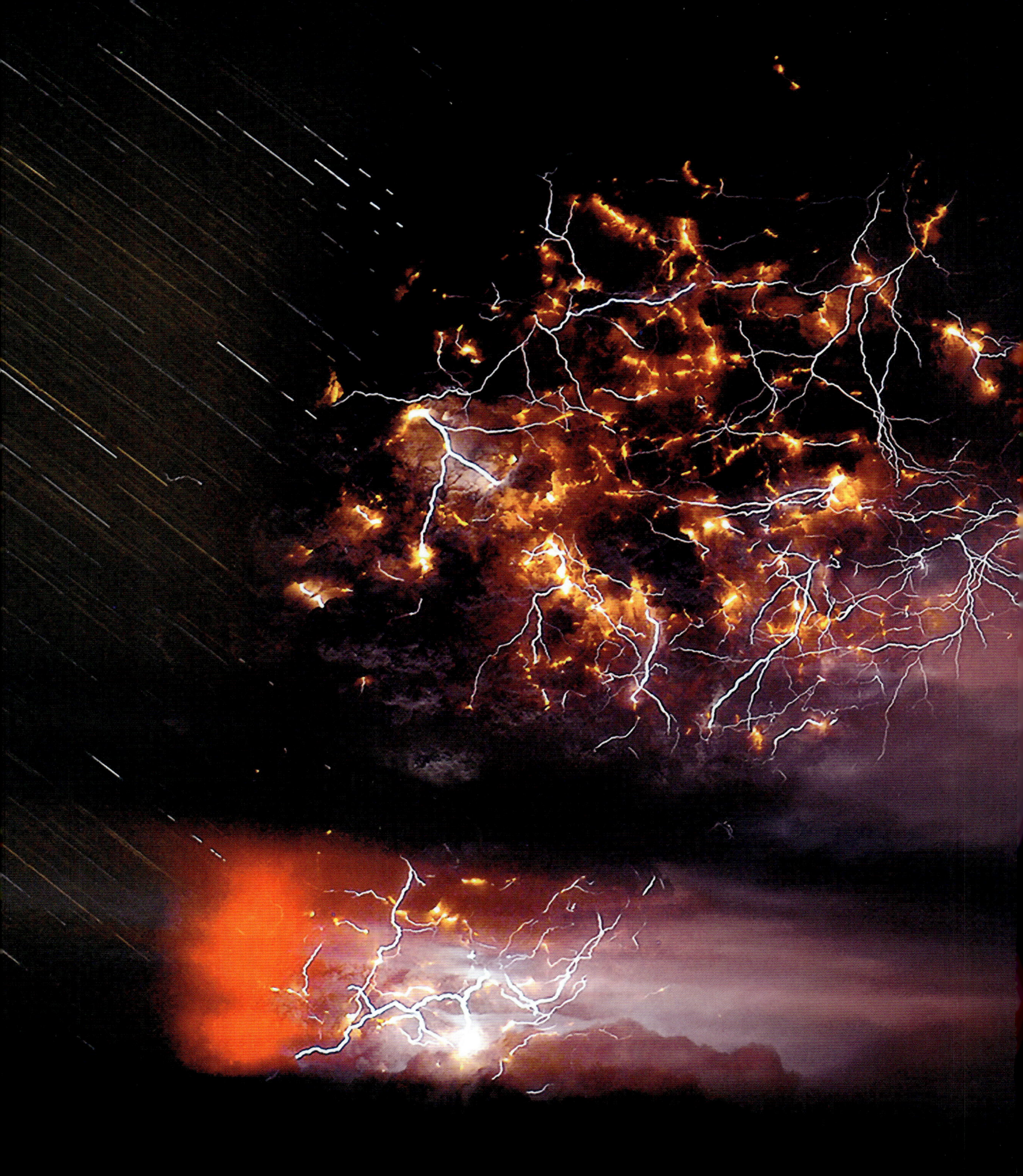

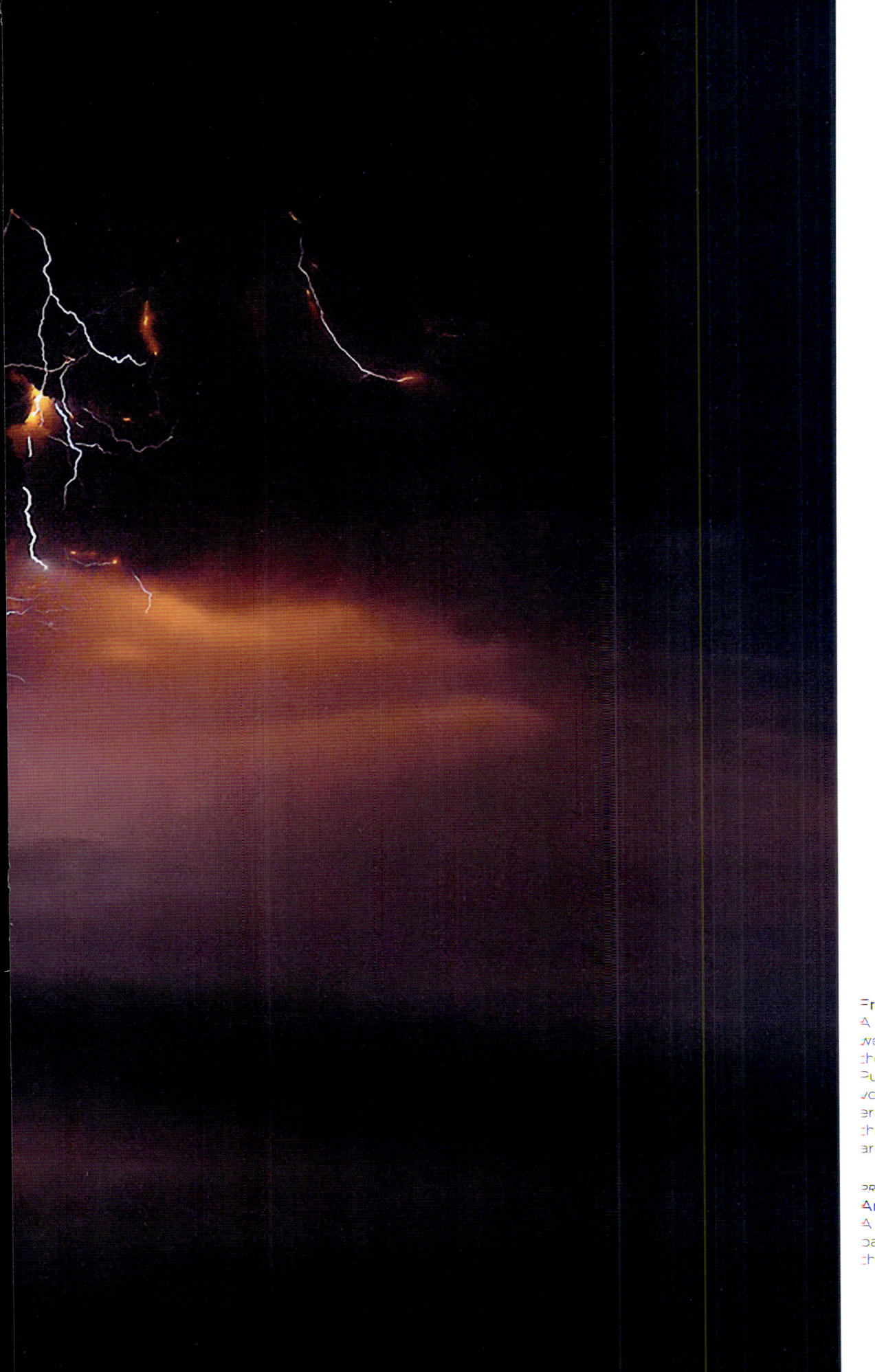

Francisco Negroni
A long exposure reveals a web of lightning formed by the ash plume of Chile's Puyehue-Cordón Caulle volcanic complex. This eruption was so powerful that it sent an ash cloud around the globe in 2011.

PREVIOUS PAGES:
Antonyus Bunjamin
A long exposure sketches the path of a ballet dancer across the stage.

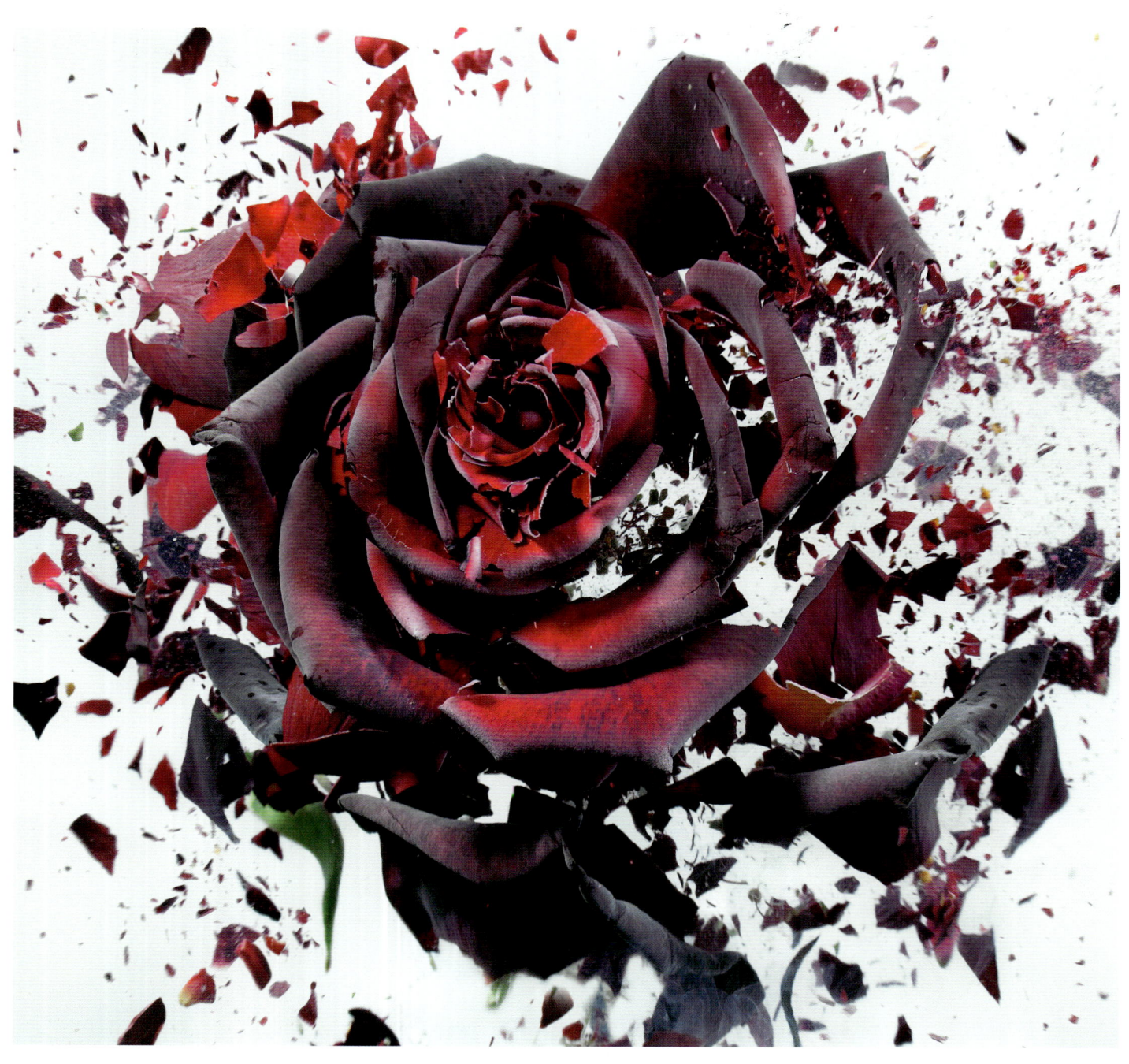

Martin Klimas
A rose, dipped and frozen in liquid nitrogen, explodes when blasted with air.

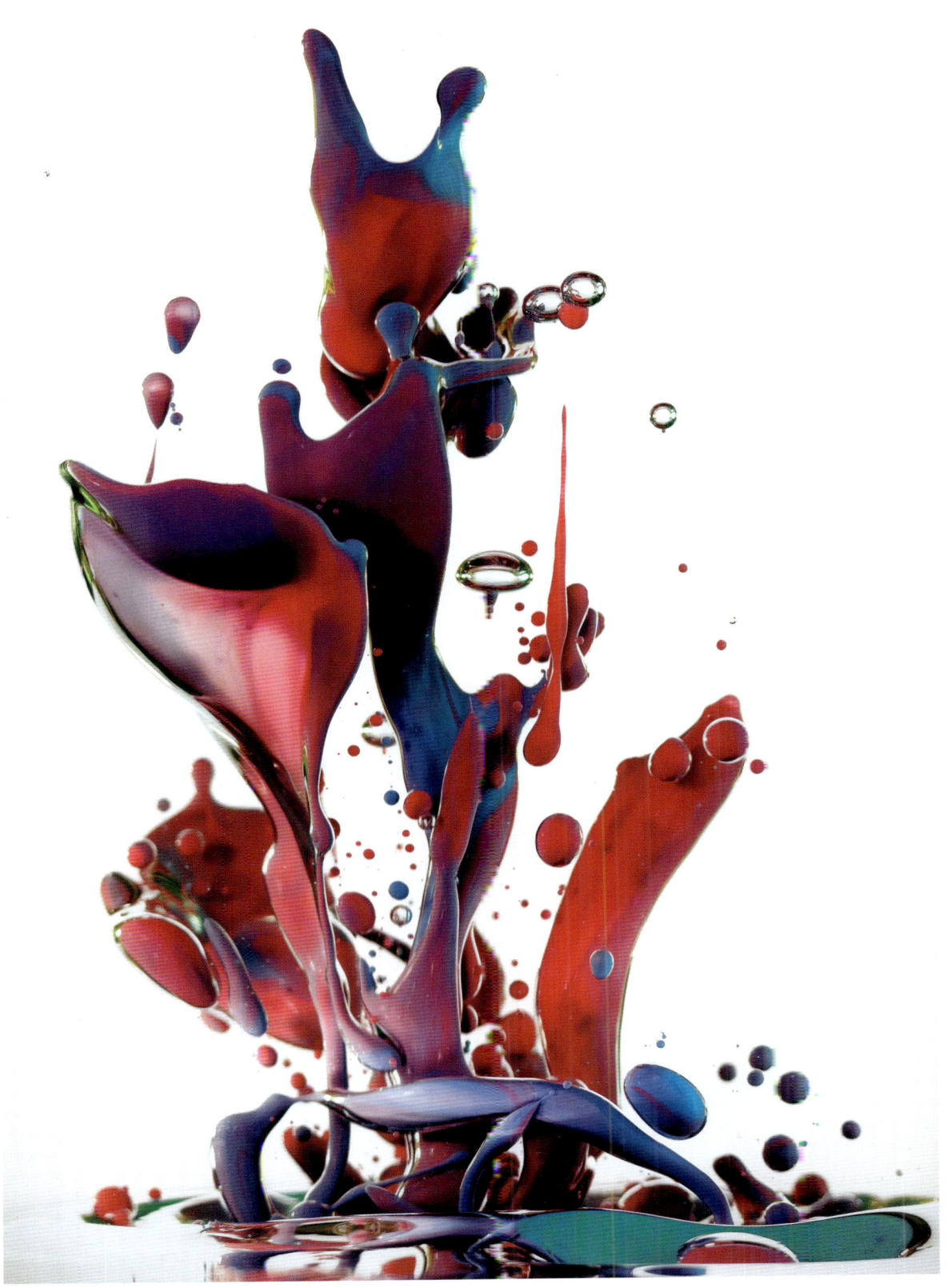

Alberto Seveso
Ink mixes with oil at high speed.

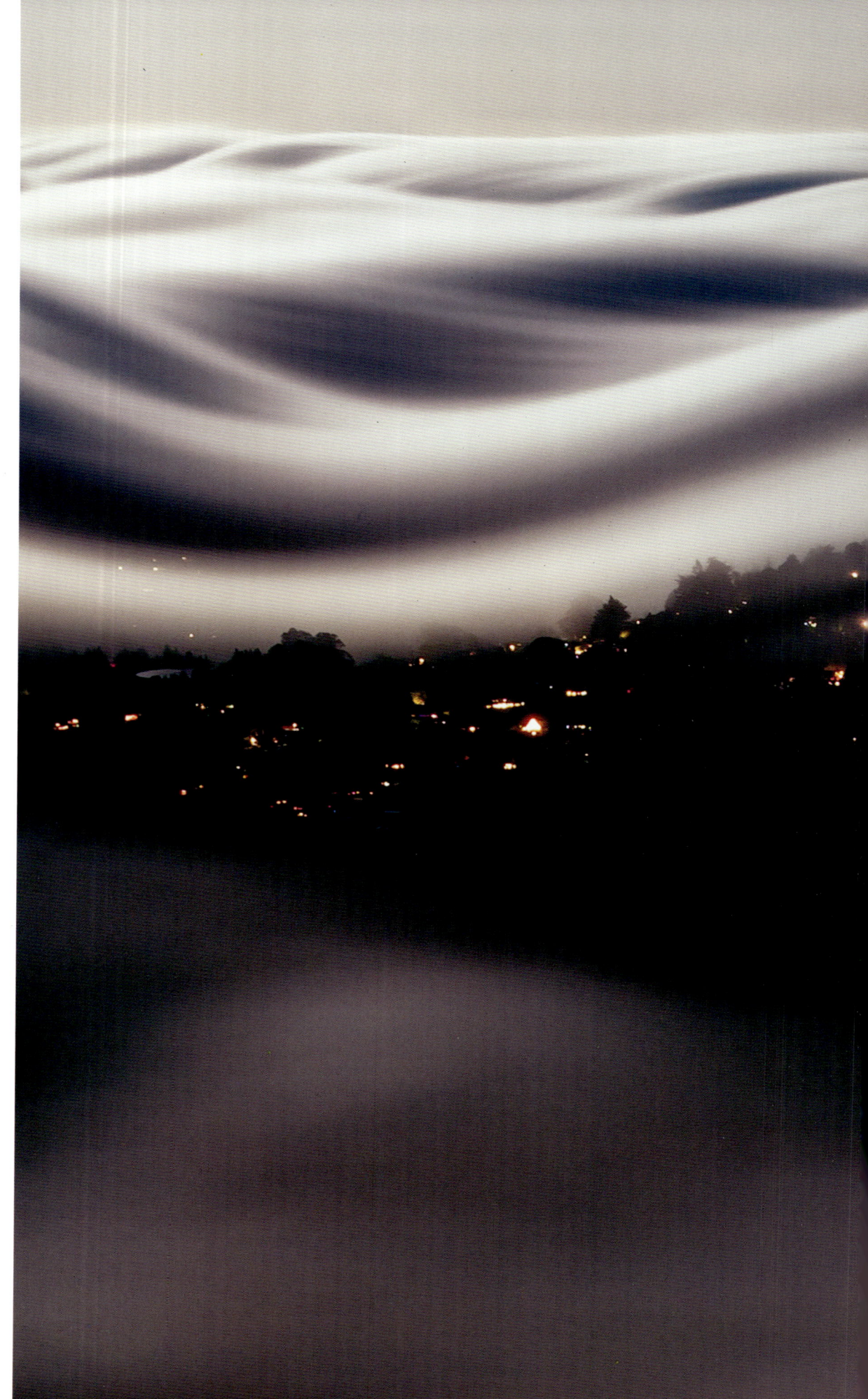

Lorenzo Montezemolo
The full moon illuminates fog as it rolls through the hills surrounding San Francisco Bay in this long exposure.

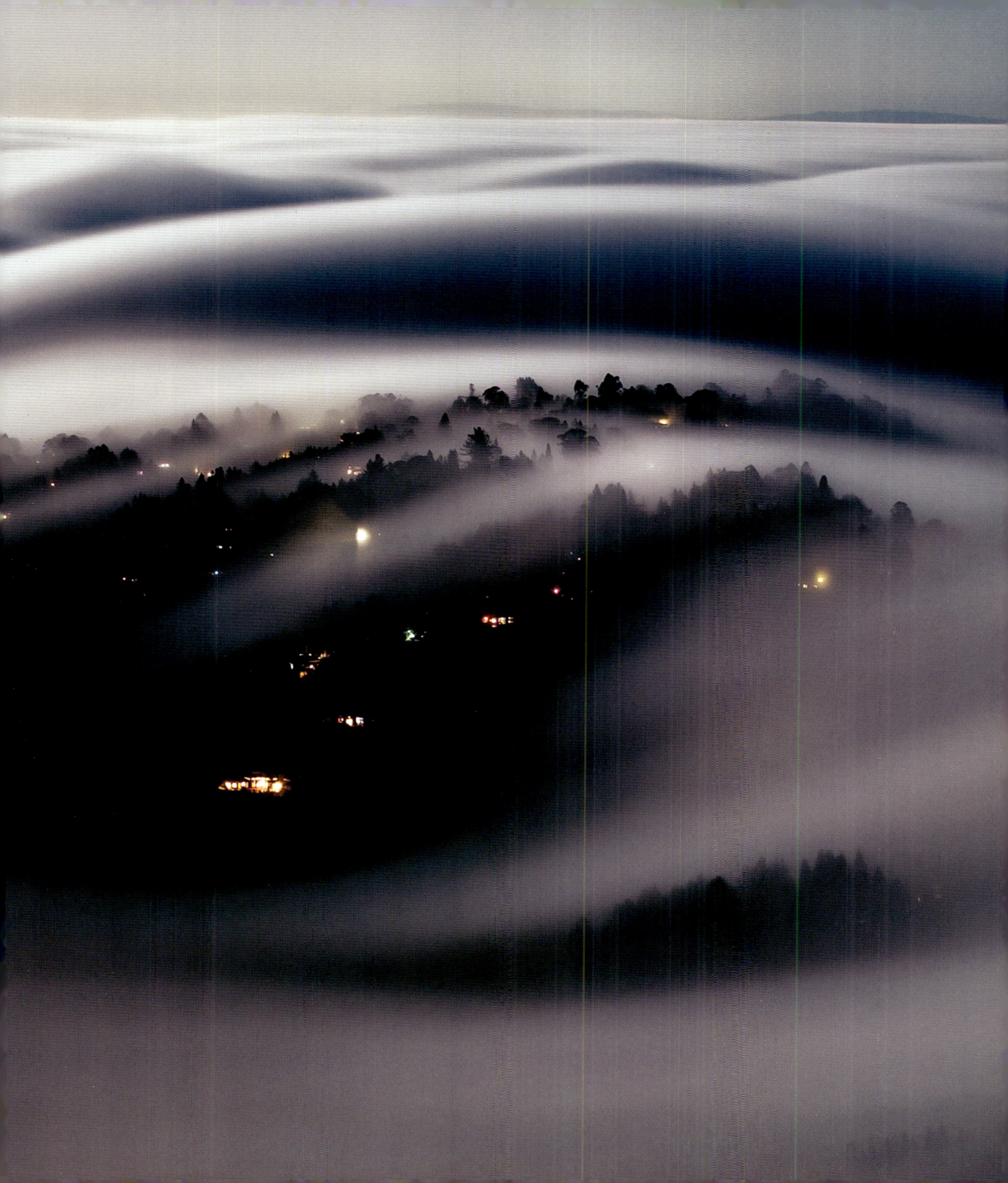

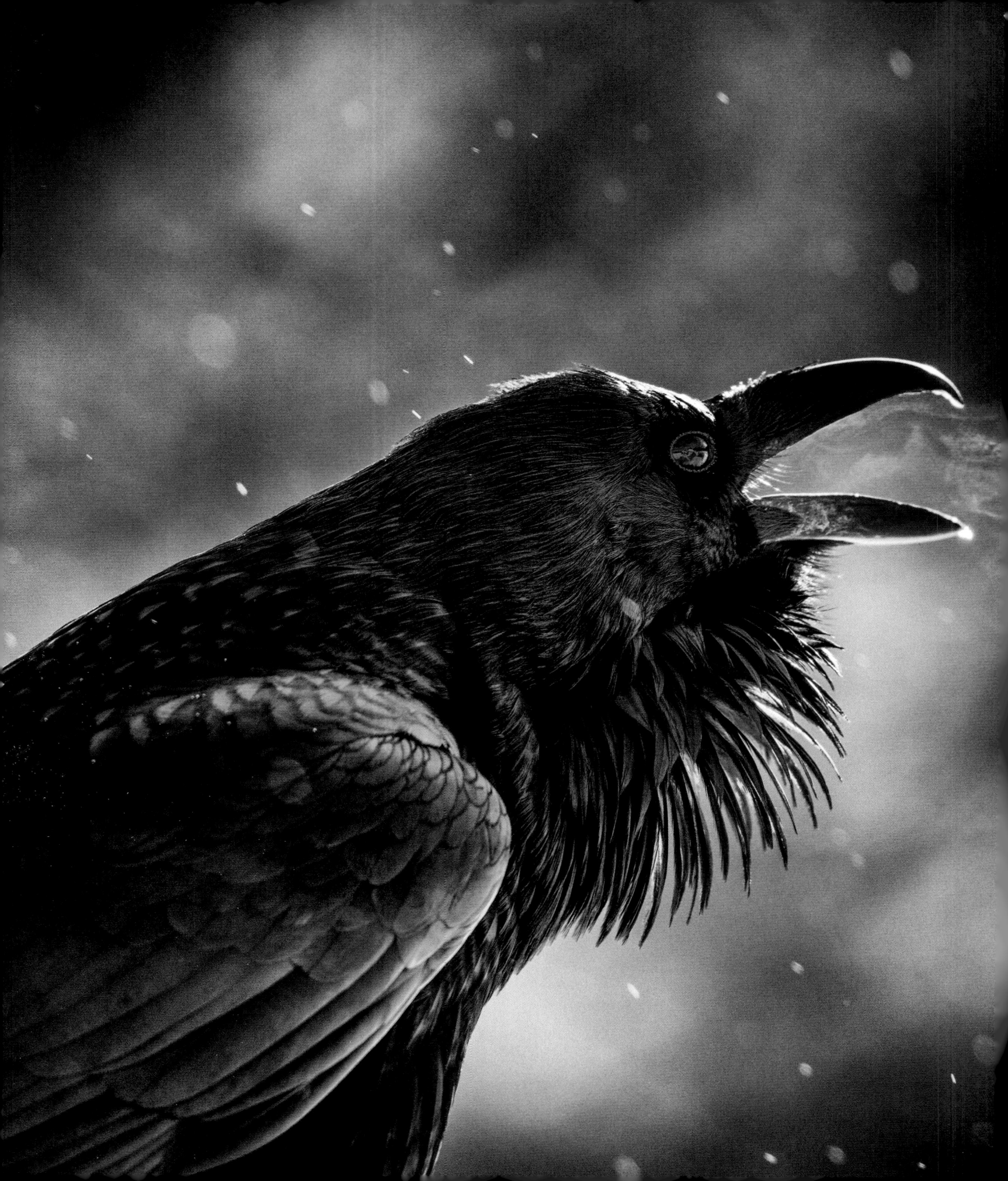

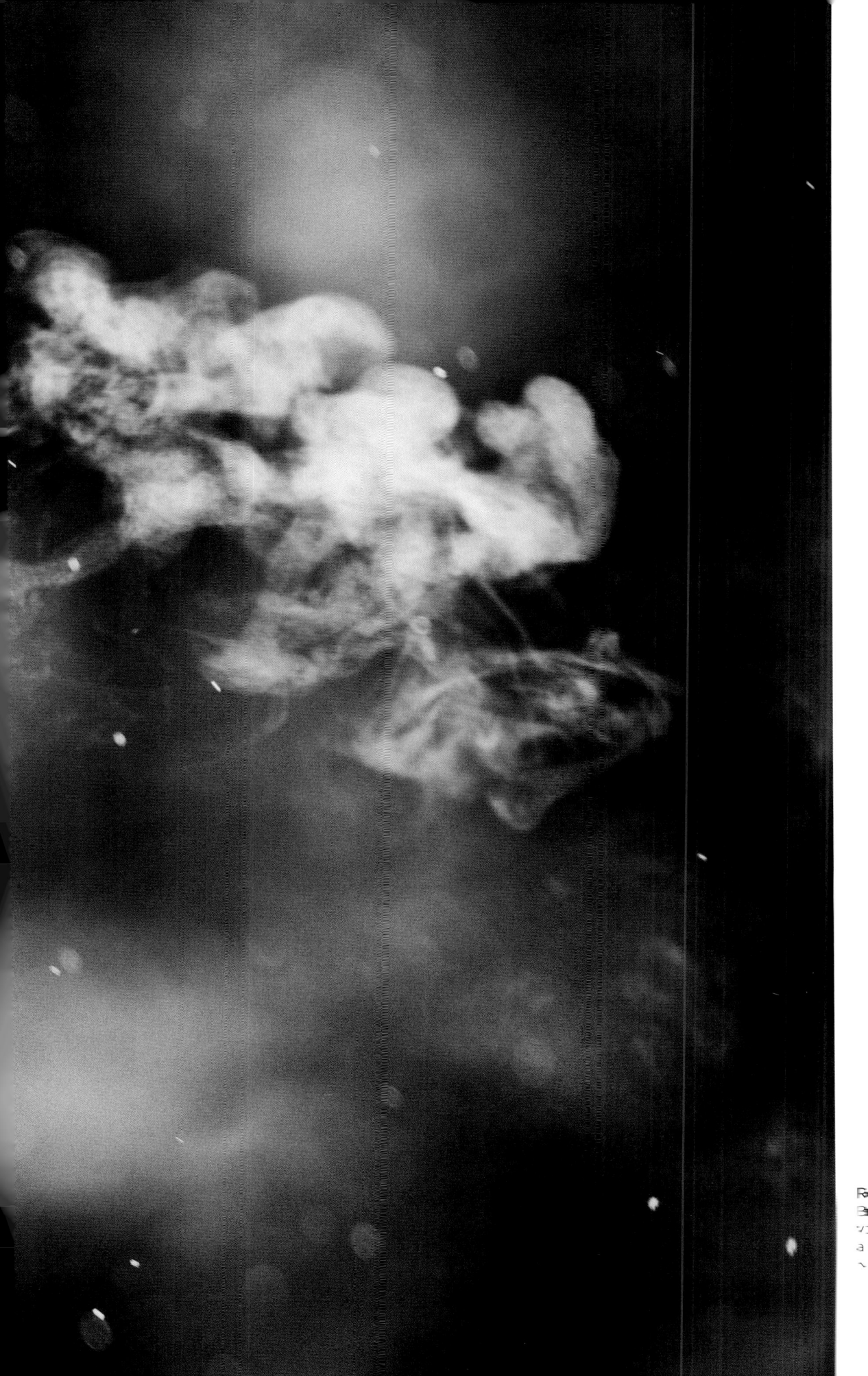

Ronan Donovan
Breath condenses as a raven vocalizes in the frigid morning air in Wyoming's Yellowstone National Park.

WITH A LITTLE TRICKERY,
A PHOTOGRAPH CAN PORTRAY
AN EXPERIENCE OF TIME
OTHERWISE UNSEEN.

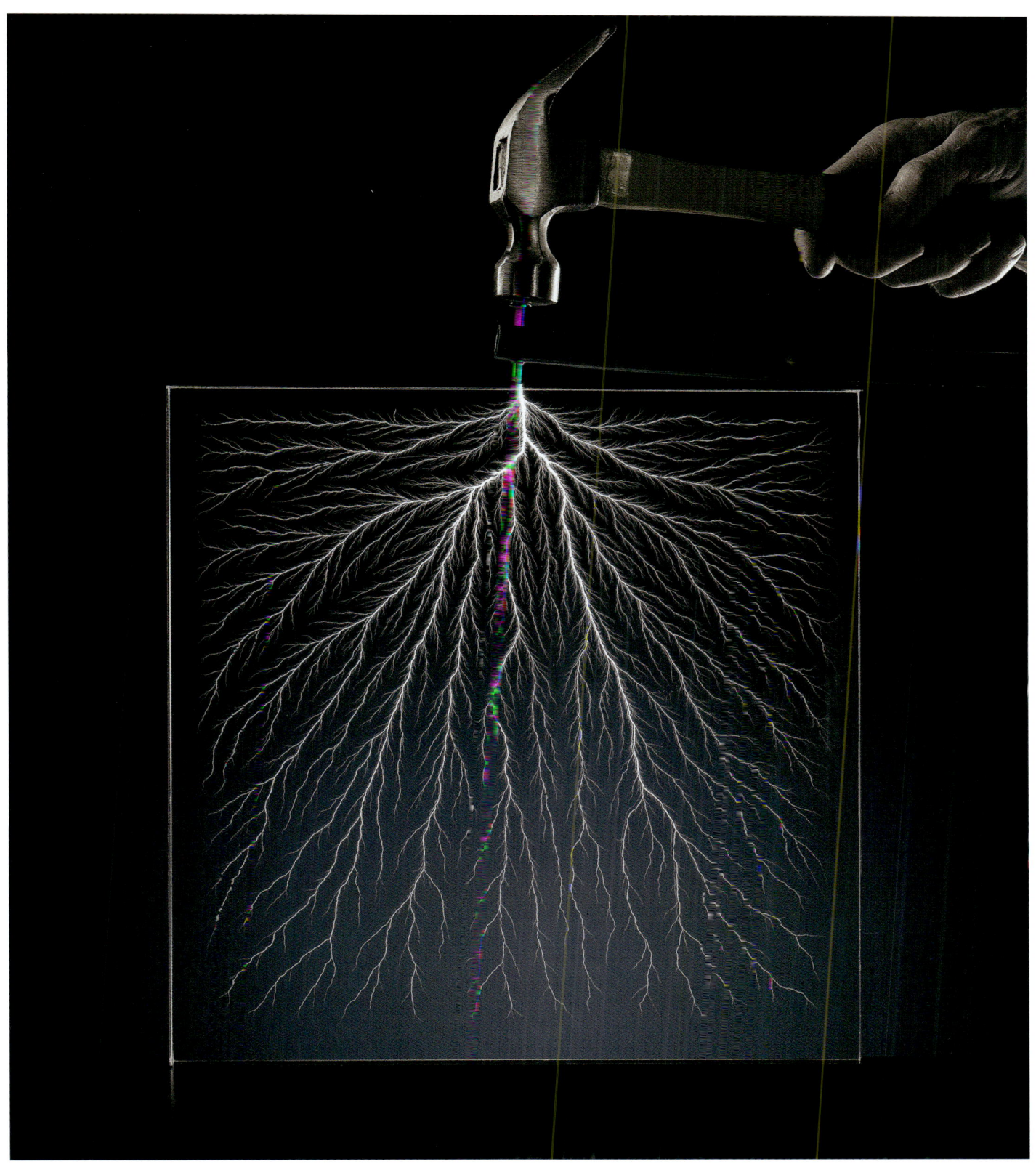

Mike Walker

A single hammer strike releases electricity stored within a charged sheet of acrylic,
permanently etching this lightning-bolt pattern, called a Lichtenberg figure, into the plastic.

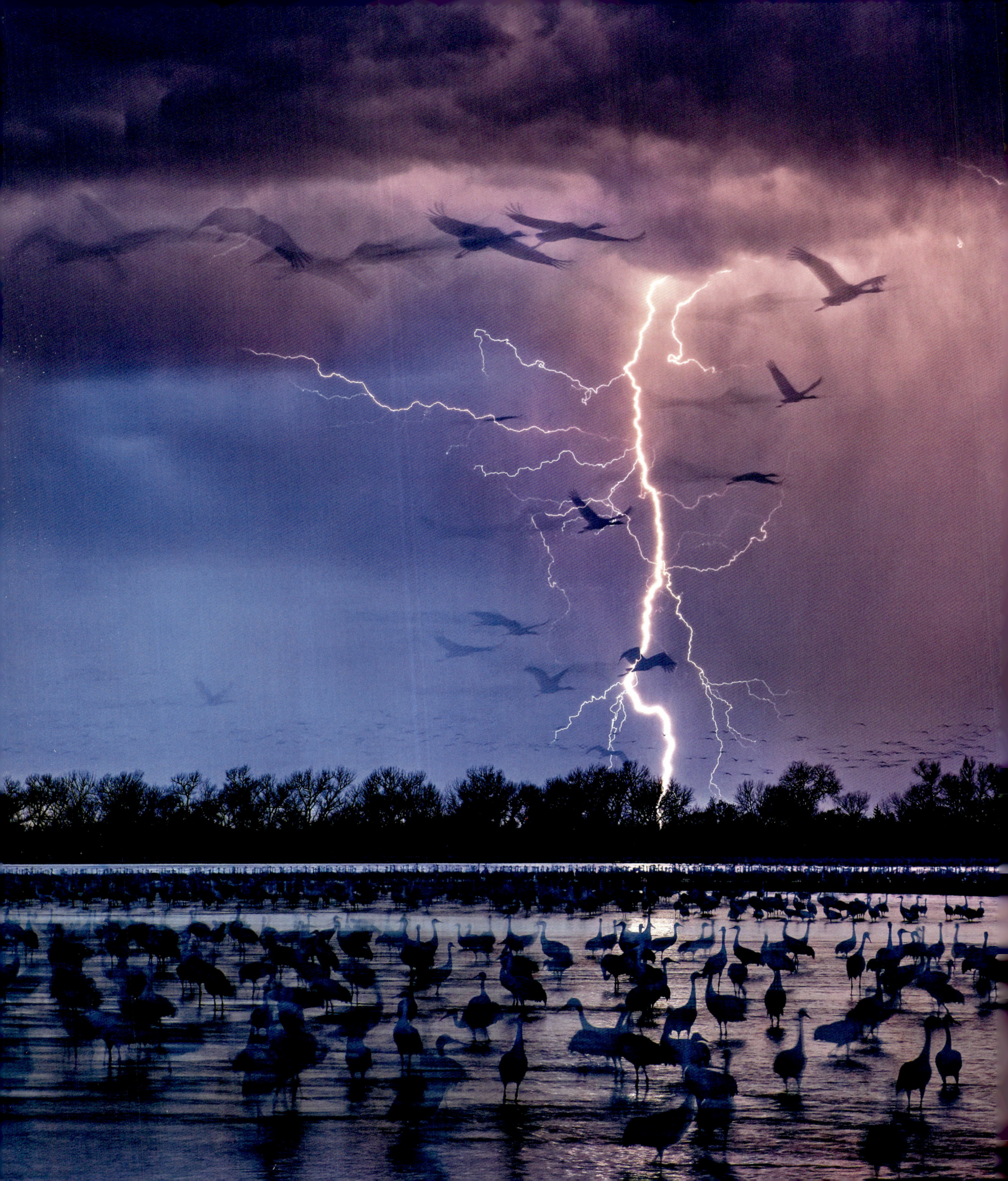

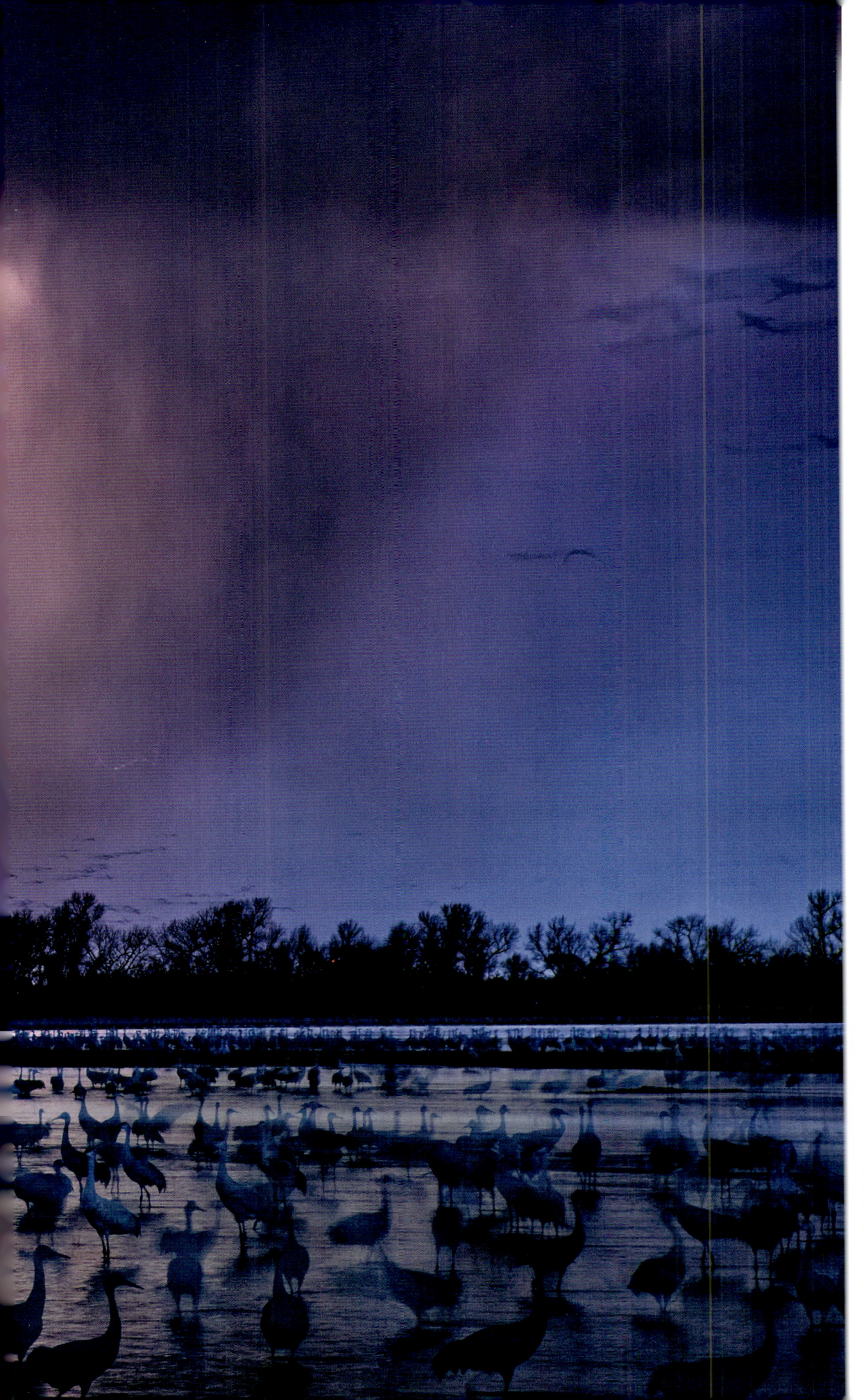

Randy Olson
A flash of lightning backlights
sandhill cranes as they fly
along Nebraska's Platte River,
where more than 400,000
cranes stop during their
annual migration.

INFLUENCES
EDGERTON

Dr. Harold "Doc" Edgerton (1903–90) was a visionary inventor whose work inspired generations of photographers, myself included. His groundbreaking images revealed movement too fast for our eyes to see.

Edgerton was instrumental in developing the electronic flash, the most critical component of my photographic tool kit today. By building a device that could produce pulses of light in rapid succession, Edgerton created a new way to visualize motion. In his image of a tennis player, for example, the repeated exposures give us a sense of movement over time. As a young photographer, I was inspired by the way Edgerton pioneered new methods for visualizing the world. He became my hero.

When I took on the challenge of photographing hummingbirds in flight, I knew I could learn a trick or two from "Papa Flash," as Edgerton was affectionately called. But I didn't want to merely copy his work. I asked myself, How would

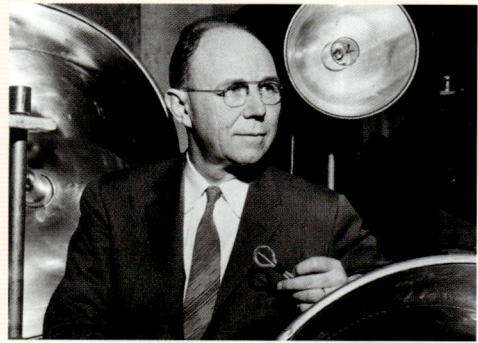

ABOVE: Dr. Harold Eugene Edgerton
RIGHT: Edgerton's 1949 stroboscopic study of a man hitting a tennis ball

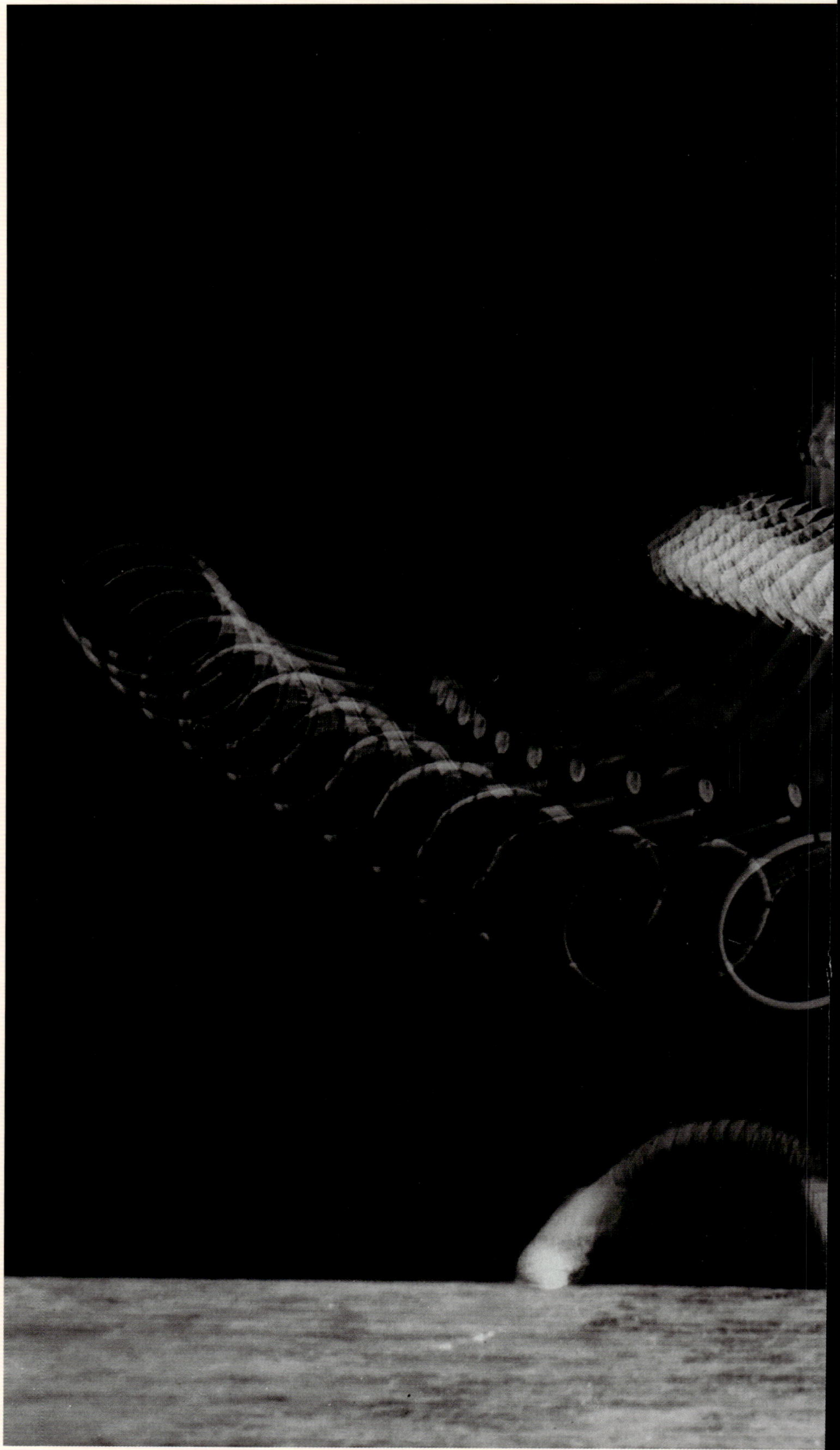

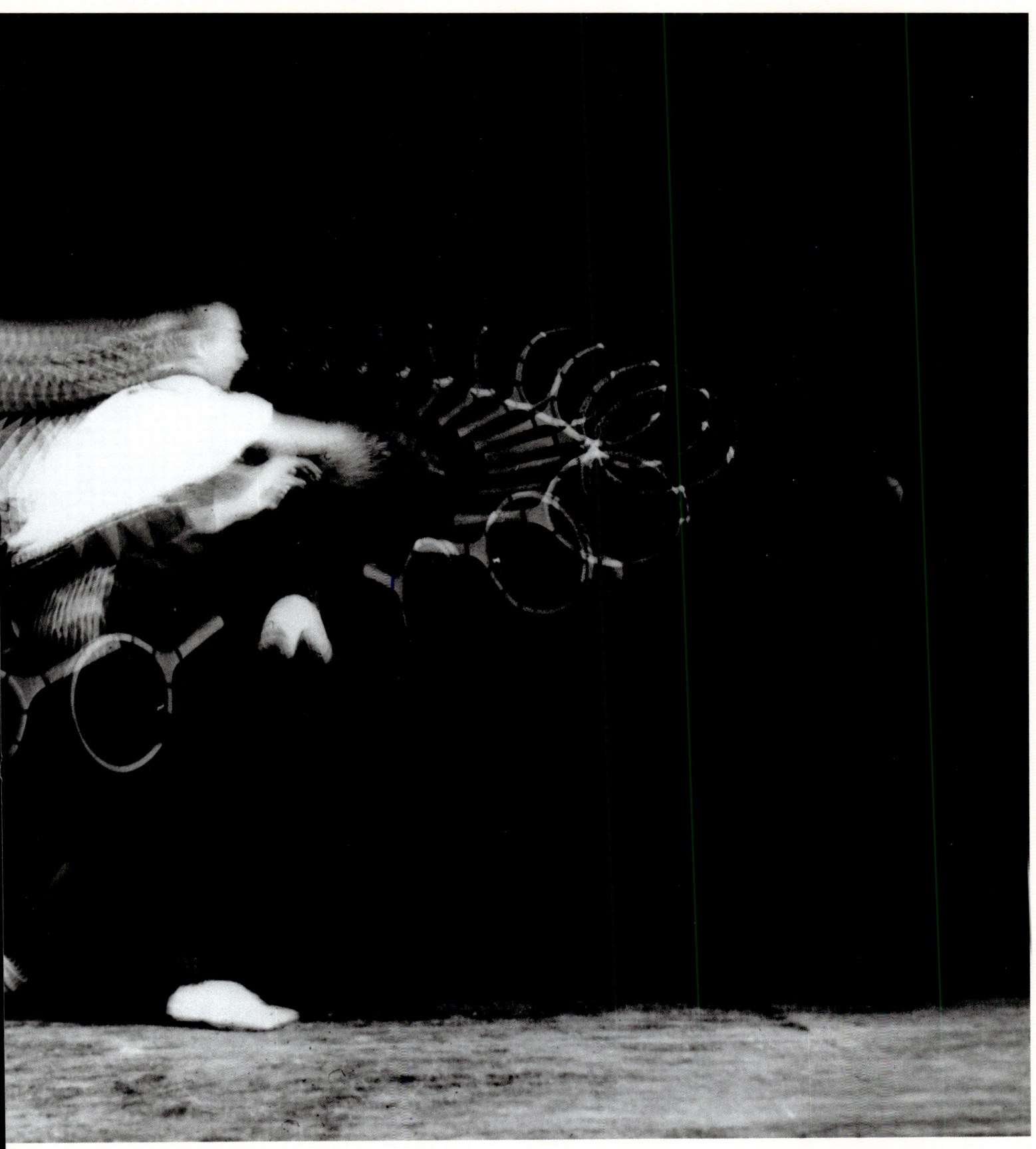

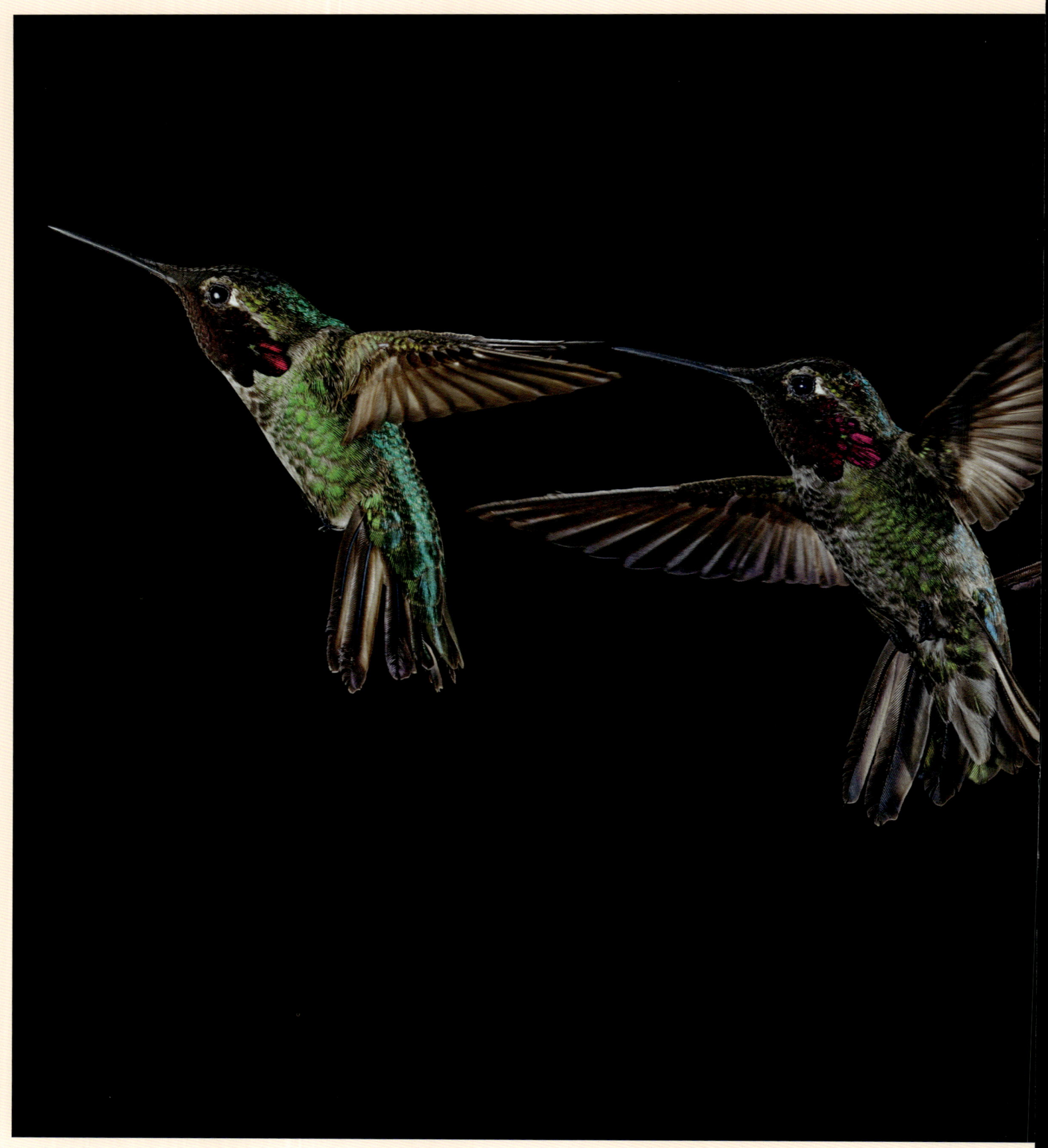

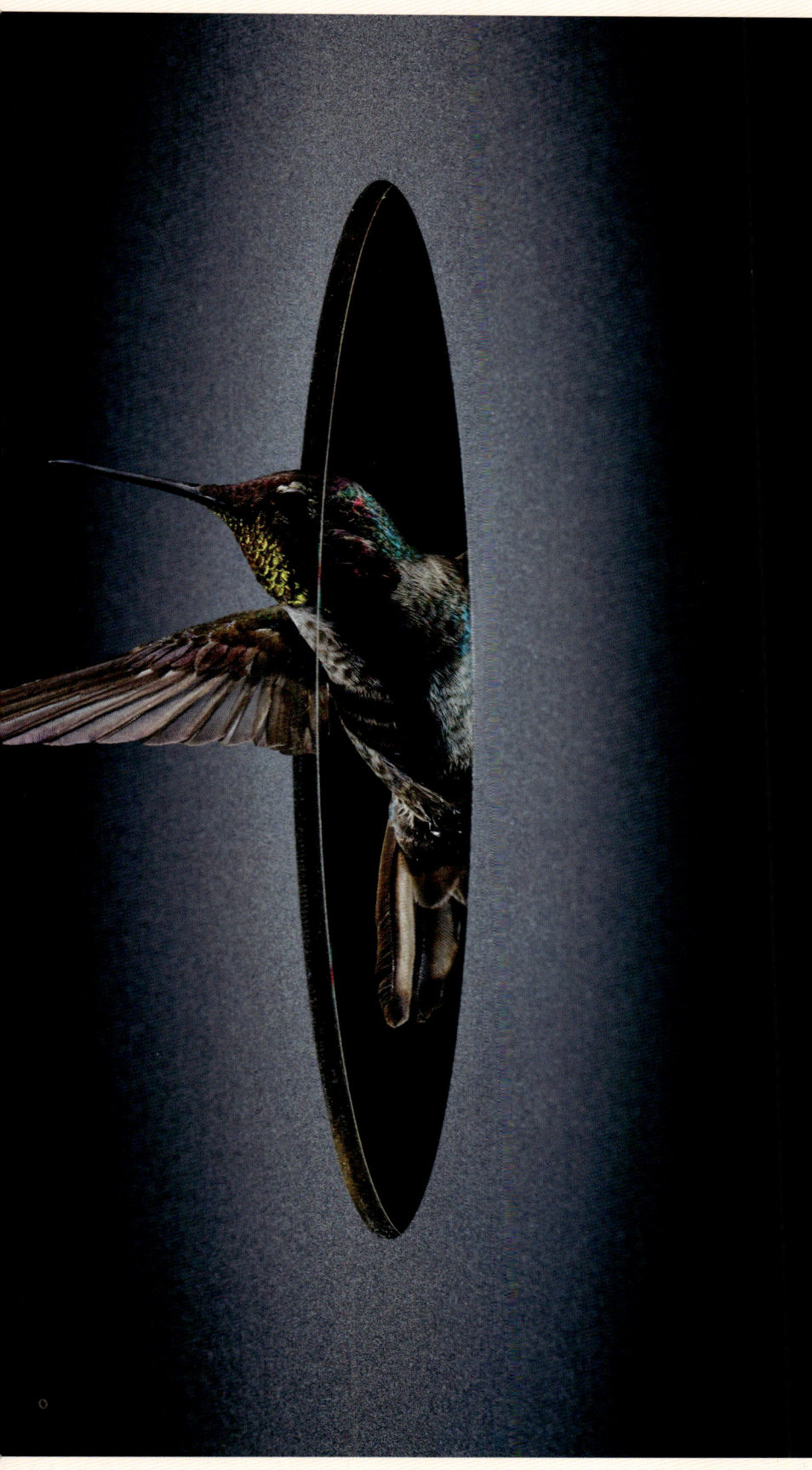

Doc Edgerton approach this task differently if he were alive today?

I realized there was a trade-off in his stroboscopic technique. The approach resolves fast motion, like the swing of the tennis racket in the image on pages 160–61. But when those flashes of light repeatedly hit part of a scene that is relatively static, like the tennis player's body, the excessive light obscures the details. Could I improve the technique by controlling the timing and brightness of each individual flash within a rapid sequence? Would that allow me to illustrate movement while preserving detail?

I decided to test my idea while visiting Marc Badger at the University of California, Berkeley, who was studying how hummingbirds manage high-speed maneuvers through confined spaces. I re-created one of Marc's experiments by building a flight cage consisting of two chambers connected by a narrow window. As a bird flew back and forth through that opening, I programmed a series of flashes to fire with just enough delay to ensure that the sequential positions of the hummingbird never overlapped. I also increased the intensity of the light from the first exposure to the last in order to give the final position of the bird the greatest visual weight.

Years after this hummingbird story was published in *National Geographic*, I received a message that read "Your photographs remind me of my grandfather's work." The sender's name was Sylvia Edgerton. I could not have imagined a finer compliment.

Anand Varma
An Anna's hummingbird performs a sideways shimmy to fit through a narrow, oval-shaped hole.

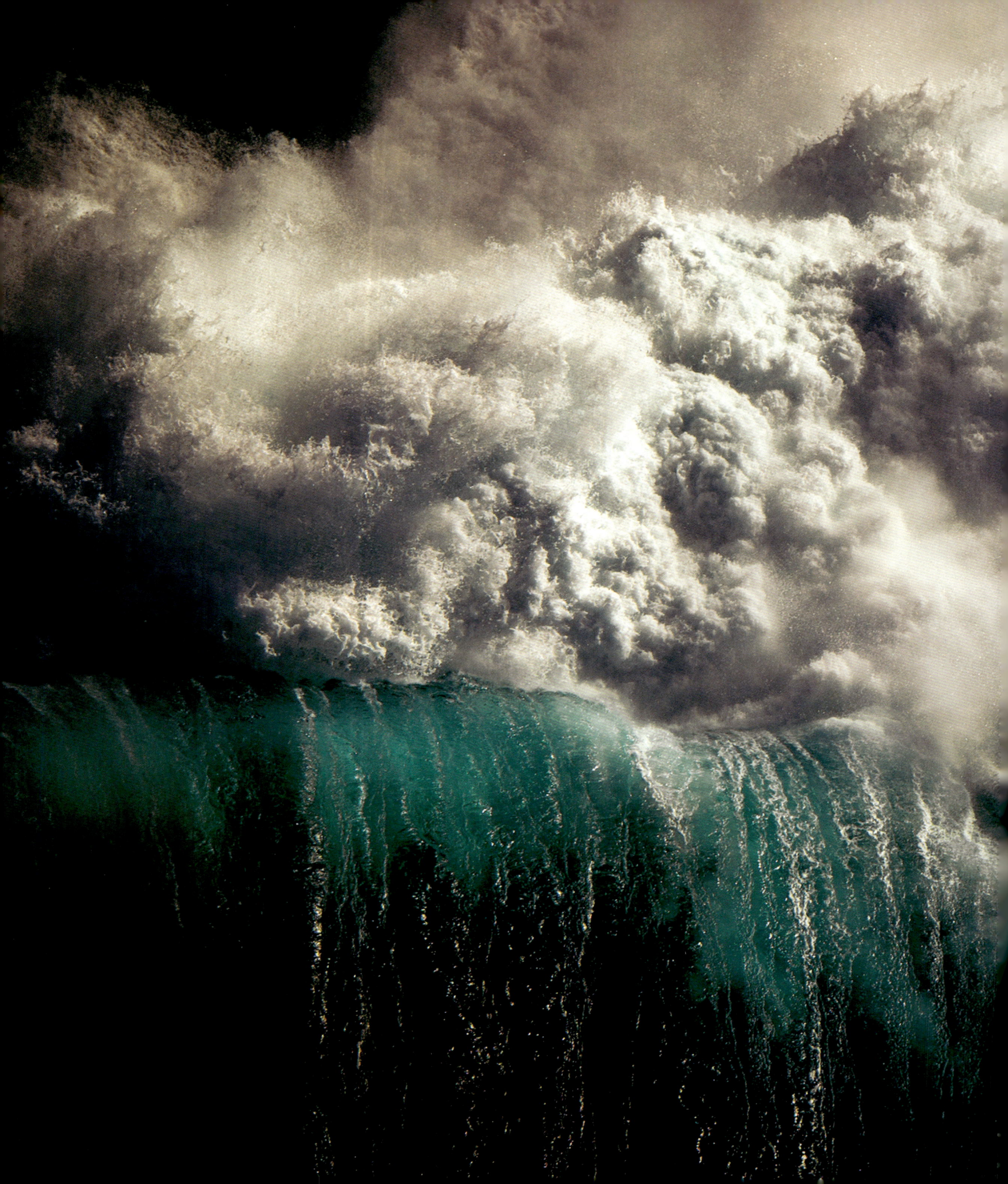

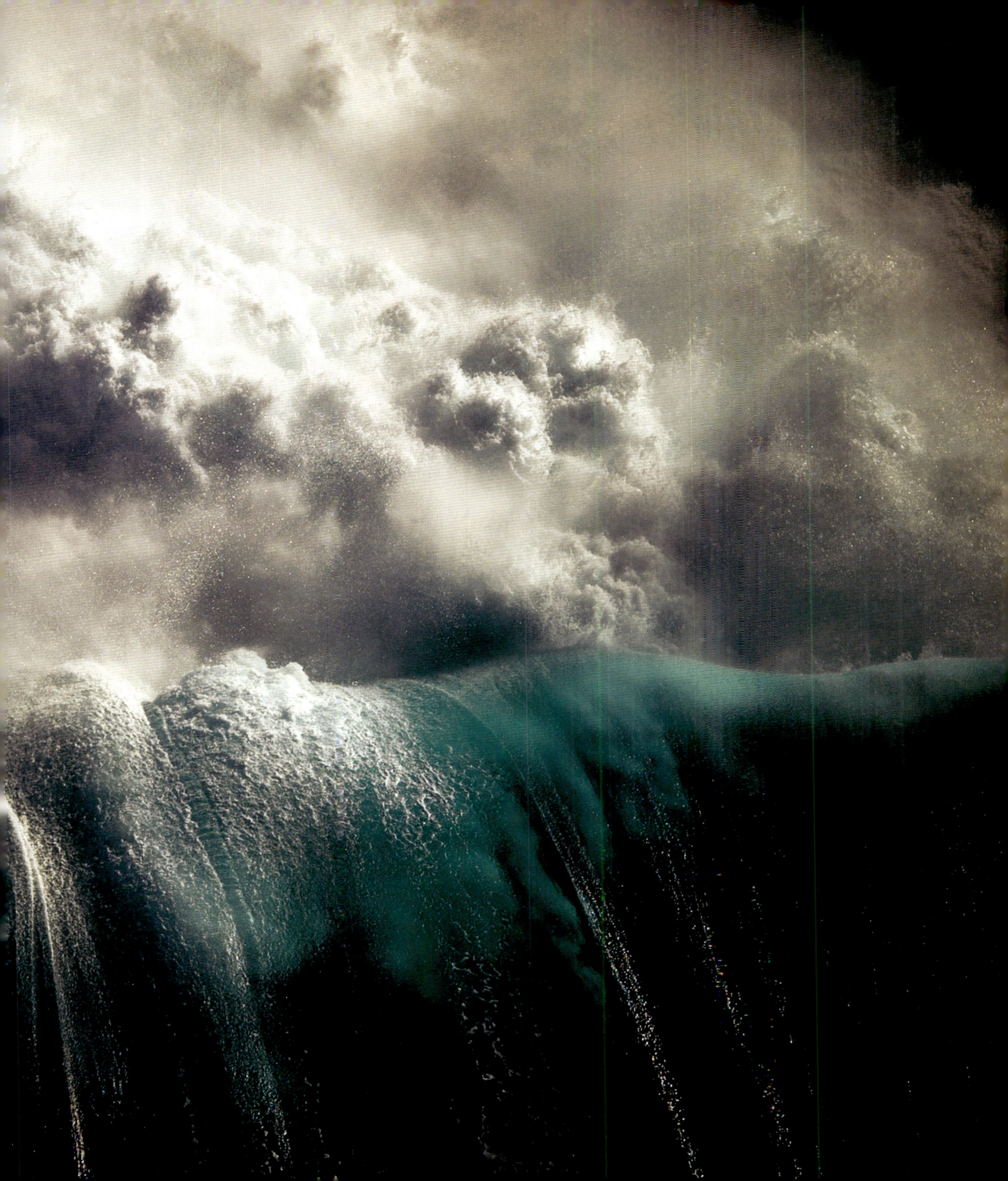

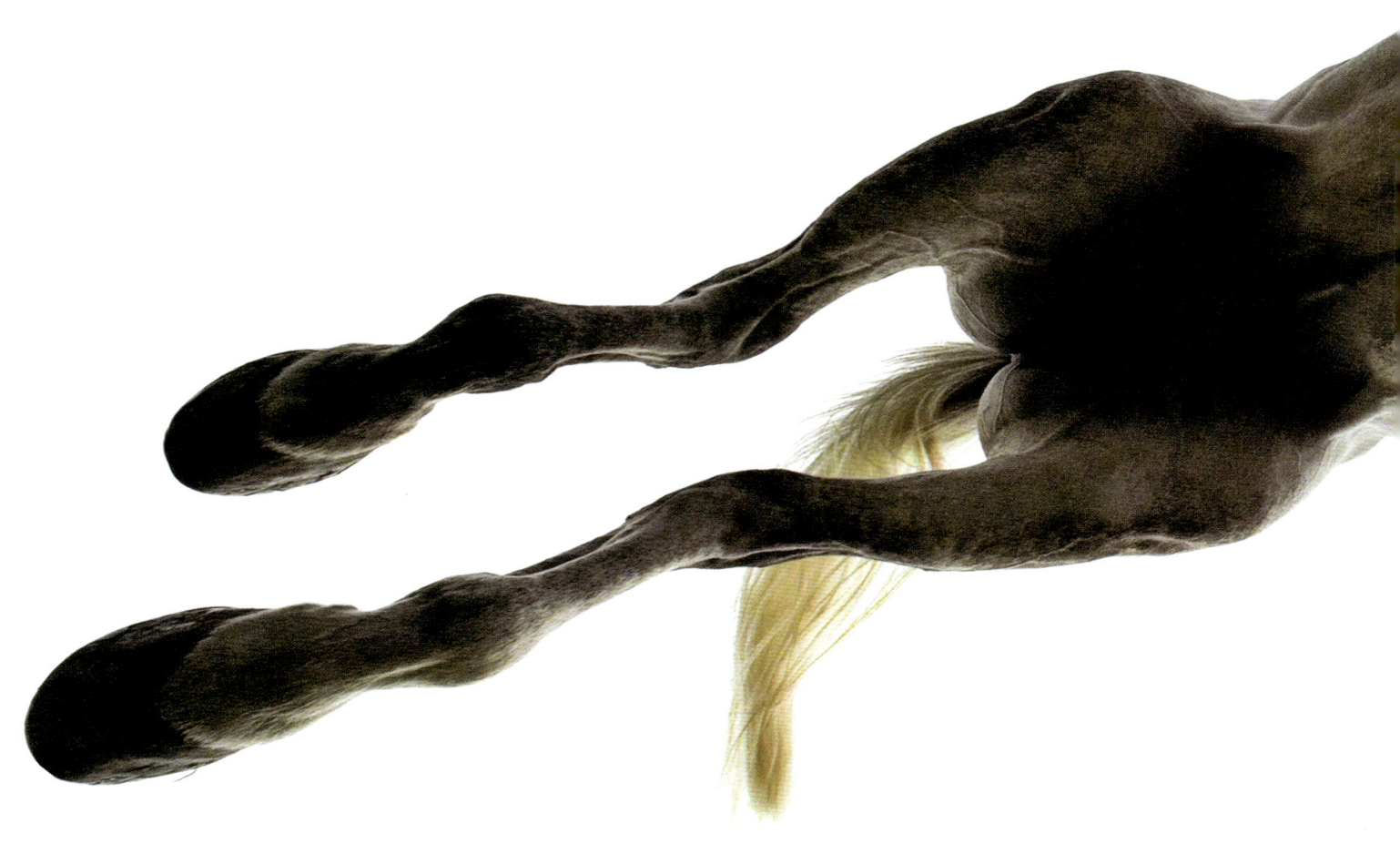

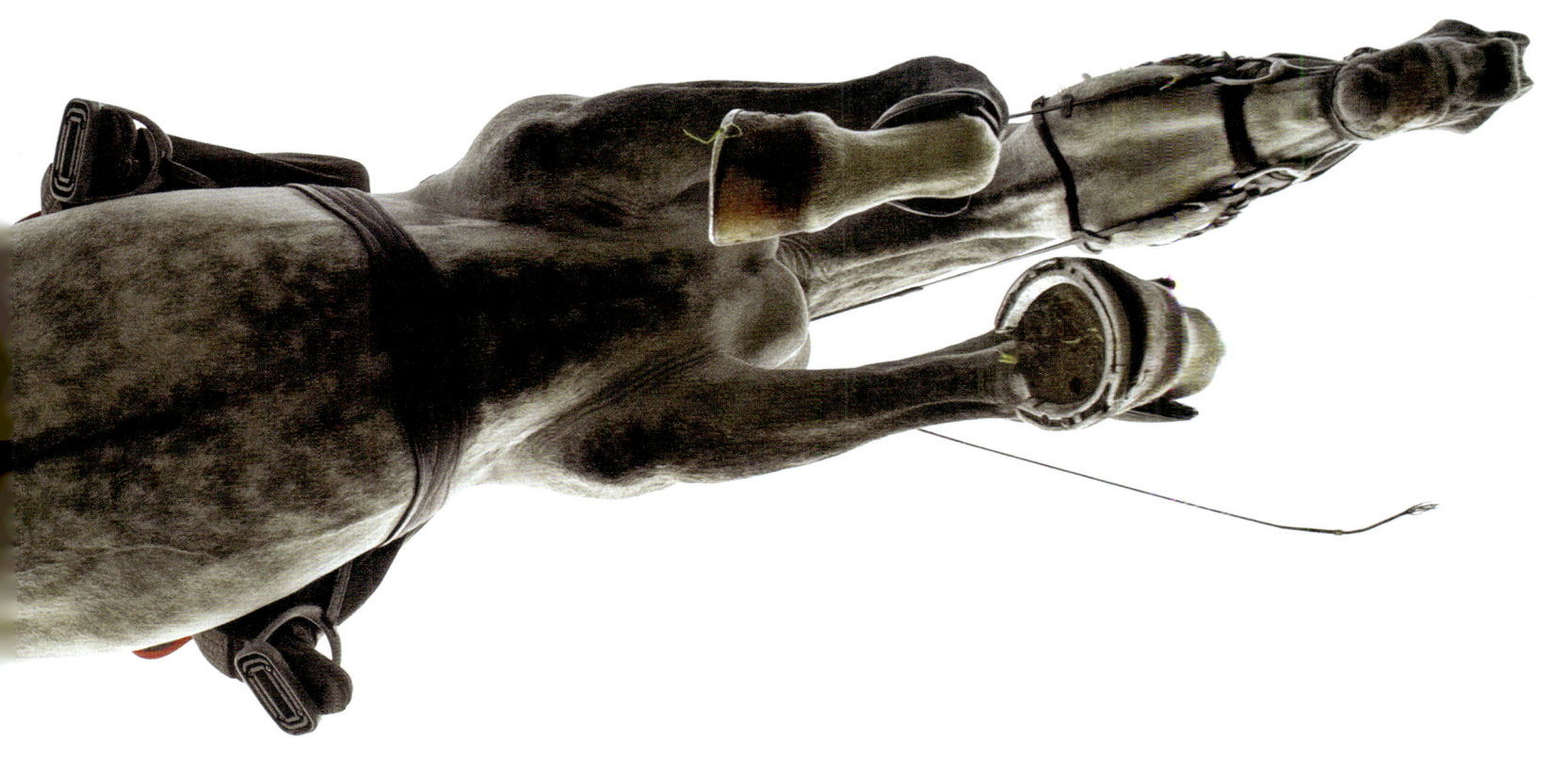

Tim Flach
A fast exposure freezes
a horse mid-jump.

PREVIOUS PAGES:
Ray Collins
An aerial view of a wave
crashing

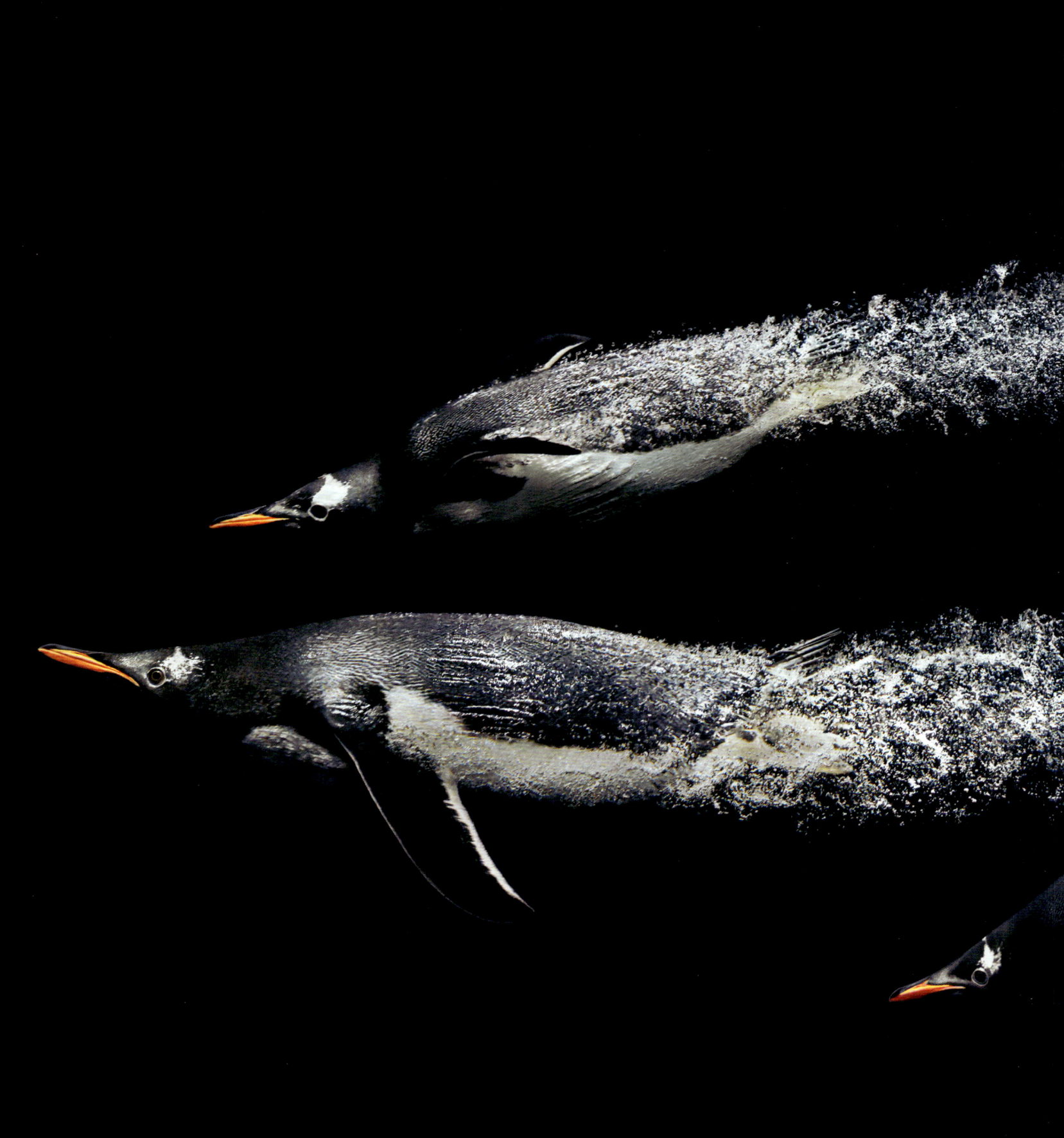

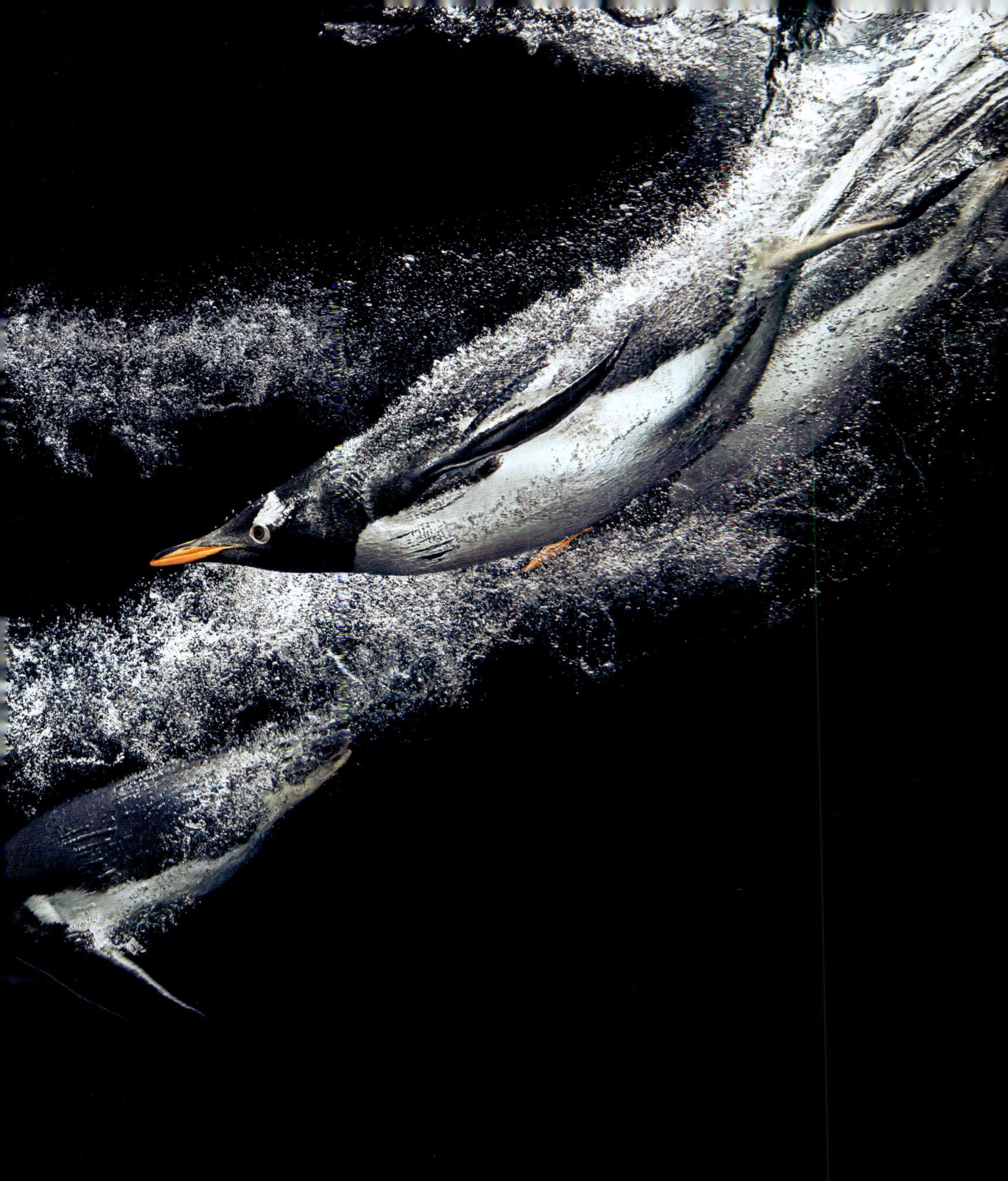

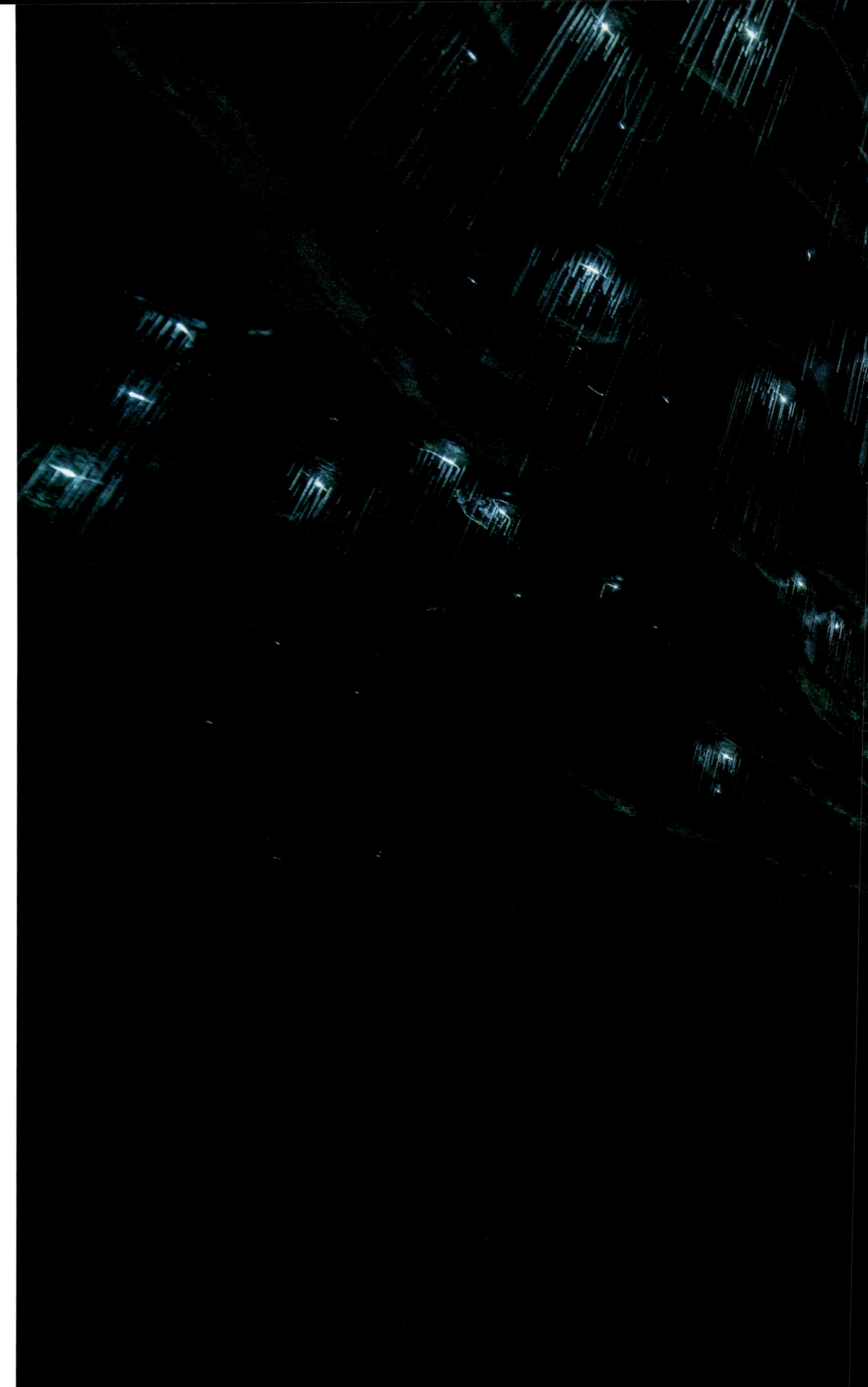

Shaun Jeffers
Bioluminescent glowworms
hang sticky tendrils of mucus
from the roof of a cave in
New Zealand. The light
attracts flying insects,
which get ensnared as prey
for the worms.

PREVIOUS PAGES:
Tim Flach
Diving gentoo penguins leave
behind trails of bubbles.

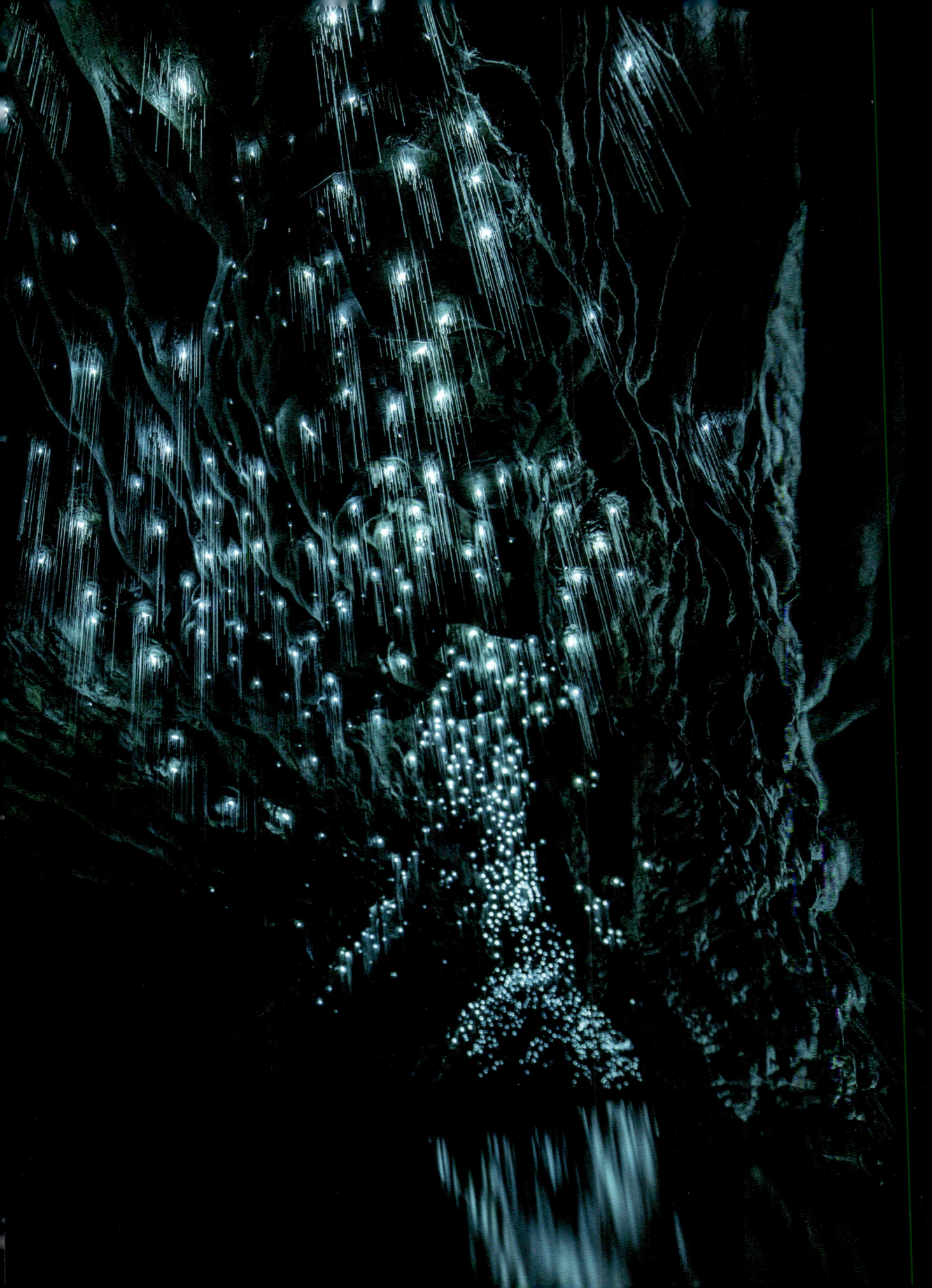

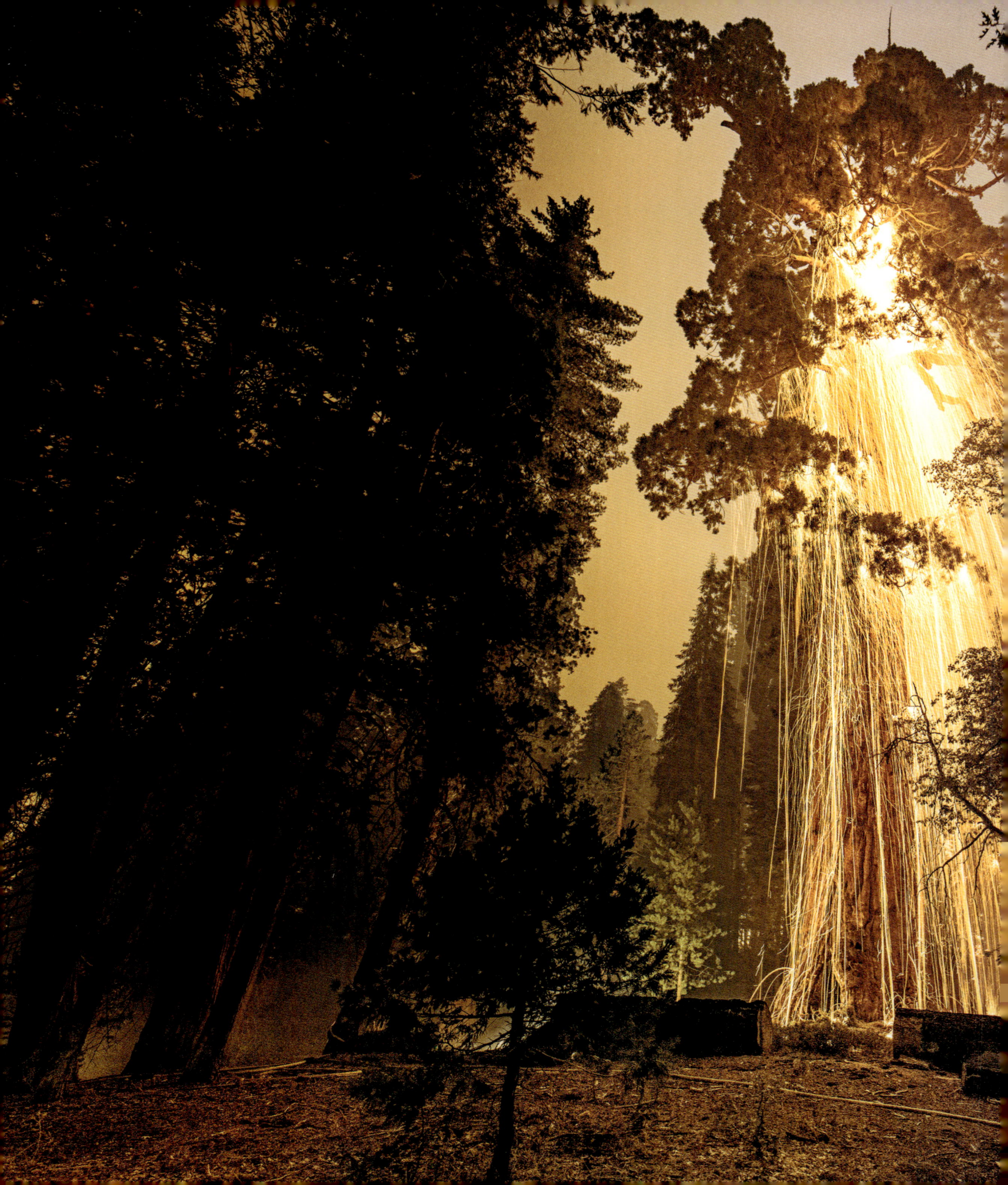

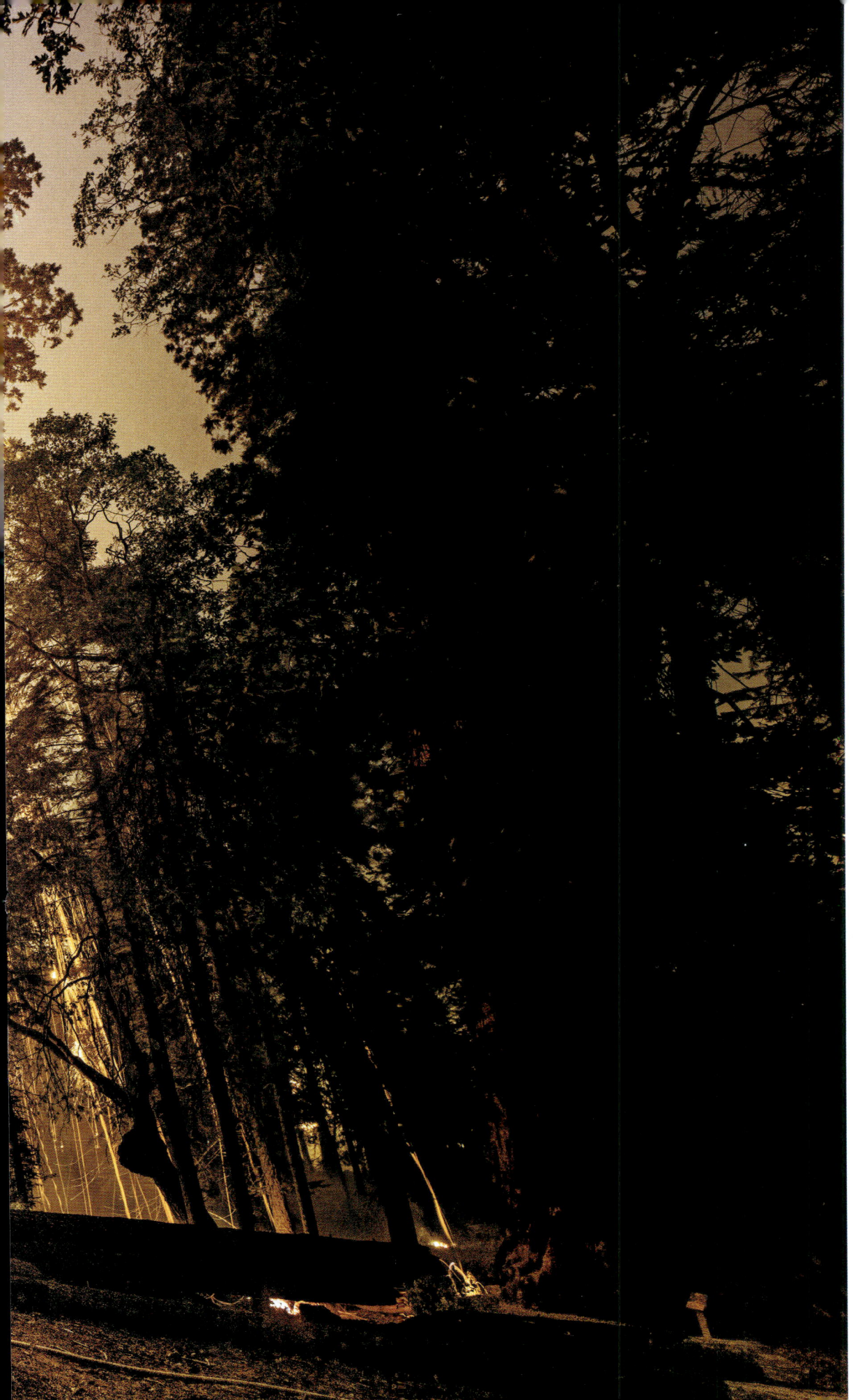

Jeff Frost
Embers rain down from the crown of a giant sequoia tree during the 2021 Windy fire in California.

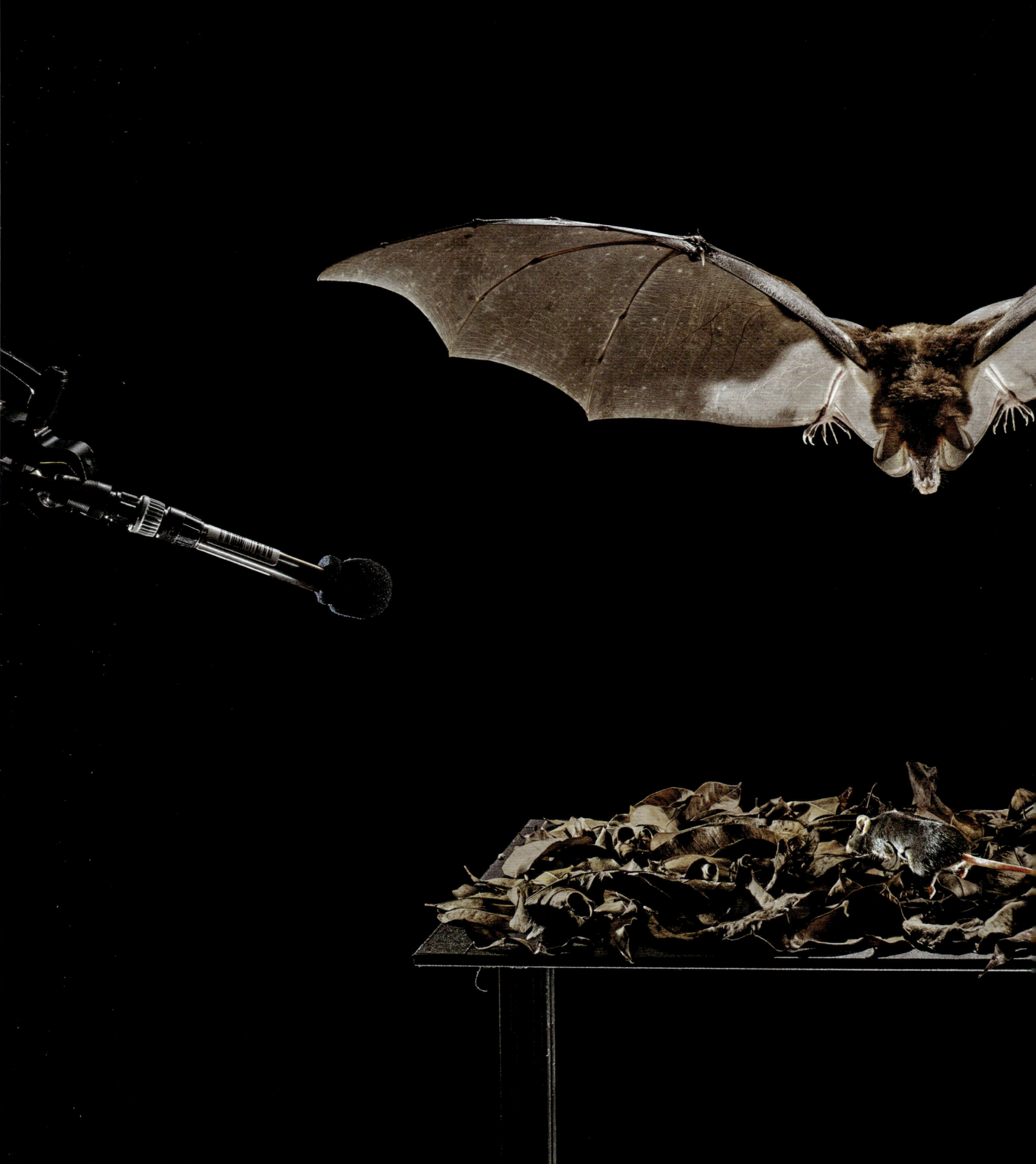

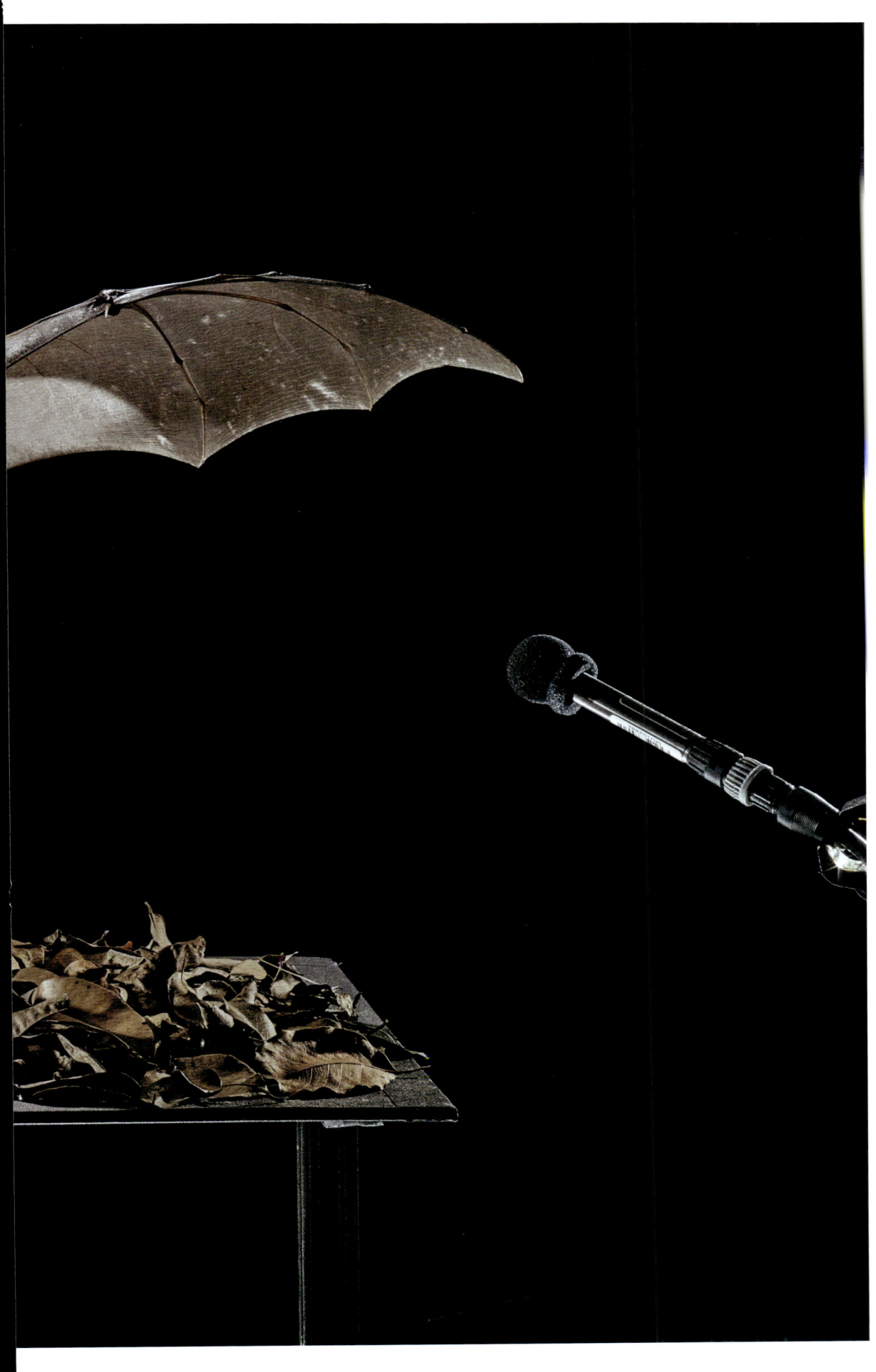

Anand Varma
A spectral bat pounces on a mouse in a laboratory flight cage. Ultrasonic microphones listen for the bat's echolocation signals.

FOLLOWING PAGES:
Stephen Wilkes
Some 670,000 white flags installed on the National Mall in Washington, D.C., in 2021 symbolize U.S. lives lost to COVID-19. This composite image shows the memorial over the course of 30 hours.

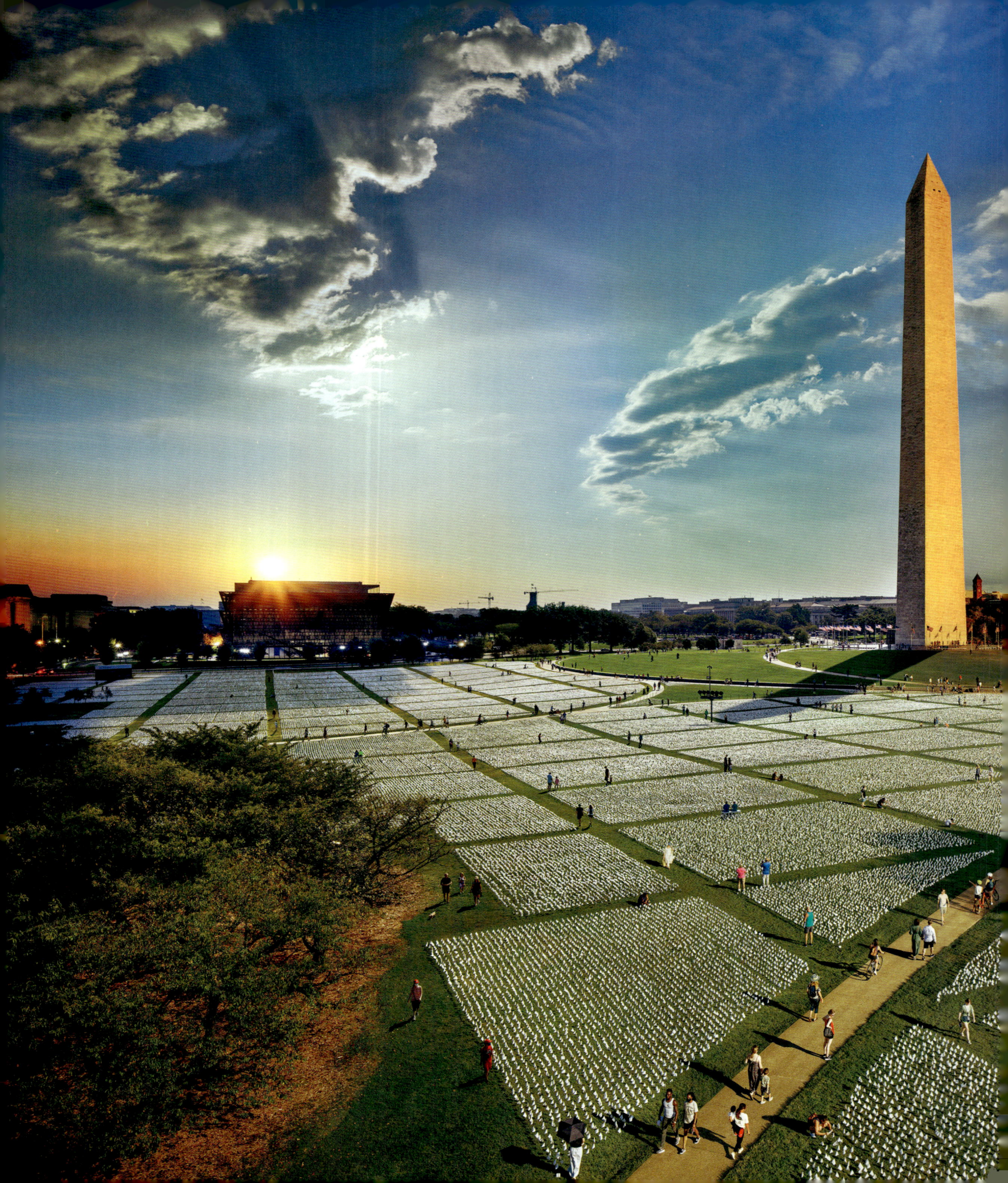

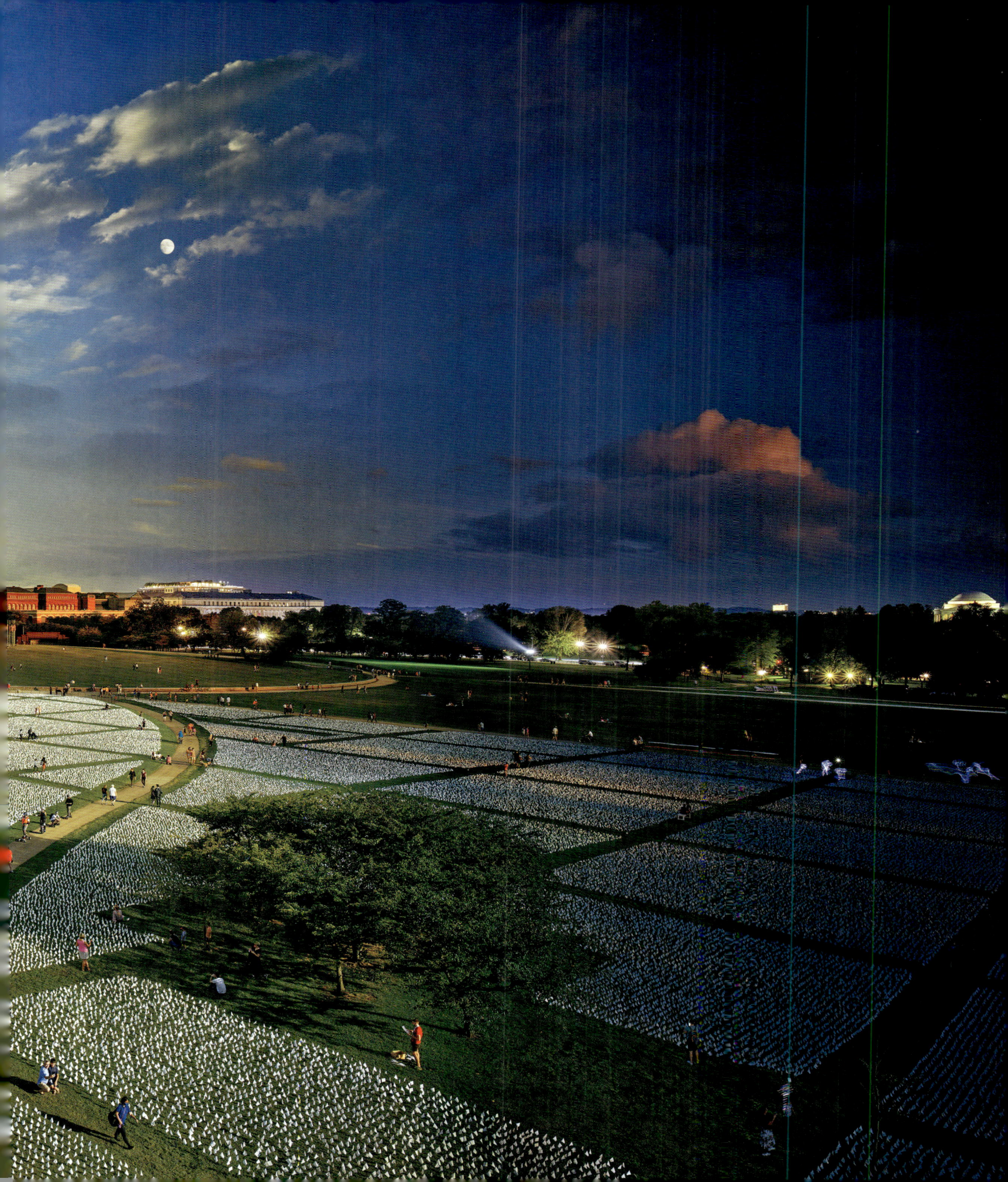

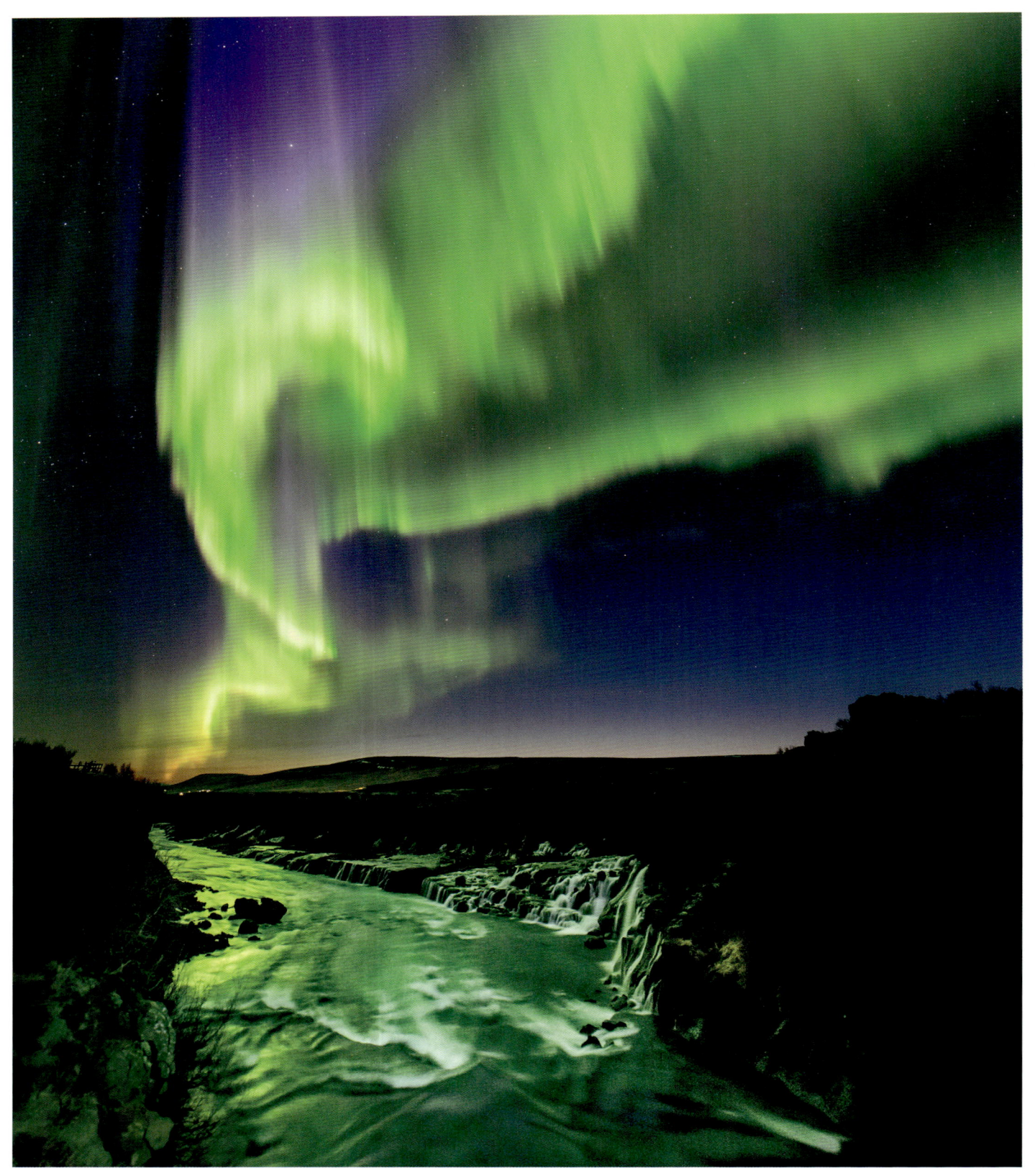

Babak Tafreshi
An aurora lights up the sky above the Hvítá River in Iceland.

EACH OF THESE IMAGES
STRETCHES OR COMPRESSES
TIME BEYOND WHAT OUR
NAKED EYES CAN DETECT.

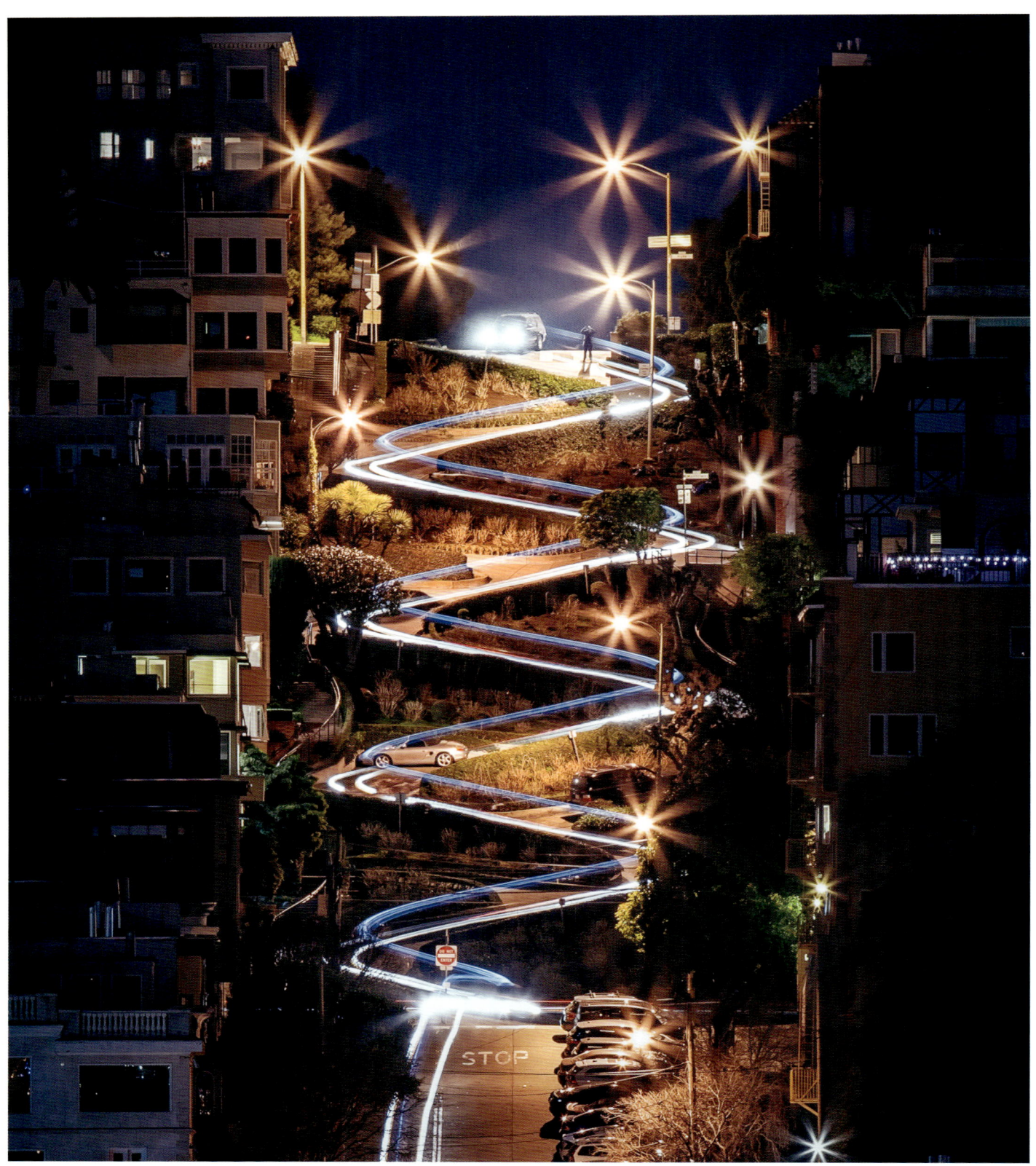

Michael Shainblum
Car headlights trace the legendary hairpin turns of
San Francisco's Lombard Street in this long-exposure image.

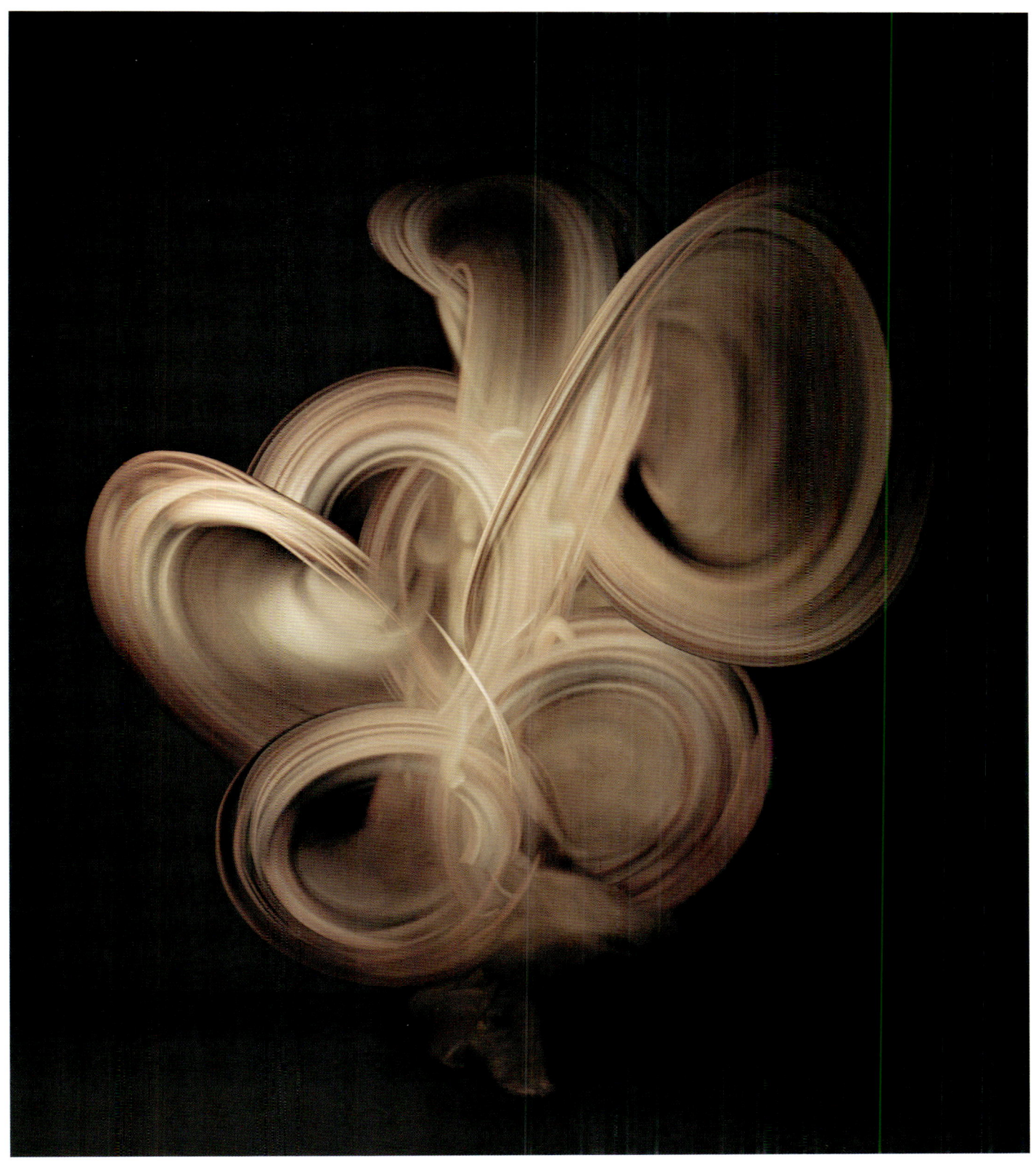

Shinichi Maruyama
A nude dancer's motion through time is revealed by a composite of 10,000 images captured over the course of five seconds.

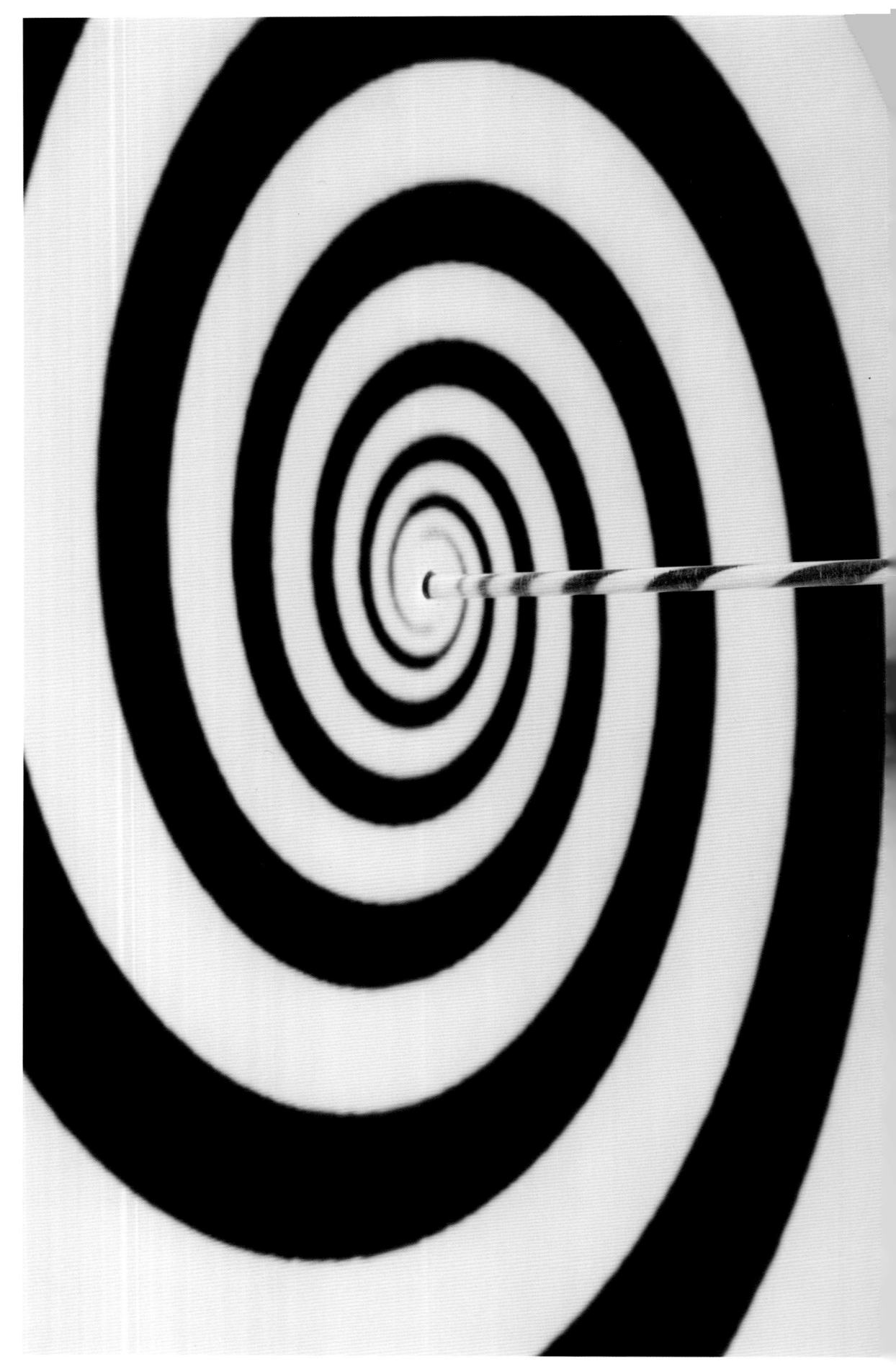

Anand Varma
An Anna's hummingbird sips nectar before an optical illusion in order to test its visual perception, in an experiment designed by the Altshuler Lab at the University of British Columbia.

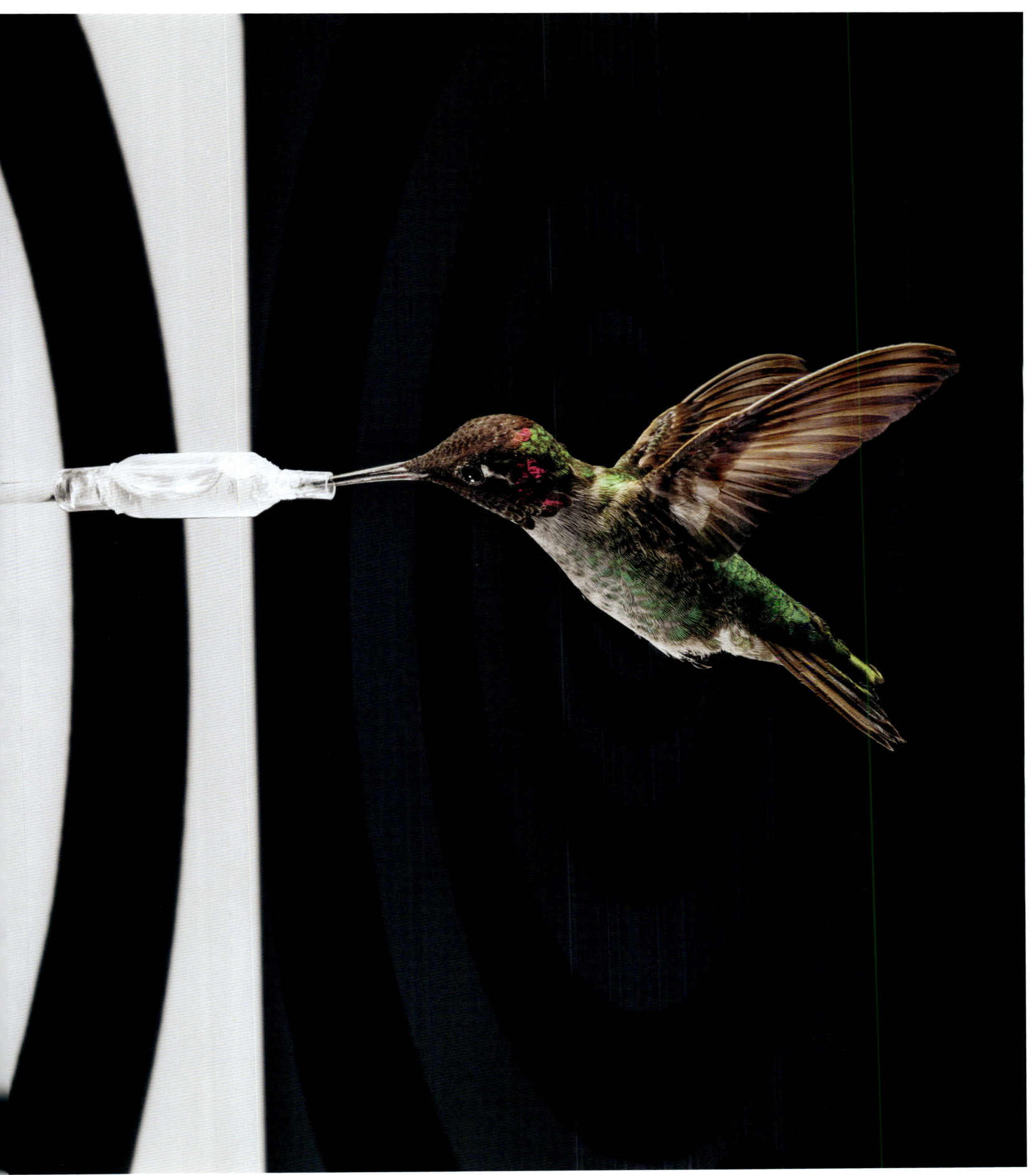

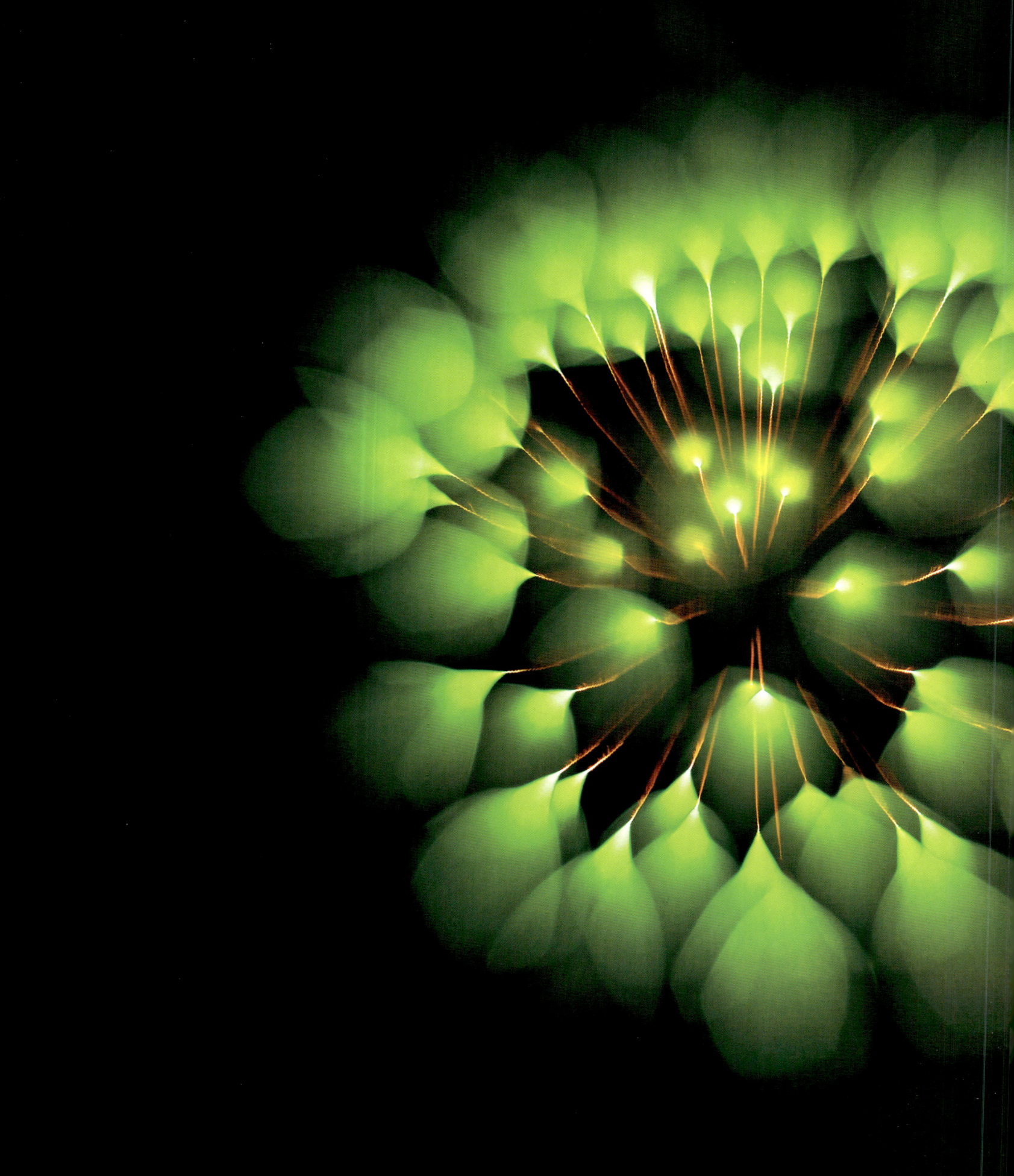

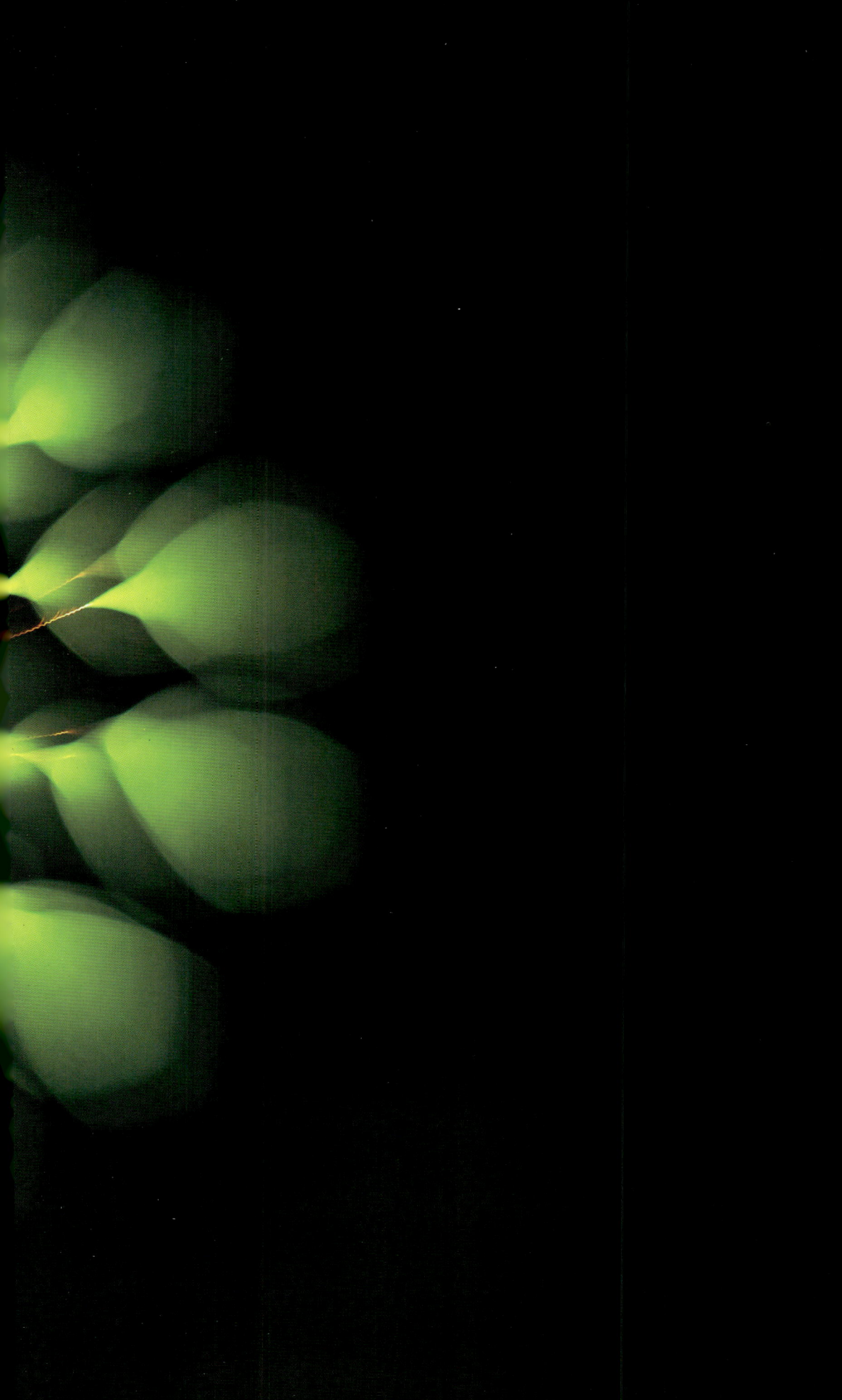

David Johnson
A unique view of fireworks created by adjusting the camera's focus as it captures the bursts of light

FOLLOWING PAGES:
Sriram Murali
Synchronous fireflies flash as part of a mating ritual that involves billions of individuals overall. The Anamalai Tiger Reserve in Tamil Nadu, India, helped create this image.

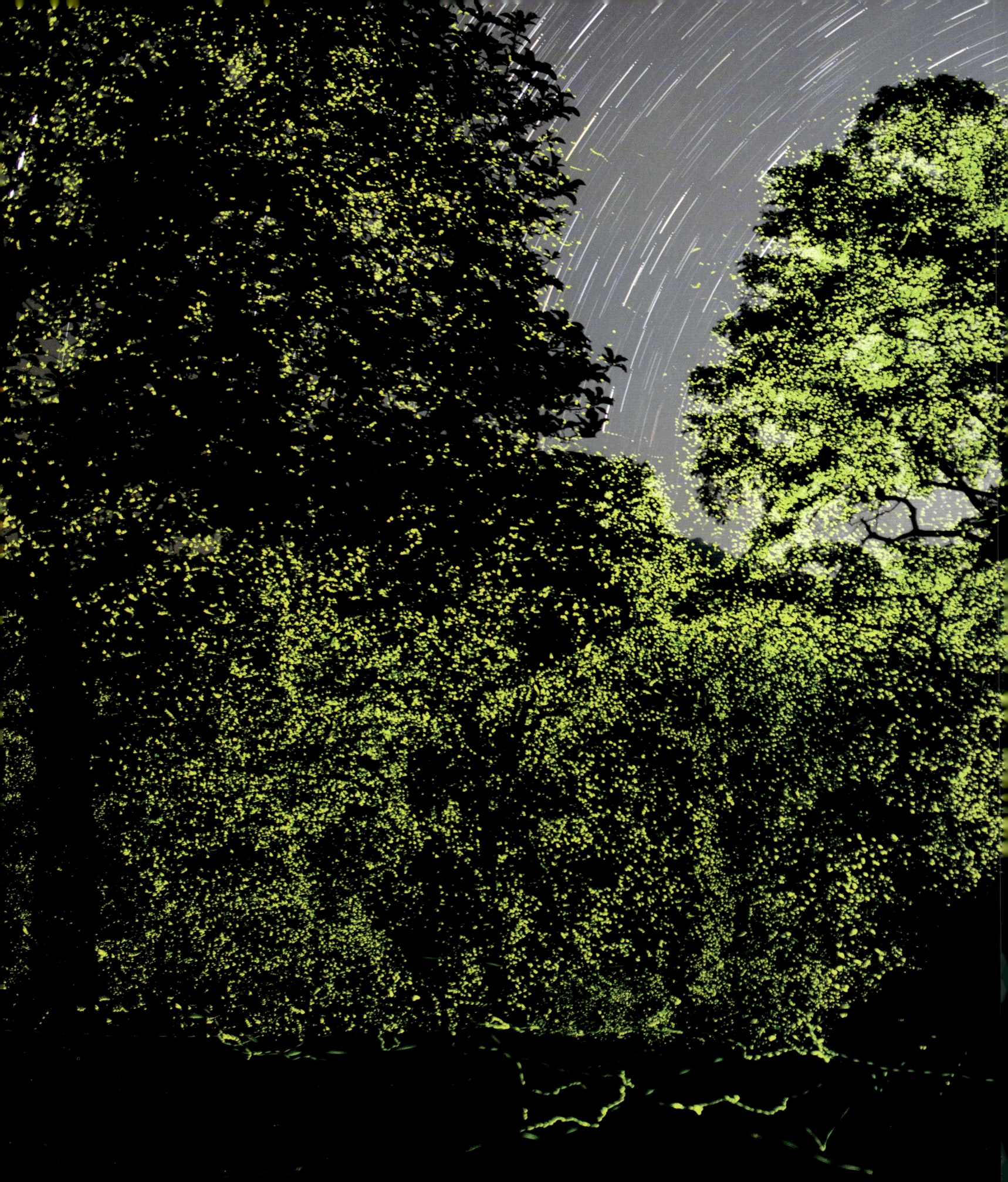

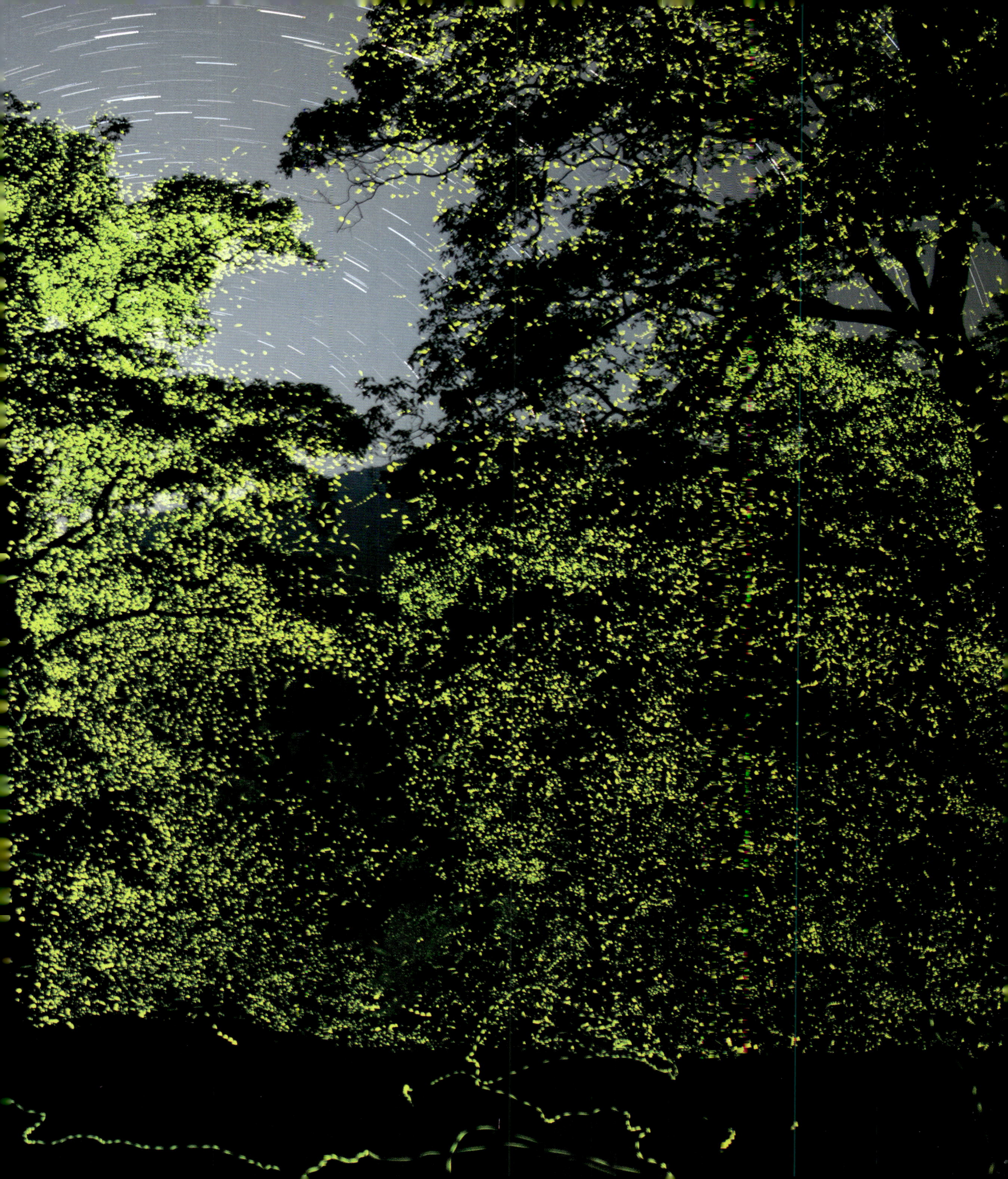

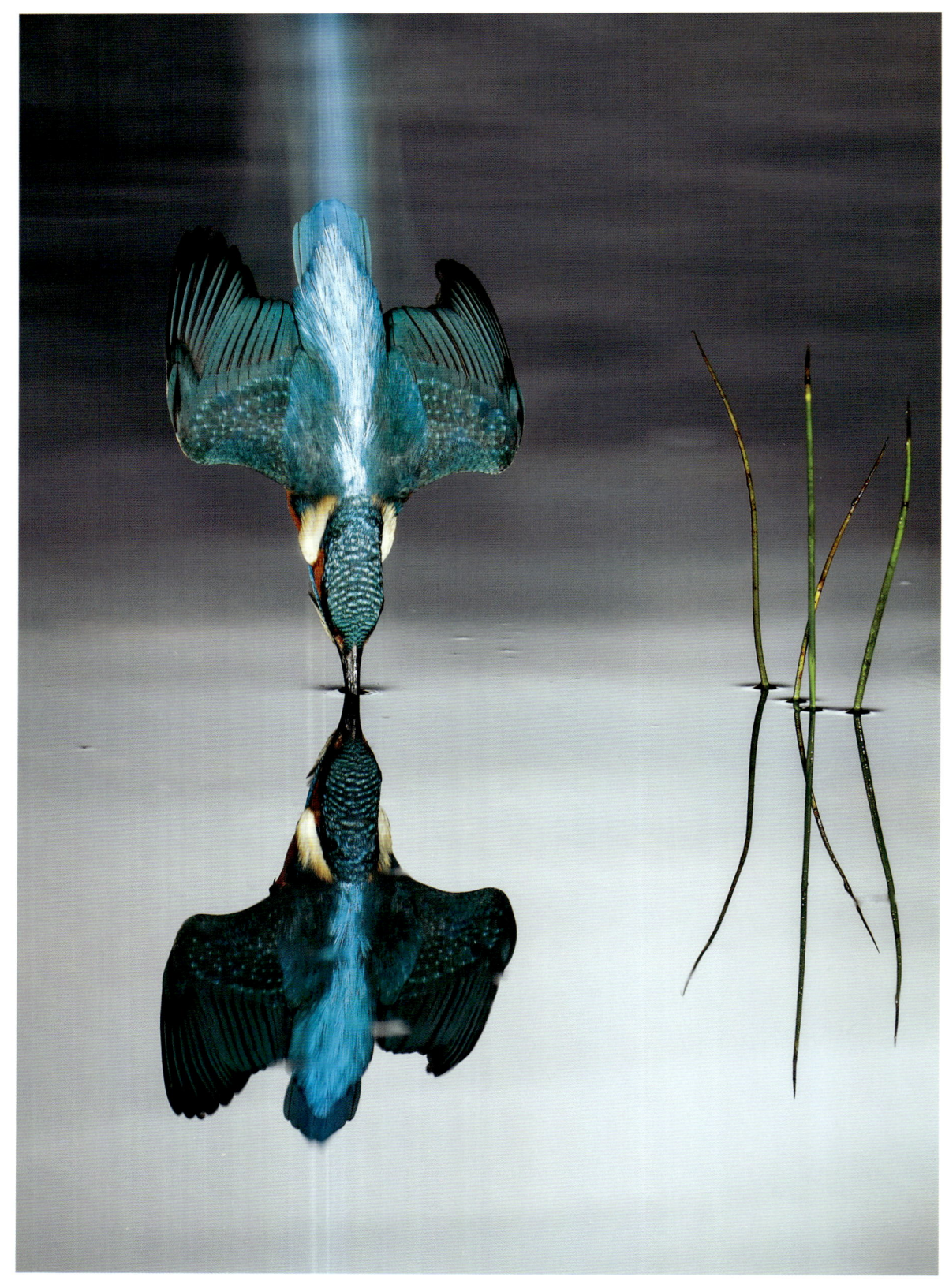

Mario Céa
A kingfisher pierces the surface of a pond as it dives after a fish.

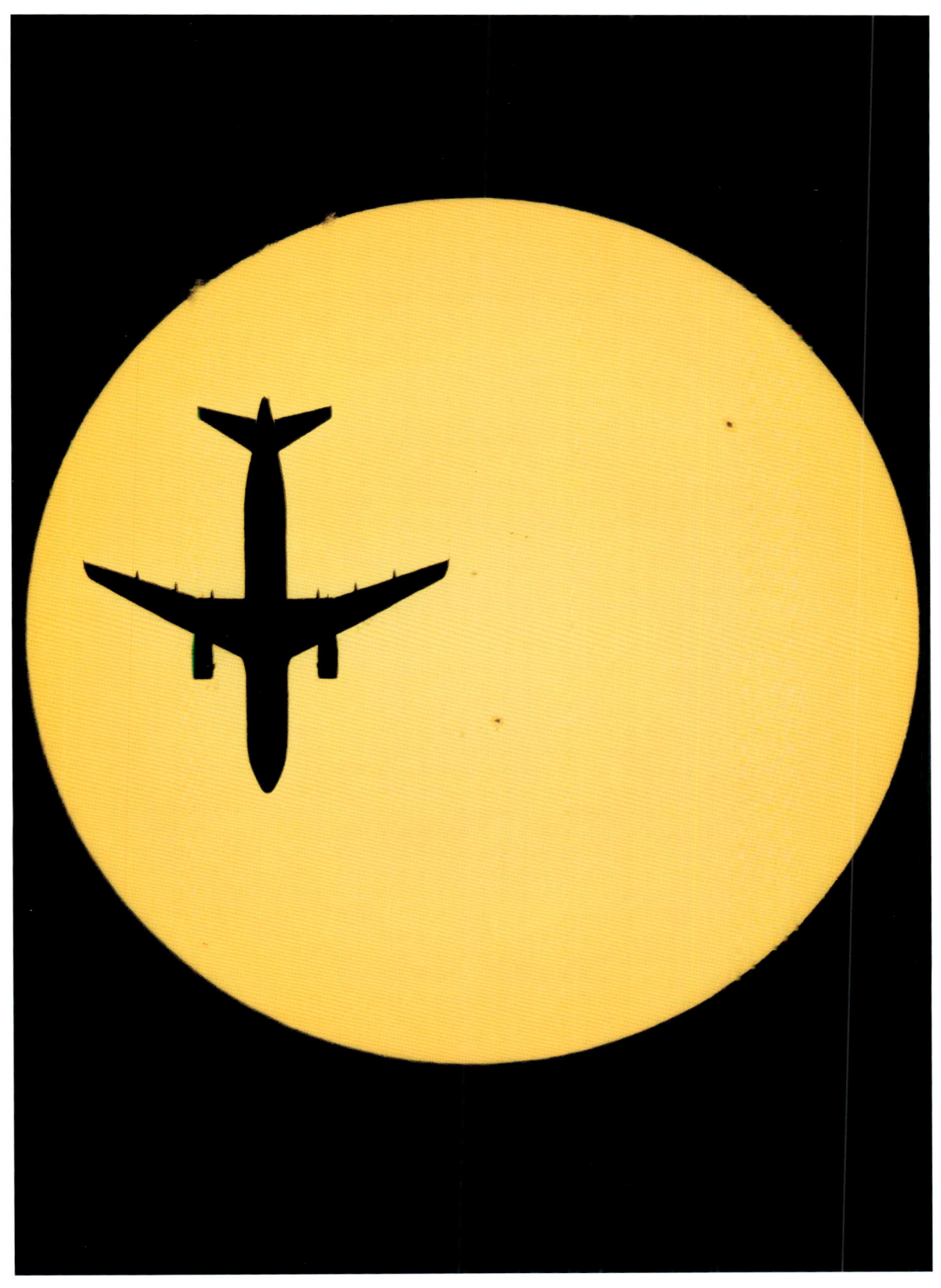

Michael Seeley
The sun silhouettes an airplane.

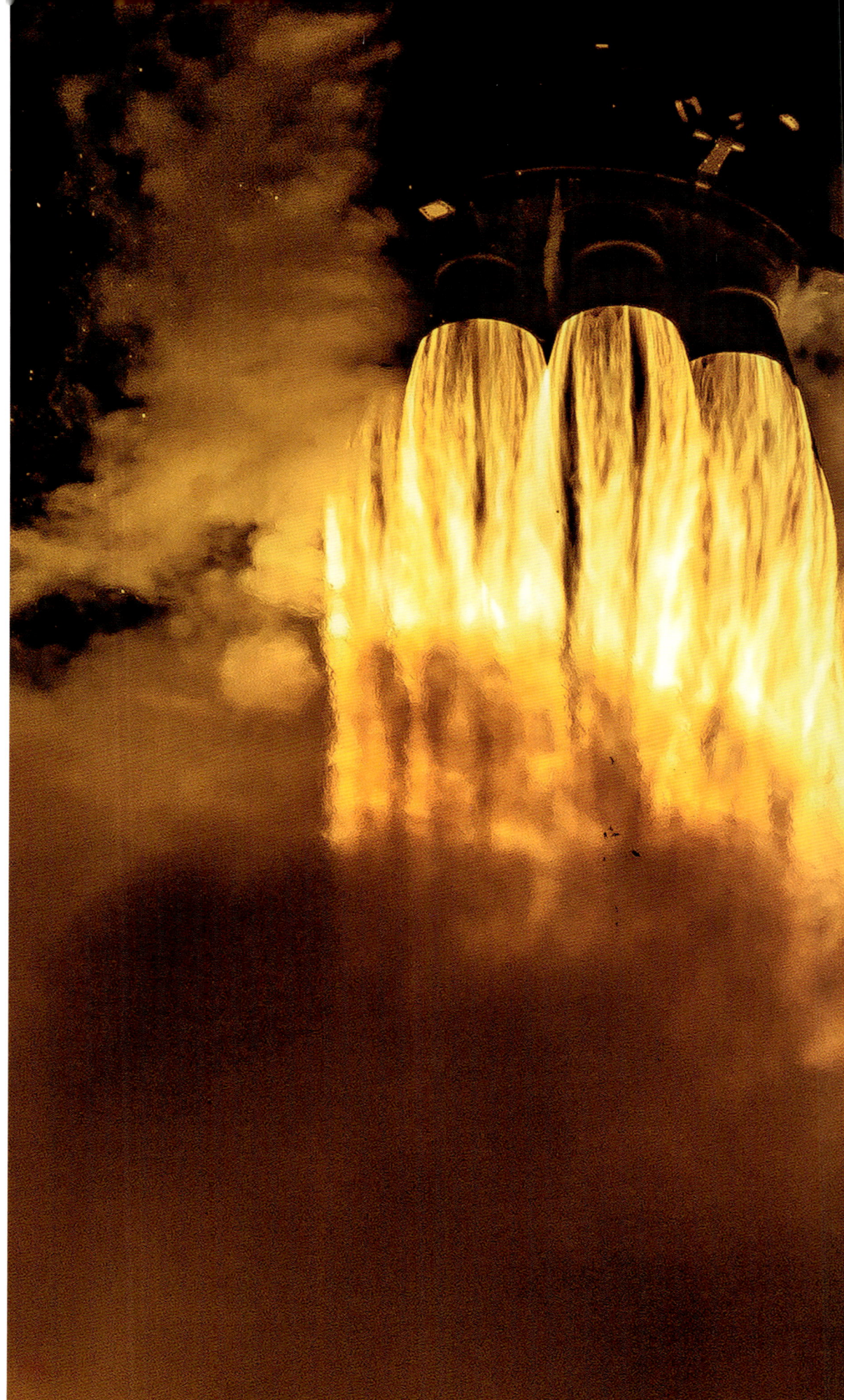

SpaceX
Twenty-seven Merlin engines propel the Falcon Heavy rocket skyward from Florida's Kennedy Space Center, launching the Arabsat-6A satellite into orbit.

FOLLOWING PAGES:
Sergio Tapia
Rust-colored water flows through clumps of green algae in a reservoir in southern Spain.

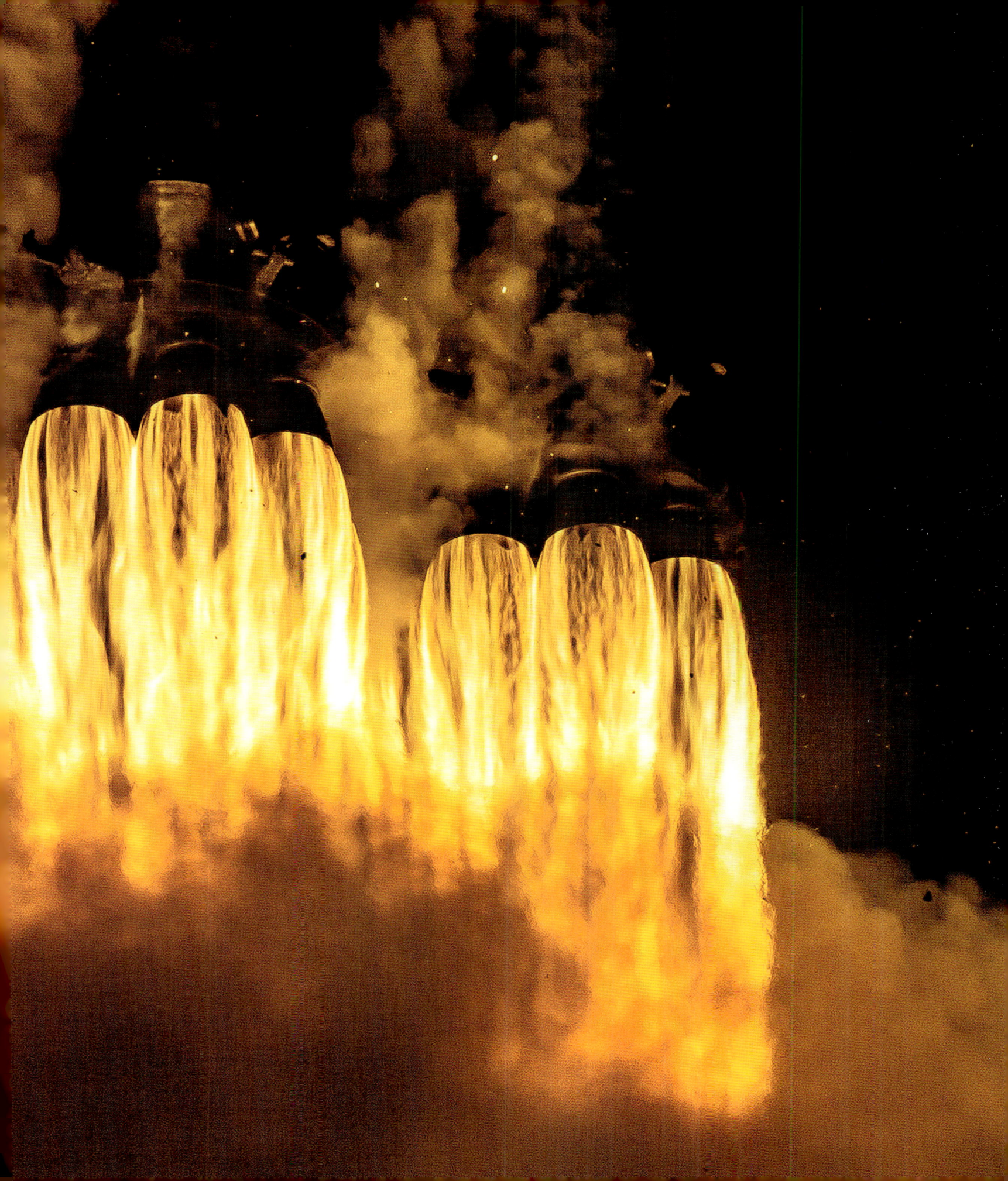

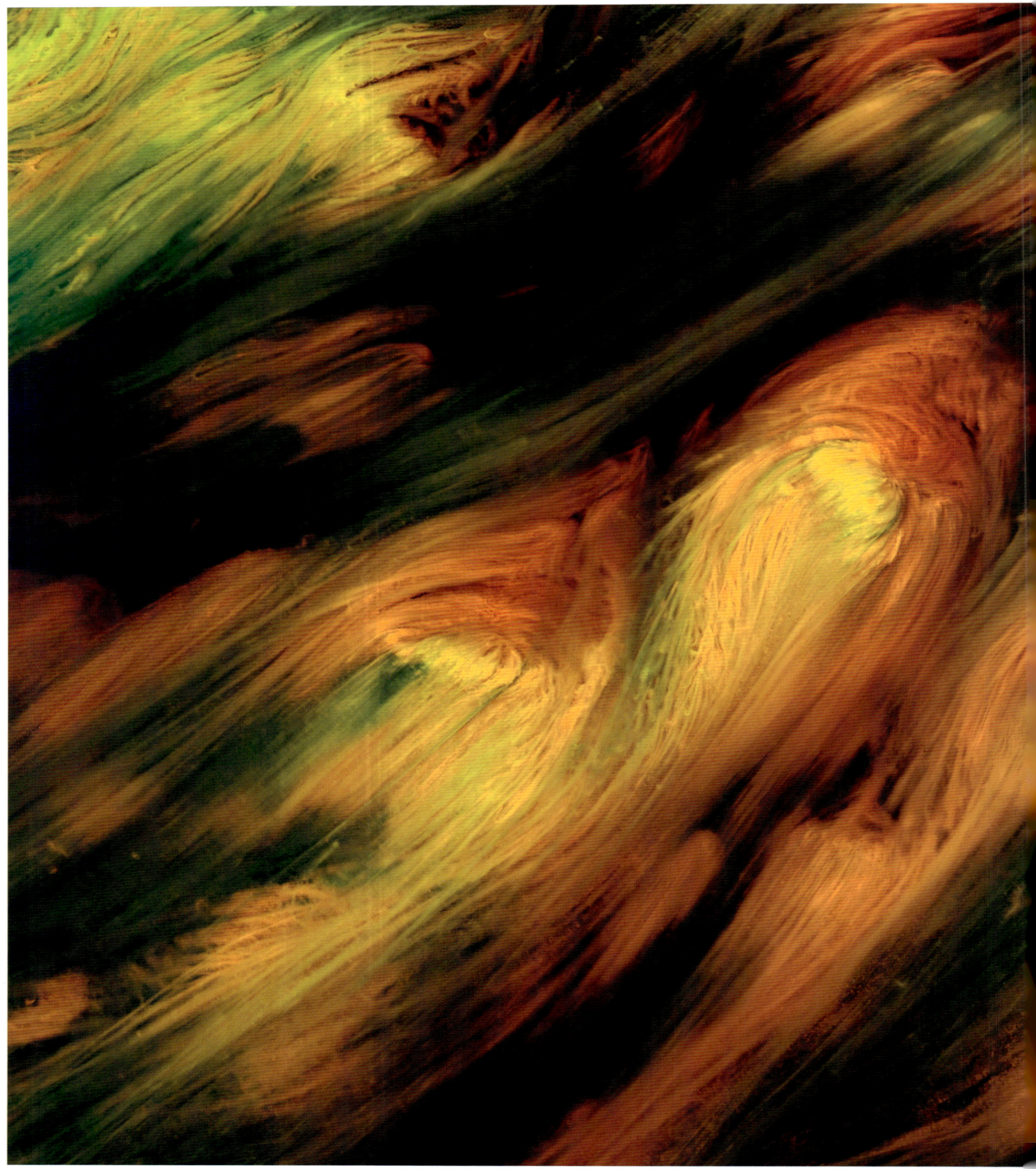

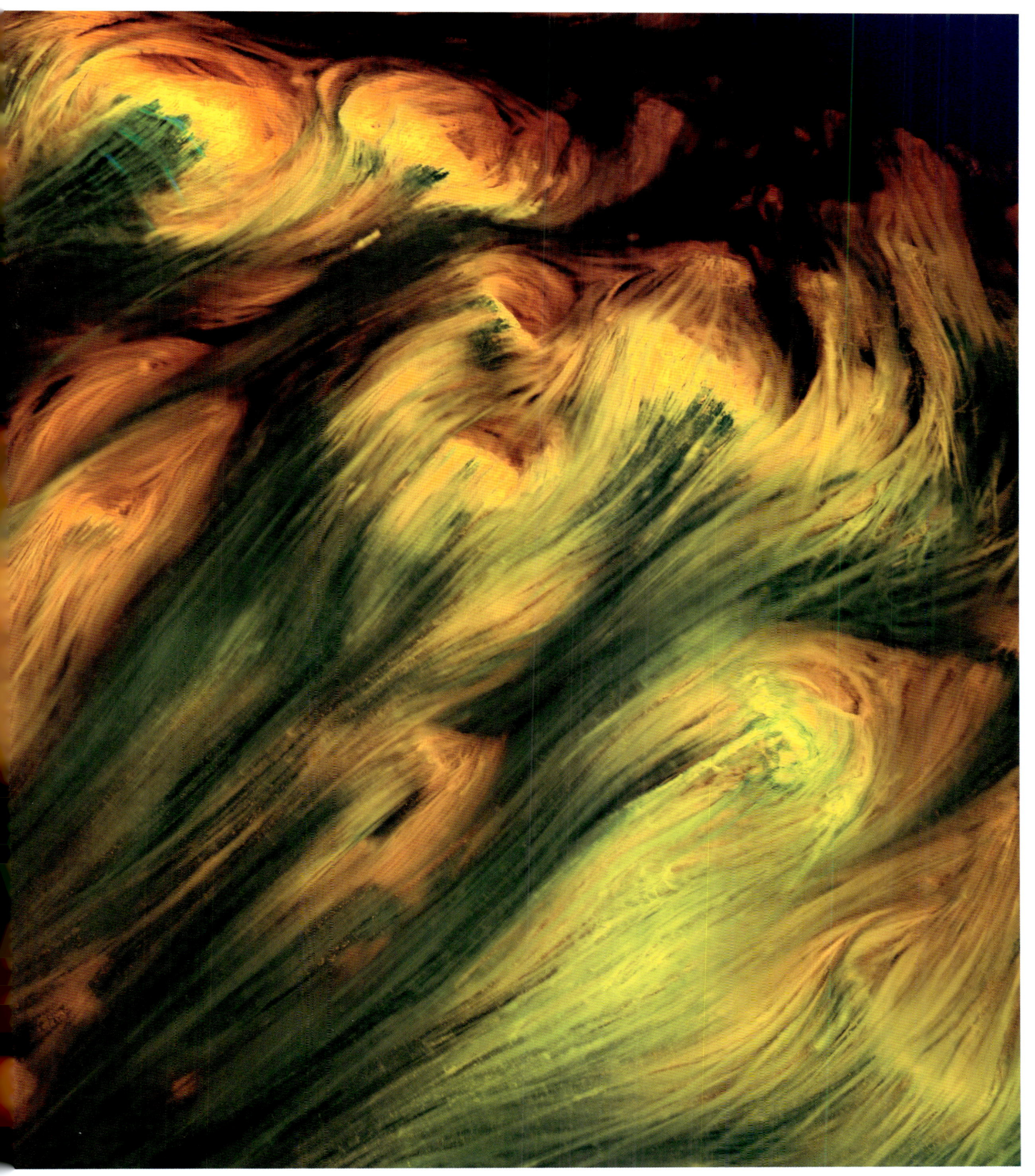

JEN GUYTON

Jen Guyton is a National Geographic Explorer with more than a decade of experience working on wildlife and conservation projects in Africa. She received numerous honors for her work documenting the recovery of Mozambique's Gorongosa National Park, becoming a Fulbright–National Geographic Storytelling Fellow in 2019 and the first winner of the Murie Family Conservation Award in 2020. At Gorongosa, she was a specialist in mammals, surveying antelope, mice, rats, shrews, and 45 species of bats.

ANAND VARMA: How did you become interested in photography?

JEN GUYTON: My parents bought me a film camera when I was 12. It was a silly little point-and-shoot, but I loved it. In general, *National Geographic* magazine has always been influential for me. I was literally looking at the pictures in the magazine before I could read the words. So as a teenager, when people asked me what I wanted to be, I always said, "A National Geographic photographer." But when I finished high school, photography school was too expensive, so it remained this nebulous thing that I dreamed about but didn't know how to accomplish. I followed my other passion, wildlife, and ended up studying conservation and doing fieldwork. I had gotten an SLR [single-lens reflex camera] for my high school graduation, and I started shooting when I was out in the field. To be honest, I think I knew pretty early on that I didn't want to be a scientist. It was the wildlife that grabbed me. My heart was in photography and storytelling.

AV: Was there a turning point when you shifted from science to photography?

JG: In 2013, I was given a Young Explorers Grant from National Geographic, and I started meeting real National Geographic photographers. I could finally see a path forward for that as a career. I was living in Gorongosa National Park, working on my Ph.D. in ecology, so I had access to tons of subject matter. Gorongosa National Park was a civil war battleground in Mozambique in the 1980s. During the conflict, 90 percent of the large mammals declined. In the 1990s, after the war, the Carr Foundation started restoring the park and protecting its ecosystems so the animals could come back. I could go out into the park and photograph whatever I wanted.

AV: How did you come up with the idea for your composite images?

JG: I wanted to see if I could strap a camera down so that you'd be immersed in the way that an ecosystem functions over time, rather than standing back from it. The project was about resource hot spots—what are the places that attract a lot of animals for one reason or another? Why are animals congregating here, as opposed to over there?

AV: What is your process for creating these images?

JG: I'll set up my camera to shoot photos at an interval or as a camera trap using a passive infrared trigger. Passive infrared basically scans the scene for changes, so if it senses movement and warmth, it'll trigger. Over some period of time, I'll gather a few hundred images, all taken with exactly the same framing. Then I pick one of those to be my background image, based on its lighting, before layering more of the images with the animals on top of it using Photoshop.

AV: What are the main challenges of this technique?

JG: Gosh, there are a lot [*laughs*]. Light is the main challenge, because you need consistent light across your selected period of time. Otherwise it's going to look weird when you composite it together, and I don't want people to be like, "Oh, that's obviously photoshopped." Then, all the normal challenges of camera trapping are compounded because you're trying to capture a continuous period of time. Camera trapping is such a headache because you're not there with the camera, so anything can

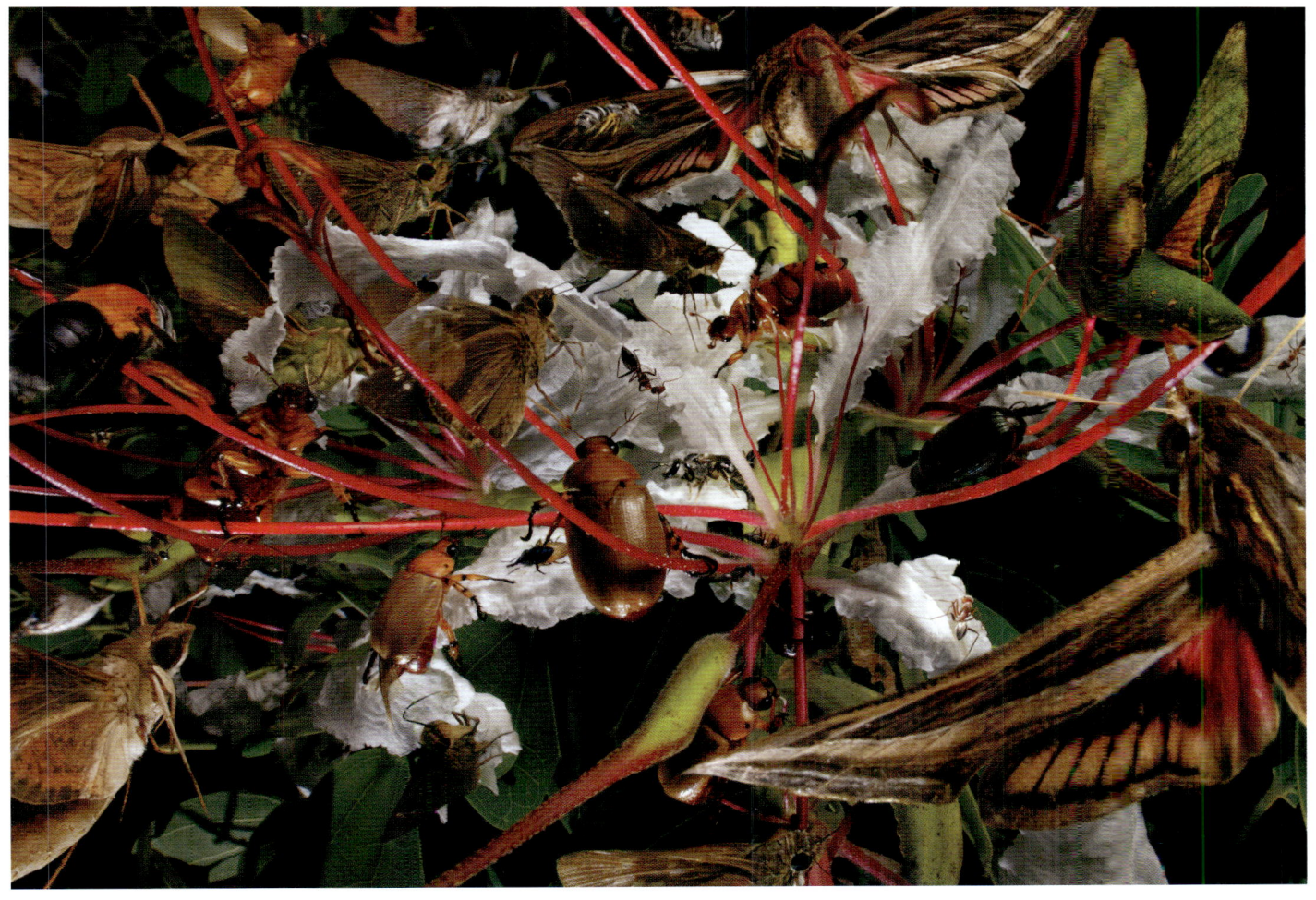

Jen Guyton
Beetles, ants, and hawk moths visit a night-blooming flower over the course of several hours in Mozambique's Gorongosa National Park.

FOLLOWING PAGES:
Jen Guyton
A pile of warthog dung attracts a menagerie of creatures over three hours' time.

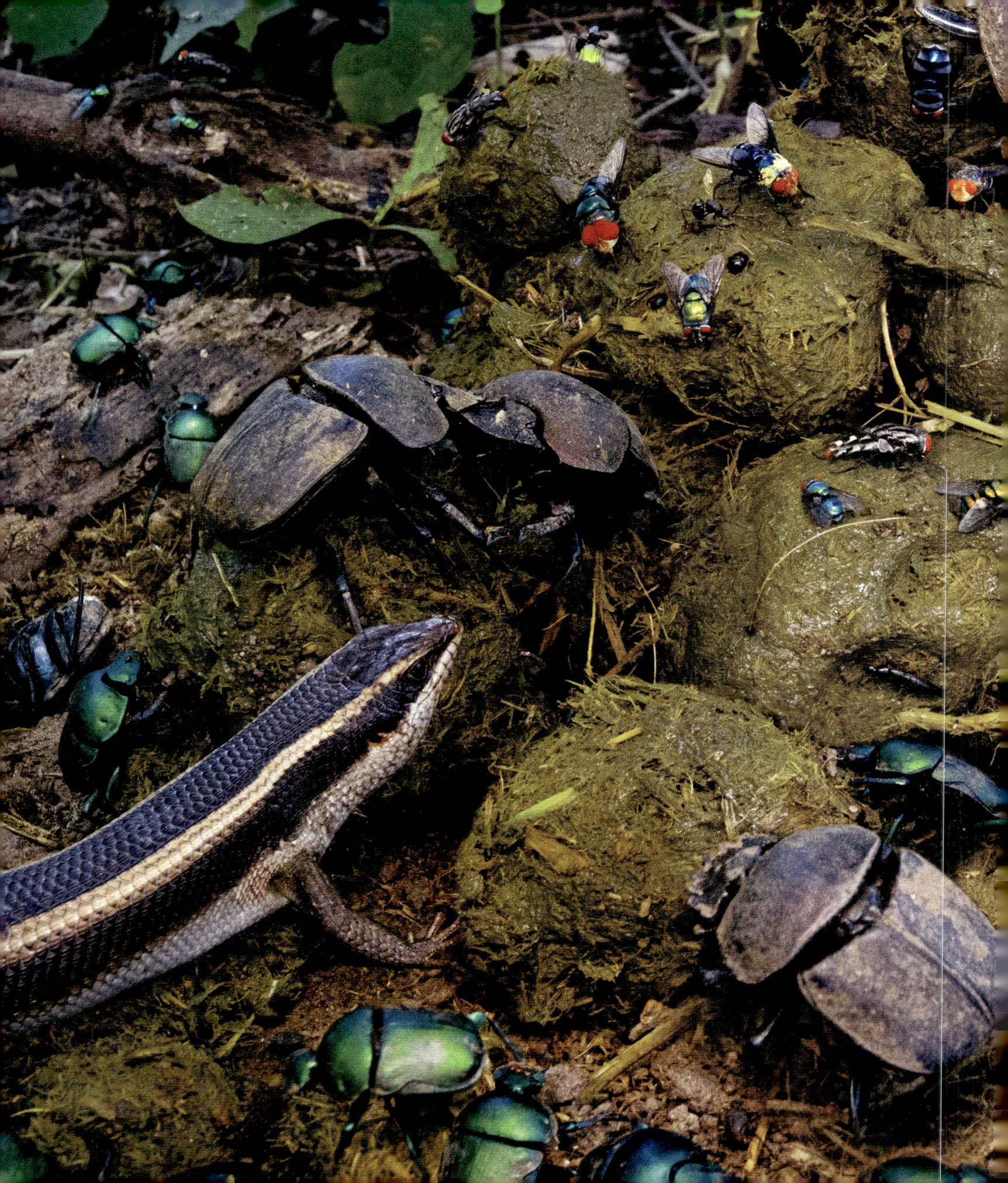

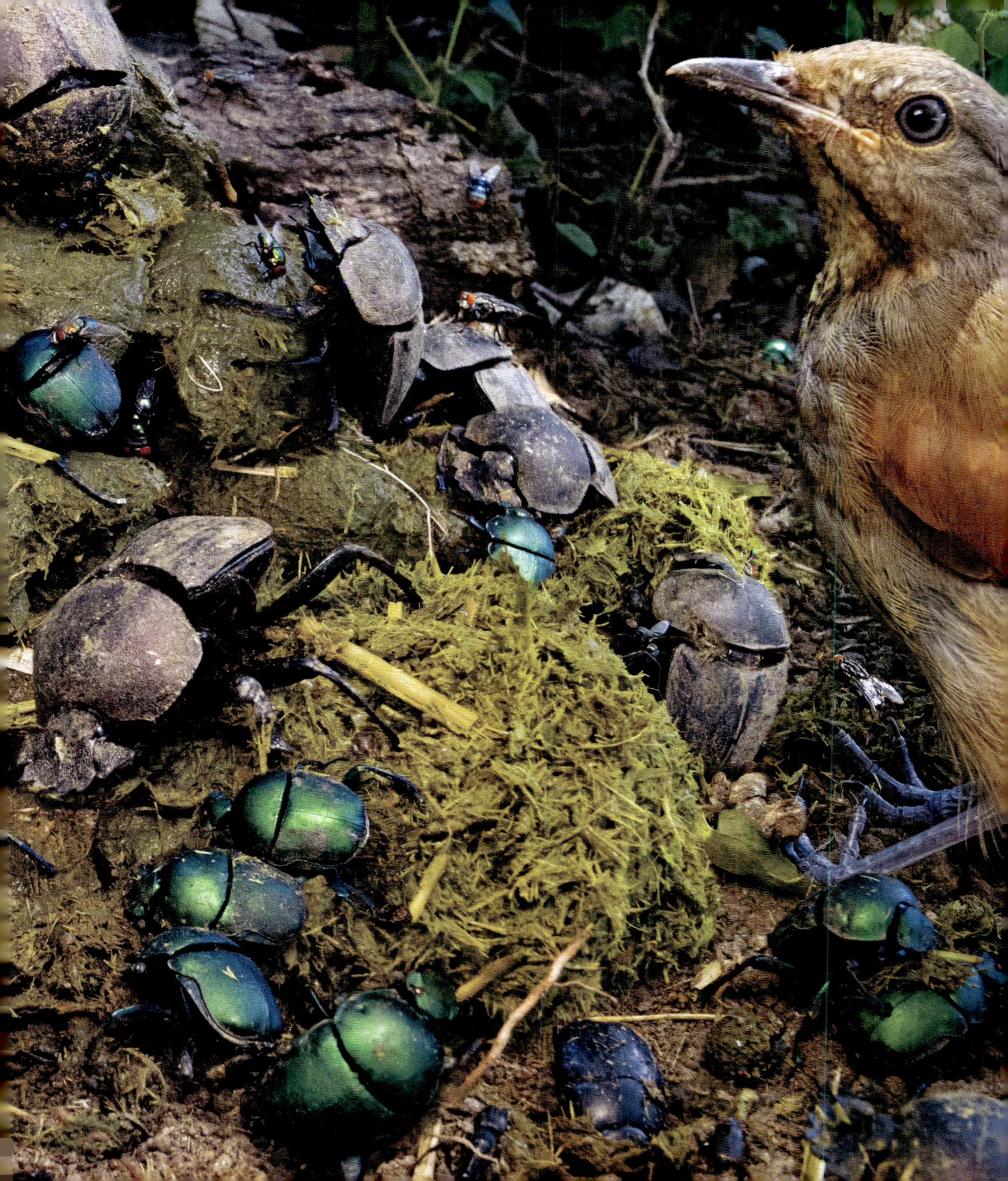

go wrong. A baboon can knock it over, the batteries can die, the lens can fog up, et cetera. That's frustrating already, but when you have an amazing two hours with a dung pile and then a baboon comes and knocks the camera over, you have to start all over. The other thing was finding the right hot spots and making sure the camera was pointed in the right direction to get the most animals.

AV: Why did you choose a dung pile as a resource hot spot?

JG: I spent a lot of time brainstorming about the places, big and small, that attract a lot of animals for one reason or another. I paid attention to the ecosystem as I was moving around in it: Where am I seeing a lot of animals? In the case of the dung, it was pretty obvious. There were beetles all over it. And in the case of this particular dung pile, I kept it simple. We had a ton of warthogs that lived in camp, so I just followed them around until they pooped [*laughs*].

AV: How did you determine what length of time to run the process for?

JG: I made a lot of rules because I wanted to make sure I was representing reality as much as possible. I decided that each composite needed to show a continuous period of time—and I included at least one of every species that visited during that time period, with no repeats of the same individual. And I didn't move anything. Wherever the animal is in the frame is where it was in the original image. With the dung, I took pictures on an interval from as soon as the dung dropped to when it was completely gone—about three hours.

AV: Was there anything that surprised you once you compiled

the images into that composite?

JG: Honestly, how beautiful it was. You take a pile of shit and turn it into this beautiful, big, jeweled mountain of ... I don't know, insects? That's what I love about it. You don't think that poop can be beautiful, but if you look at it the right way, it can.

AV: What made you want to focus on the night-blooming flower?

JG: I was walking past this bush at night, and I passed over it with my headlamp and saw these beautiful bright white flowers, and tons of moths all over the bush. So I set up my camera to take images at an interval and left so I wouldn't disturb the scene. But it took me a few nights to get the composite, and thank god I got it on the last night, because these flowers were dying pretty rapidly. The first night I had a monkey sit on the branch and break it, and then these beetles made it tricky because they were eating the flower. So as the scene went on, there was less and less and less flower [*laughs*]. It was probably the hardest one to composite.

AV: What inspired the termite-mound composite?

JG: Termite mounds are fascinating habitats because they have their own microclimates. They're cooler than the surrounding area, they tend to be wetter, and the vegetation on them tends to be more nutritious because of all the nutrients the termites are bringing in. Because of all those aspects, the mounds attract a lot of large mammals, especially the bushbuck antelope, nyala antelope, and kudu antelope. You'll often see the antelope standing on top of the mound because it makes a good vantage point for predators. I knew there

would be a lot of animals moving through the area, but it took me many tries to get a good composite. The time span for the composite had to be about two weeks to see the full range of animals using it. But I thought this one was particularly good because of the monitor lizard living in a hole under the mound, the genet cat living in the big tree—it would crawl down those lianas every night. And then I got two monkey species and two antelope species. I used a camera trap with a passive infrared trigger and set my camera to the deepest possible depth of field to try and get as much in focus as possible.

AV: Do you have any interesting stories from this project?

JG: Yes. I really wanted to do a carcass. But I had to find a carcass that was fresh enough for vultures to still be using it. On my last week in Gorongosa, I found one that hadn't even been opened yet—I actually had to open it with a knife, which was gross. Then I set up my camera for the vultures, but the first night no vultures came down at all. I think they were freaked out by my camera. They're really smart and really cautious. So I tried putting some mud and leaves on top of the camera to disguise it a bit better, but still, they didn't come down the second day. By this time, the carcass was starting to smell *really* bad. Finally, I took off the glass filter from the front of my camera trap, because it's reflective, and the vultures came down the next day, and I got a couple frames. But when I went to check my trap, it had been opened and the parts strewn everywhere and stomped on. I suspected I knew who was responsible. My colleague had his own camera traps set up nearby, so we checked the photos on it. There was a big bull elephant walking

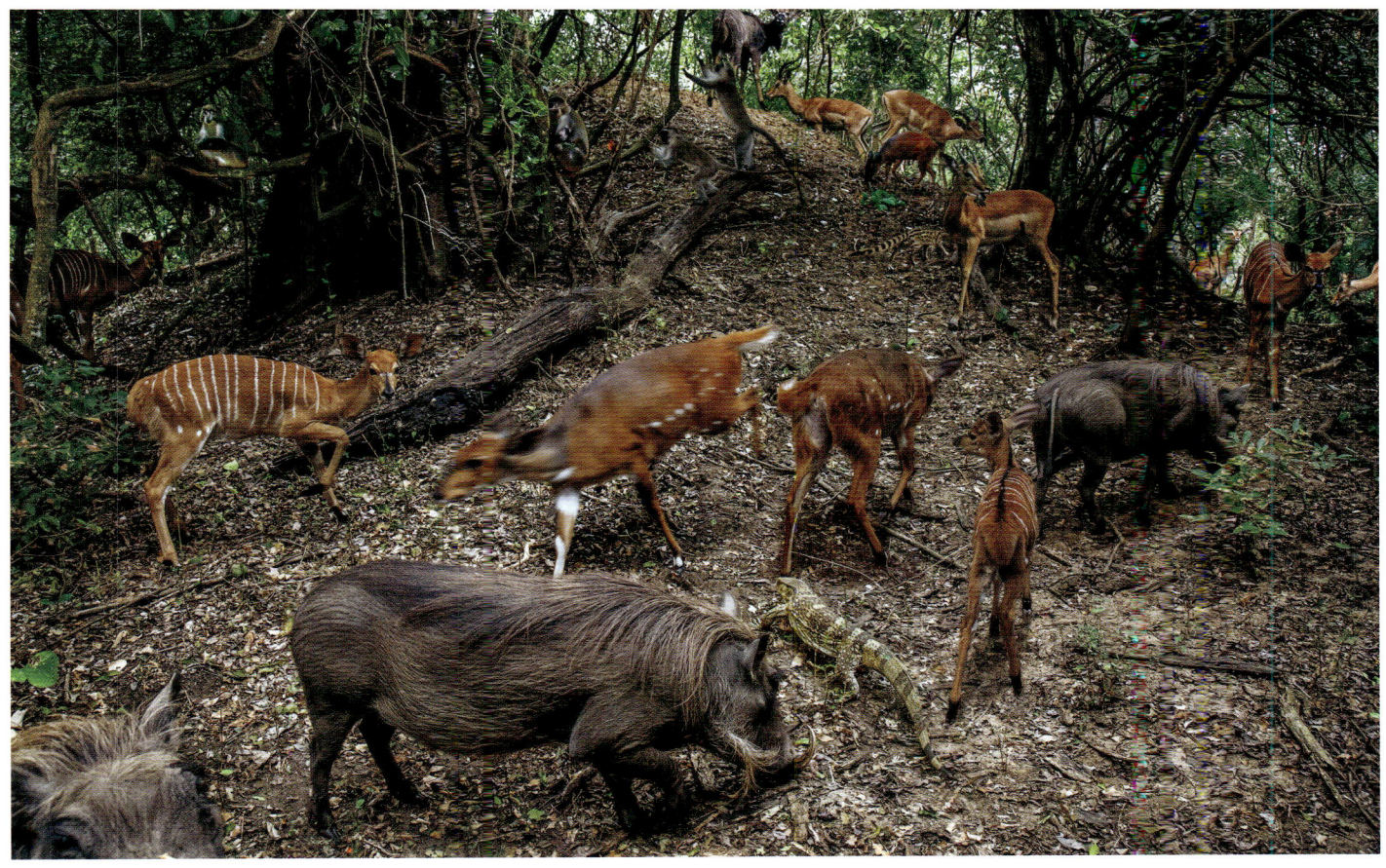

Jen Guyton

The diversity of wildlife that visits a Gorongosa termite mound over the course of two weeks: warthogs, two monkey species, two antelope species, a genet, and a monitor lizard

past his camera with my camera in his trunk [*laughs*].

AV: Oh wow!

JG: So that one failed miserably, and I never managed to get my carcass photo, sadly. But the elephants were a pain in the butt. Actually, they had completely destroyed another one of my traps as well, to the point where they even took the tripod and pulled apart every section of the legs.

AV: Do you know why elephants do that?

JG: I think there are two possible explanations. One is that they really don't like people, especially in Gorongosa, where they were persecuted so badly during the civil war. A lot of the elephants there are old enough to actually remember the war. They're even known to chase and headbutt cars. It might be that they smell our scent on the camera trap and just get angry and trash it. But the other possible explanation is that they're really curious, and if there's something new in their environment, they're going to explore it. Elephants explore things by tearing them apart sometimes. So, as with any type of nature photography, it can be hard to predict. And since I needed a very specific set of things to happen for these images to work, it didn't always come together. I spent a year on this project, and I got five great images and maybe a couple more that are pretty good.

AV: What do you want your audience to take away when they see these photographs?

JG: I hope they're inspired by how beautiful they are, and I hope it helps them to think more deeply about how ecosystems work. You can get a cool image of a lion, but that doesn't necessarily make you think about what was at that watering hole before or after the lion was there, and how different animals use space or time in unique ways. I hope these images can be an entry point into thinking more about community living and how animals interact with each other.

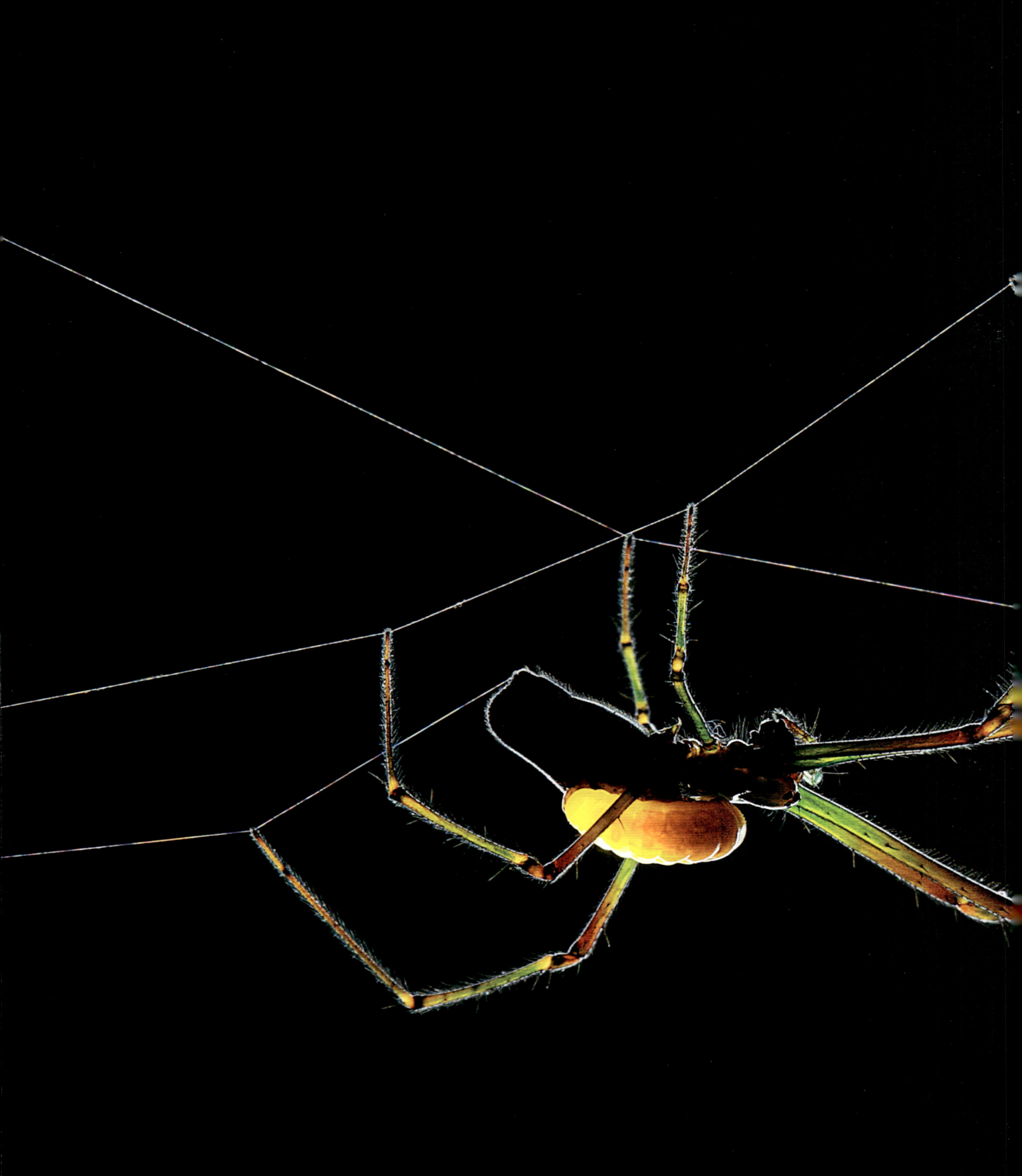

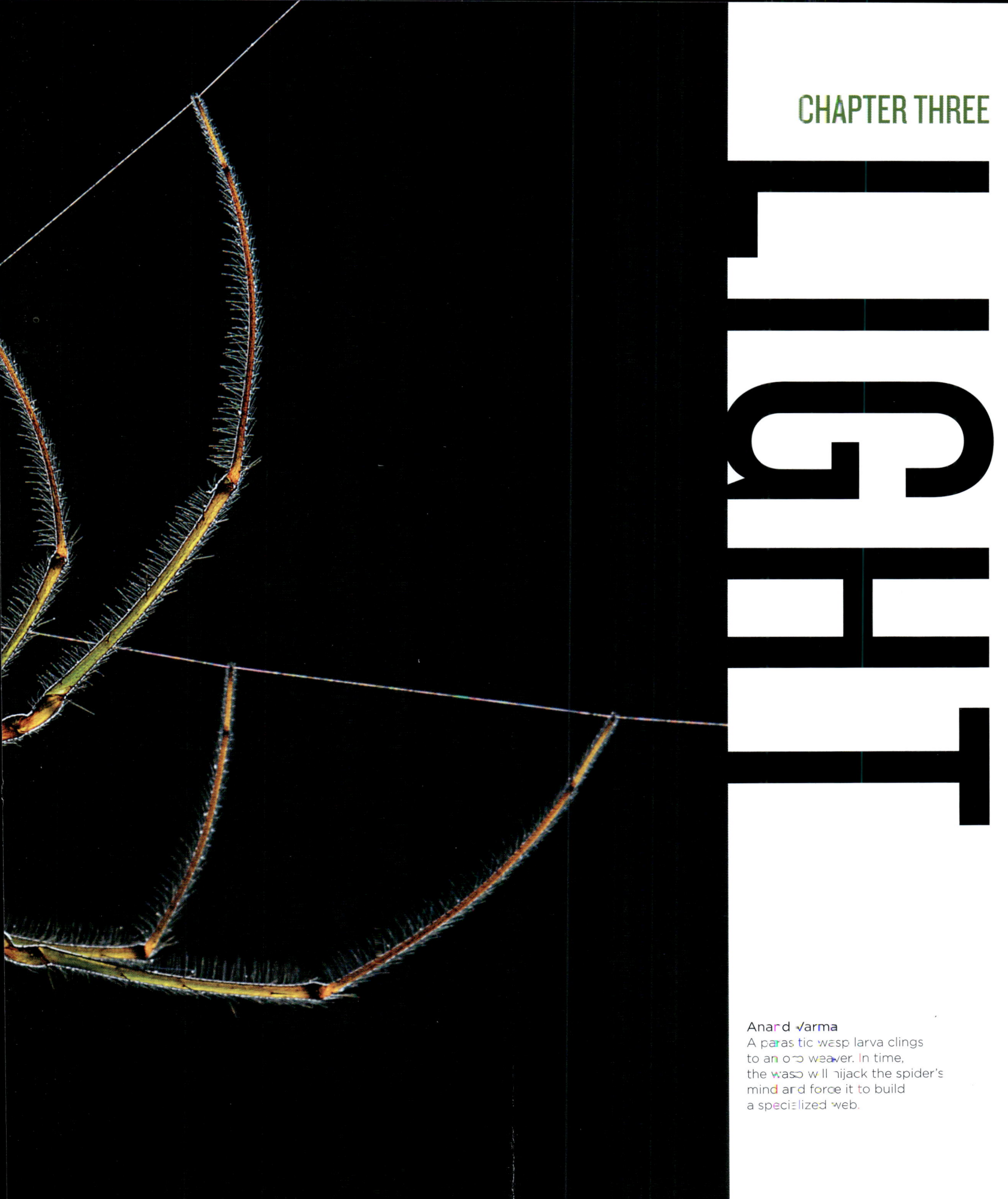

Anard Varma
A parasitic wasp larva clings to an orb weaver. In time, the wasp will hijack the spider's mind and force it to build a specialized web.

CHAPTER THREE

LIGHT

PARASITES ARE a tough sell. When I proposed a story about mind-controlling parasites to *National Geographic* magazine, my photo editor Todd James defined my task succinctly: "Your job is to take a photograph that stops a reader in their tracks and makes them want to read the caption." That was no simple task. How could I convince skeptical readers that these devious little beasts deserved their attention?

I started with the techniques that had produced reliable results for me in the past. Get close and magnify details. With enough proximity, some surprising features usually emerge. Tiny hairs make a bee's face look furry. Glossy reflections make a beetle appear metallic. But in this case, the parasites just seemed more revolting the closer I got. Manipulating size was not helping my cause. By the time my fieldwork started, I had no clear plan for how to succeed.

I arrived in Costa Rica with the goal of photographing a parasitic wasp that attacks a spider. With the help of researchers from William Eberhard's lab at the University of Costa Rica, I collected infected spiders from the surrounding forest and brought them back to my hotel room, carefully stashed in plastic bins to hide them from the housekeeping staff. I began a tedious trial-and-error process, searching for some flattering angle from which to portray them. I soon gave up and instead started experimenting with the lighting. By varying the position and direction of the light source, I stumbled on a hidden secret. When illuminated from just the right angle, the spider's body became translucent and began to glow. Its delicate strands of silk shimmered with color, and tiny hairs along its legs stood out against the dark background. Light revealed brilliant details that even my best close-up lenses had missed.

Every photograph is a collaboration with light. But the images in this chapter show how photographers manipulate light in specific ways to reveal hidden patterns. Prasenjeet Yadav used a headlamp to illuminate the intricate designs of a living root bridge in India (page 226). Stephen Orlando attached LEDs to a kayak paddle to trace the path of its movement across the water (page 238).

PHOTOGRAPHERS SHOW US HOW LIGHT CAN BE HARNESSED WITH A CAMERA TO RESHAPE OUR PERCEPTION OF THE WORLD.

Photographers do more than just position light in clever ways. Cameras can detect colors that are invisible to the human eye, and photographers use these capabilities to broaden our sensory experience. When you see a rainbow, red appears at the top of the arch and violet at the bottom. Present and yet hidden from our view are the infrared colors that exist past the red and the ultraviolet colors that appear beyond the violet. With an infrared-sensitive camera, Paolo Pettigiani reveals concealed layers of a landscape (page 272). Craig Burrows's ultraviolet photography shows the secret structures of flowers (pages 218 and 244).

Whether using traditional illumination in innovative ways or special techniques that go beyond the reach of our senses, these photographers show us how light can be harnessed with a camera to reshape our perception of the world. ∎

FOLLOWING PAGES:
Reuben Wu
Light from a drone defines the angular lines of a rock face at the Hoel Tryfan slate quarry in Wales

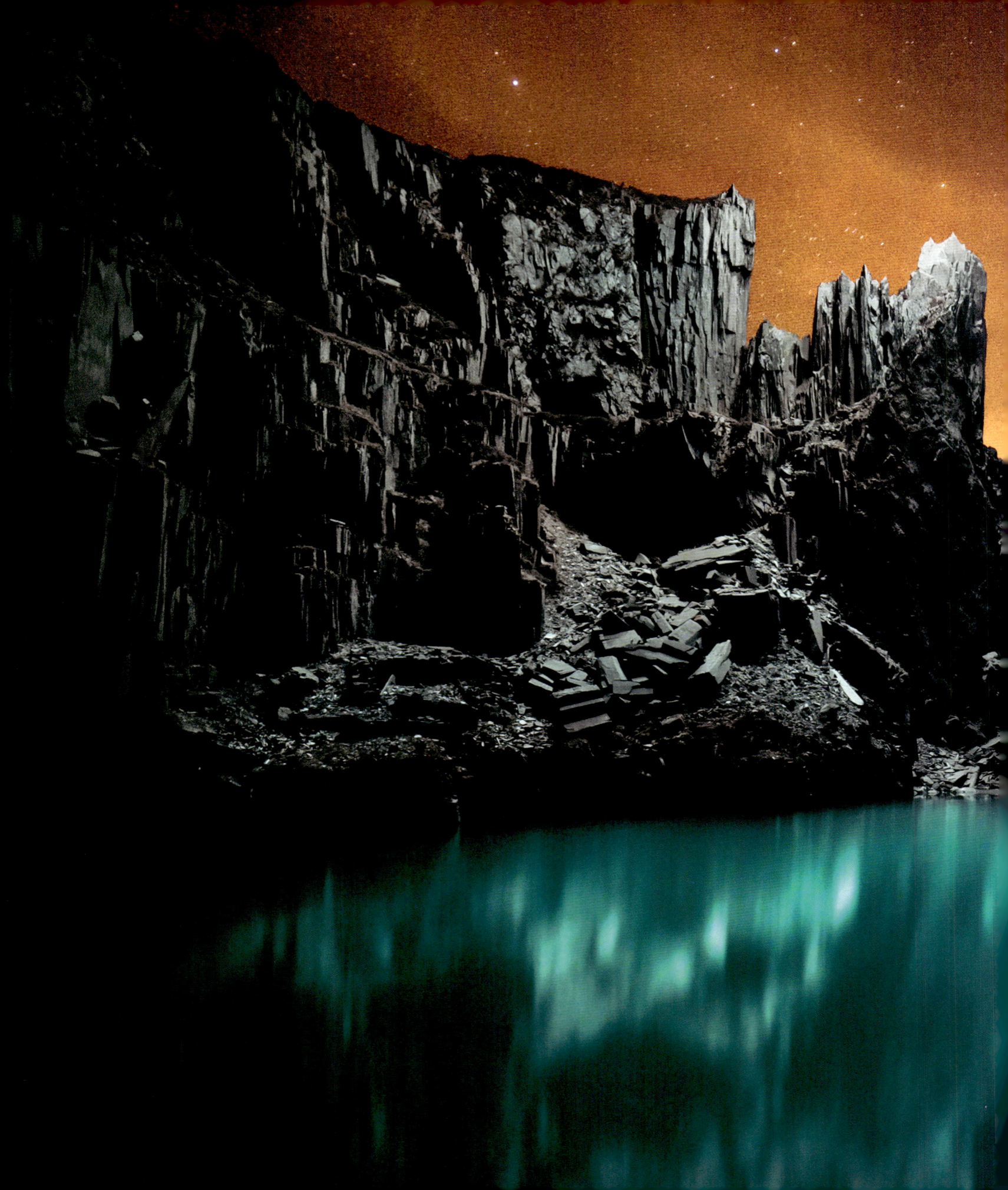

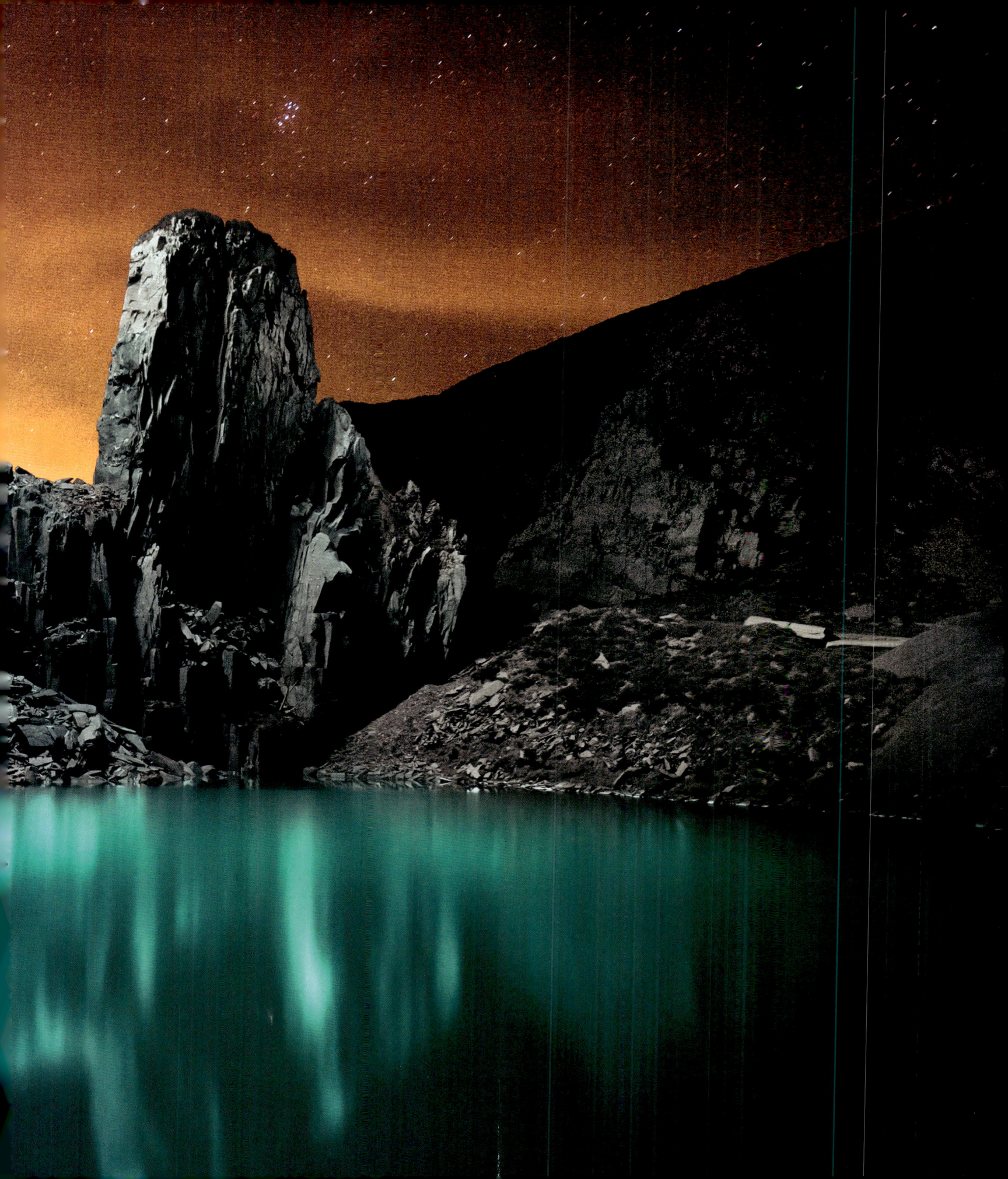

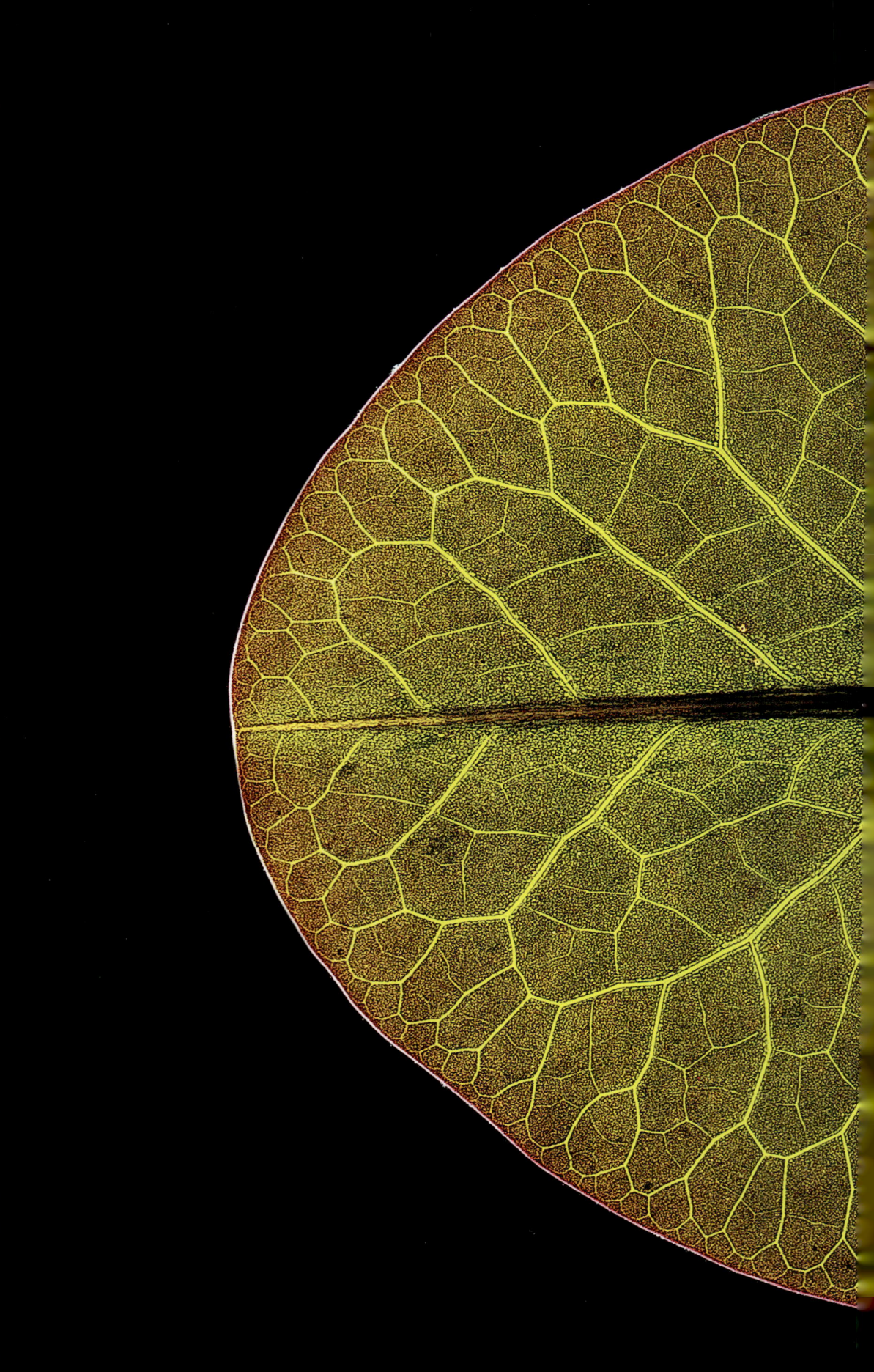

Anand Varma
Light behind a blue water lily
pad illuminates the delicate
pattern of its veins.

FOLLOWING PAGES:
Nathan Myhrvold
Crystals of vitamin C dissolved
in water and illuminated with
polarized light

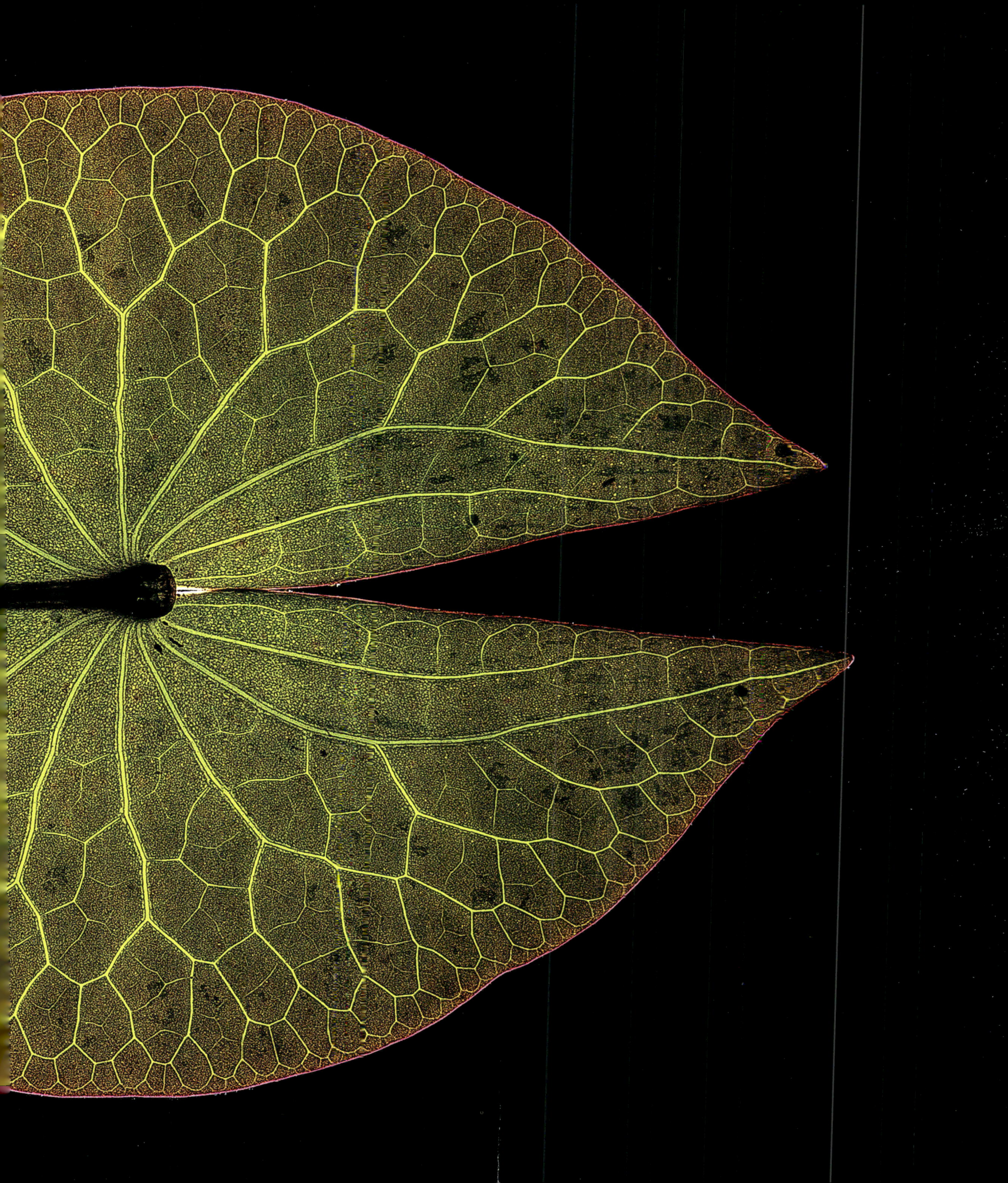

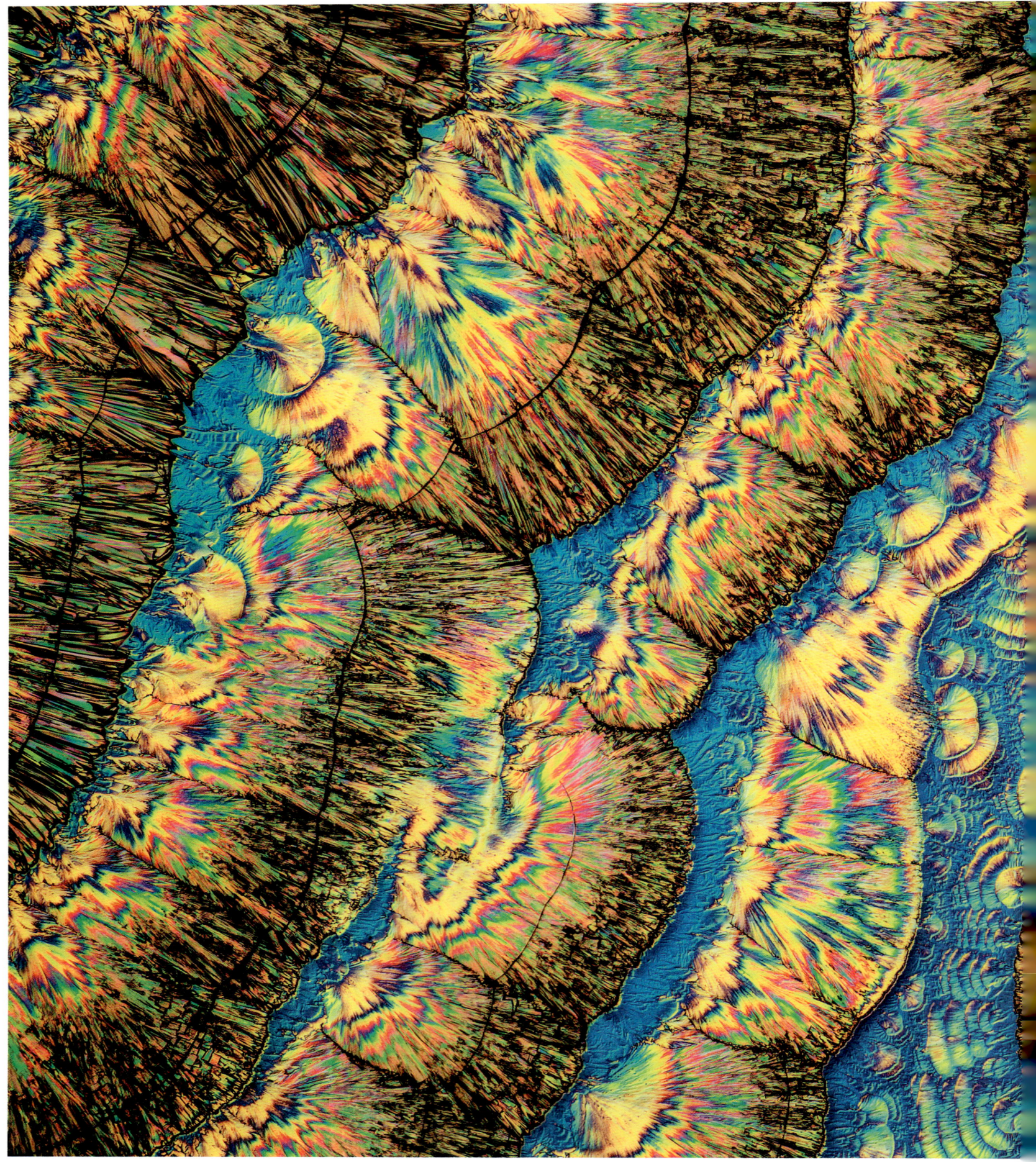

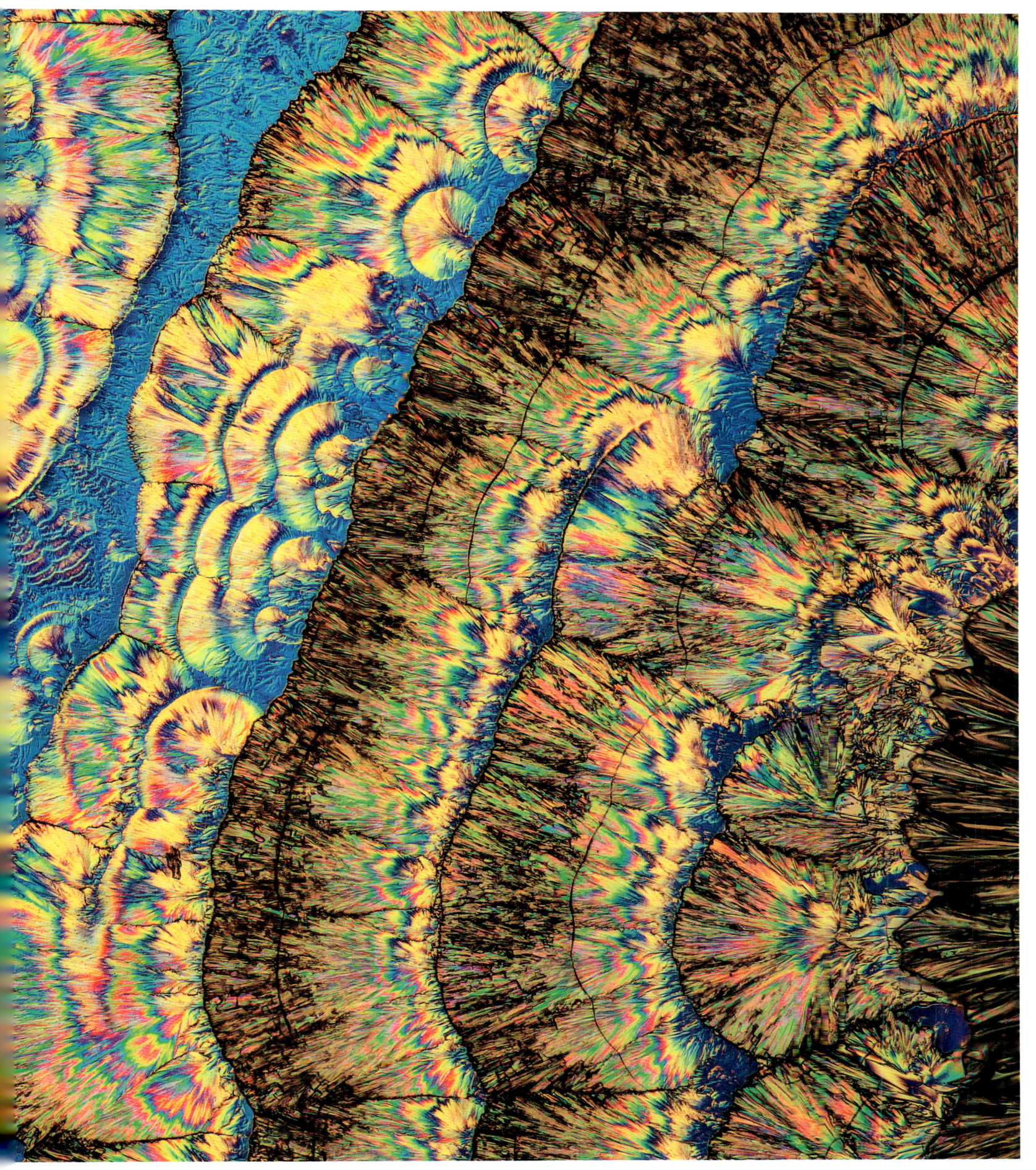

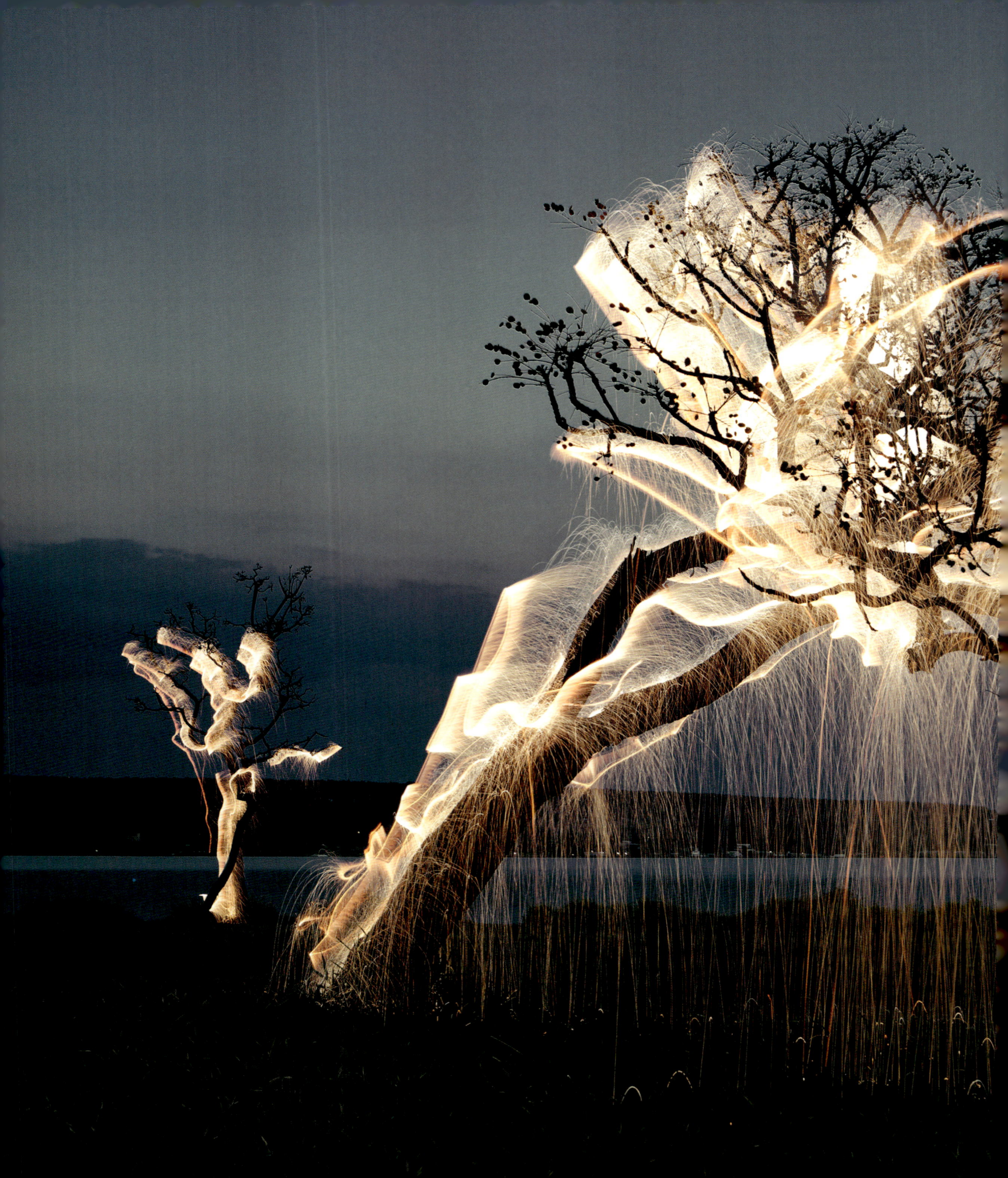

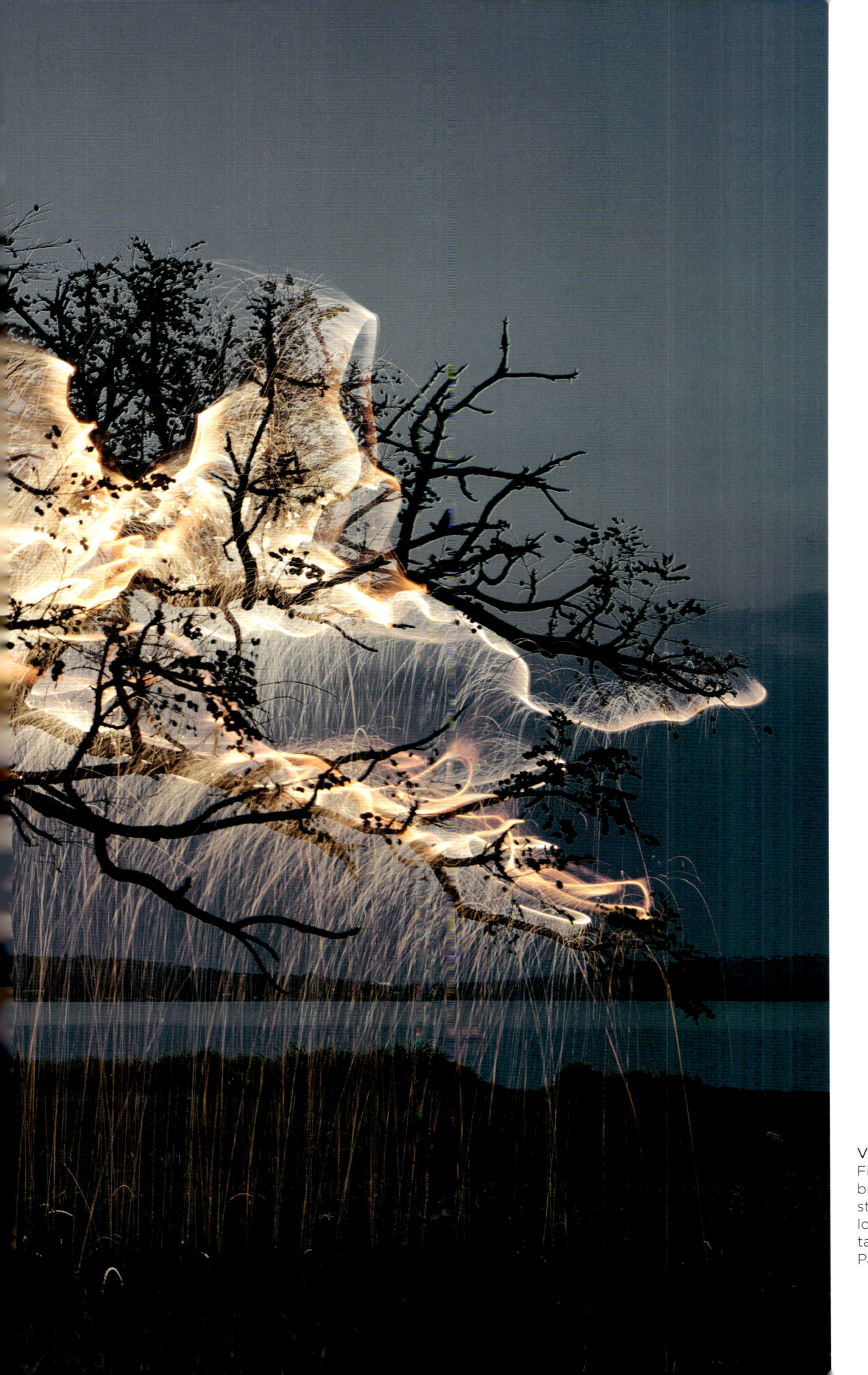

Vitor Schett
Fireworks traced along
branches highlight the
structure of trees in this
long-exposure composite
taken near Brazil's Lake
Parancá

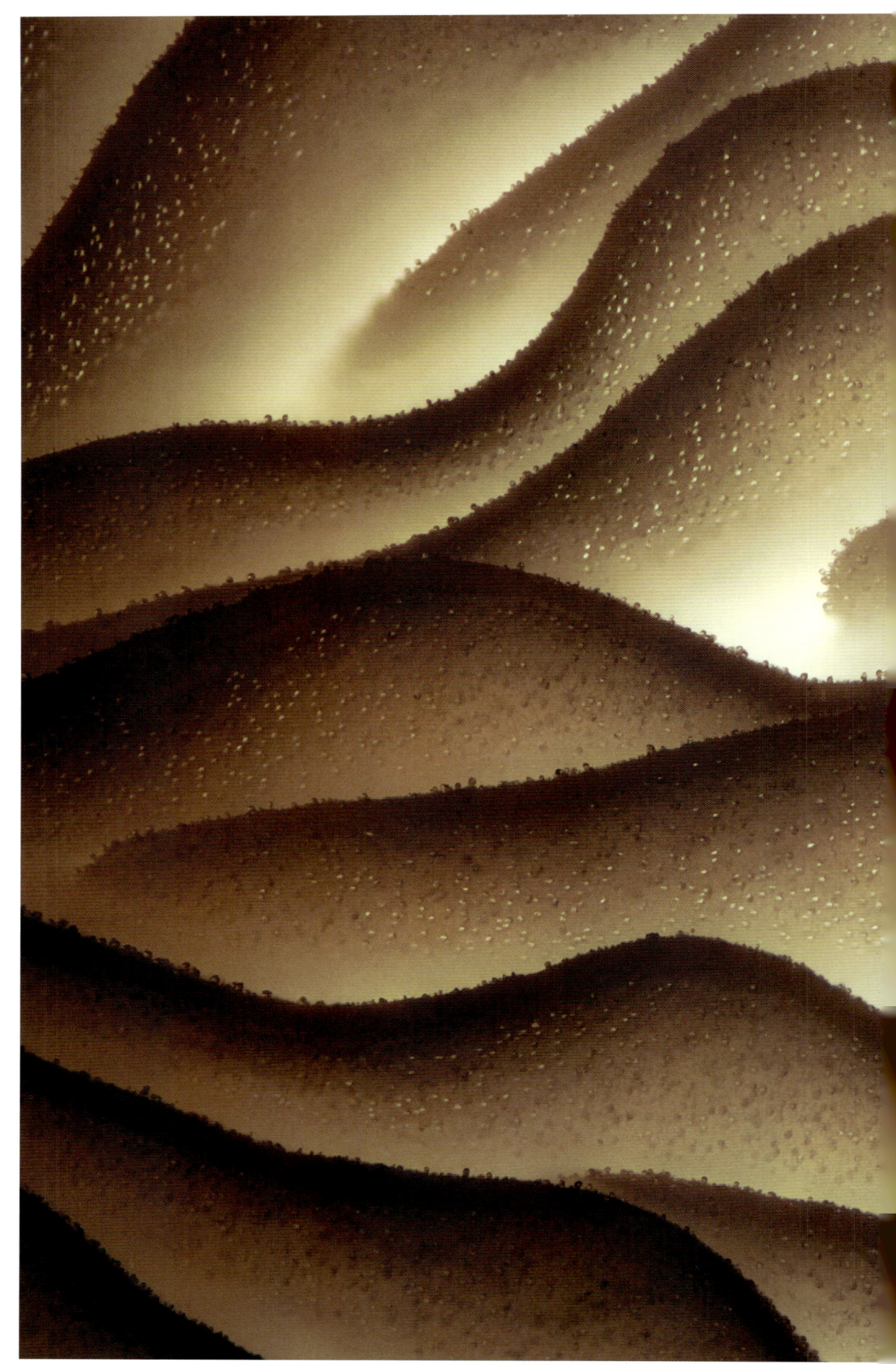

Charles Krebs
Light shining through the magnified gills of a mushroom reveals their textured ridges.

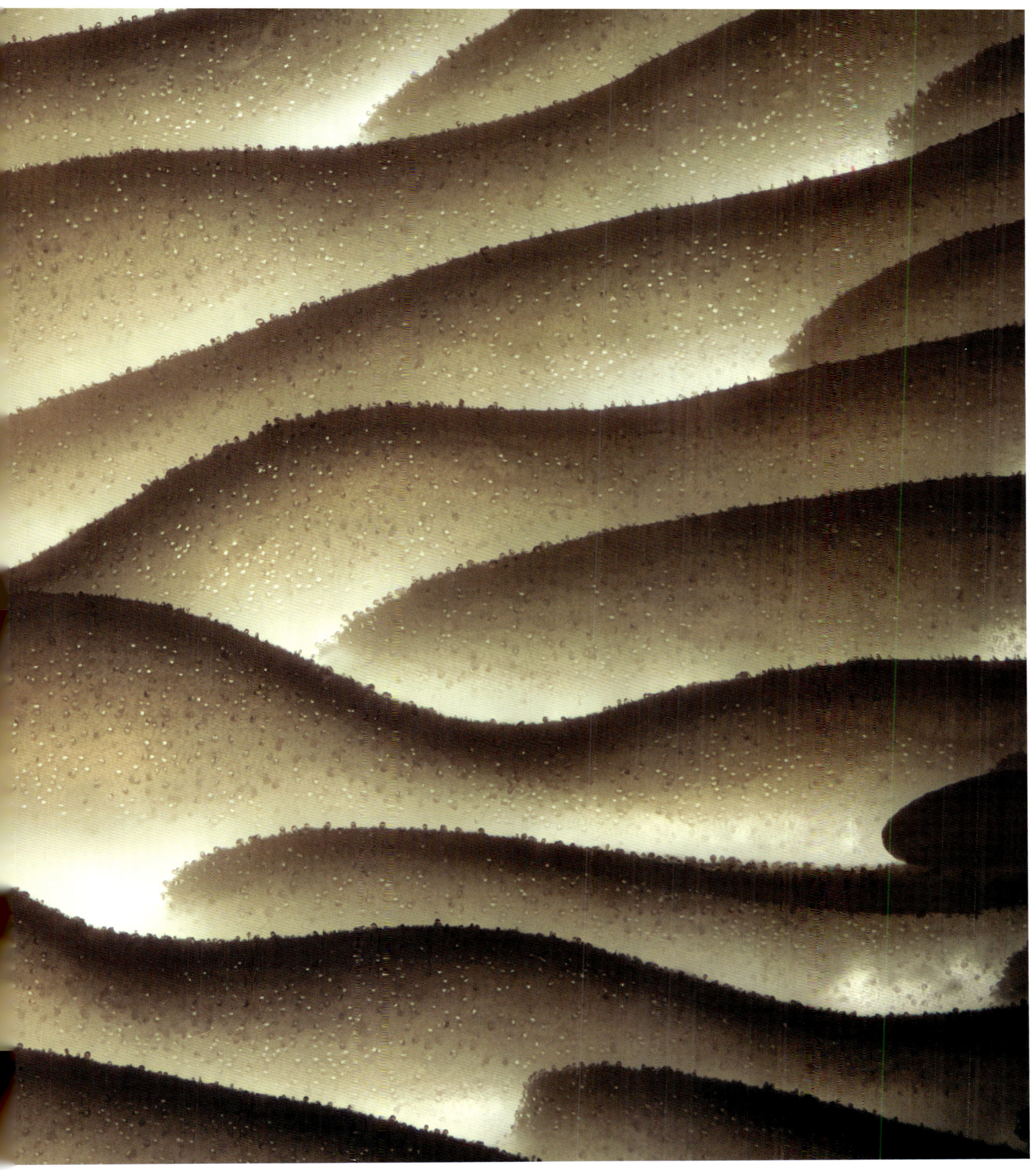

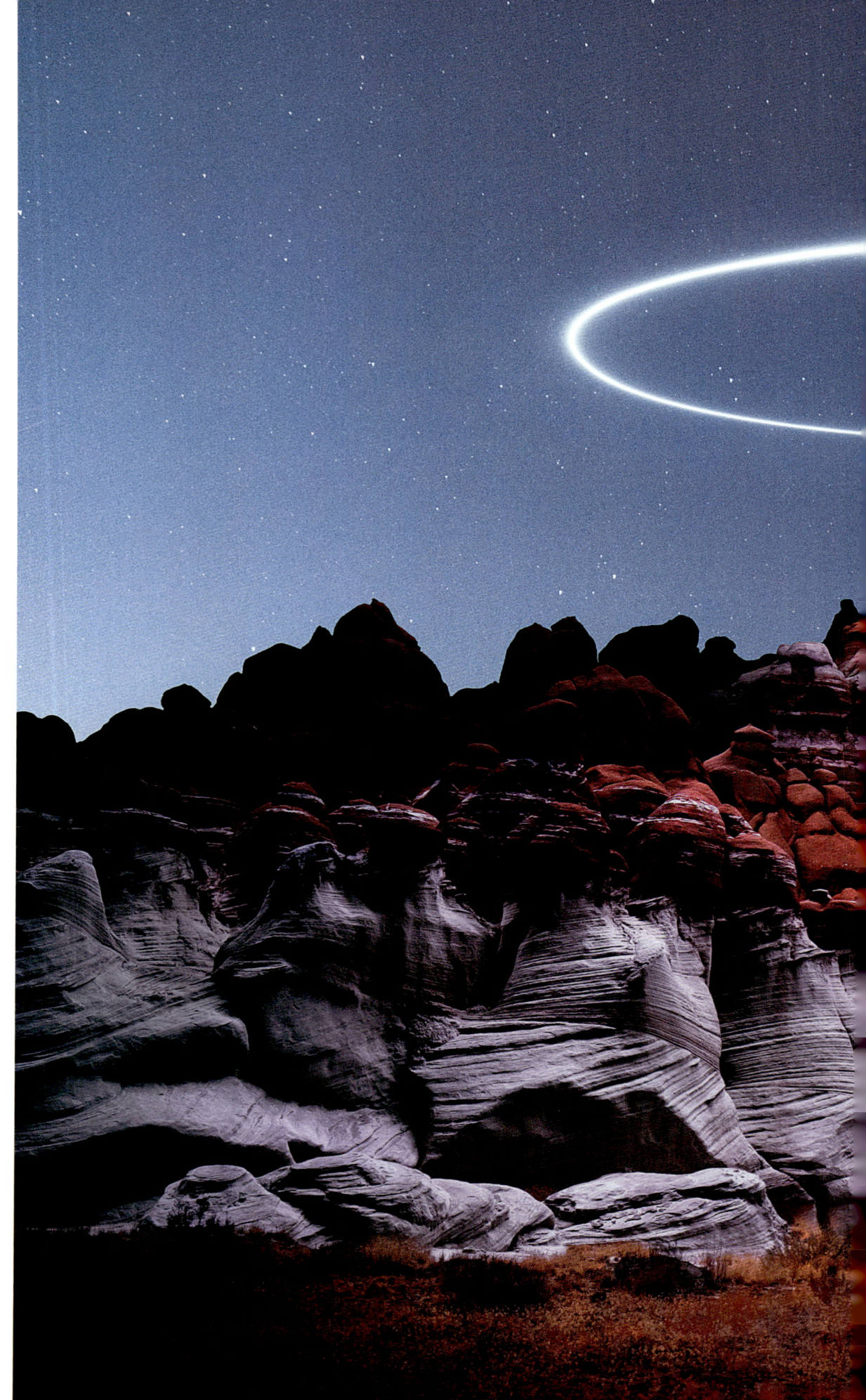

Reuben Wu
A drone carrying a light paints
a halo above a mountain peak
in the Arizona wilderness.

FOLLOWING PAGES:
NASA
The Sombrero galaxy,
28 million light-years from
Earth, represented in a mosaic
of six images captured by
the Hubble Space Telescope

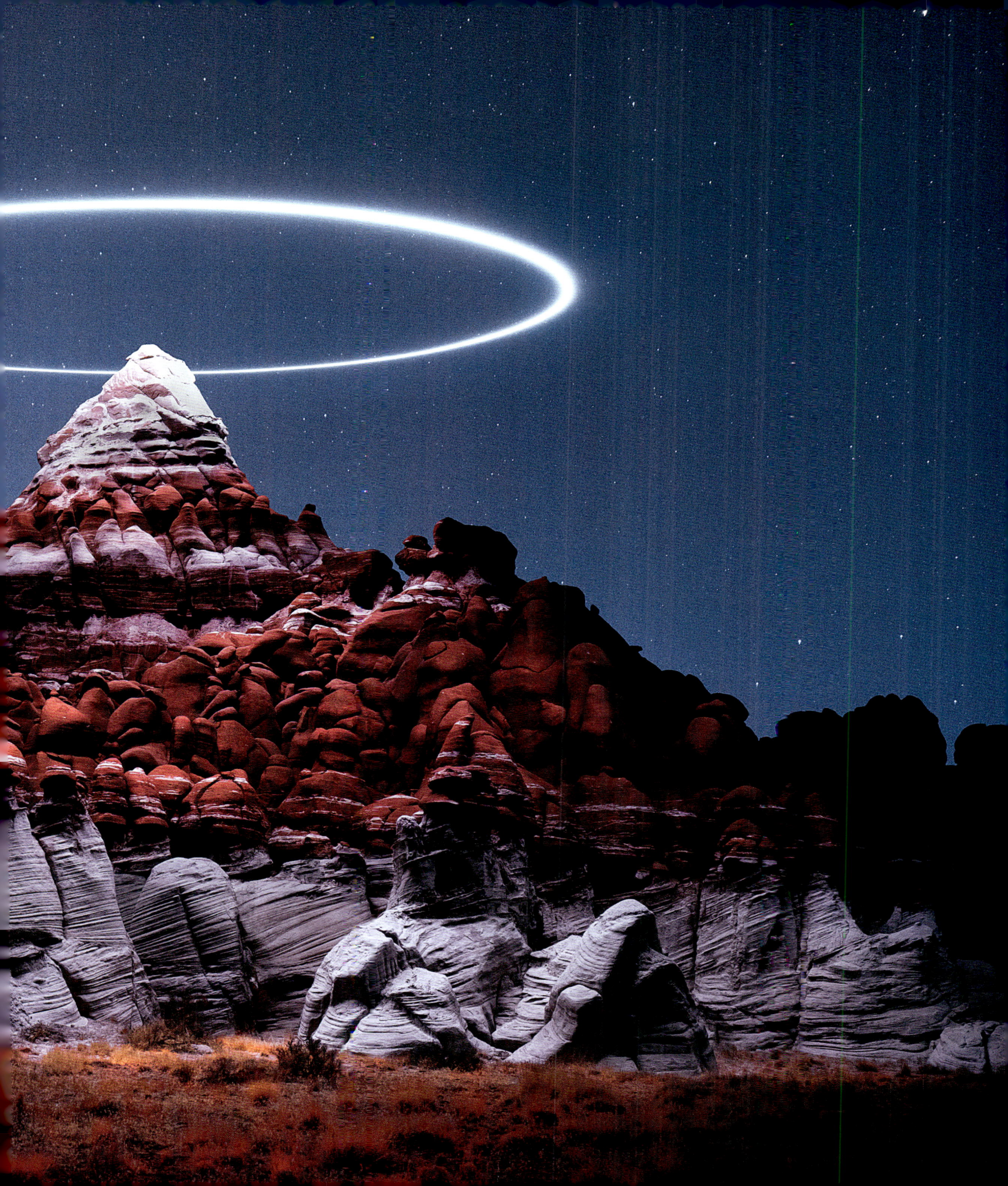

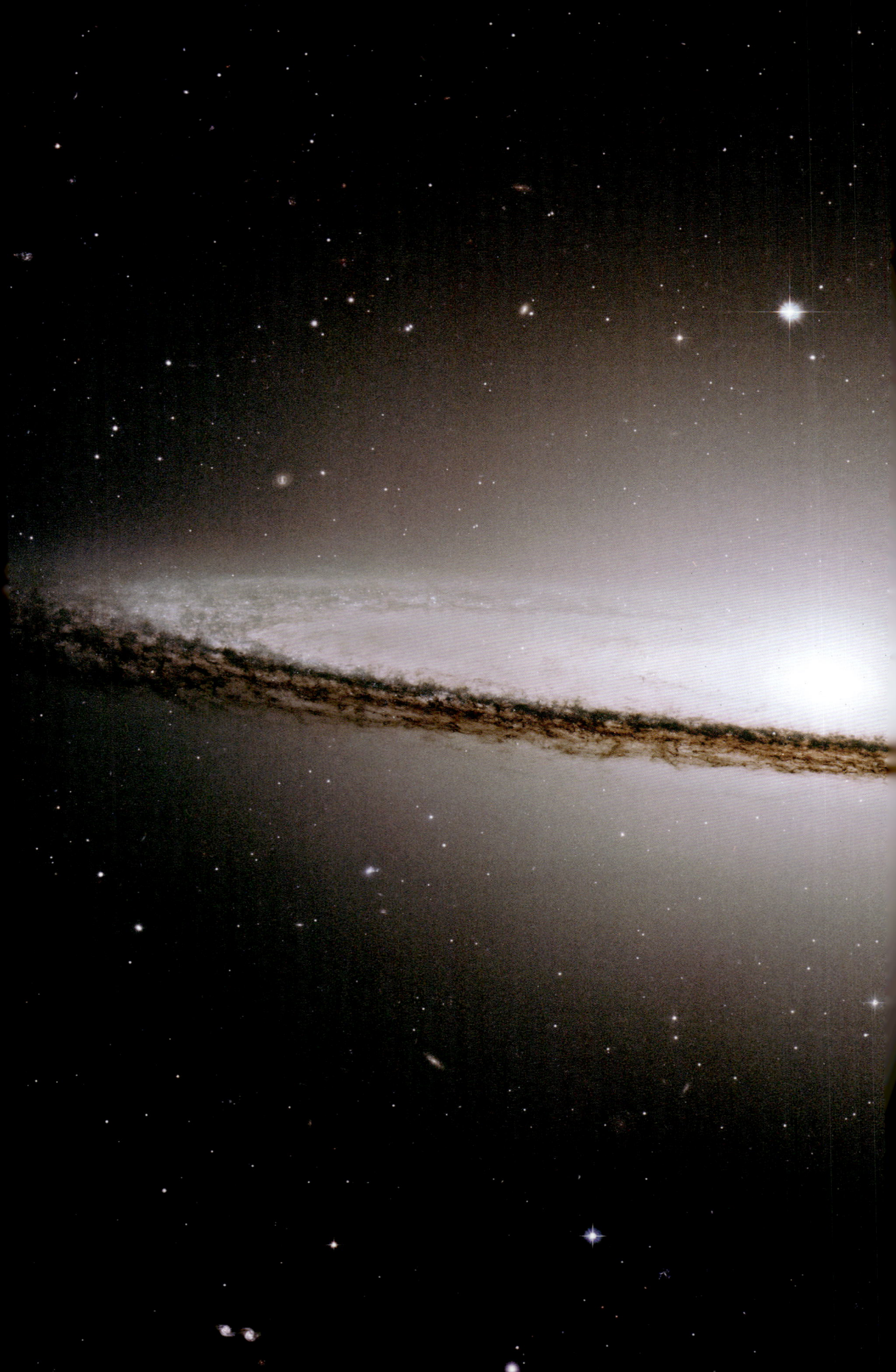

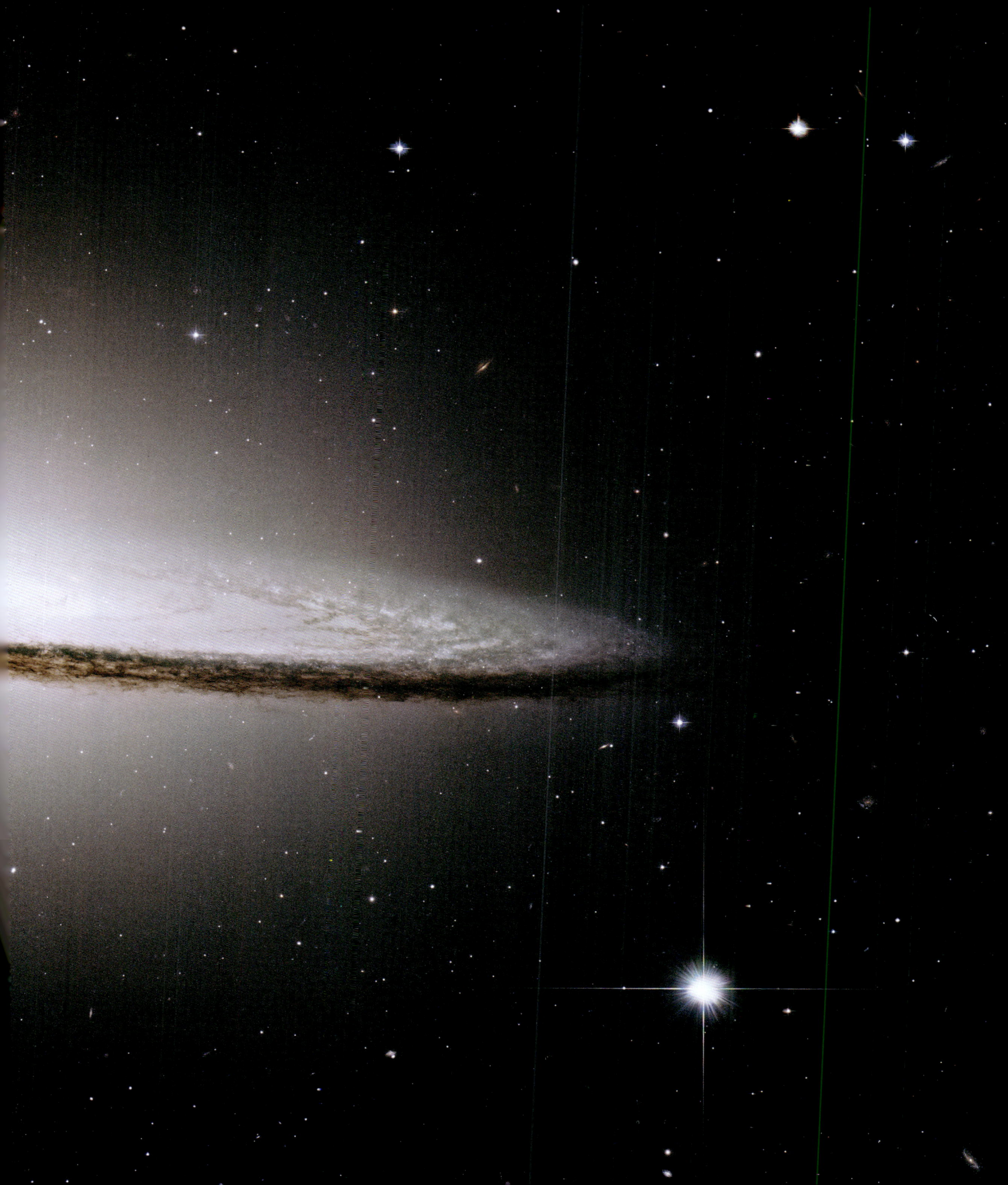

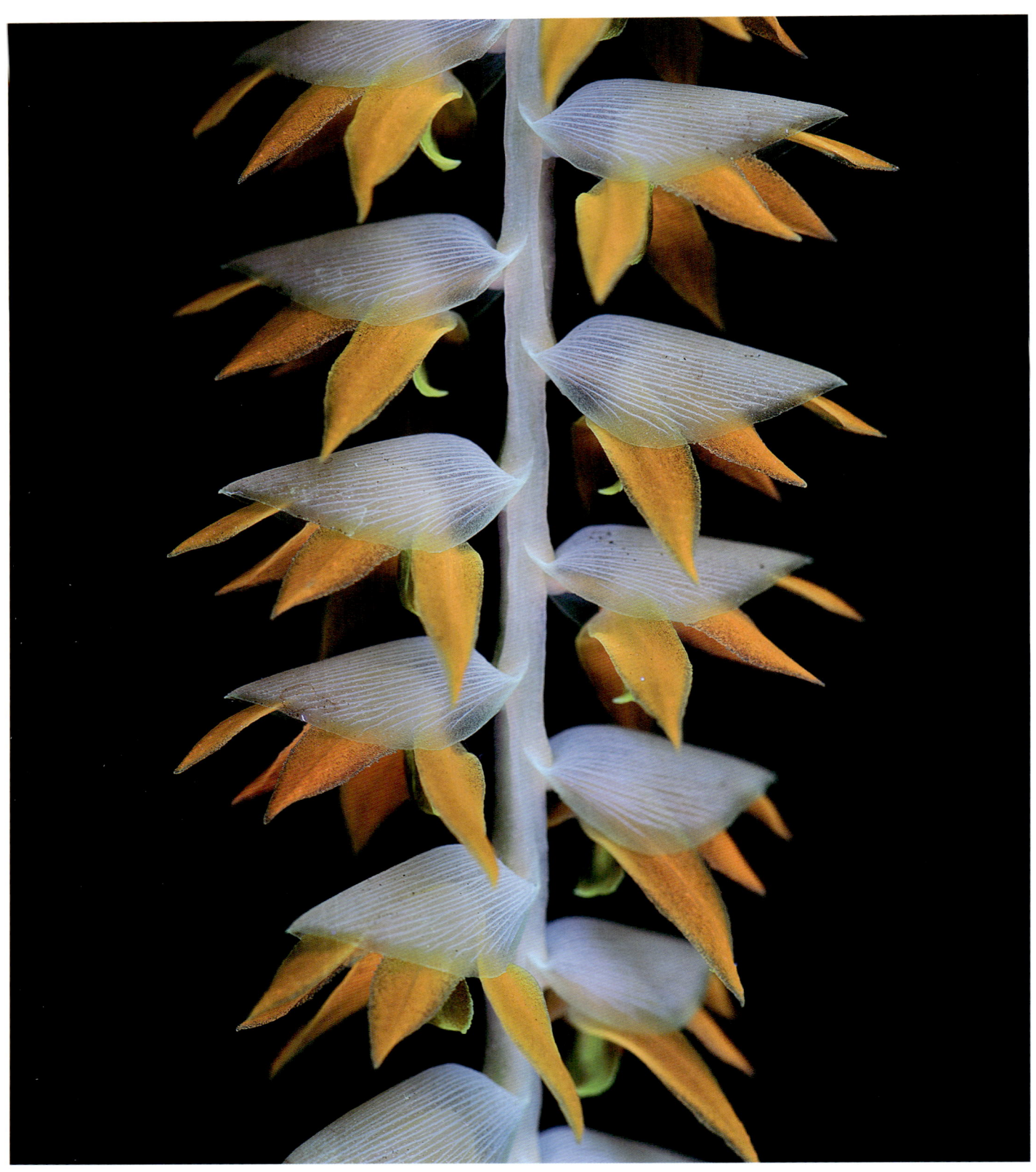

Craig P. Burrows
Hidden details of this orchid species appear
when illuminated with ultraviolet light.

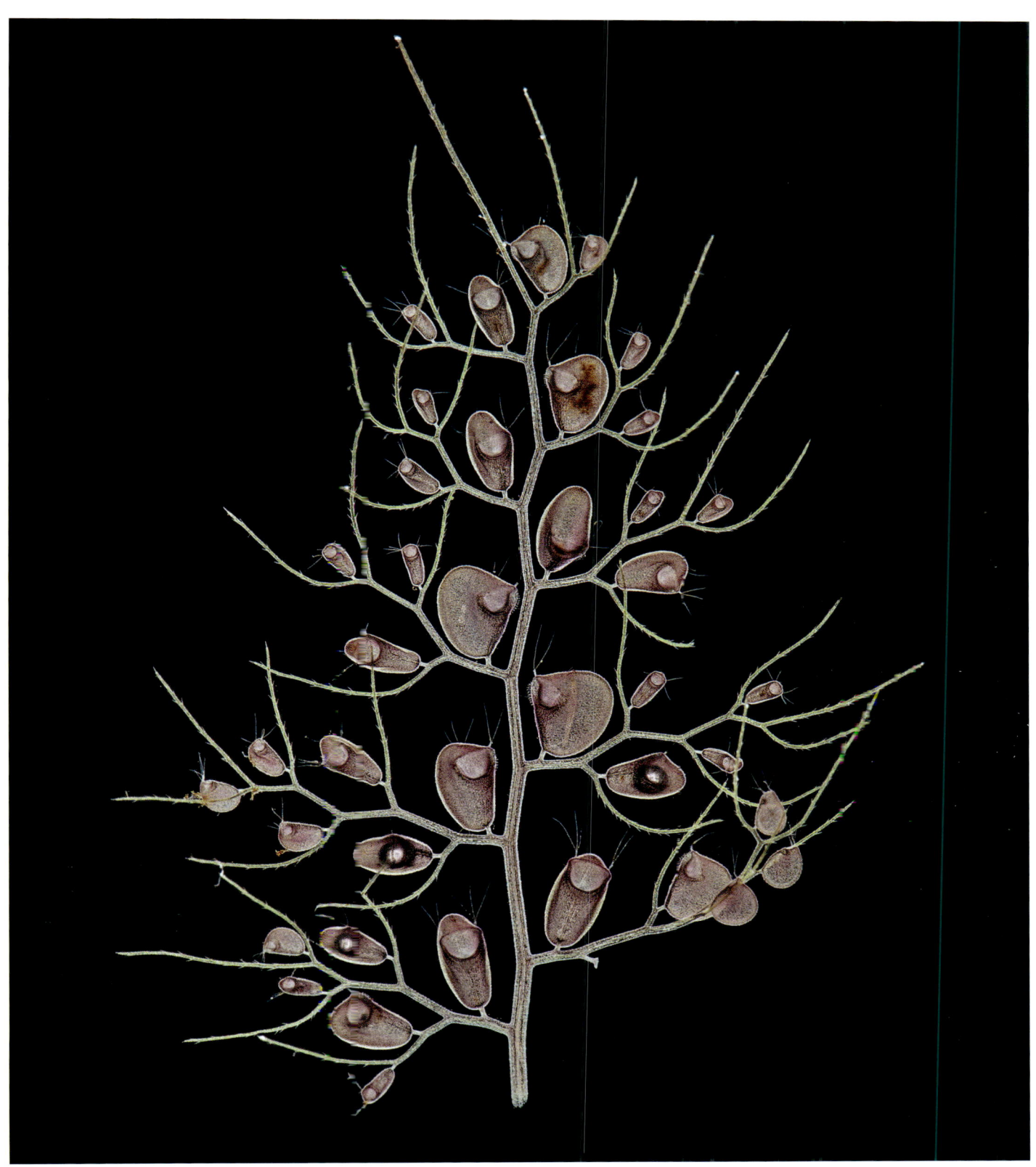

Anand Varma
A carnivorous aquatic plant called a bladderwort,
found in the Okavango Delta of Botswana

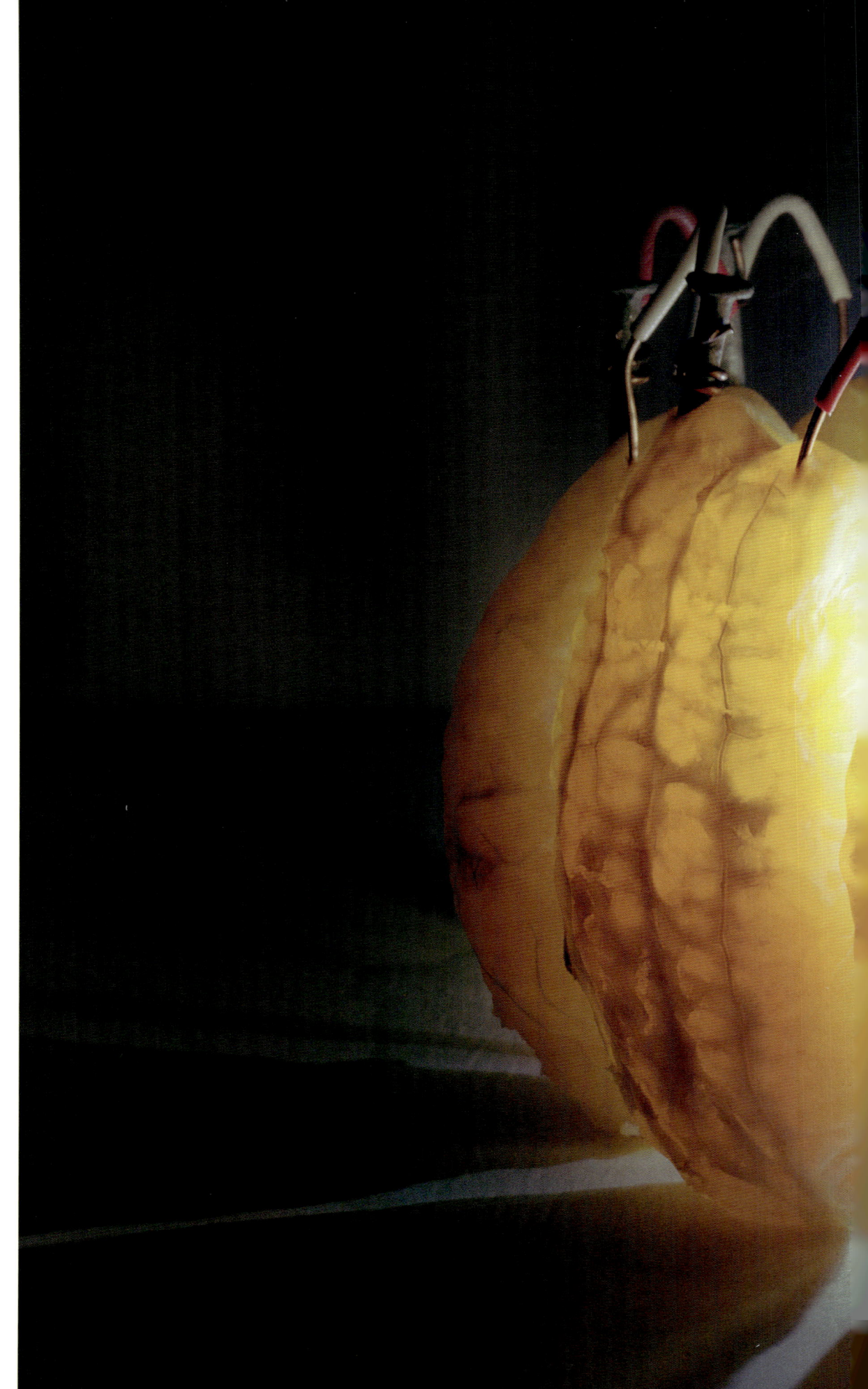

Caleb Charland
Nai s inserted into this orange react with citric acid to generate electricity. Copper wires transmit this power to an LED, which illuminates the orange from inside.

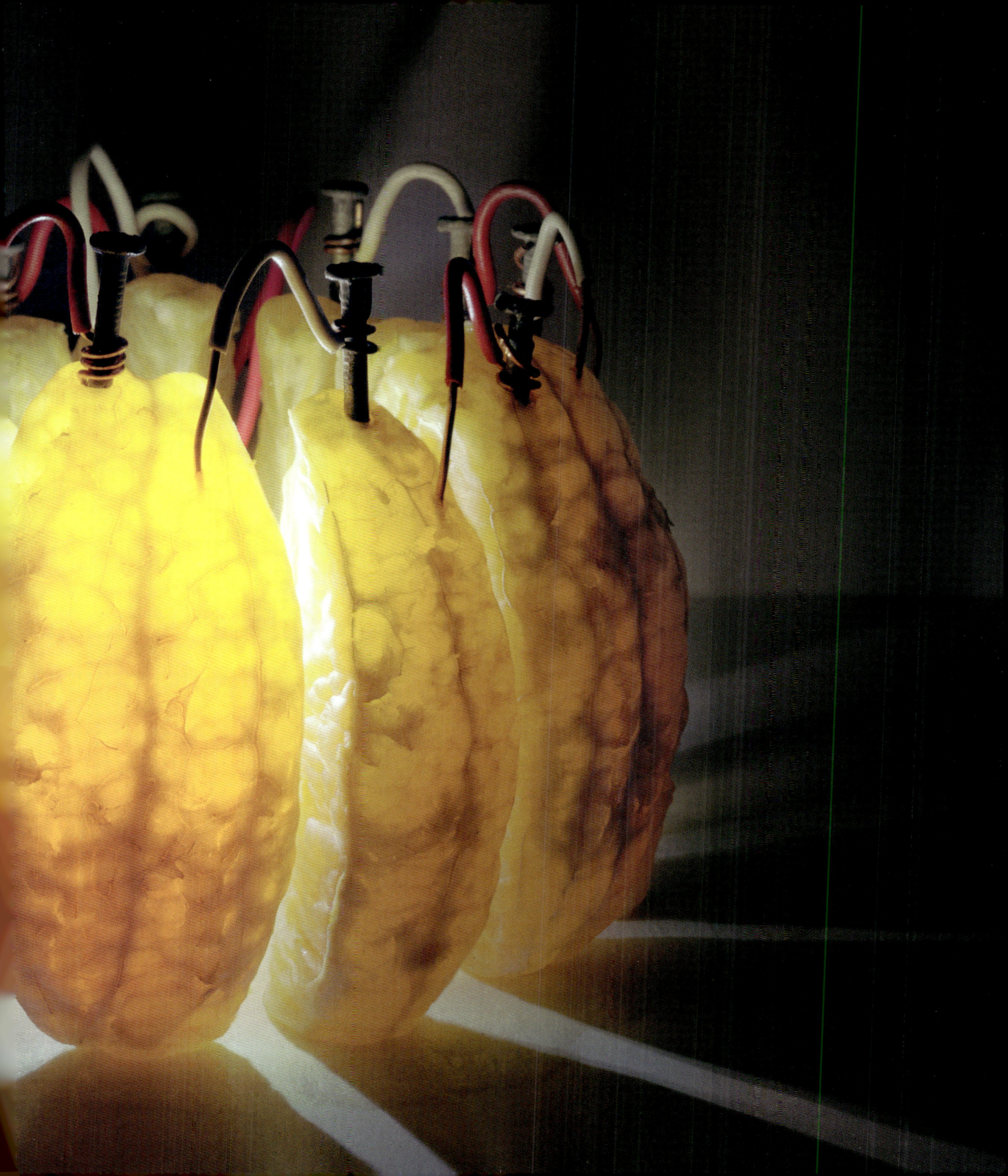

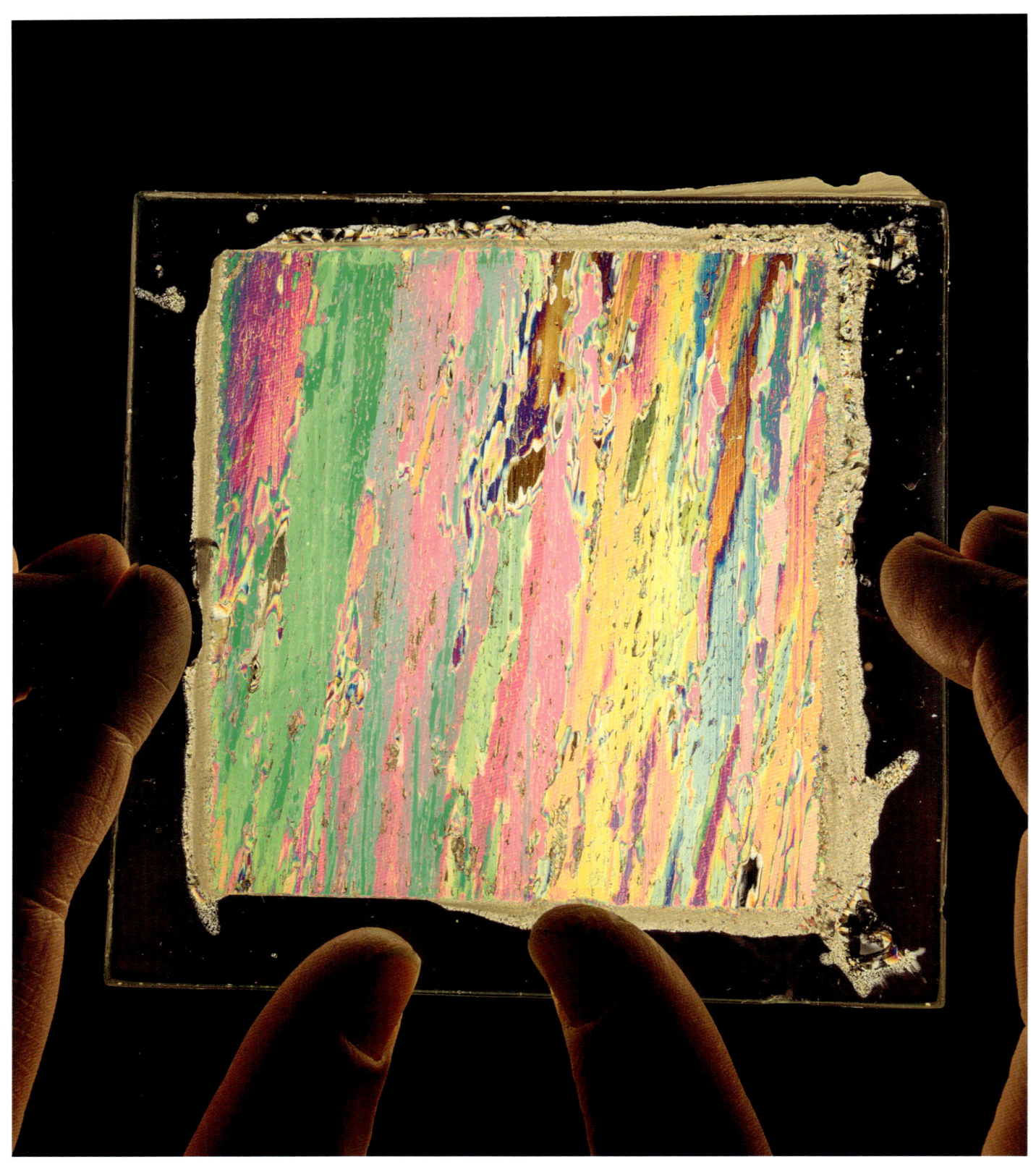

Nick Cobbing
This thin cross section from an ice floe reveals its complex crystalline
formations when photographed with polarized light.

EVERY PHOTOGRAPH IS A COLLABORATION WITH LIGHT. THESE IMAGES SHOW HOW PHOTOGRAPHERS MANIPULATE LIGHT IN SPECIFIC WAYS TO REVEAL HIDDEN PATTERNS.

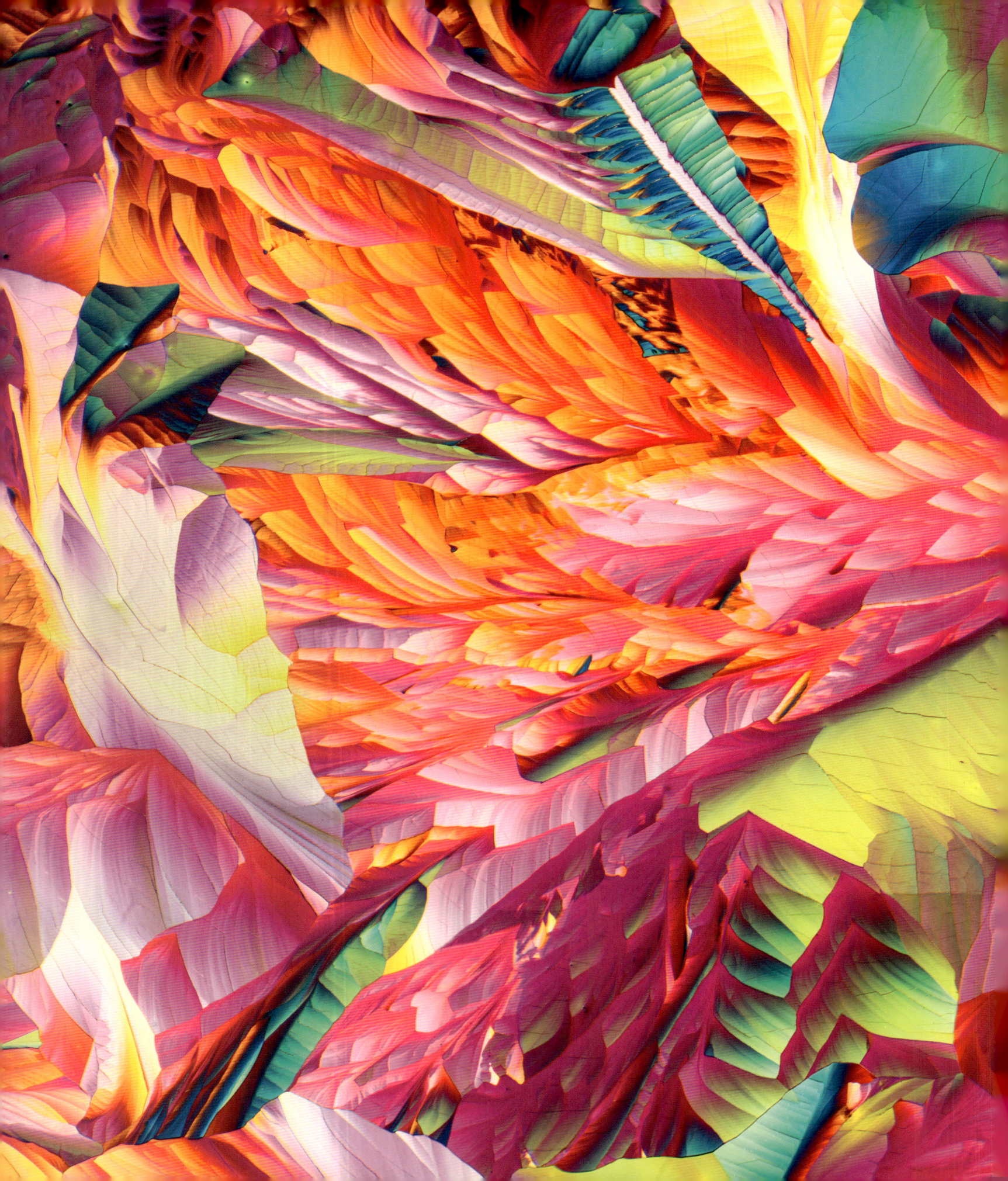

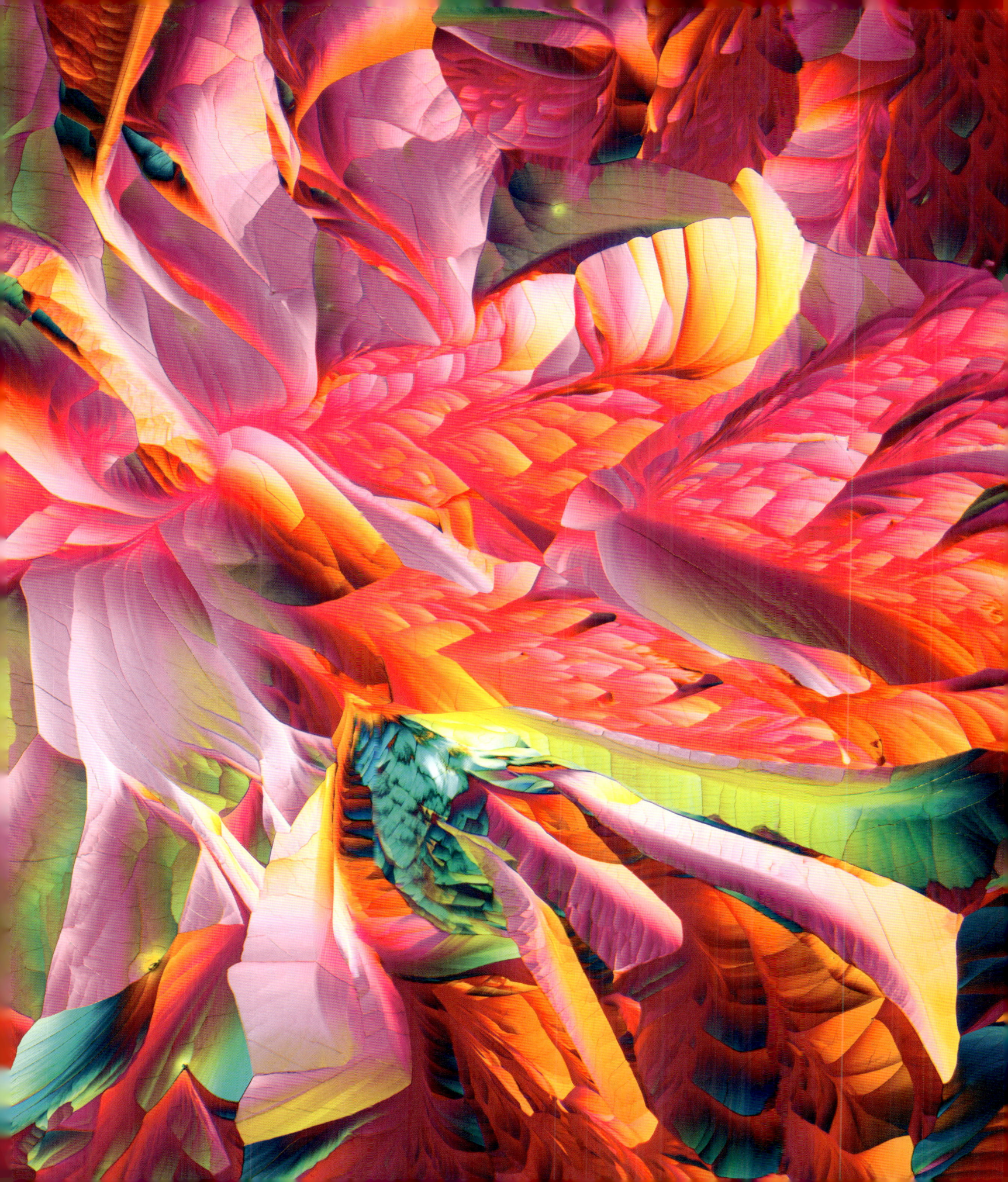

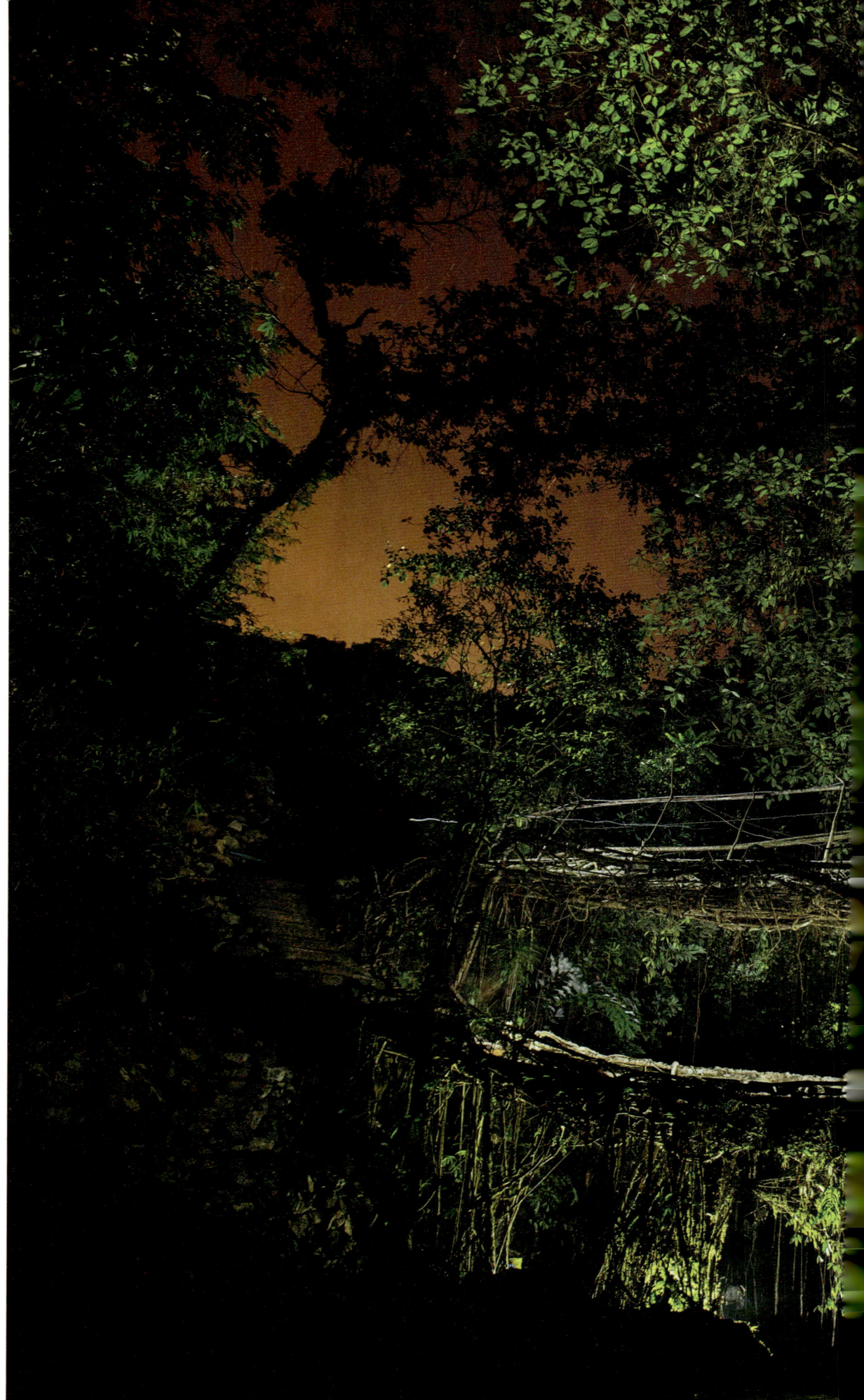

Prasenjeet Yadav
To illuminate this living root bridge, the photographer wore a headlamp and walked each span of the structure while his camera recorded a long exposure.

PREVIOUS PAGES:
Justin Zoll
Kaleidoscopic crystals of l-glutamine and beta-alanine amino acids revealed with polarized light

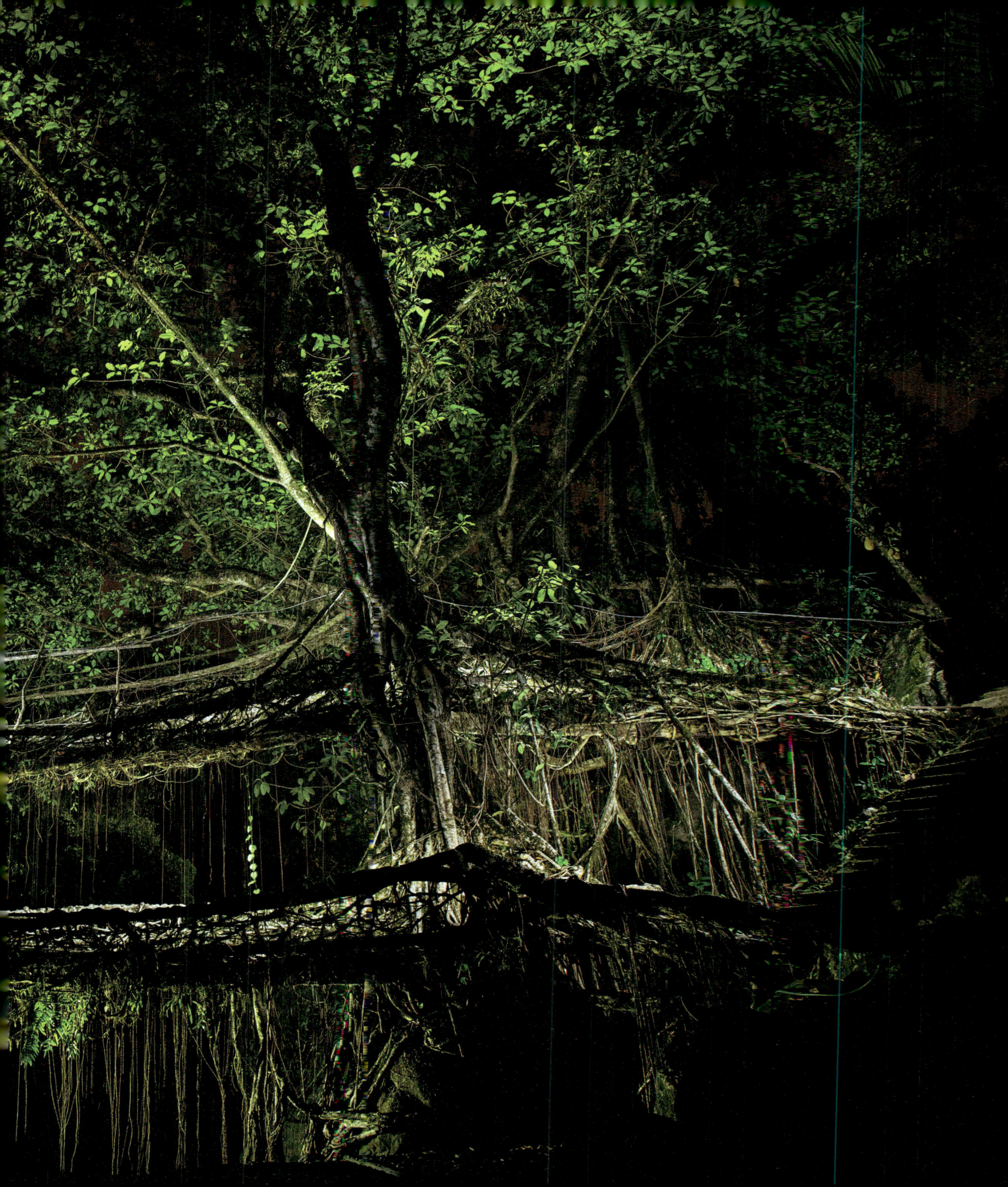

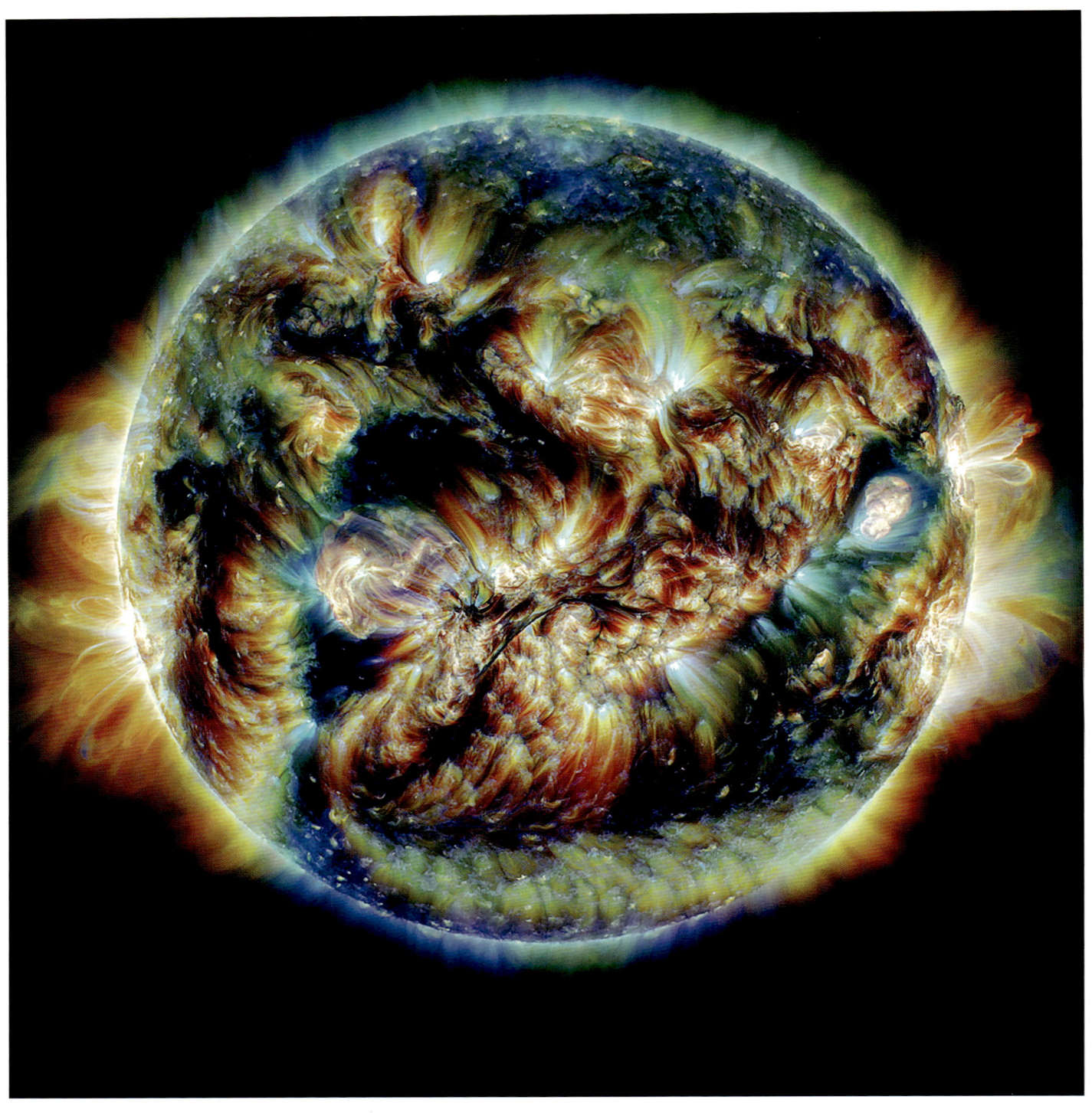

Hassan Hatami
This portrait of the sun, taken by NASA's Solar Dynamics Observatory, was created by combining thousands of images at multiple wavelengths of visible and ultraviolet light, allowing us to see the elaborate dynamics of the sun's atmosphere.

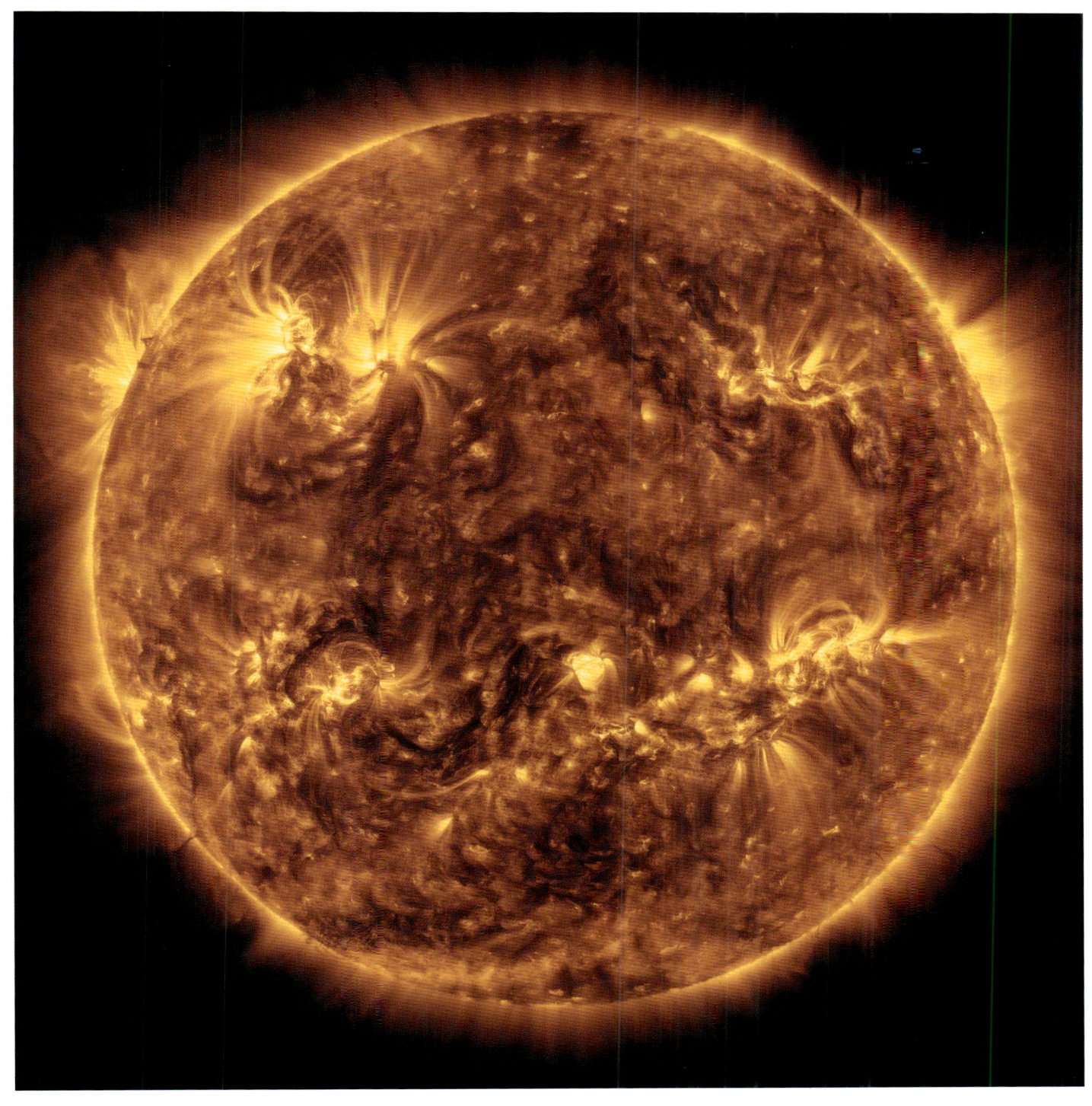

Emil Kraaikamp/ESA and NASA
A specialized suite of telescopes, the Extreme Ultraviolet Imager,
captured this composite image of the sun.

INFLUENCES
DALLAS NAGATA WHITE

While hiking in the rain along an active lava flow near Kalapana, Hawaii, Dallas Nagata White decided to try out an idea. A fashion photographer by trade, she set up a tripod to take a self-portrait with her husband, Ed. A typical portrait photographer would have illuminated the scene from the front, but instead she positioned a flash directly behind them. Ed spontaneously dipped her for a kiss, captured in the photograph you see here. The clever placement of light illuminates the falling rain while drawing your attention to the key action of the scene: the couple kissing.

I discovered this image while browsing the contest web page for *National Geographic Traveler* magazine back in 2012. It grabbed my attention, and I immediately realized that I could use Dallas's lighting technique for the project I was working on at the time—photographing parasites.

I had been given very firm constraints for this project. "You get one picture to tell the story of each parasite," my editor, Todd James, told me, clearly leaving no room for negotiation. This created an enormous challenge for one subject I wanted to photograph in particular, a parasitic barnacle called a rhizocephalan that attacks sheep crabs.

Dallas Nagata White
The photographer and her husband on a rainy night, next to a lava flow near Kalapana, Hawaii

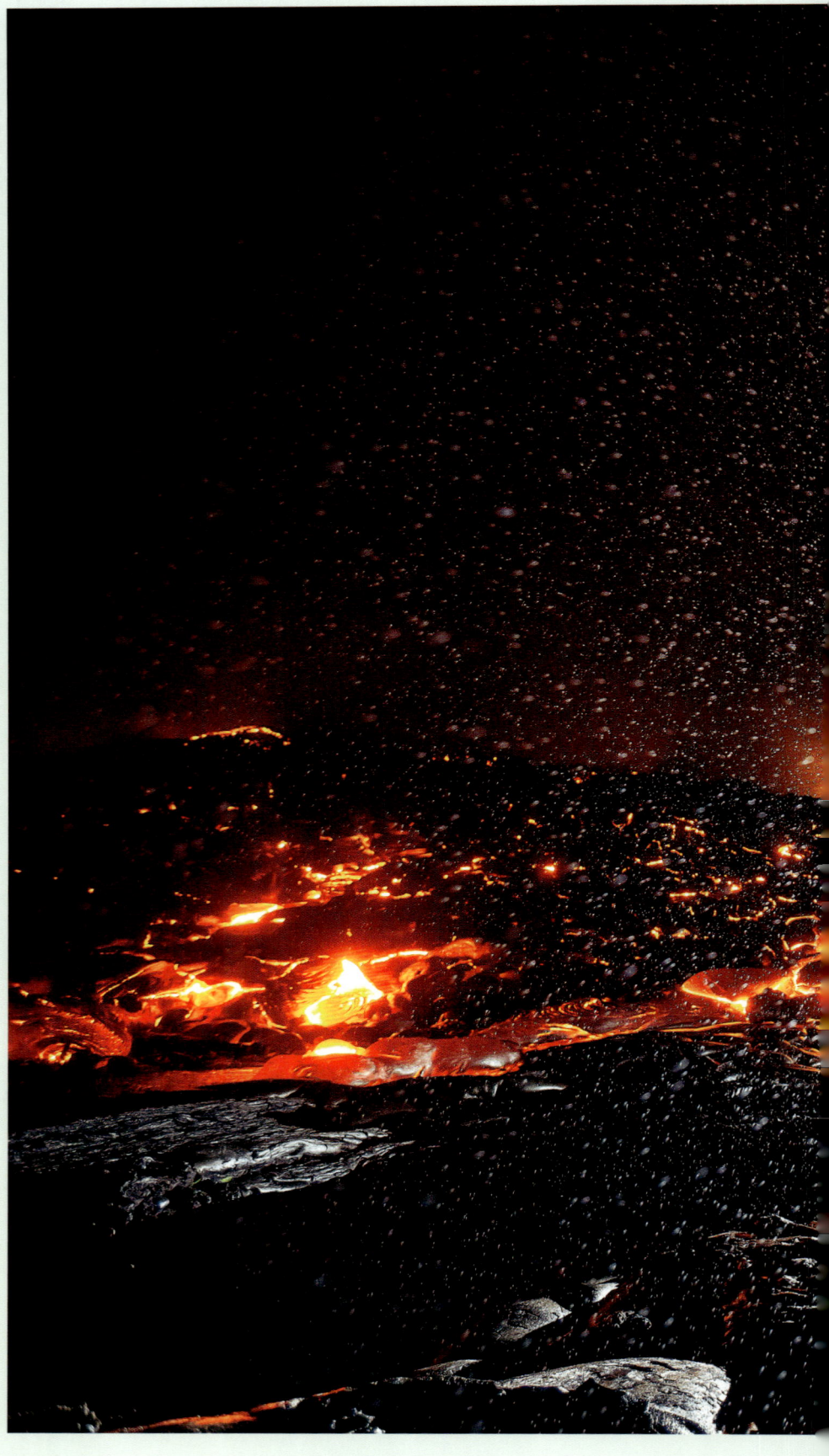

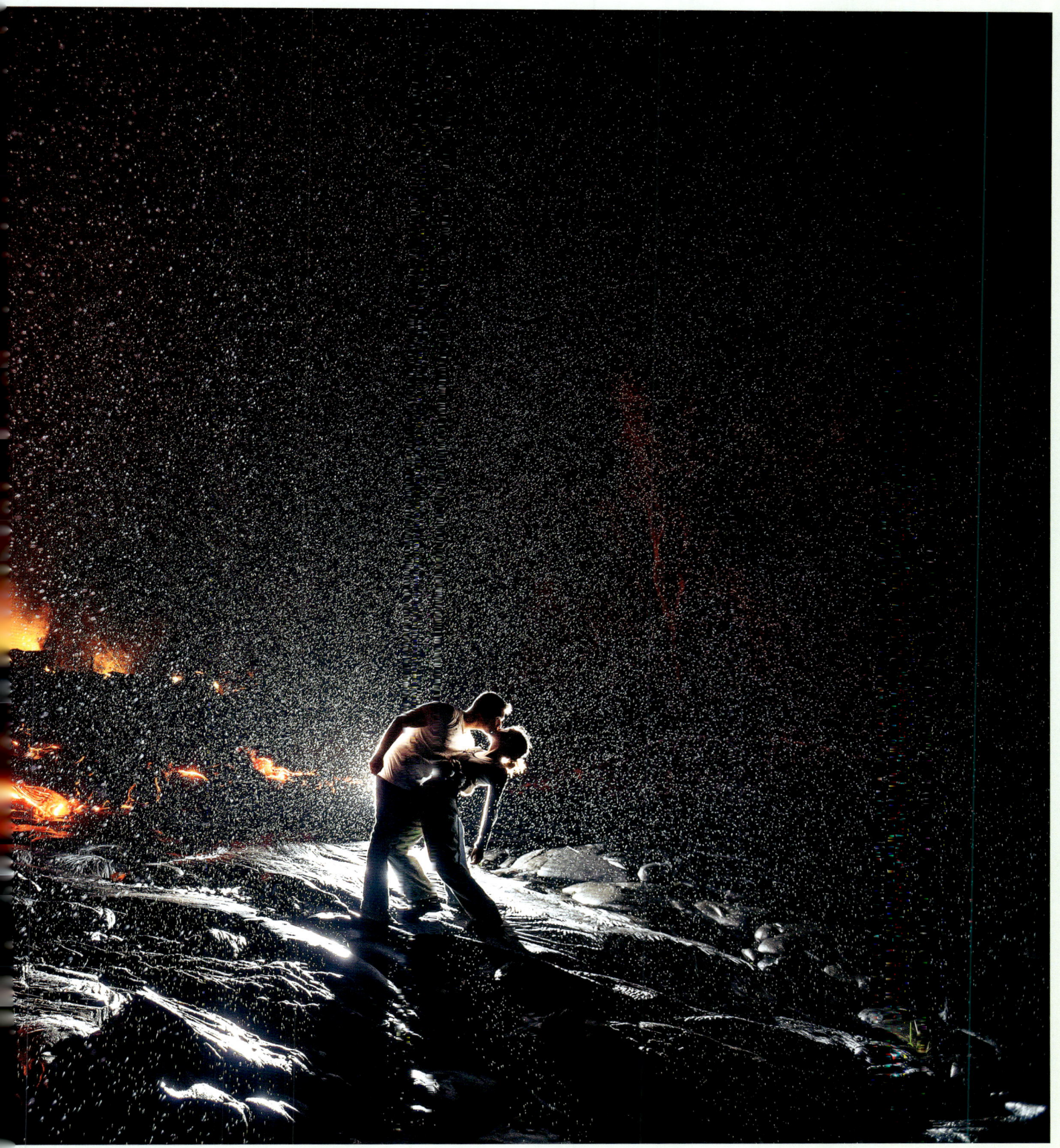

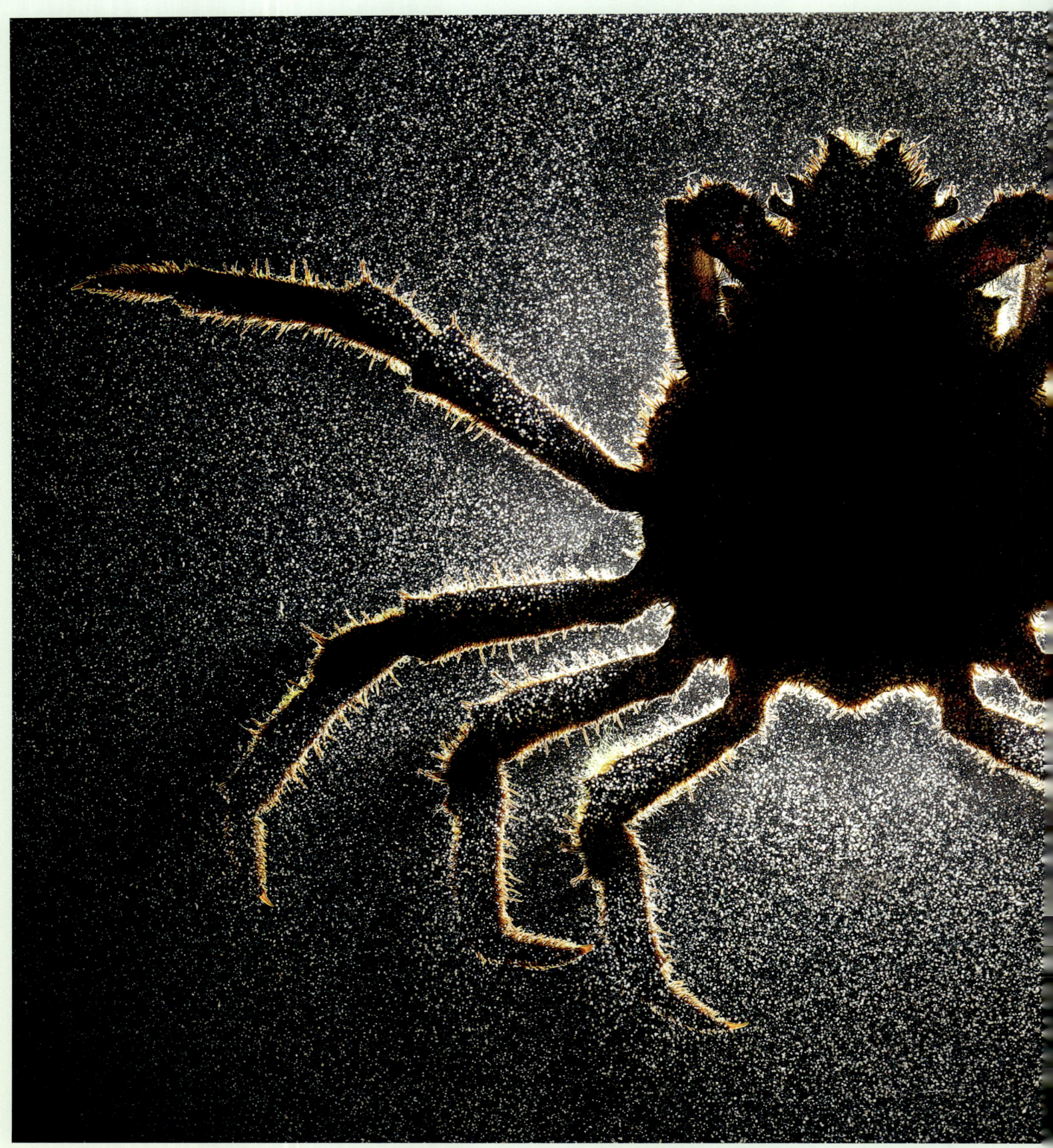

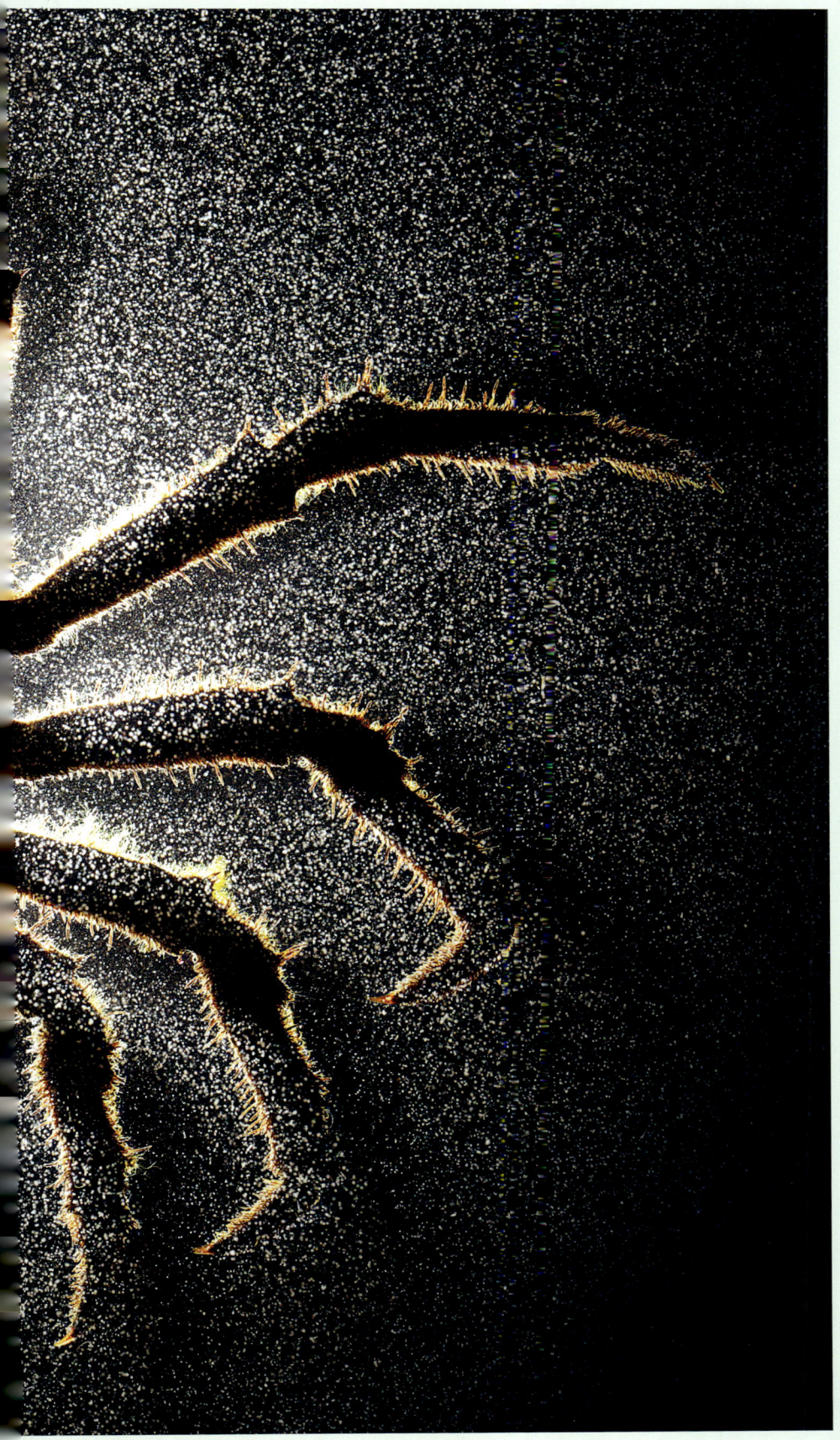

The remarkable thing about this parasite is what it does to the crab it infects. If the crab is male, the parasite forces it to become a female. The abdomen of the feminized crab widens, providing a "womb" for the barnacle to fill with its own eggs. Nurtured by the crab, whose maternal care instincts have been activated by the parasite, a new batch of eggs hatches every few weeks. With each hatching, tens of thousands of baby barnacles disperse to hunt for new crabs to infect.

The problem for me in photographing this process was that the parasite is thousands of times smaller than the host it infects. I was stumped: How could I juxtapose the two creatures in a single image? Dallas's image gave me the solution. I drove down to the University of California, Santa Barbara where Armand Kuris ran a parasitology lab that studied these parasites. I put a custom aquarium to house an infected crab and waited for the parasites to emerge.

When they did, I placed a light behind the crab, inspired by how Dallas had placed the light behind herself on the volcano. Just as her light shone against the minuscule rain droplets, my light revealed the microscopic barnacle larvae in the aquarium, making visible what otherwise would have been hidden from view.

Anand Varma
Tens of thousands of parasitic barnacle larvae emerge from a sheep crab.

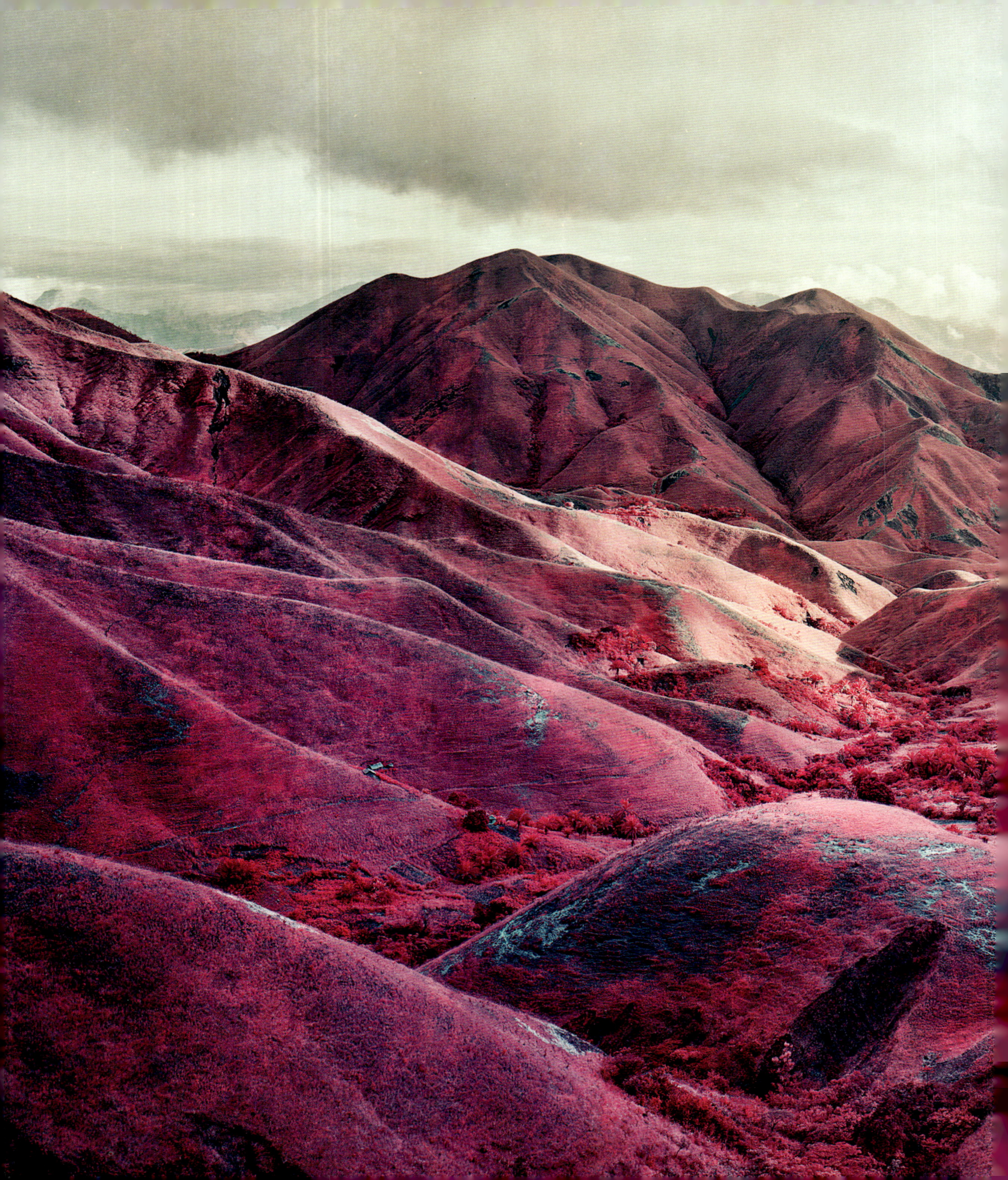

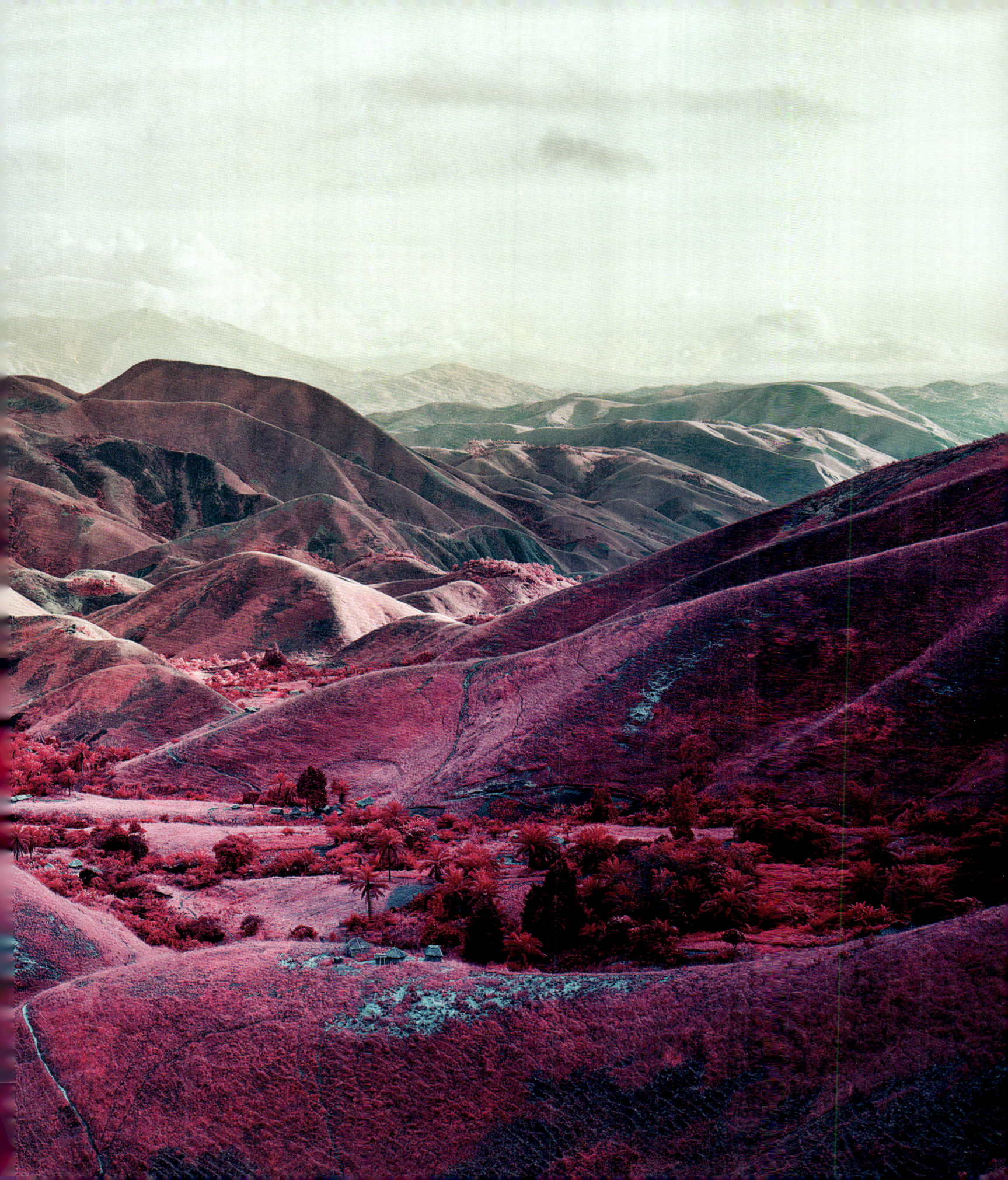

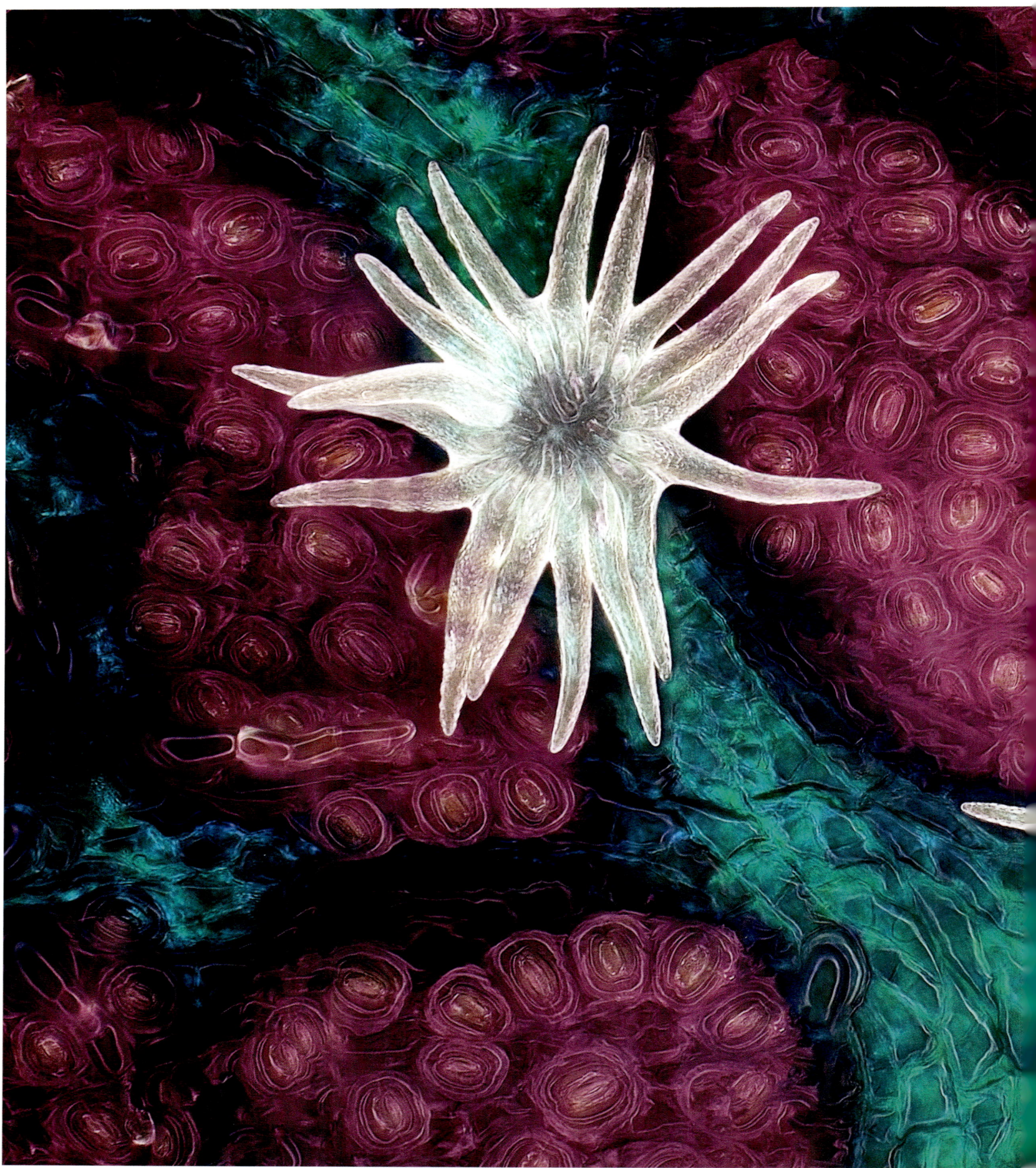

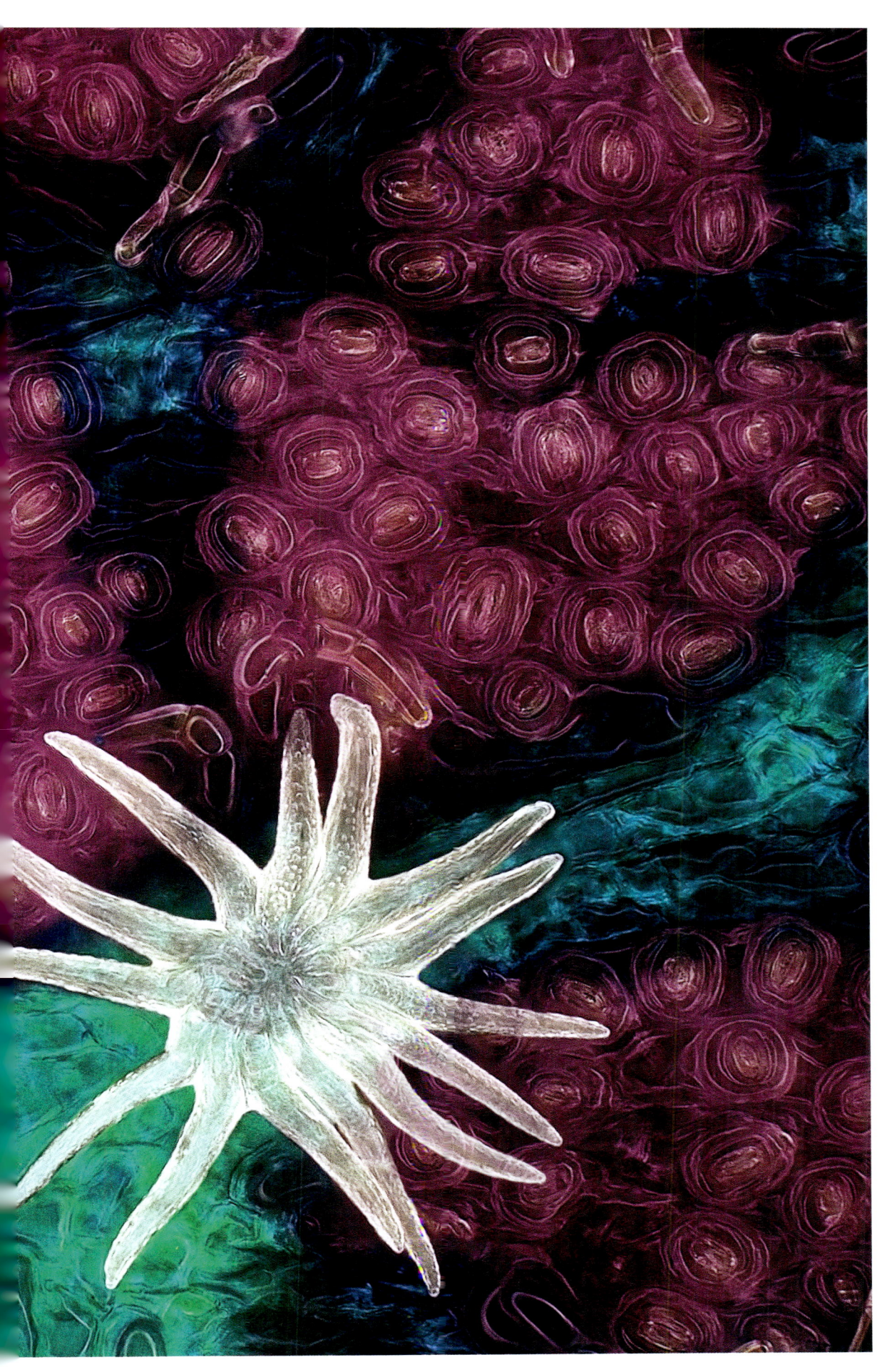

Jason Fik
By stacking roughly 200 magnified images of a southern live oak leaf, the photographer created a revealing portrait of the leaf's anatomy, including trichomes (white bristles) and stomata (purple pores).

PREVIOUS PAGES:
Richard Mosse
A camera sensitive to infrared light shows vegetation patterns within the folds of South Kivu's mountains in the Democratic Republic of the Congo.

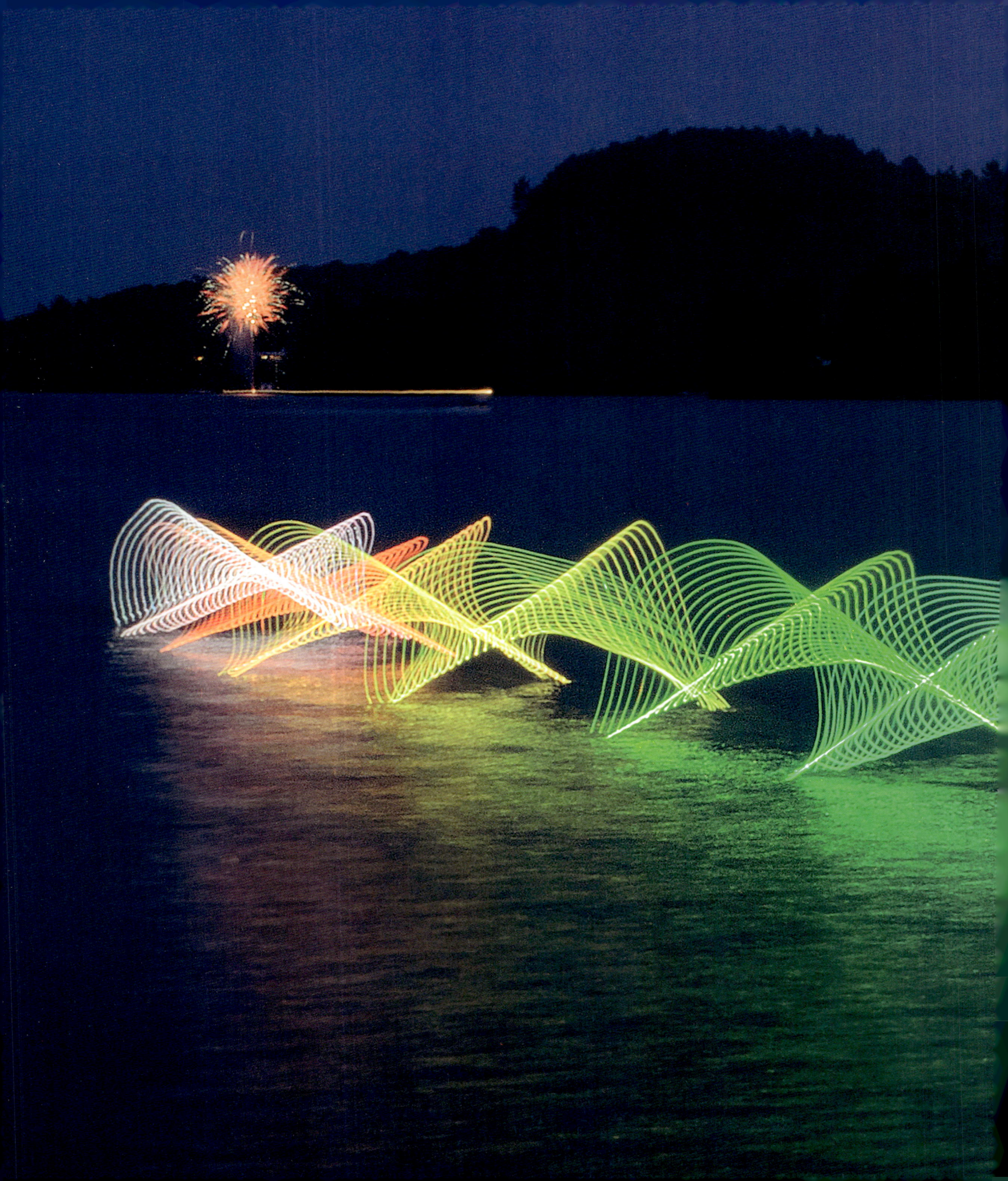

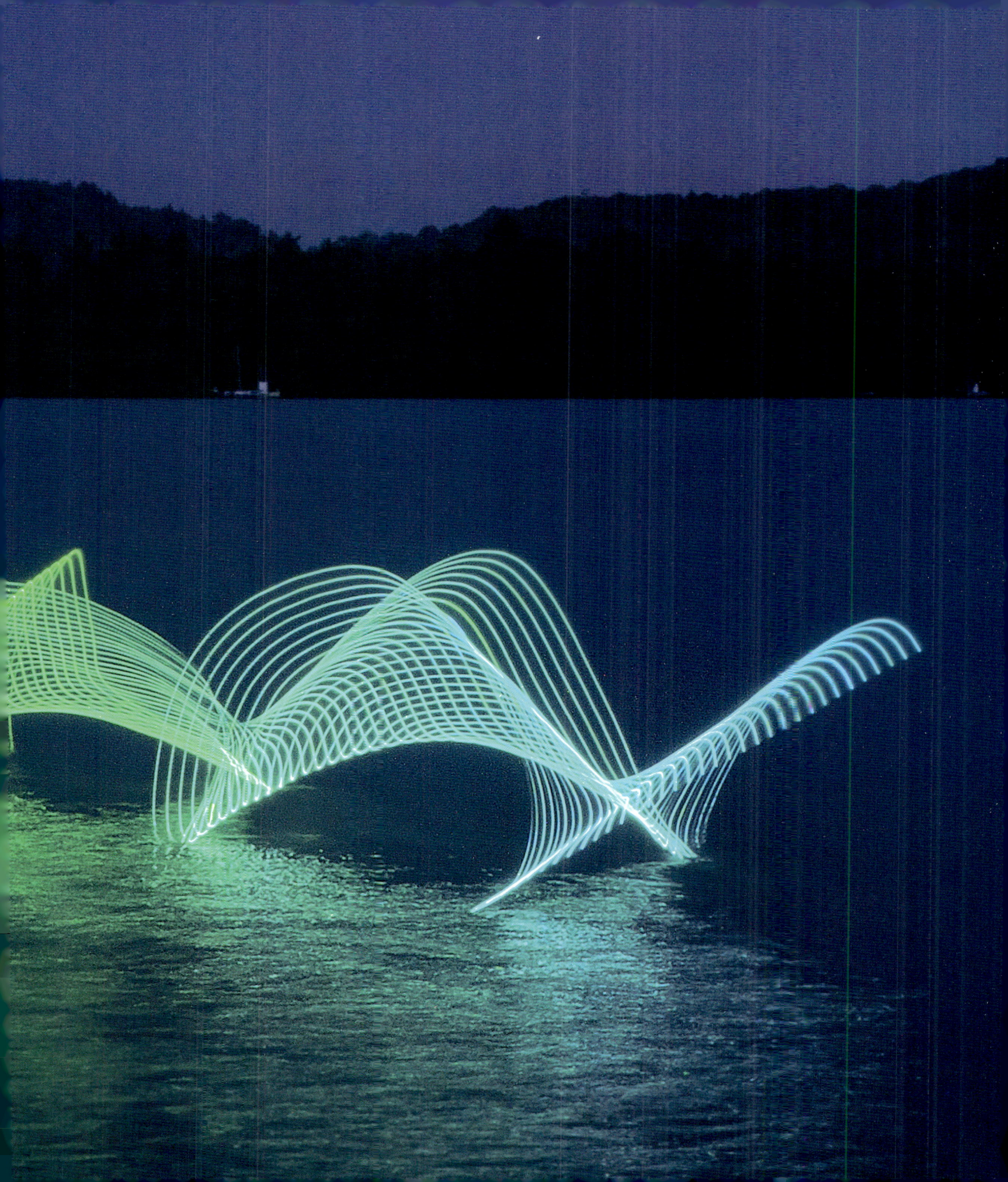

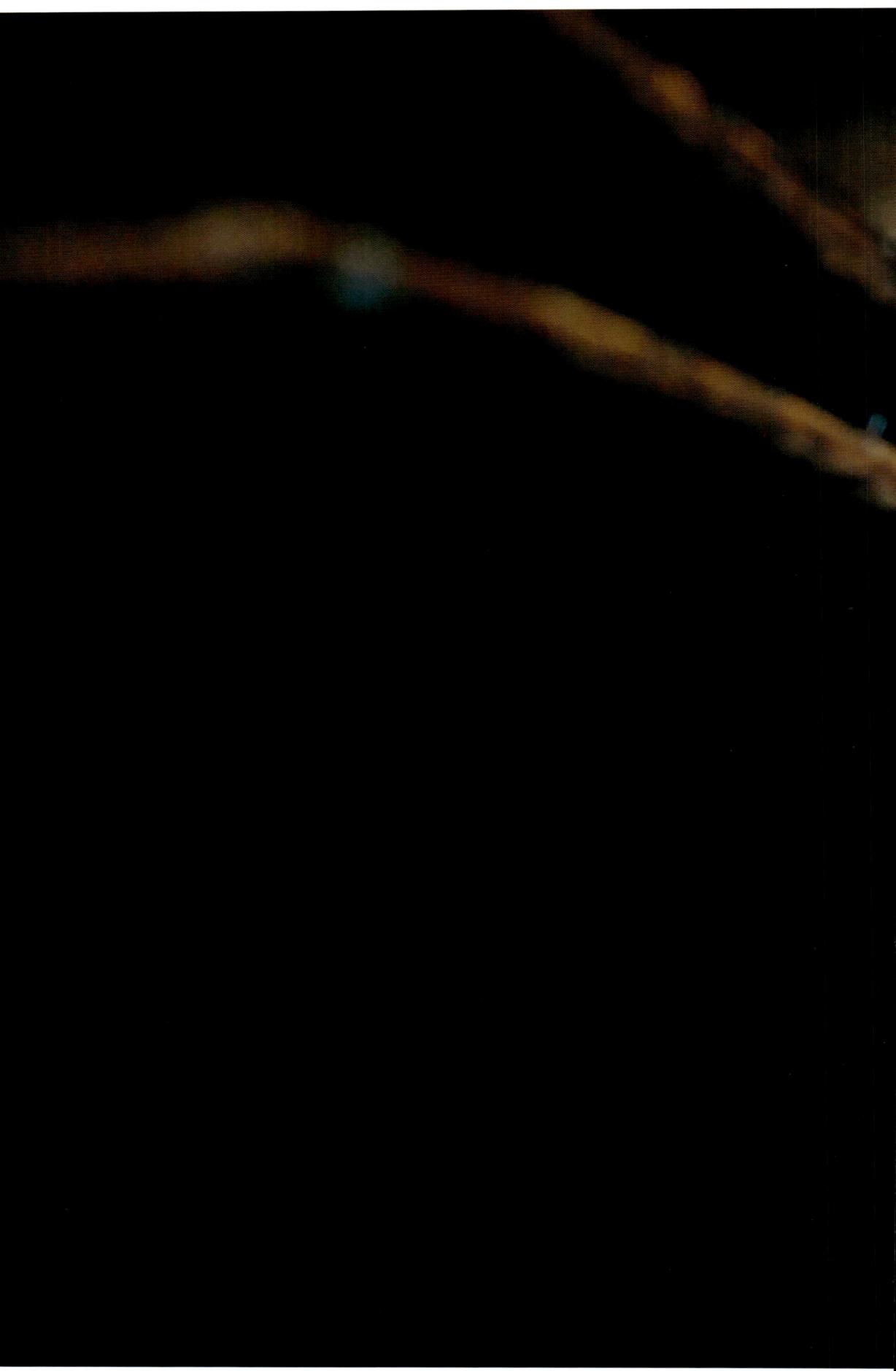

Nicky Bay
Ultraviolet light gives this arachnid's features an ominous glow appropriate to its common name: the ogre-faced spider.

PREVIOUS PAGES:
Stephen Orlando
Color-changing LEDs attached to a paddle trace the strokes of a kayaker in Ontario, Canada.

FOLLOWING PAGES:
Jim Obester
When illuminated by blue light, the clear tentacles of a tube-dwelling anemone fluoresce green.

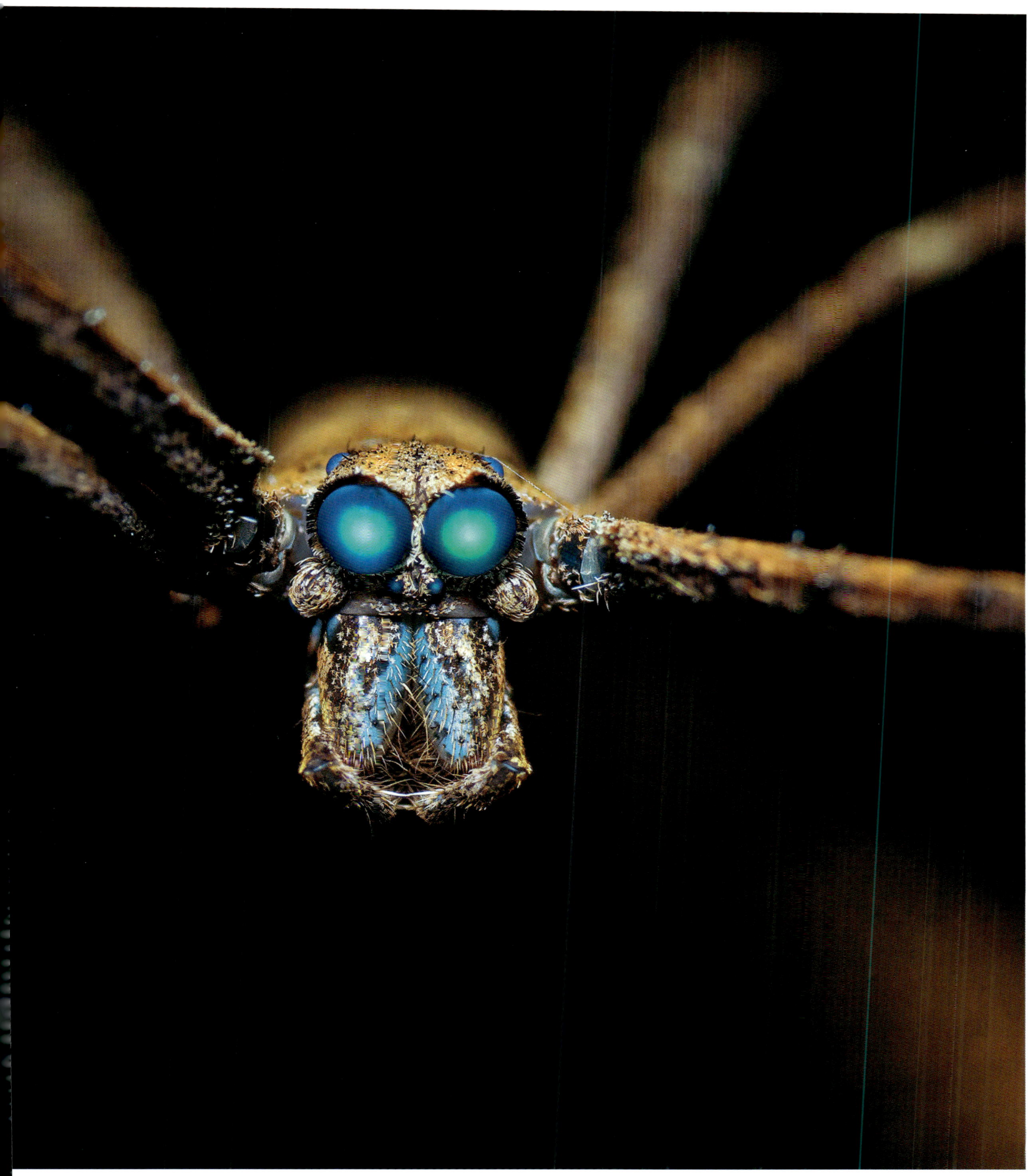

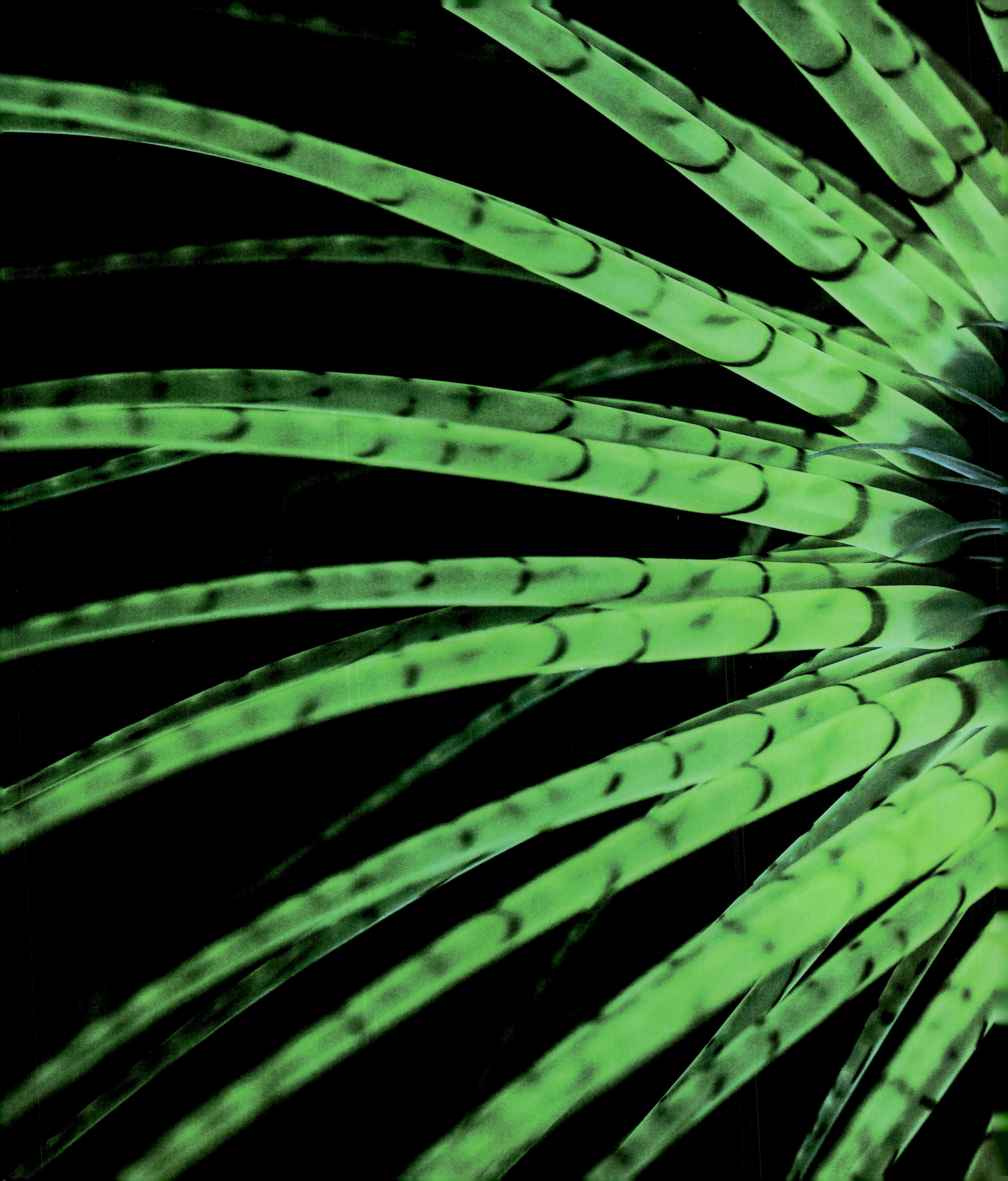

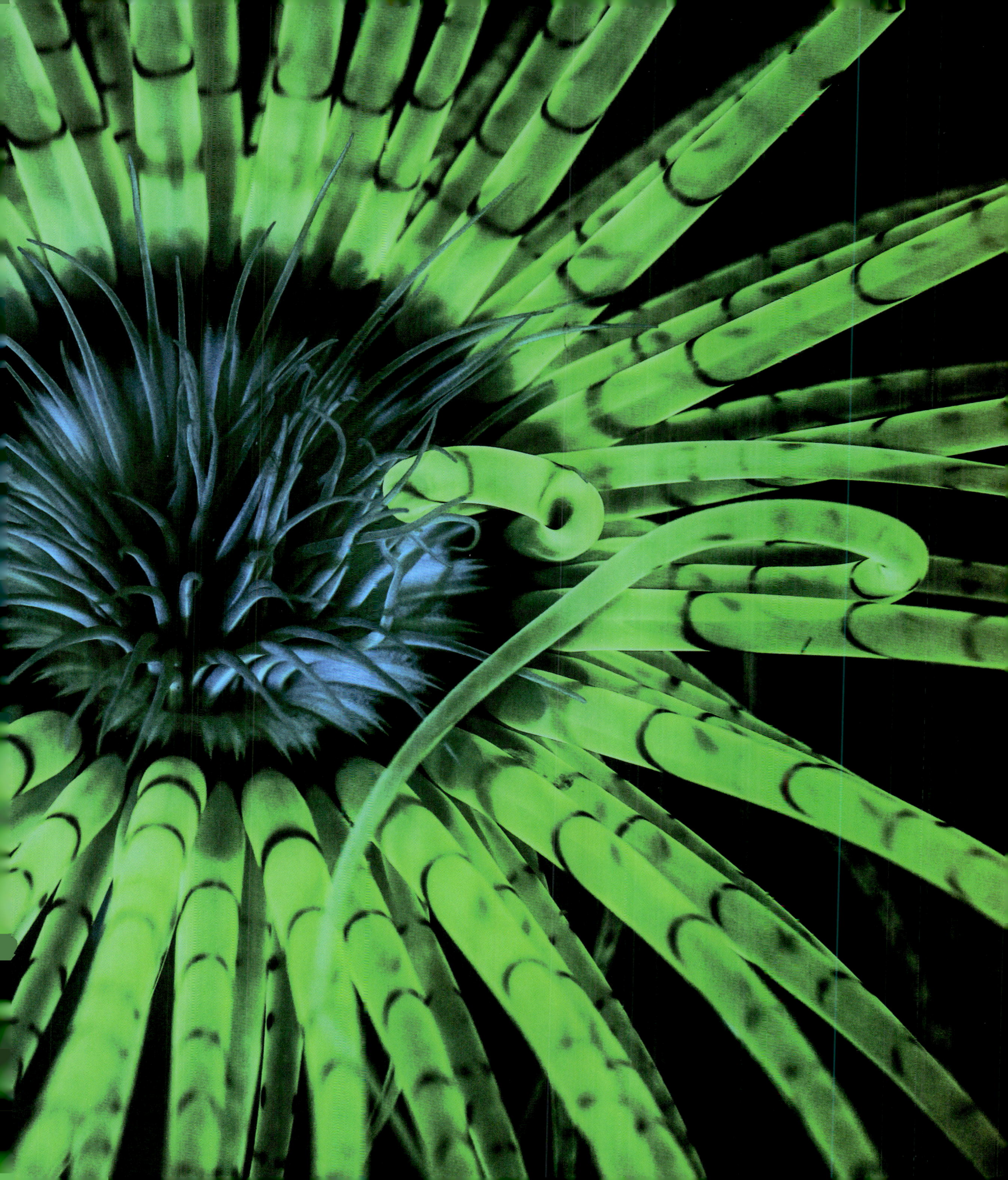

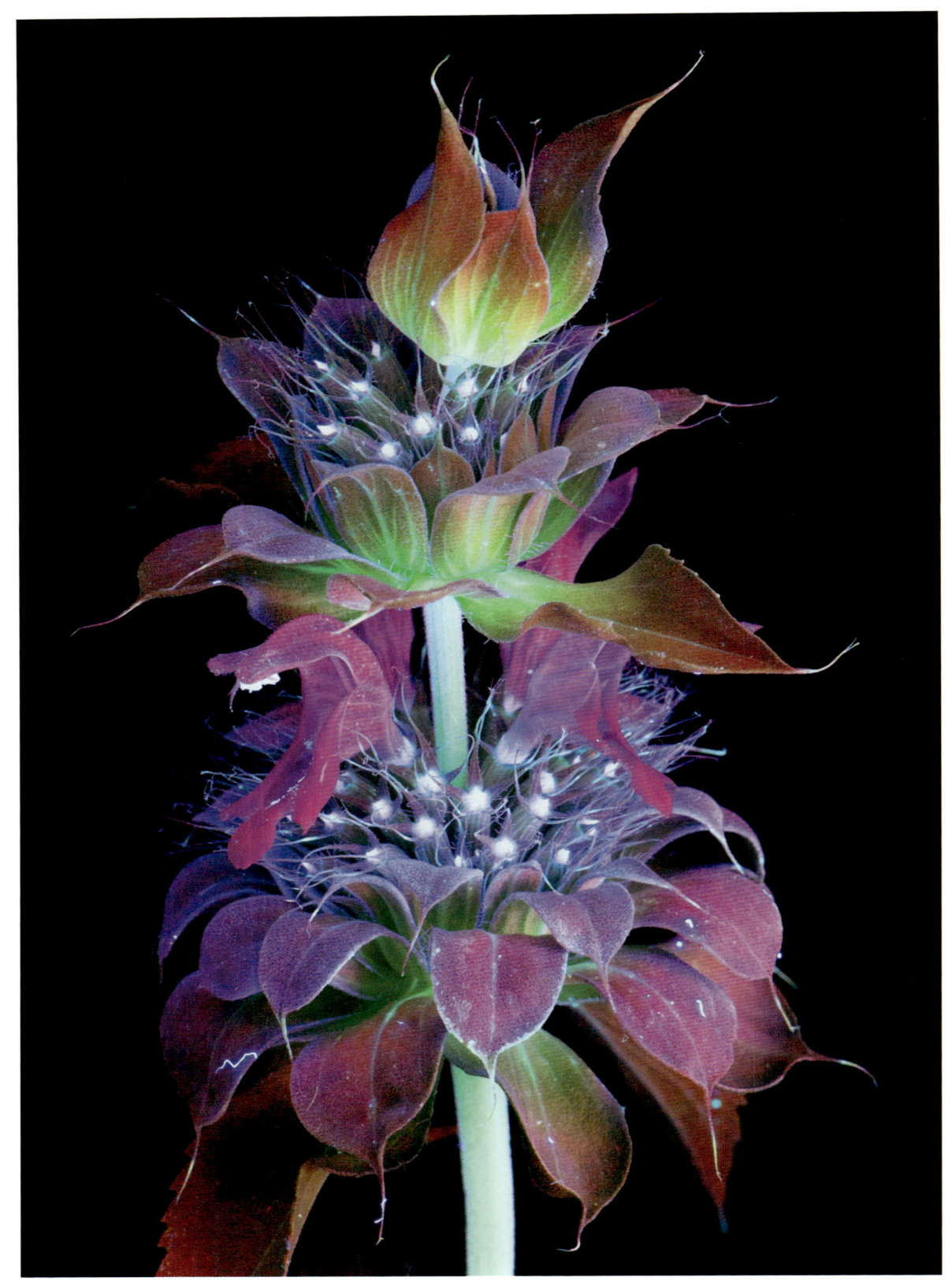

Craig P. Burrows
A bee balm flower becomes iridescent when bathed in ultraviolet light.

CAMERAS CAN DETECT COLORS
THAT ARE INVISIBLE TO THE HUMAN EYE,
AND PHOTOGRAPHERS USE THESE
CAPABILITIES TO BROADEN OUR SENSORY
EXPERIENCE.

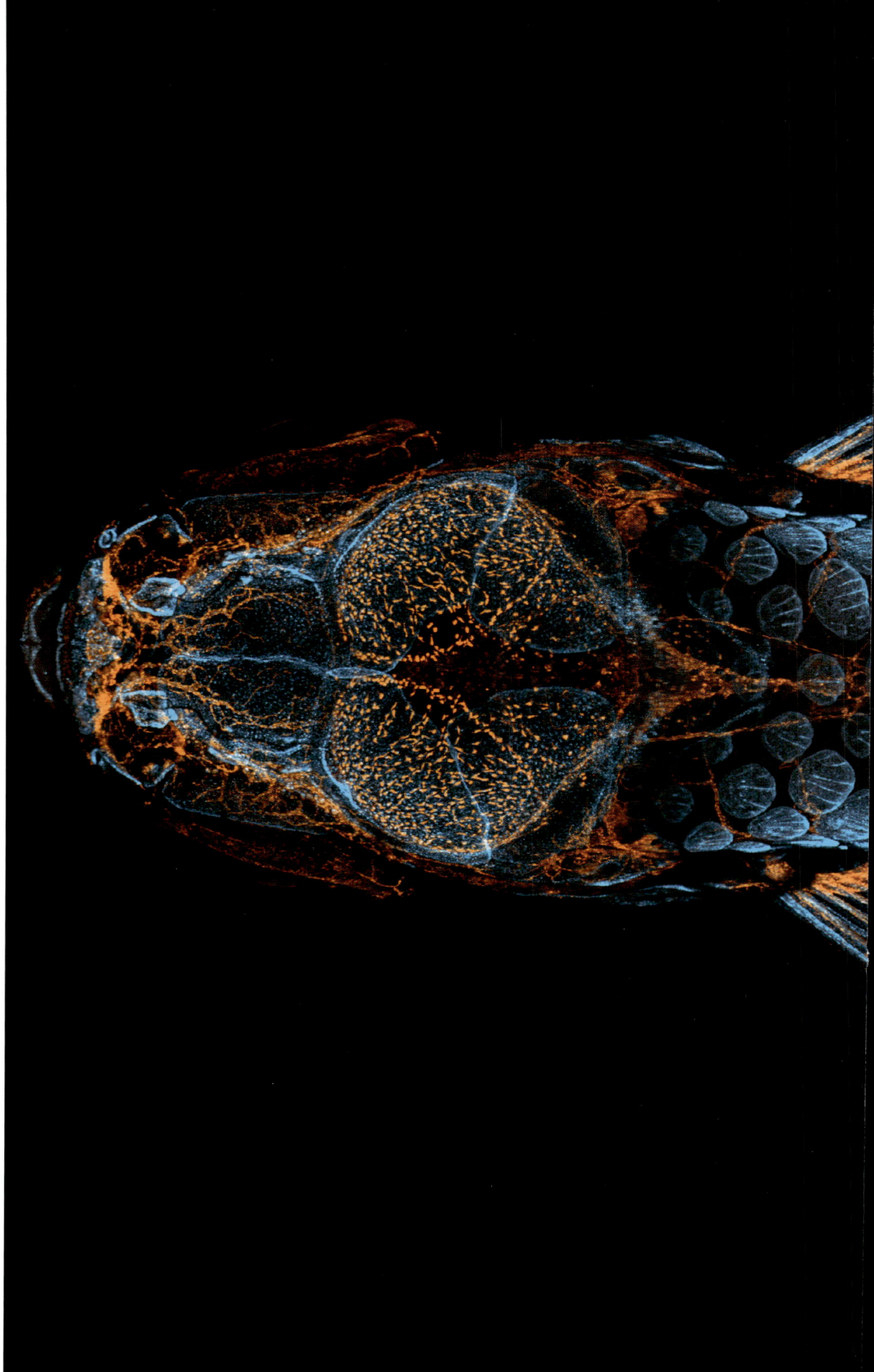

Daniel Castranova, Brant
Weinstein, and Bakary
Samasa/NIH
Fluorescent proteins provide
a detailed look at the anatomy
of a living zebrafish. Under
filtered light, the lymphatic
vessels of its immune system
glow orange while the scales
glow blue.

FOLLOWING PAGES:
Hugh Turvey
A colored x-ray turns tulips
translucent, revealing the
stamens and pistils inside
the blooms.

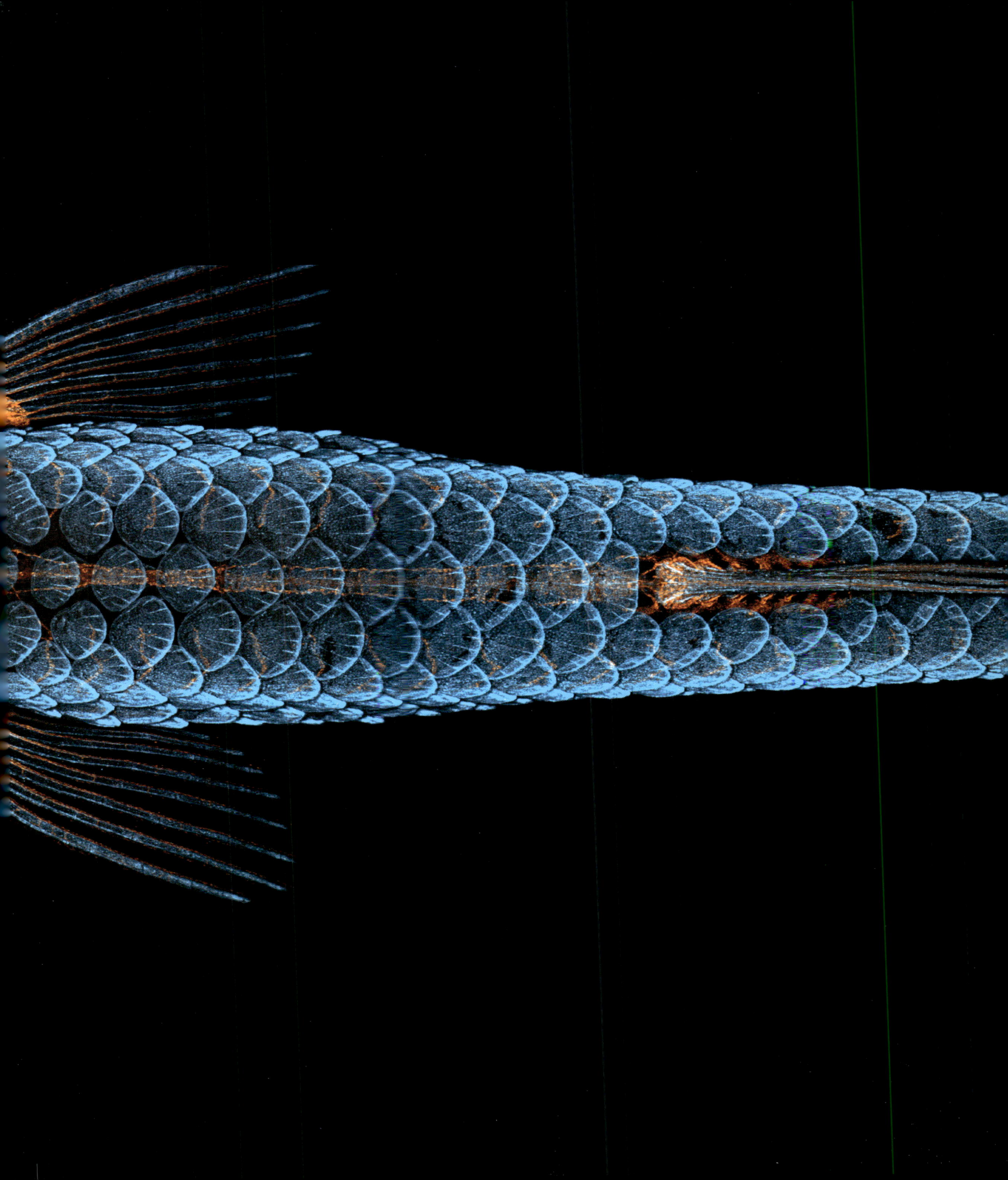

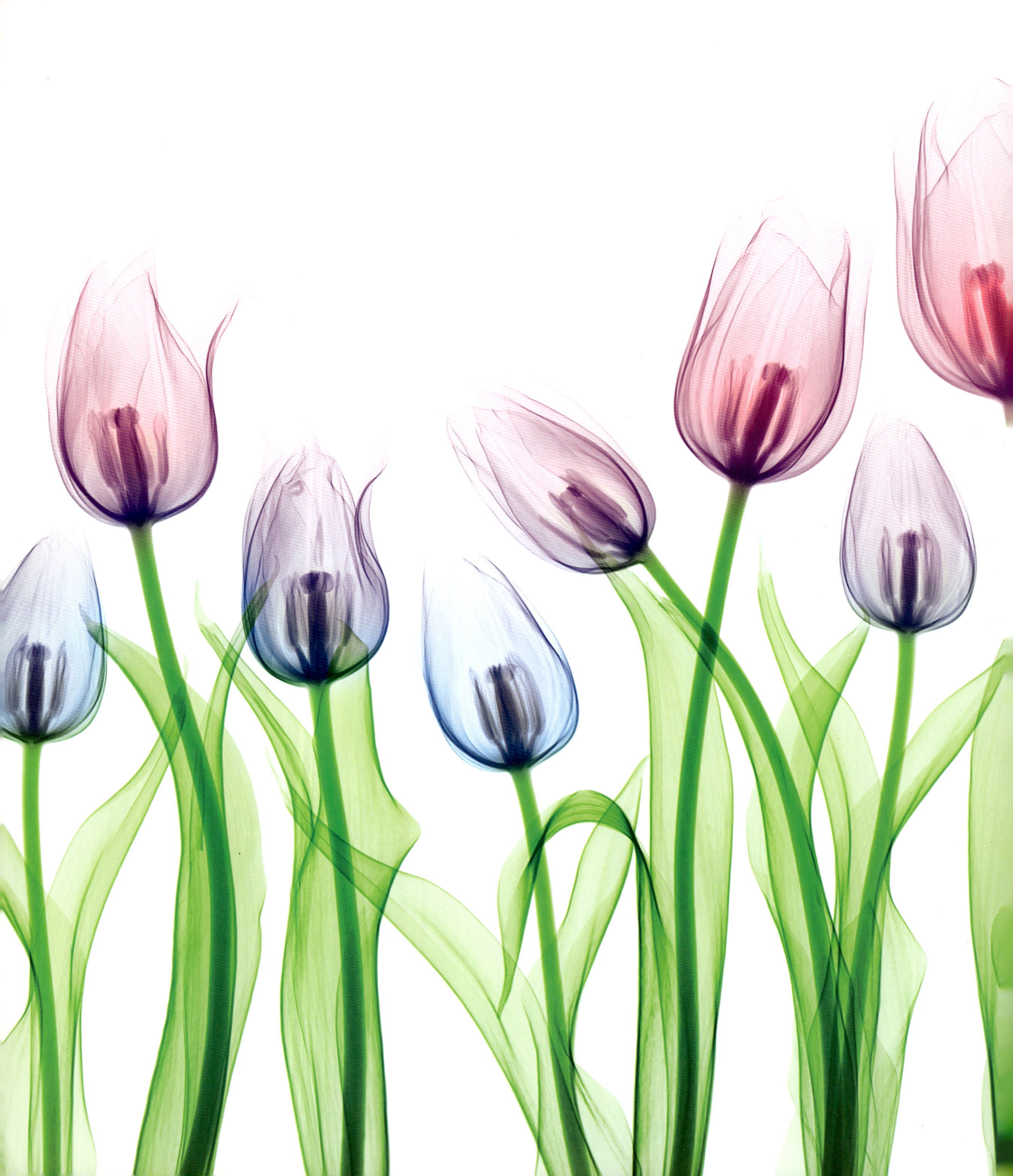

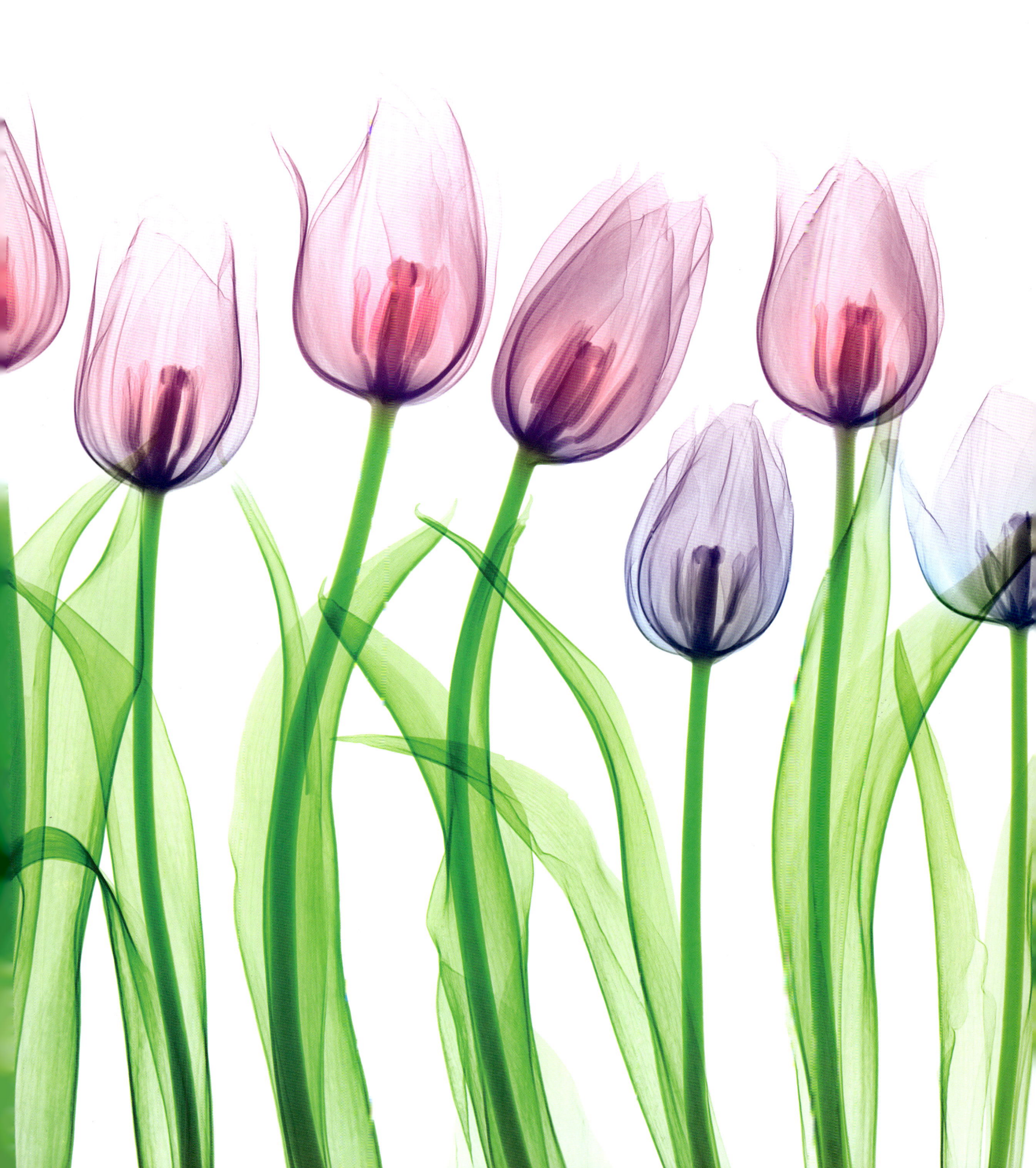

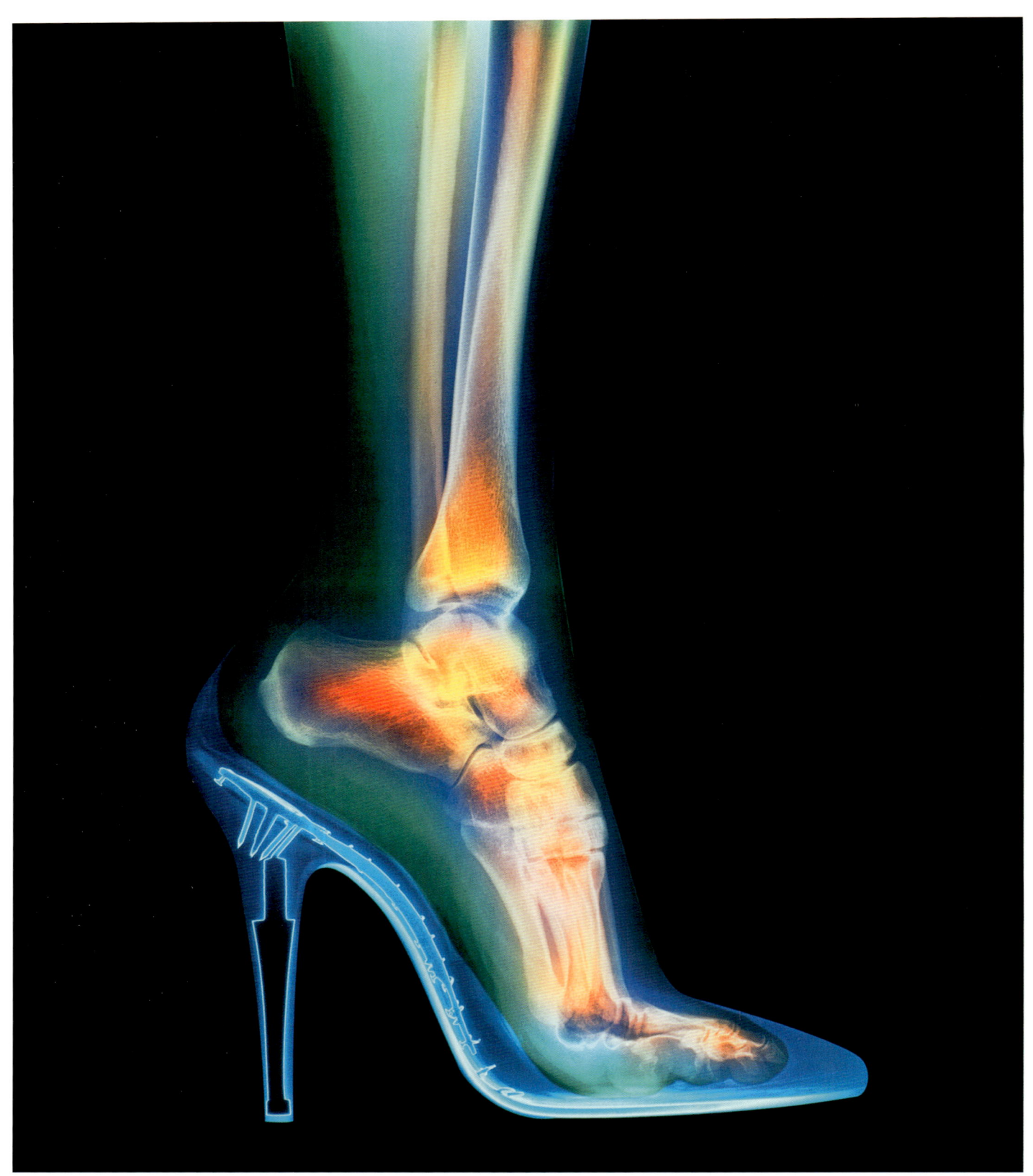

Hugh Turvey
A colored x-ray reveals the anatomical details of a woman's foot clad in a high-heeled shoe.

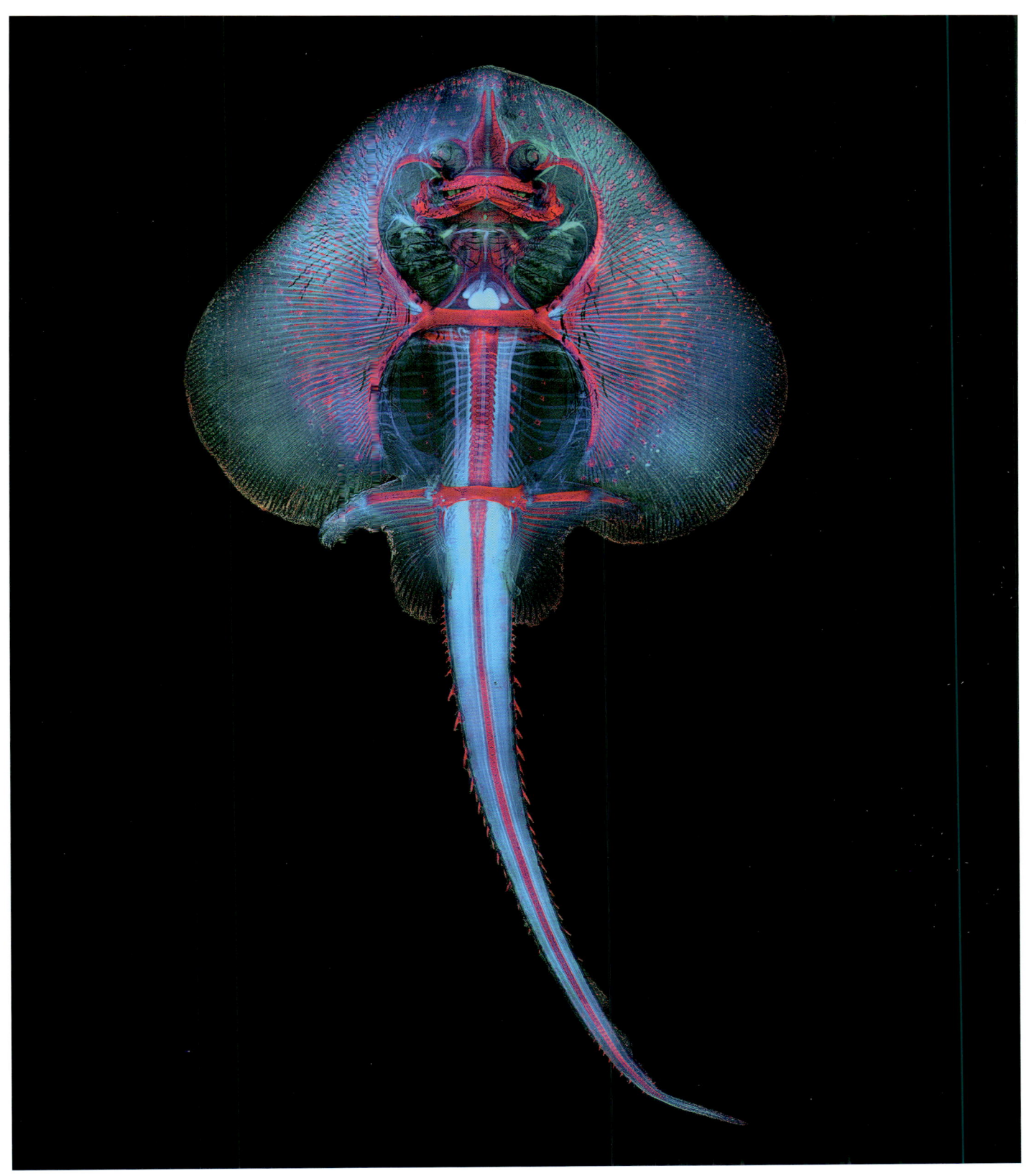

Teresa Zgoda

A skate embryo's skeletal structure shows up, thanks to fluorescent dyes that highlight bones and cartilage.

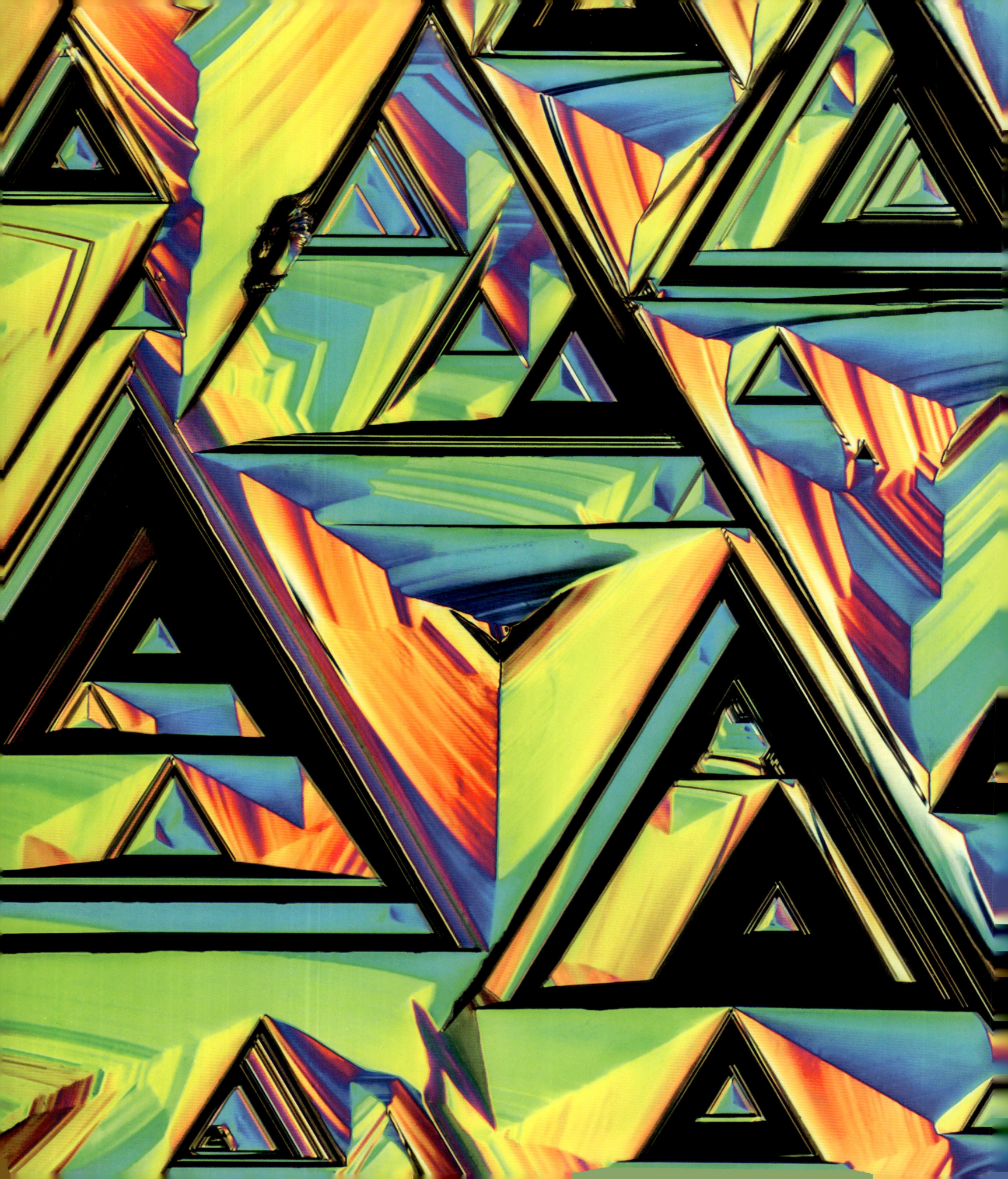

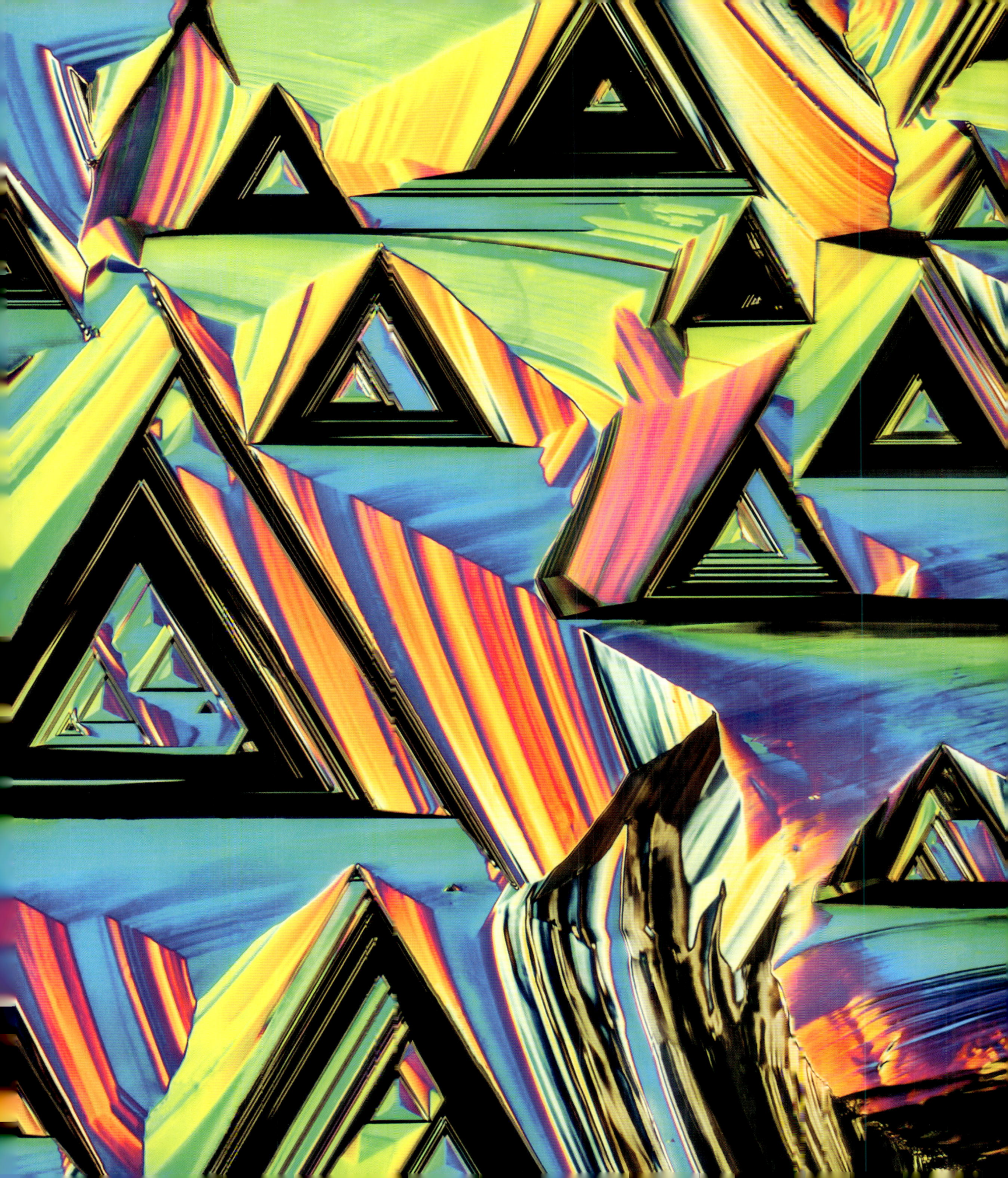

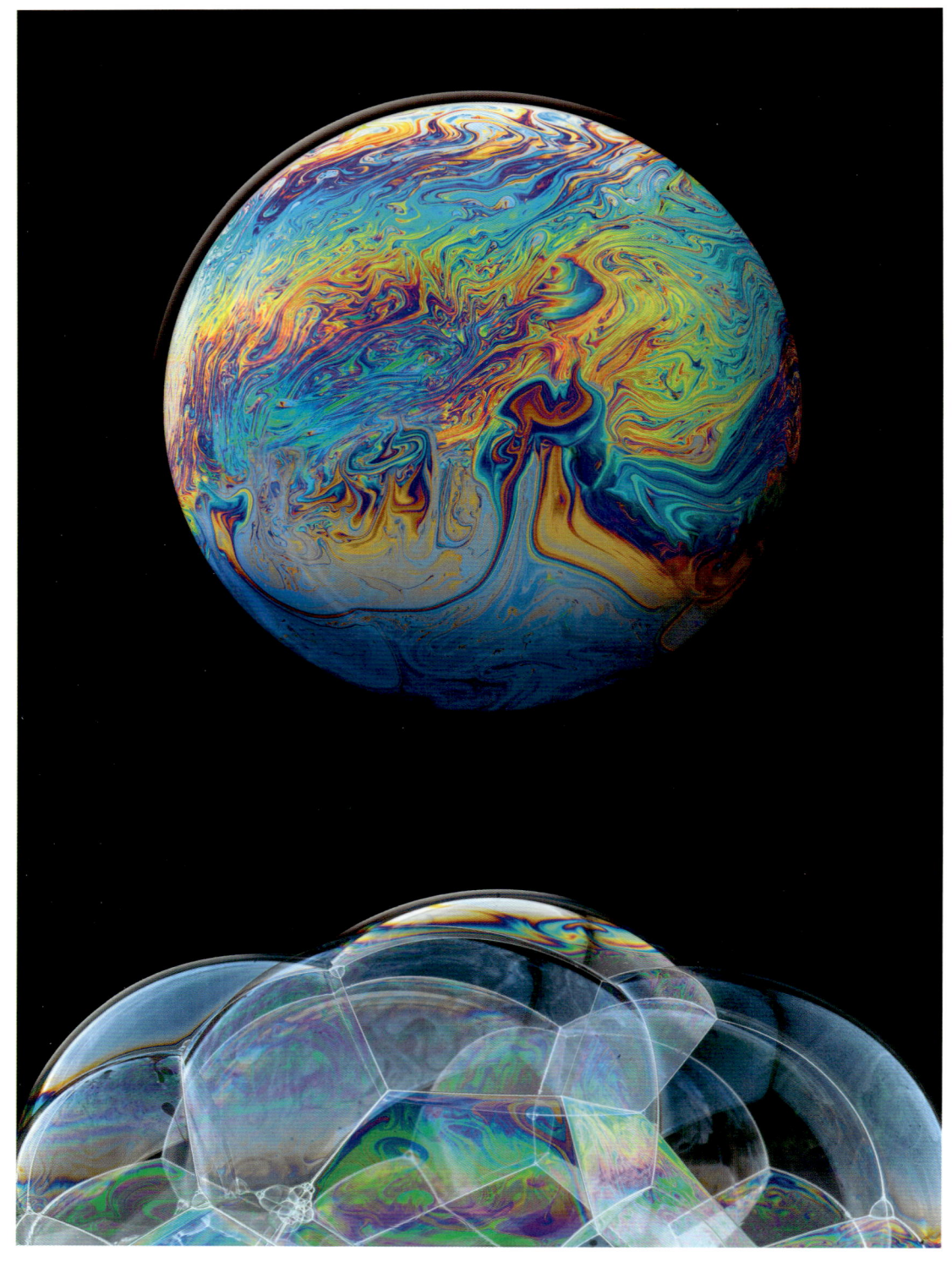

David Pearce
Diffuse lighting captures the natural iridescence of a floating soap bubble.

PREVIOUS PAGES:
Nathan Renfro
Microscopic structures called trigons become visible on a diamond
when illuminated by polarized light.

A BEAUTIFUL, BEWILDERING PHOTOGRAPH IS A WONDERFUL PLACE TO START MAKING SENSE OF THE STAGGERING COMPLEXITY AROUND US.

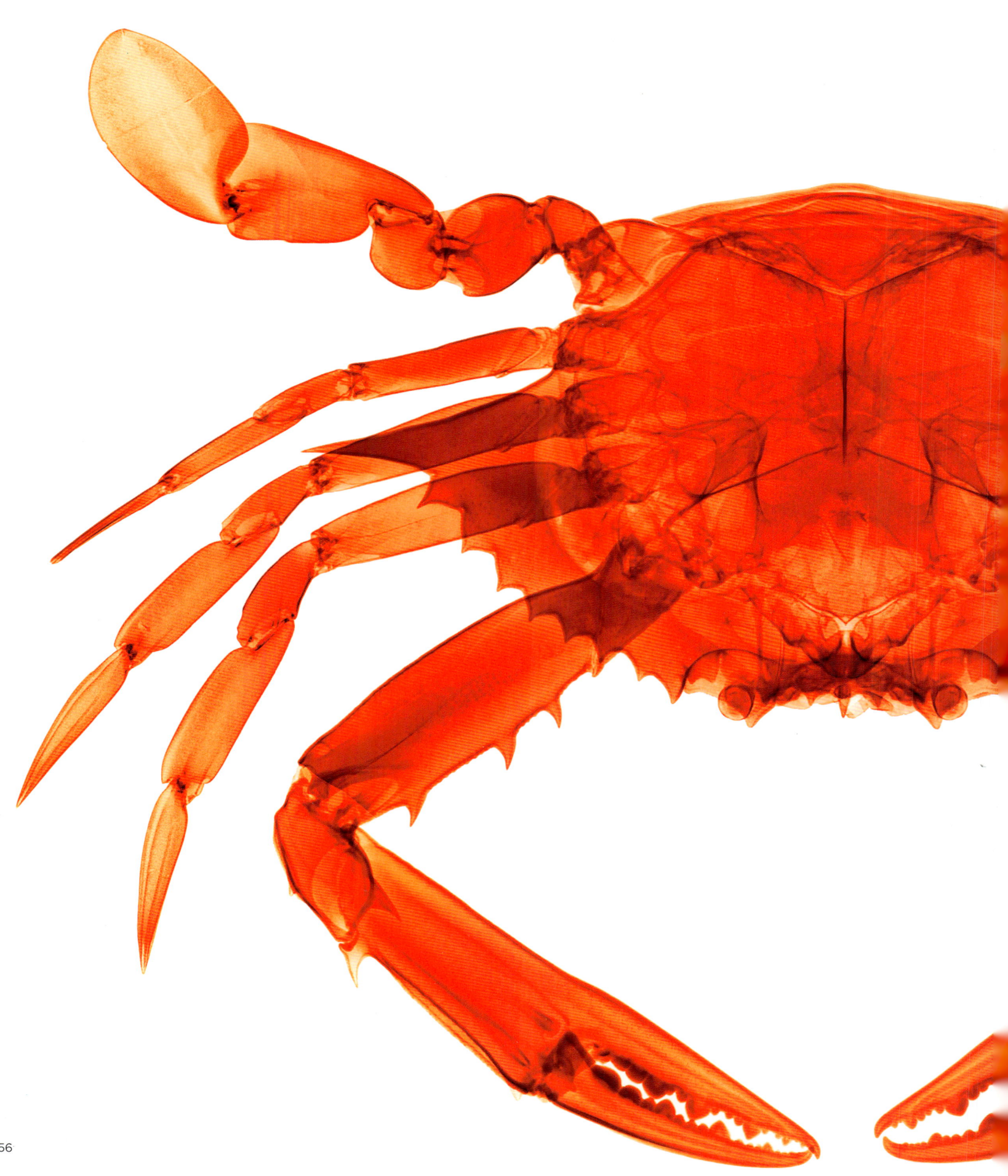

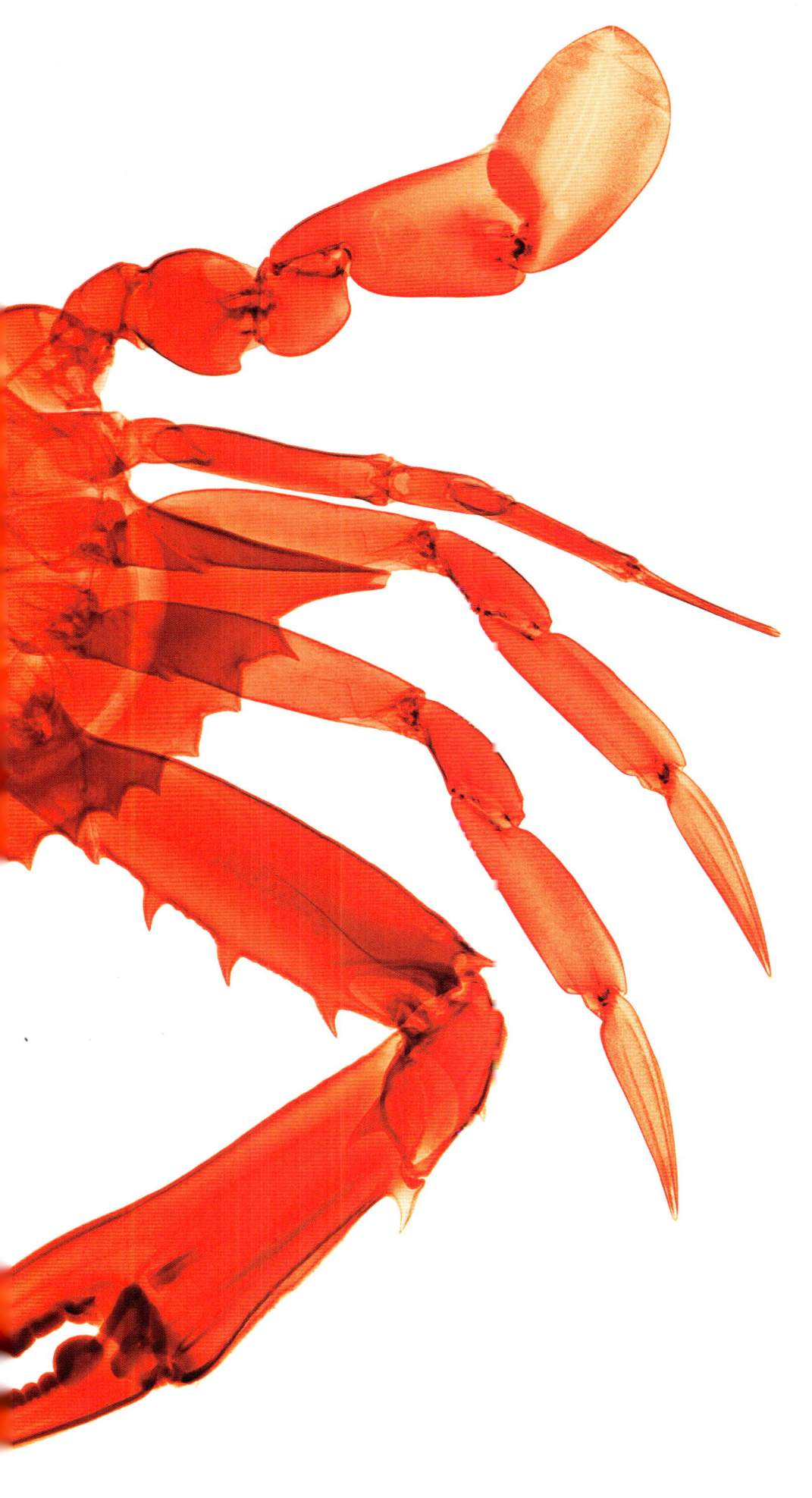

Brendan Fitzpatrick
A colored x-ray reveals
the internal structure
of a mud crab.

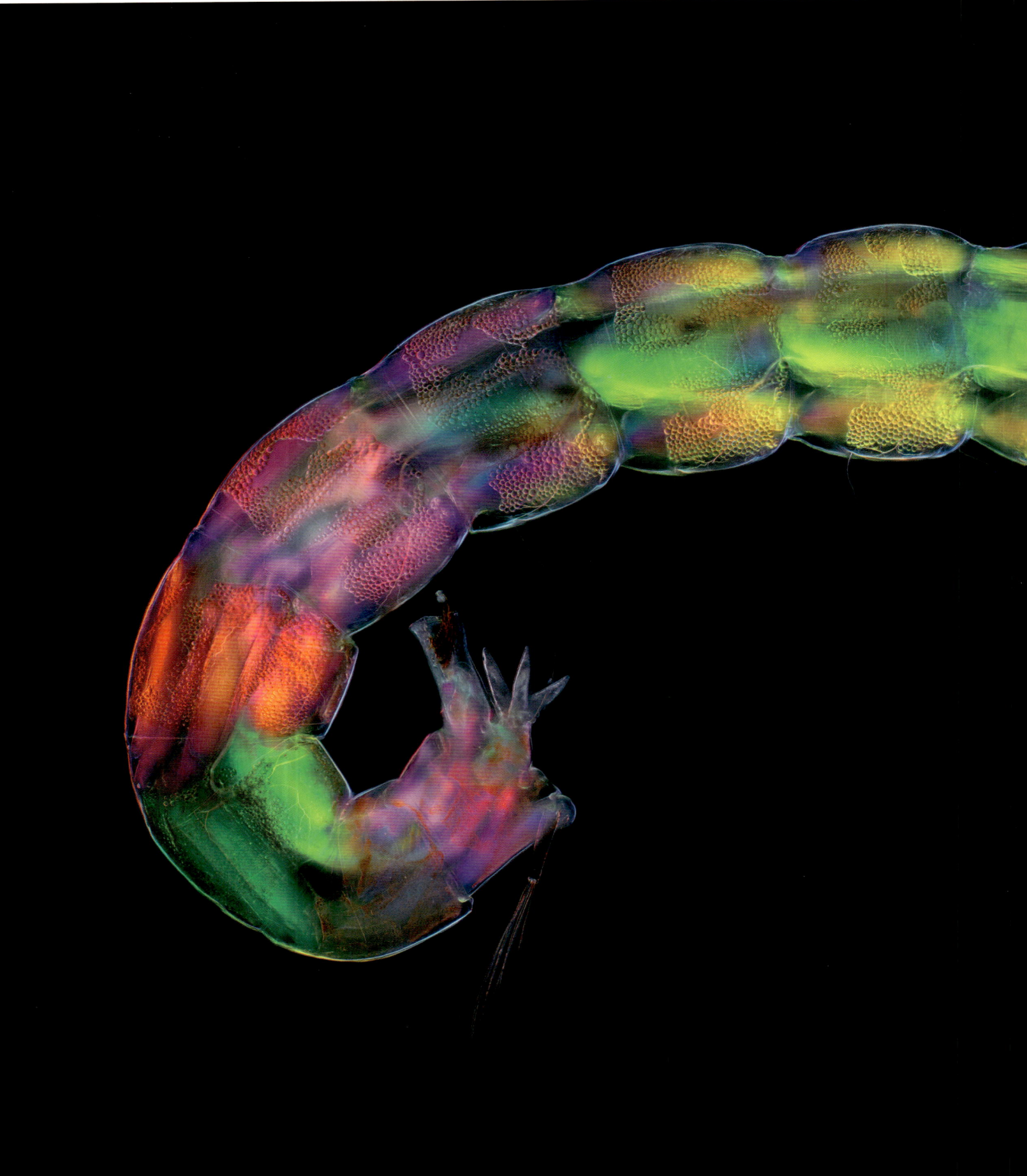

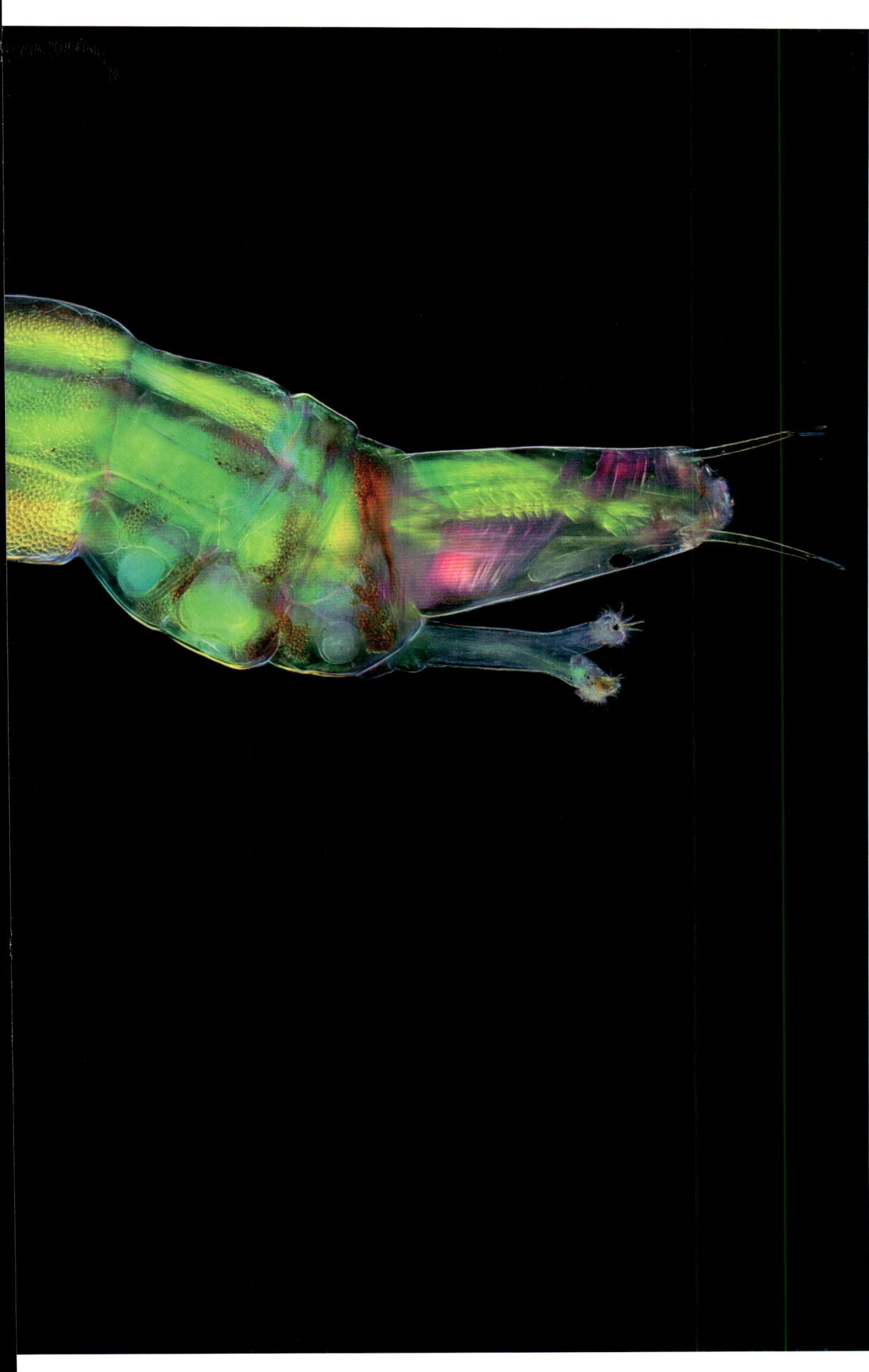

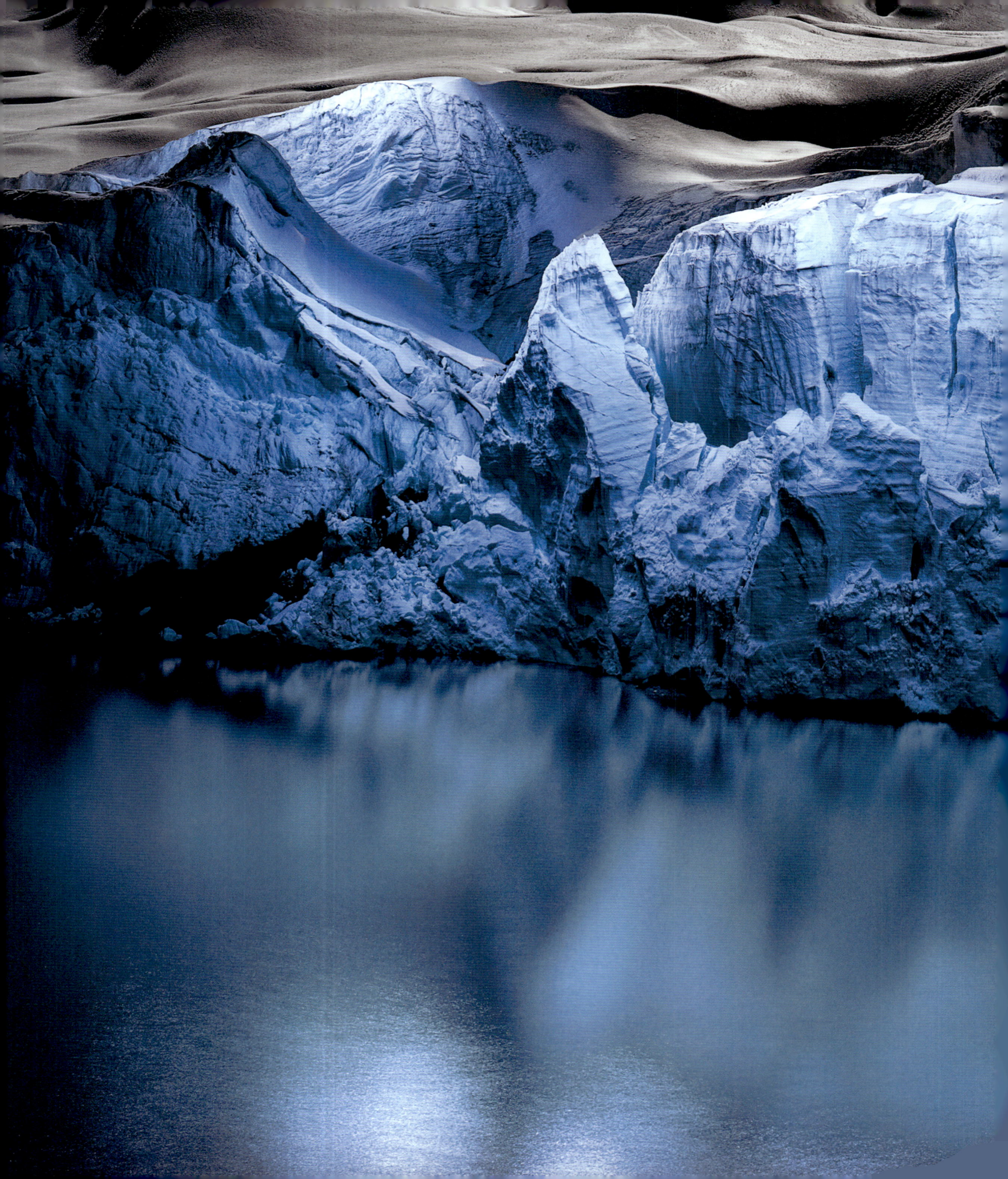

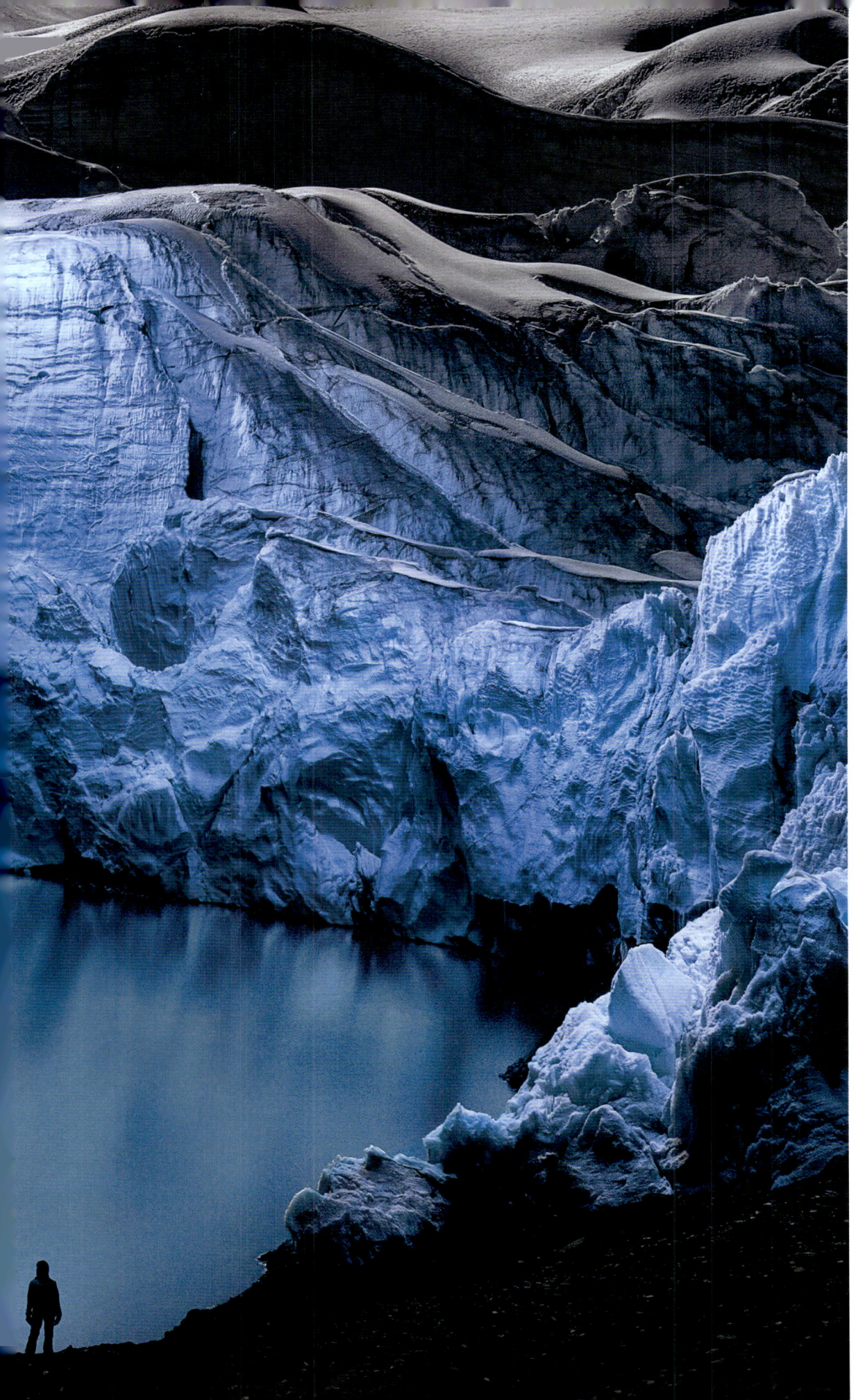

Reuben Wu
An LED mounted on a flying
drone illuminates the Pastoruri
Glacier in Peru's Cordillera
Blanca.

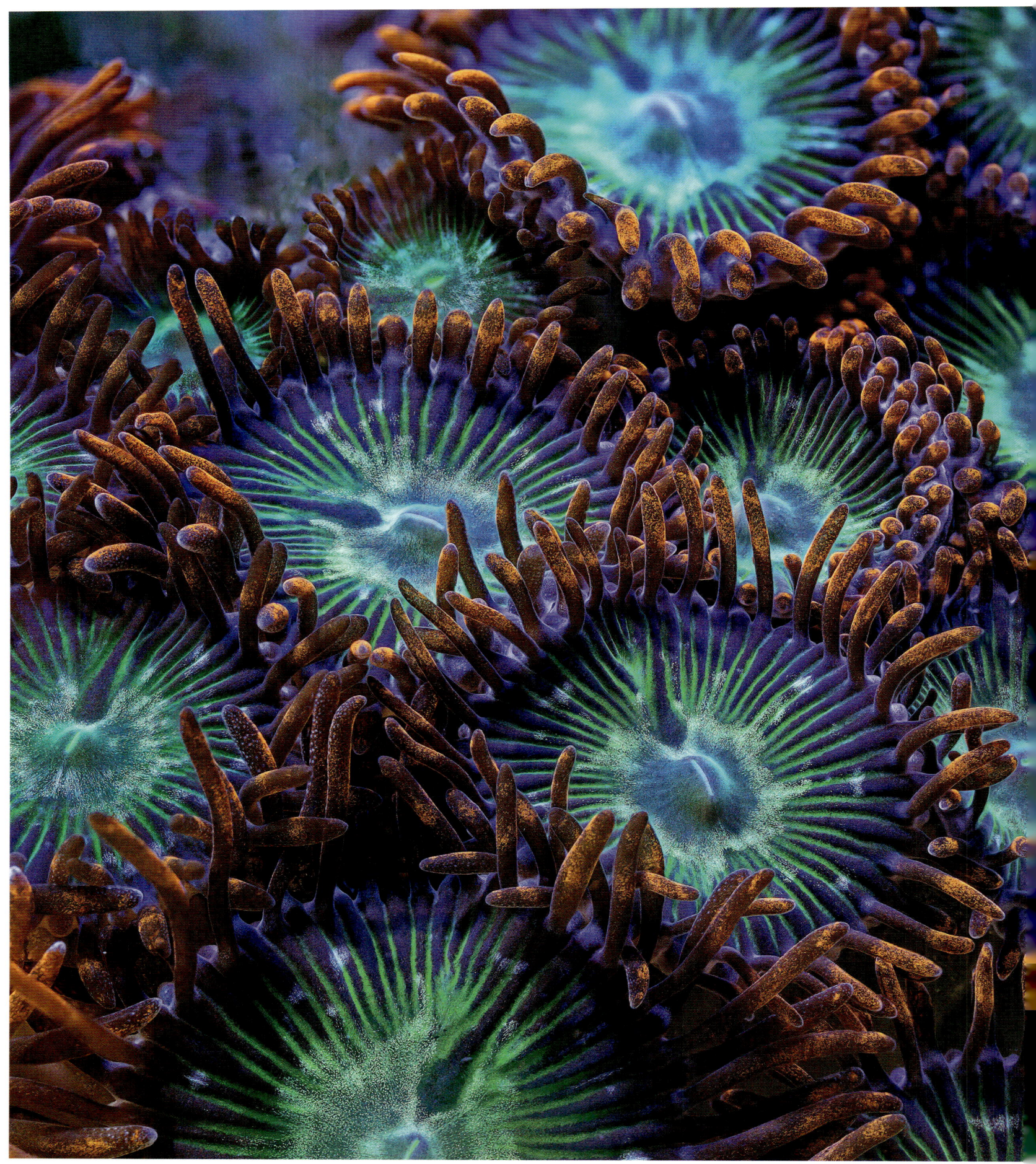

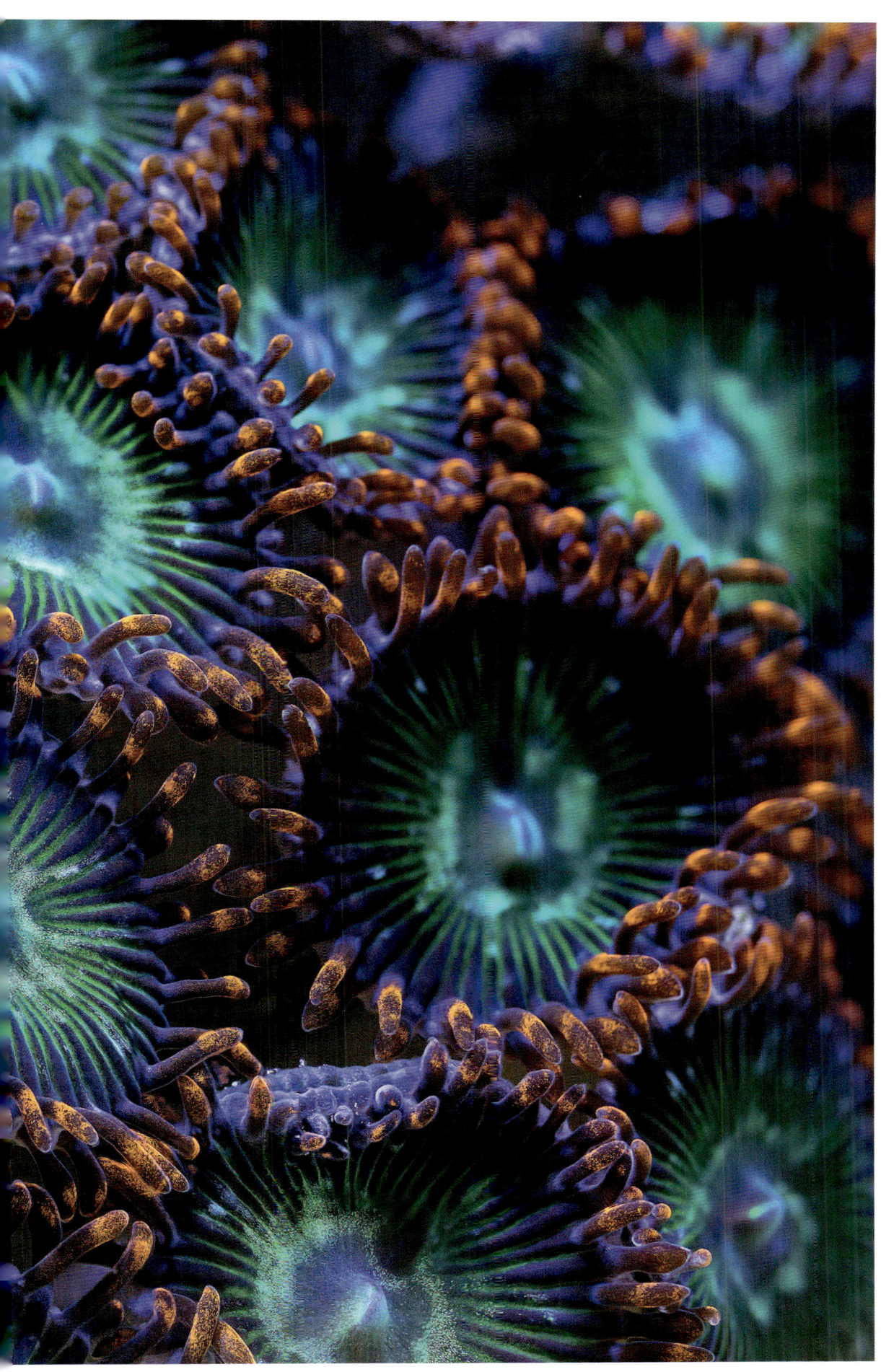

Daniel Stoupin
Zoanthid coral polyps
fluoresce under a combination
of visible and ultraviolet light.

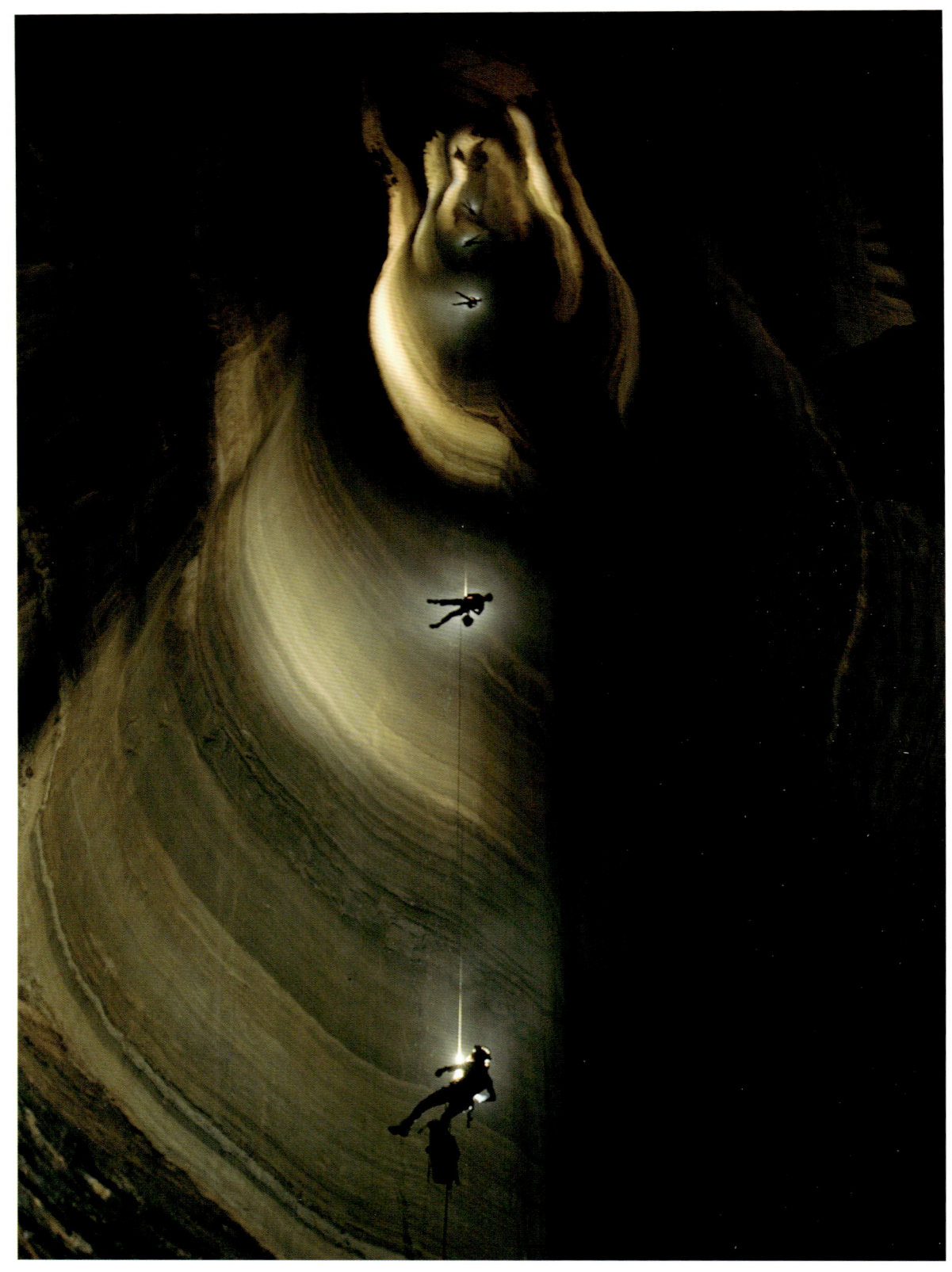

Stephen Alvarez
Cavers explore the Fantastic Pit in Georgia's Ellison's Cave, believed to be the deepest
free-fall cave pit in the United States, with an unobstructed vertical drop of 586 feet.

HIGH-CONTRAST LIGHTING HELPS GUIDE OUR ATTENTION IN A CLUTTERED SCENE BY ACCENTUATING SPECIFIC DETAILS.

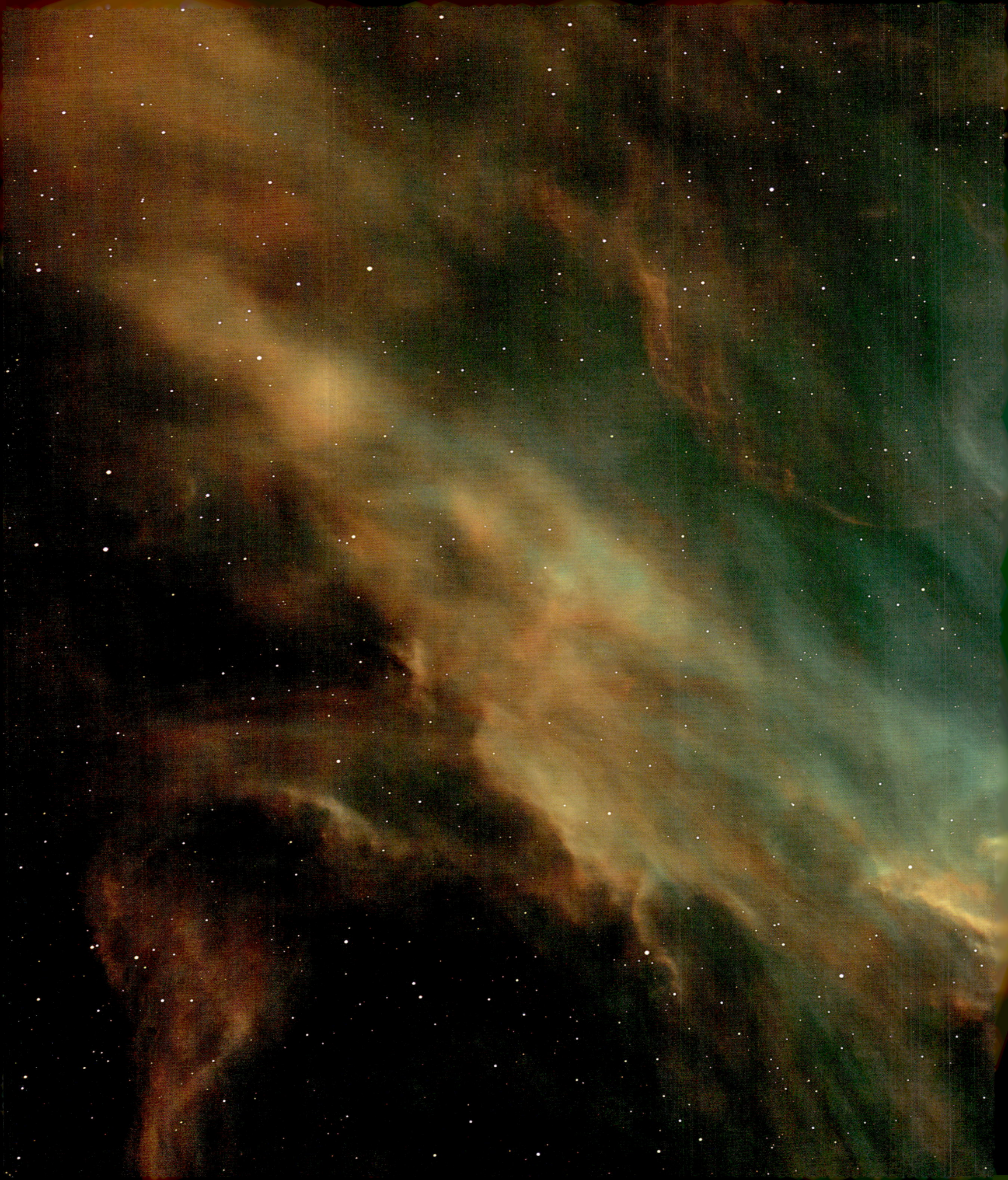

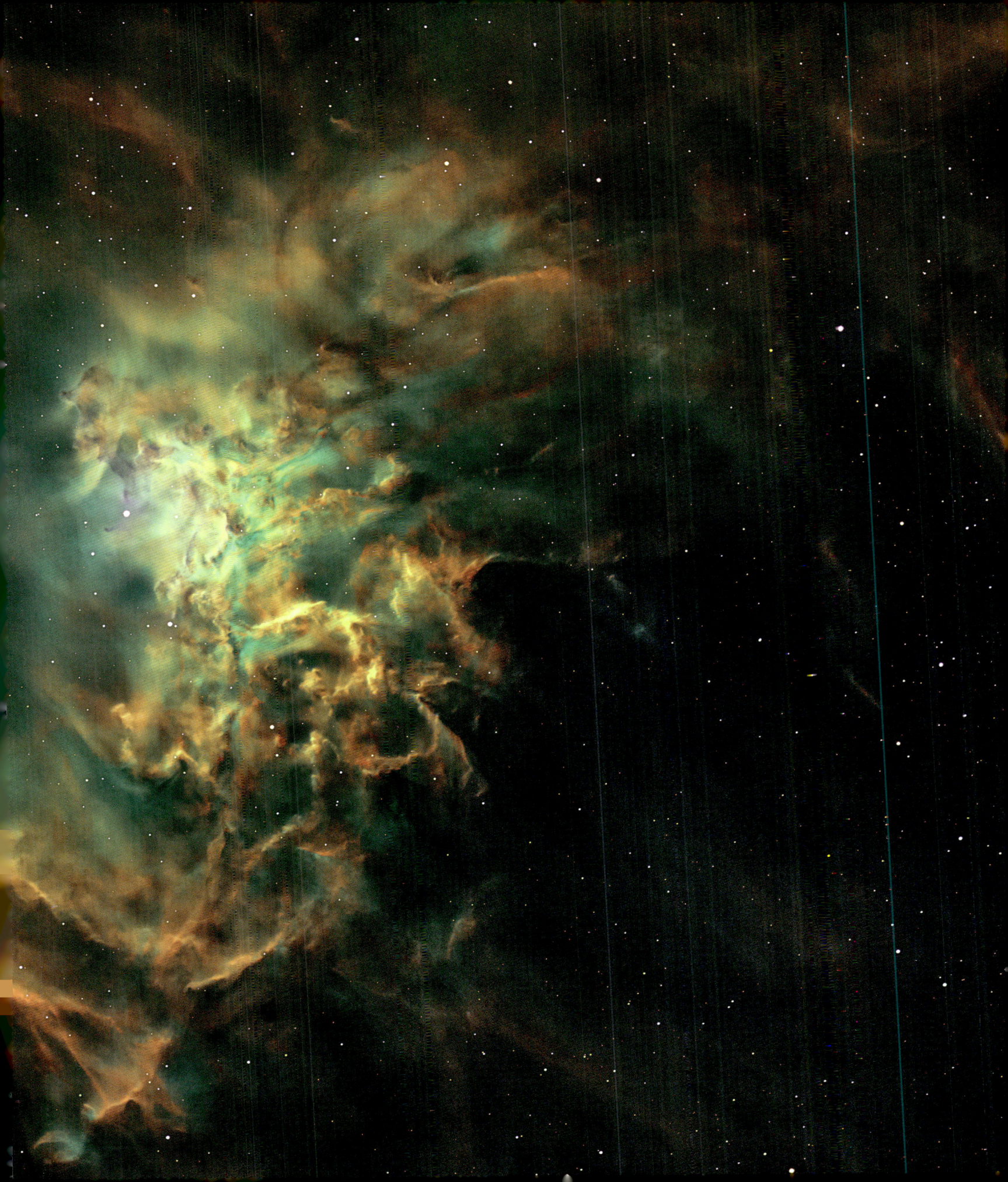

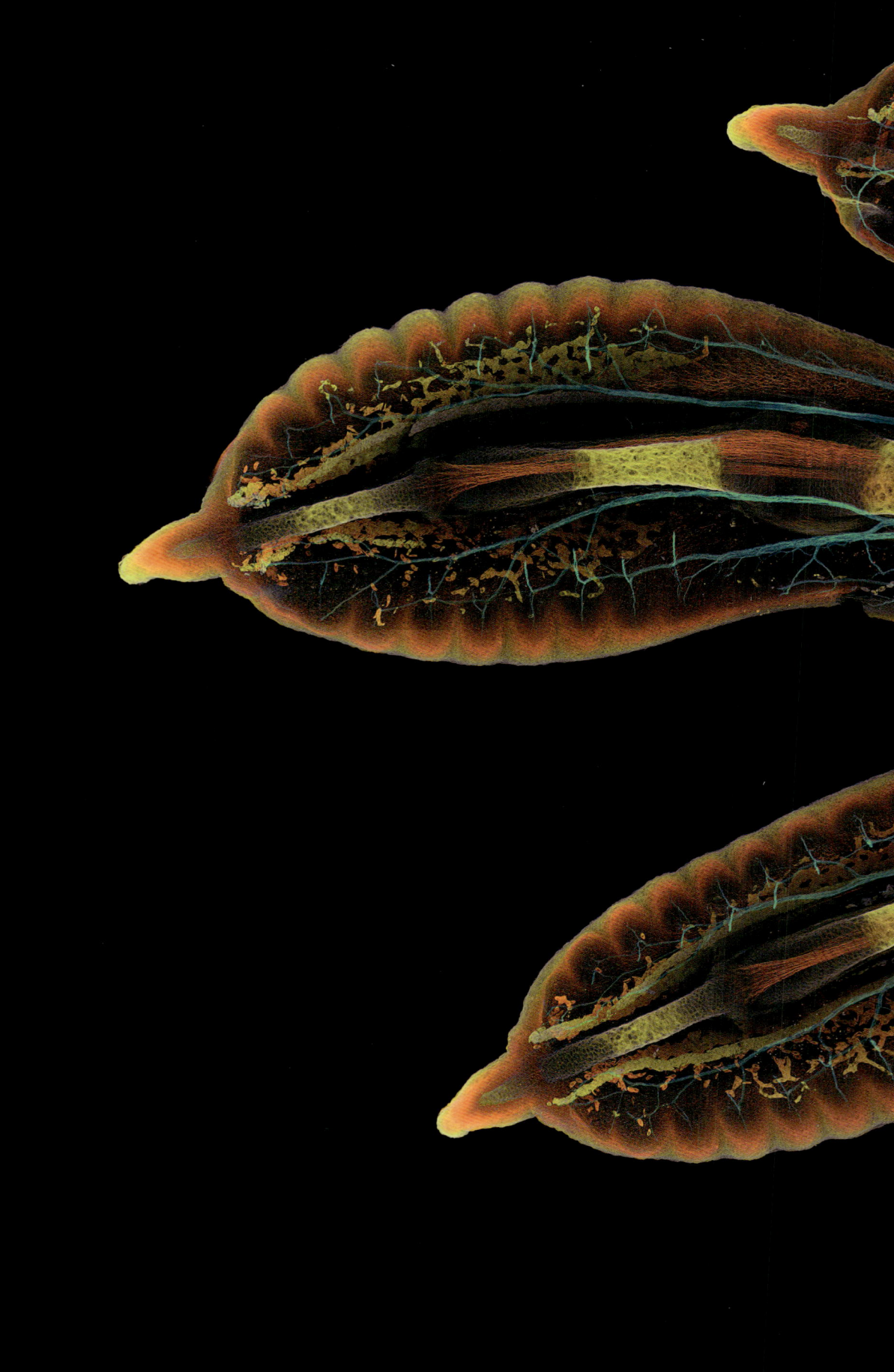

Grigorii Timin and Michel Milinkovitch
The embryonic hand of a Madagascar giant day gecko glows from the stain of fluorescent dyes. More than 75,000 images were combined to give us this single detailed view.

PREVIOUS PAGES:
Martin Cohen
A telescope fitted with three color filters produced this composite image of the Flaming Star Nebula in the constellation Auriga.

FOLLOWING PAGES:
Richard Mosse
An aerial image of a gold pit in Pará, Brazil, taken with a multispectral camera, which records wavelengths of light outside the range of human vision

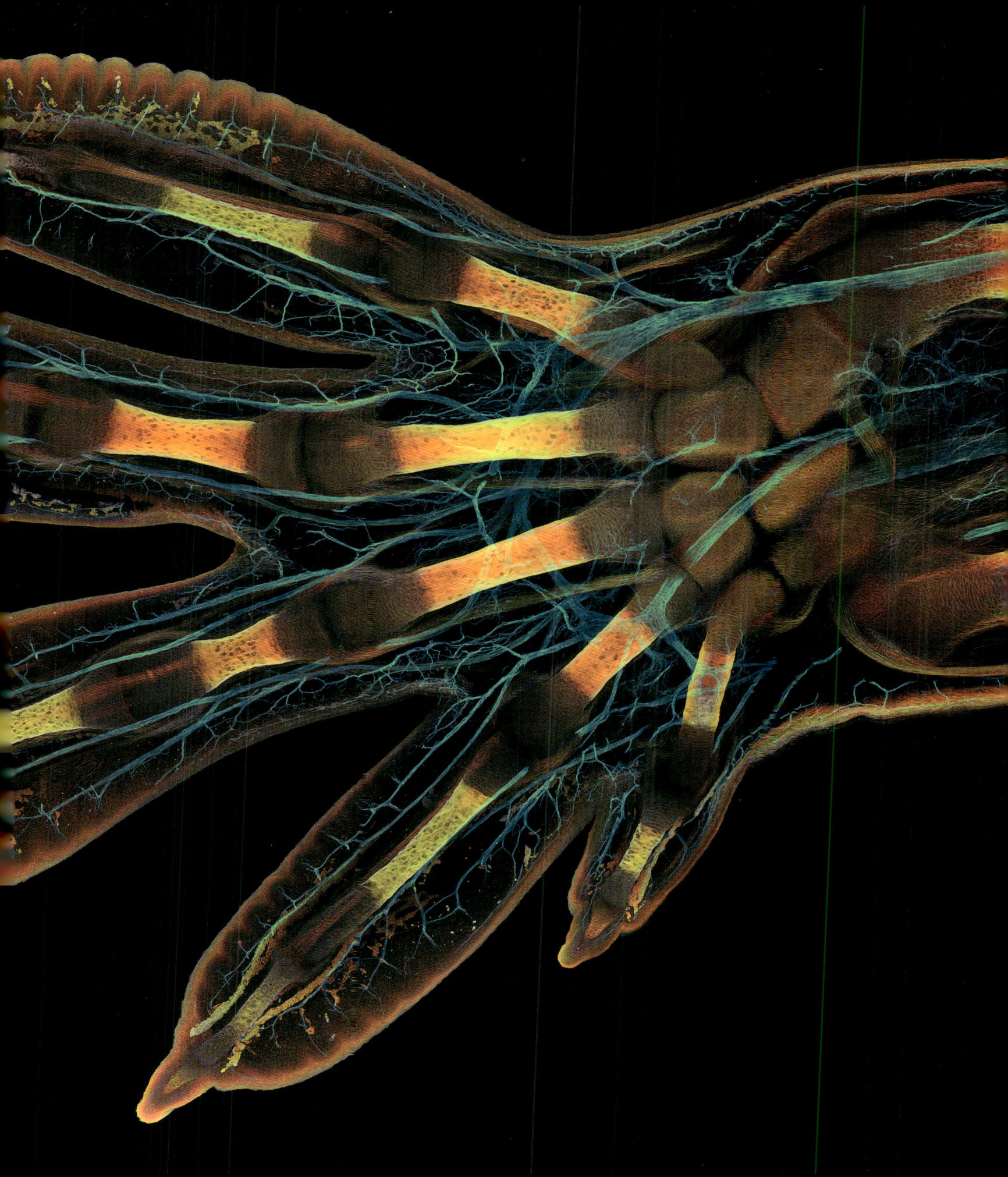

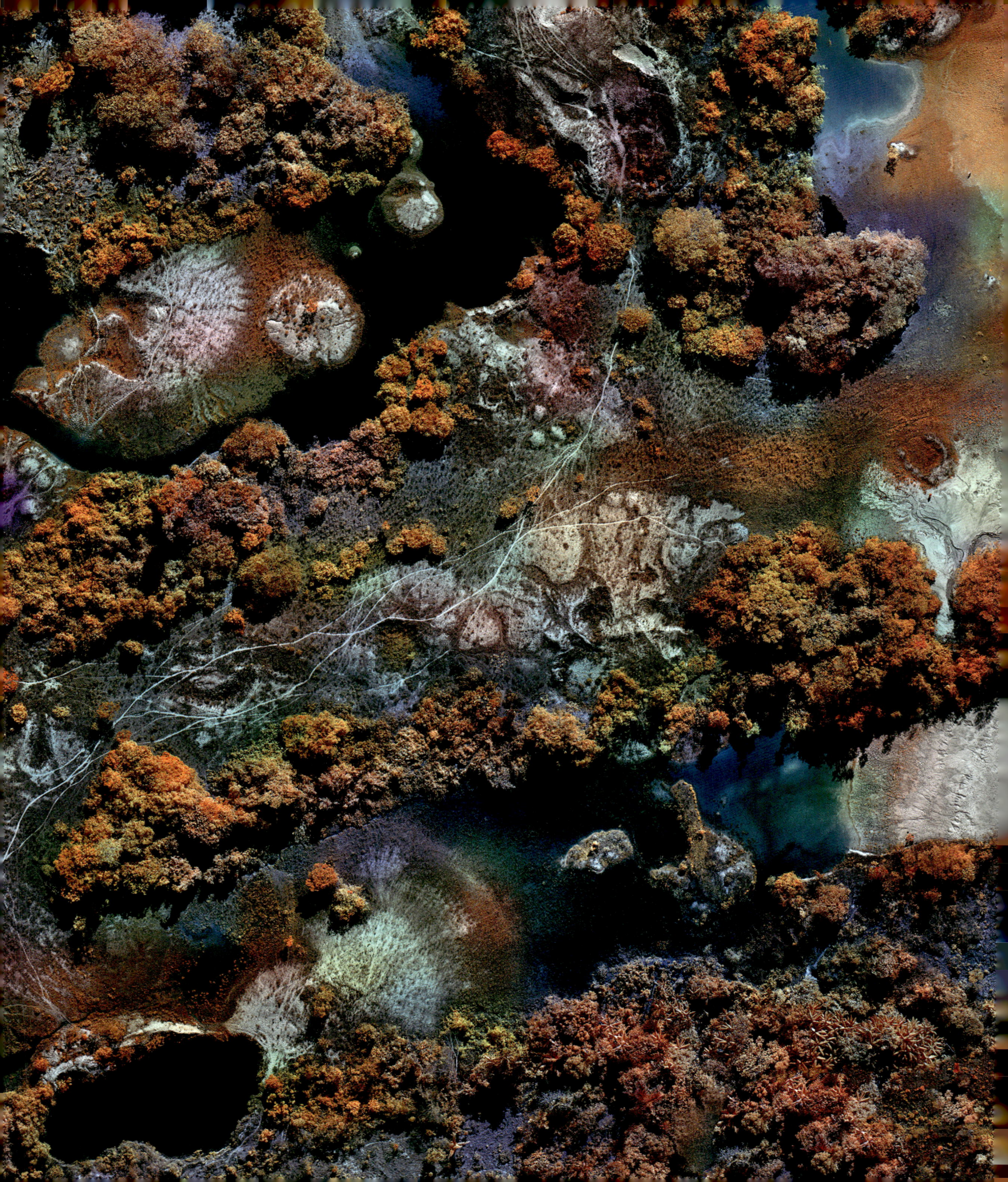

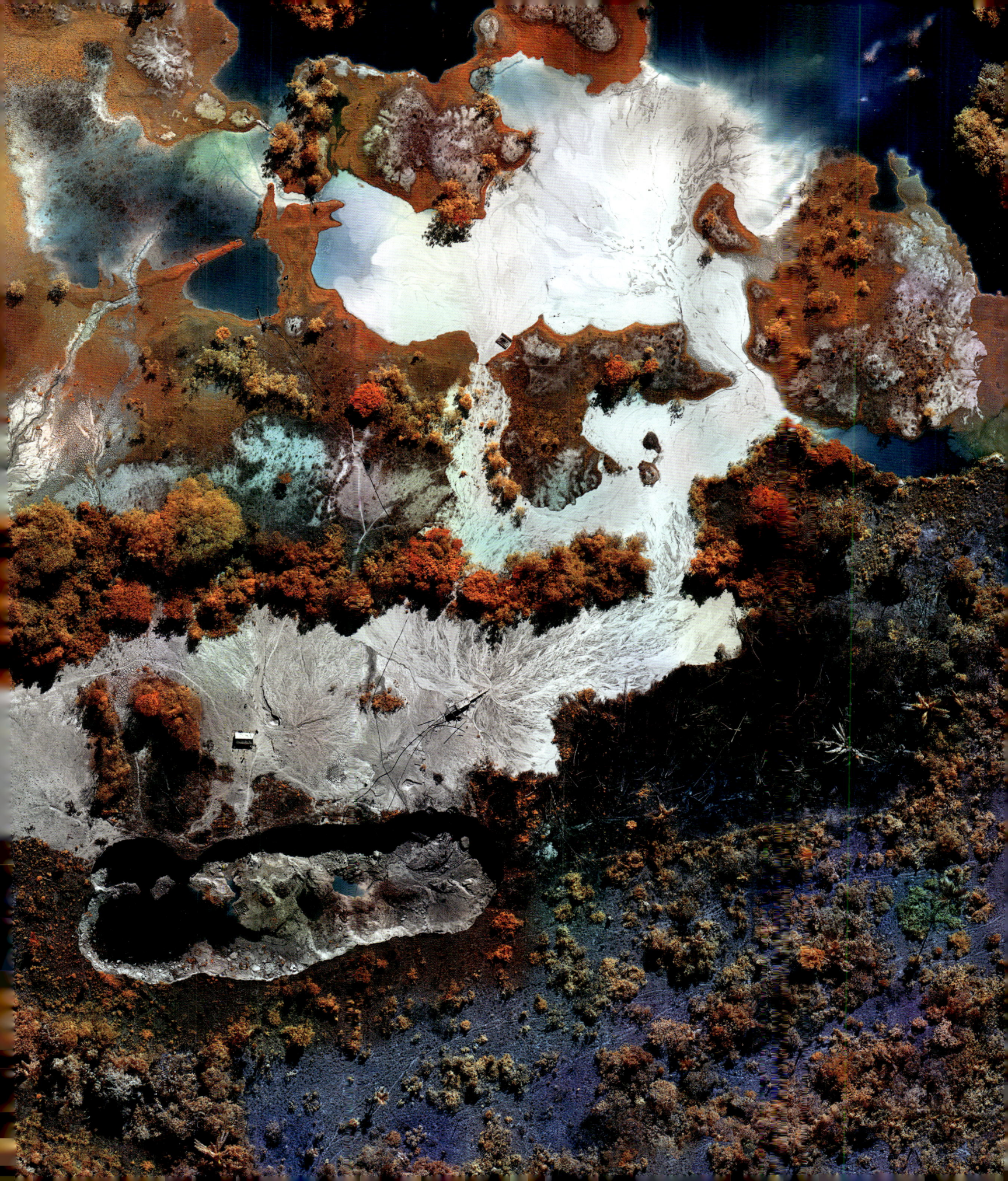

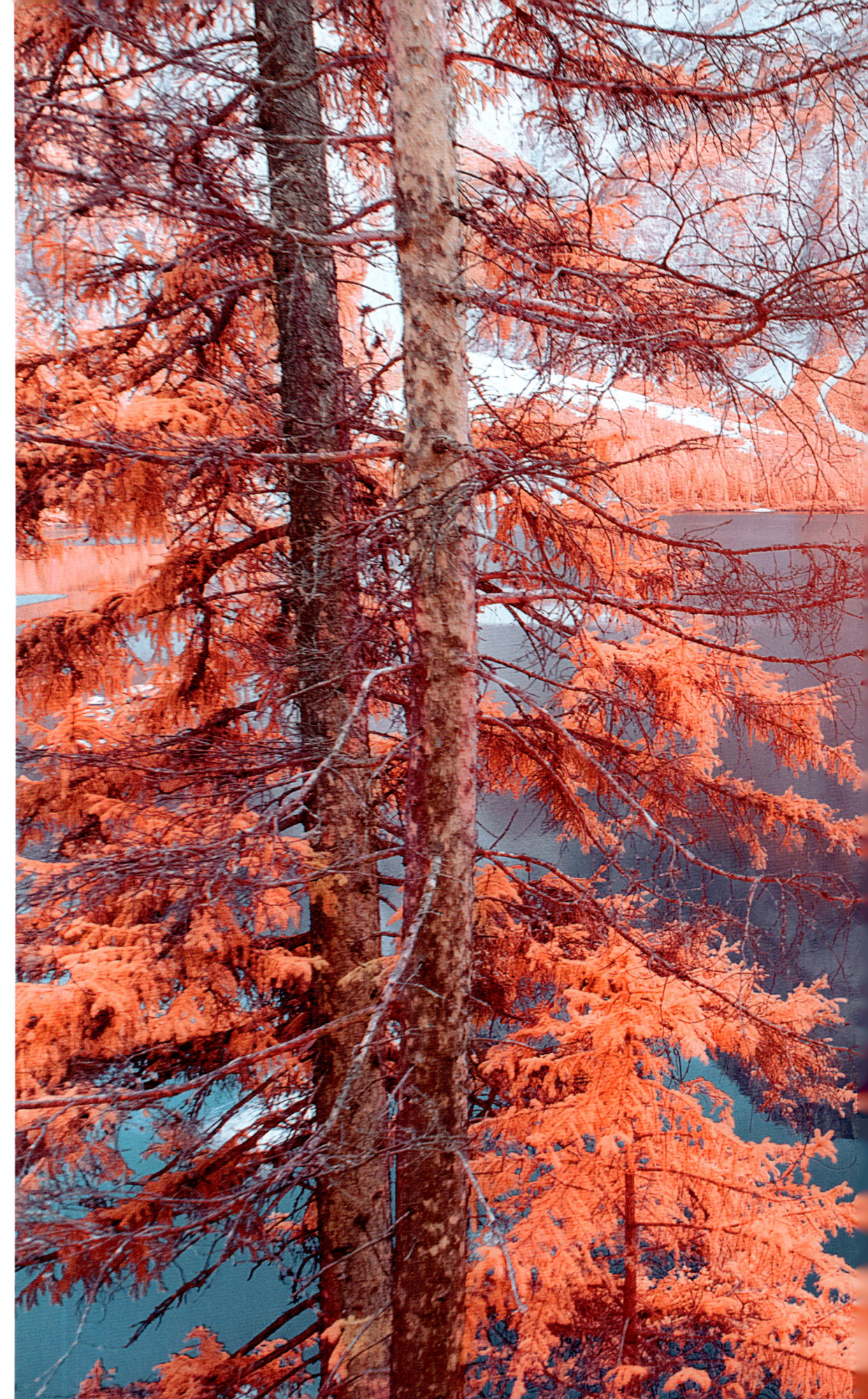

Paolo Pettigiani
In this image, an infrared camera depicts plants in pinkish hues, giving us a unique view of a landscape in Italy's Dolomite Mountains.

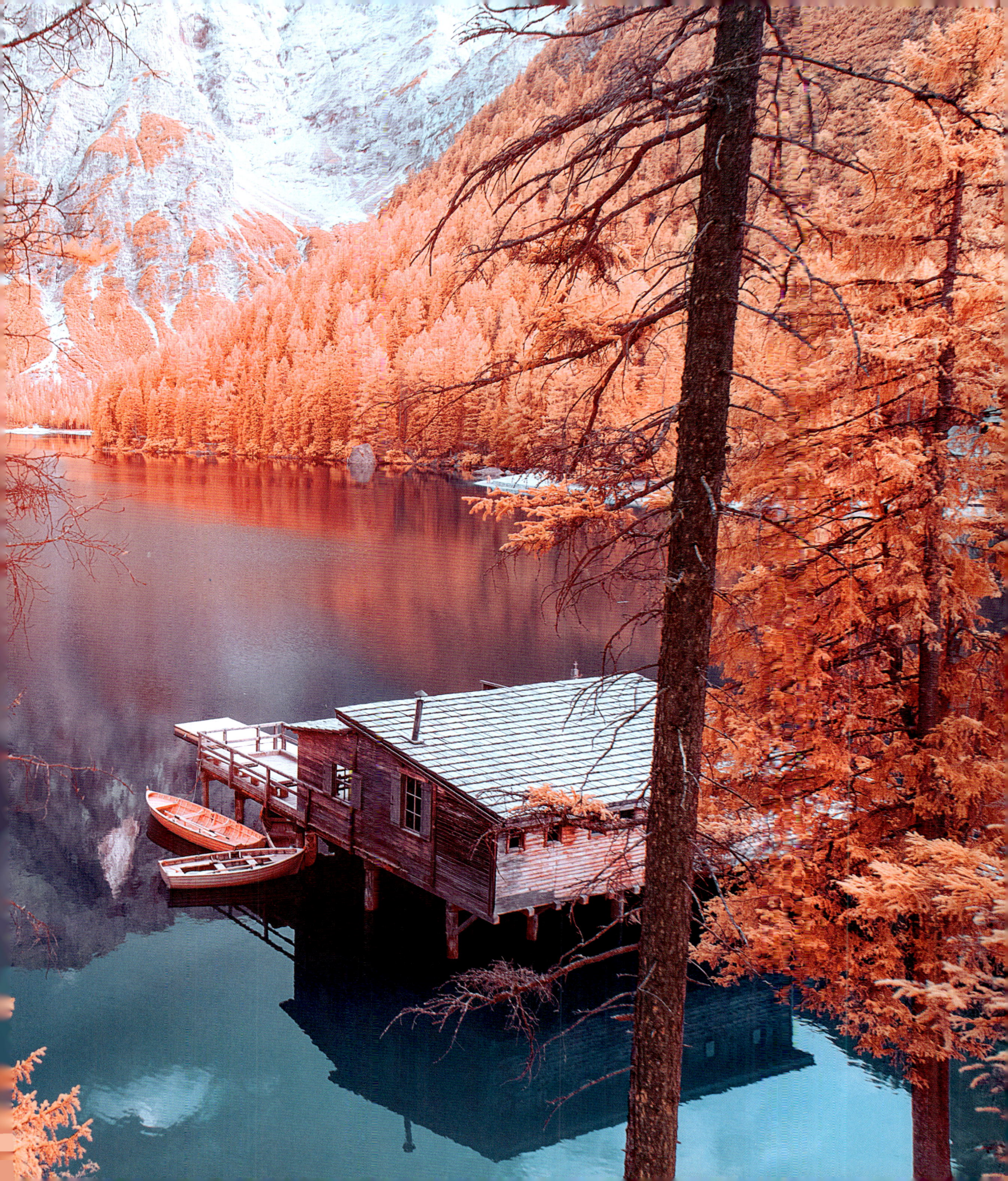

REUBEN WU

Reuben Wu is a photographer and multidisciplinary artist whose work has been commissioned by Apple, Audi, Mercedes-Benz, Samsung, and others. He is co-founder of the electronic band Ladytron, with which he has produced six studio albums, and he is a music producer and sound designer for clients including GE and IBM. His books, *Lux Noctis* and *Aeroglyphs & Other Nocturnes,* are held in the permanent collections of the Guggenheim Museum, the Metropolitan Museum of Art, MoMA, and the Art Institute of Chicago.

ANAND VARMA: What was your path to becoming a photographer?

REUBEN WU: It's a long story. I started out as a professional industrial designer in Cambridge, England. I had always been a visual person, but I was advised to steer clear of art because it was never going to make me money. During my time in industrial design, however, I co-founded a band called Ladytron as a side project—just me and a few friends making music and playing shows to have fun. But we were eventually picked up by a record label, and part of that deal was having to go on tour in the U.S. It came to a point where I had to either go full-time with music or continue with design. I ended up taking the leap and becoming a full-time musician, and I did that for the next 10 years.

Touring enabled me to travel and see the world. We were going all over the place: the U.S.A., Canada, South America, Russia, China, Asia, Australia. And while we were touring these incredible places, I decided to pick up a camera to document the world around me. It was an amazing way to explore. So travel piqued my interest in photography, but what began as a travel diary became more involved. I started to book personal time around the live shows to go take pictures. I eventually rediscovered film photography and analog cameras. Previously, I had dabbled in photography, but there was not enough around me to inspire me and keep my interest. Traveling suddenly gave it real meaning. It eventually became more rewarding than making music.

AV: Did you eventually have to choose between the two?

RW: When the band decided to take a sabbatical in 2011, I was facing that exact choice. I had no idea what I was going to do. I was either going to figure out a solo music career, or I was going to follow this dream that I wanted more than anything else—but had no idea how to do it. [In the end] I decided that photography had to be my new creative practice. For the next five or so years, I started to branch out of analog photography and learn about digital and video and time-lapse and all the things that new technologies enabled me to do, and I kept on making new work. By this time I was living in the U.S. It

had always been a dream destination for me. Especially for someone who was brought up in the U.K., poring over pages of *National Geographic,* the desert landscapes of North America felt like a fantasy. Combined with my fondness for the science fiction films of the '80s, they melded together into something I was really drawn to. In the States, I was able to travel to all of these places and make the pictures that I'd always wanted to make. It's been a really interesting seven or so years since. I started with a dream, and now I'm at a point where I'm feeling like I'm where I want to be in life, in terms of my creative output.

AV: Are there particular memories or experiences from your past that influence your work today?

RW: I grew up in an immigrant family in the U.K., and we started from very humble beginnings. All we could afford to do on holiday was to go hiking and camping. As a very introverted kid, the time I spent in these wild places, in the national parks of northwest England, was when I felt a lot of freedom, where I felt completely at ease. It was here where I developed

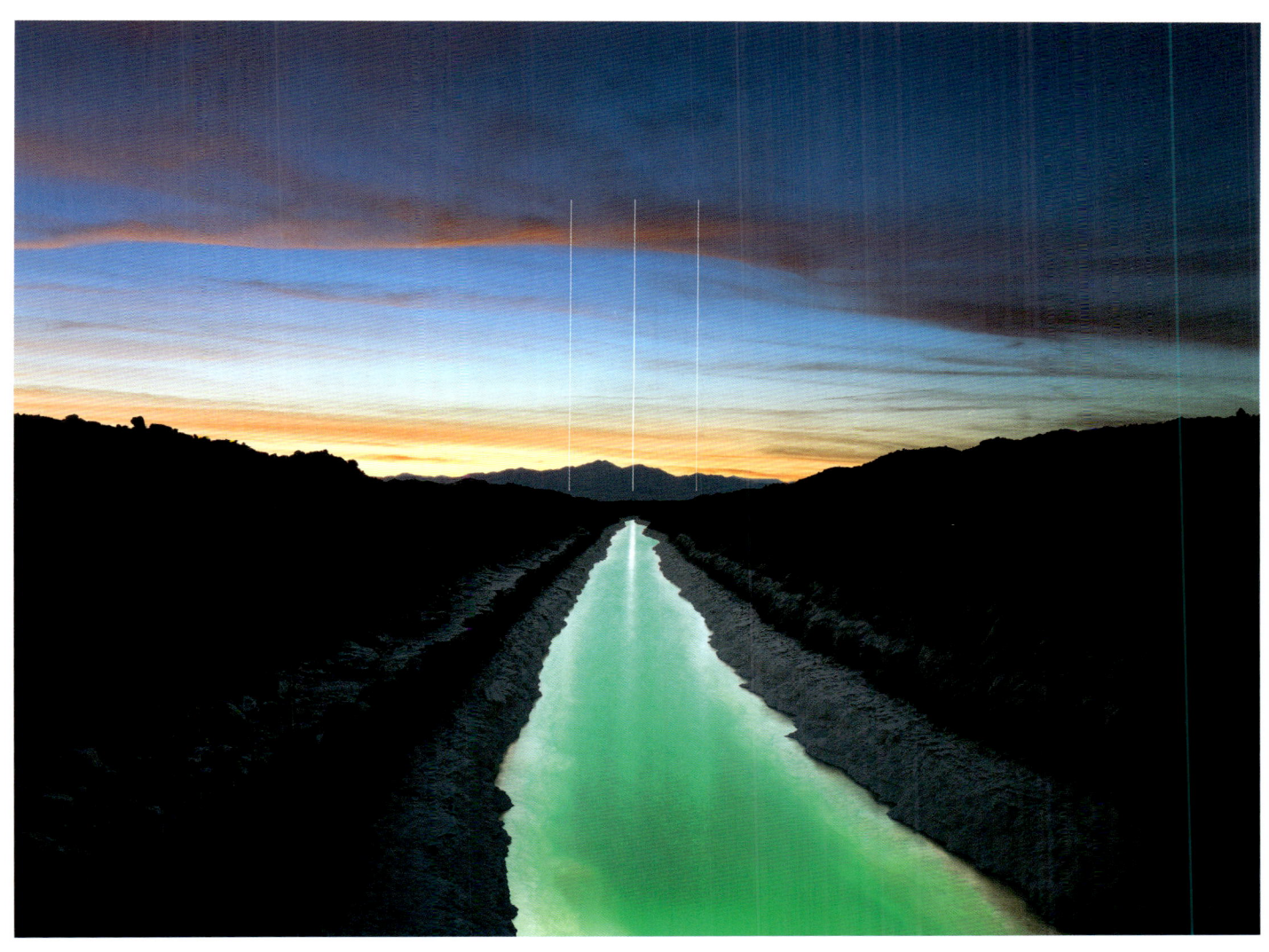

Reuben Wu
A light attached to a crone paints three vertical lines above a salt evaporation pond
in Amboy, California.

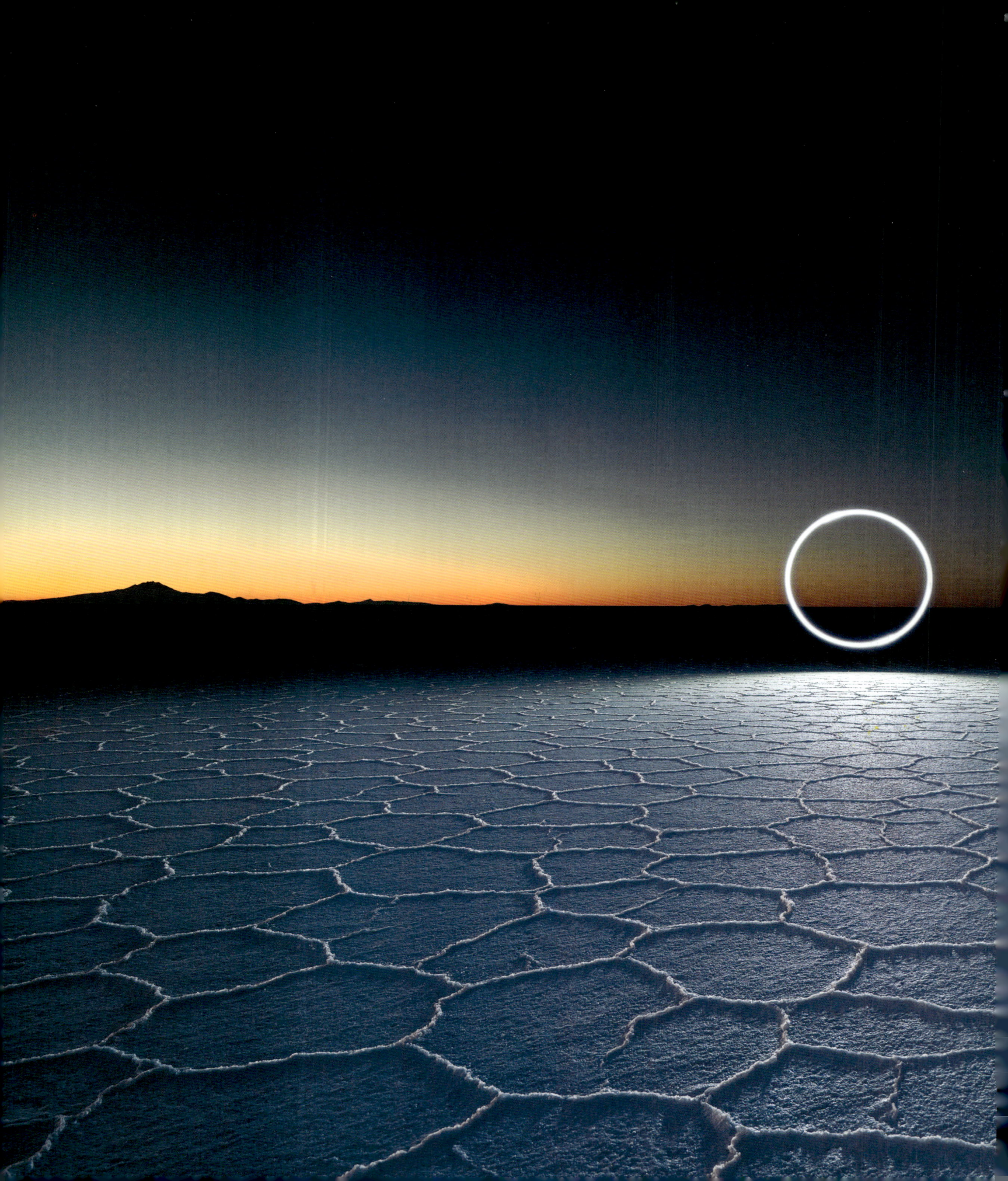

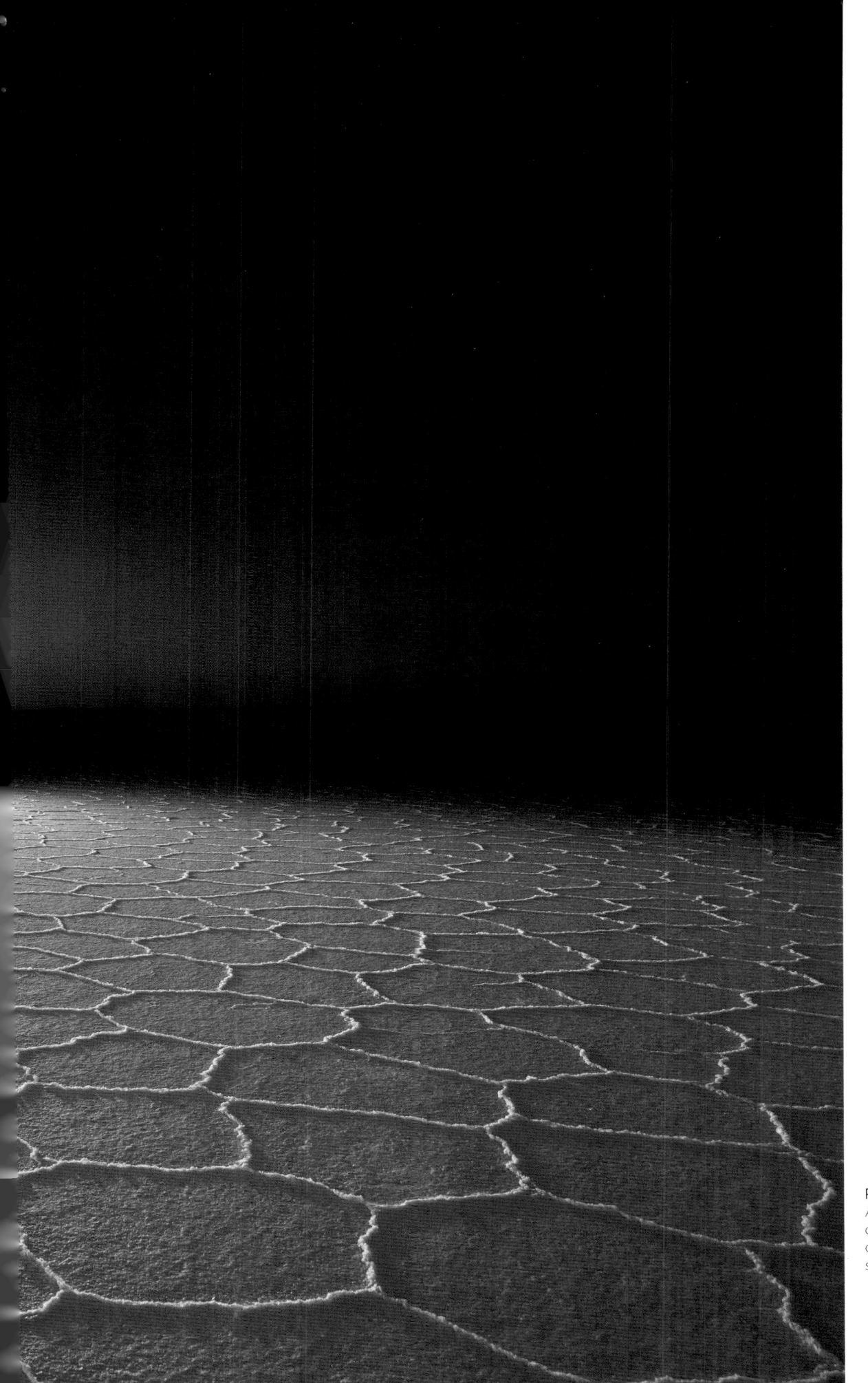

Reuben Wu
An LED connected to a drone
creates a perfect circle of light
over Bolivia's Salar de Uyuni
salt flats

277

my fascination for mountains and wild environments. I think that interest, combined with my fascination with science fiction, created a worldview for me which was very real but also very imaginative. Films like *Close Encounters of the Third Kind, Blade Runner,* and *2001* all have a realness about them. There's a familiarity about those films, and I think that has become a theme in my work.

AV: I love your music. How does it influence your photography today?

RW: The music and the photography come from the same place. Both utilize composition, both need balance, and I think storytelling is common between the two as well. A song doesn't have to have explicit meaning, and that enables people who listen to music to come to their own conclusions about the meanings. It's the same thing with photography. I don't try and say, "This is what it means." I like people to be able to pause and imagine their own stories.

AV: Your work has a very unique visual signature. Can you describe the technique you use?

RW: As a person interested in landscape photography, it was difficult to create something new. It wasn't enough for me to wait for the sunset to happen and take a picture—people have done that since the beginning of the medium. So I began to use different techniques to try and create something novel, things like long-exposure photography, taking pictures with old film cameras in places where you may not expect people to use those cameras—basically trying to combine things. For me it was easier to make a hybrid of things which are already familiar and create something novel from that. I realized that

one way of being more creative was to have an intervention with what I was taking the picture of. In studio photography, you're able to create lighting unique to the photo. I had an idea to bring it into landscape photography, to use artificial lighting in a natural landscape. That became my way of creating something new.

The juxtaposition of artificial lights in a natural and remote environment was a jarring but compelling aesthetic. I began experimenting with drones. At the time, they were pretty rubbish at taking pictures at night, so I started to put lights on them instead. Rather than the drone being a flying camera, it was a flying light beam. Then, with my camera on the ground, I realized by accident that the flight paths of the drone became apparent in long exposures. That was really interesting to me. I started to feature those and experiment with the idea of painting the sky with these robots. I later named these formations "aeroglyphs"—motifs in the sky that are viewable from the ground. It's kind of the opposite of geoglyphs, motifs on the ground which you can view from above.

AV: So your aeroglyph technique was discovered by accident?

RW: The idea in the beginning was to not have the drone in the frame, and to have no visible source of light. But I started to realize when I was working through these photographs that the flight paths were speaking to a tangible indication of a longer sense of time, which I think is important in these images. They speak to the idea of geologic time and this idea of something that's not a decisive moment anymore. It's almost a totality of decisive moments. I think it's interesting how these images are only visible

through long exposure using the camera. It's almost like a glimpse of a different dimension of time.

AV: Are there photographers or artists who have influenced your work?

RW: Gregory Crewdson's work was an inspiration for creating this element of mystery and drama out in the middle of nowhere. A lot of film directors as well: David Lynch is a huge inspiration. Alejandro Jodorowsky, Steven Spielberg. Then Caspar David Friedrich, the painter, and Georgia O'Keeffe. All these artists have been able to create their own worlds and craft a new reality.

AV: Do you think about your audience when you are taking a photograph?

RW: Not really. I think that bit comes later. When I'm taking these images, I'm in a flow state where I'm not thinking about exterior things like that. I'm thinking about where I am and what I'm trying to create. It becomes quite an all-consuming mindset. I lose track of time, particularly in these places where it's dark and remote with no other people. Later on, in the comfort of my studio when I'm looking at the images, maybe I'll think about that. But I think it's important for me to remain separate from the idea of how pictures can be received.

AV: I want to ask about one of your early aeroglyphs. It's a ring of light above what looks like a salt field. How did you take that image?

RW: That was in Bolivia at a place called Salar de Uyuni, the largest salt flat in the world. I was on a two-week road trip. We started in northern Chile, in the Atacama, and headed north into Bolivia into the high desert, where we ended

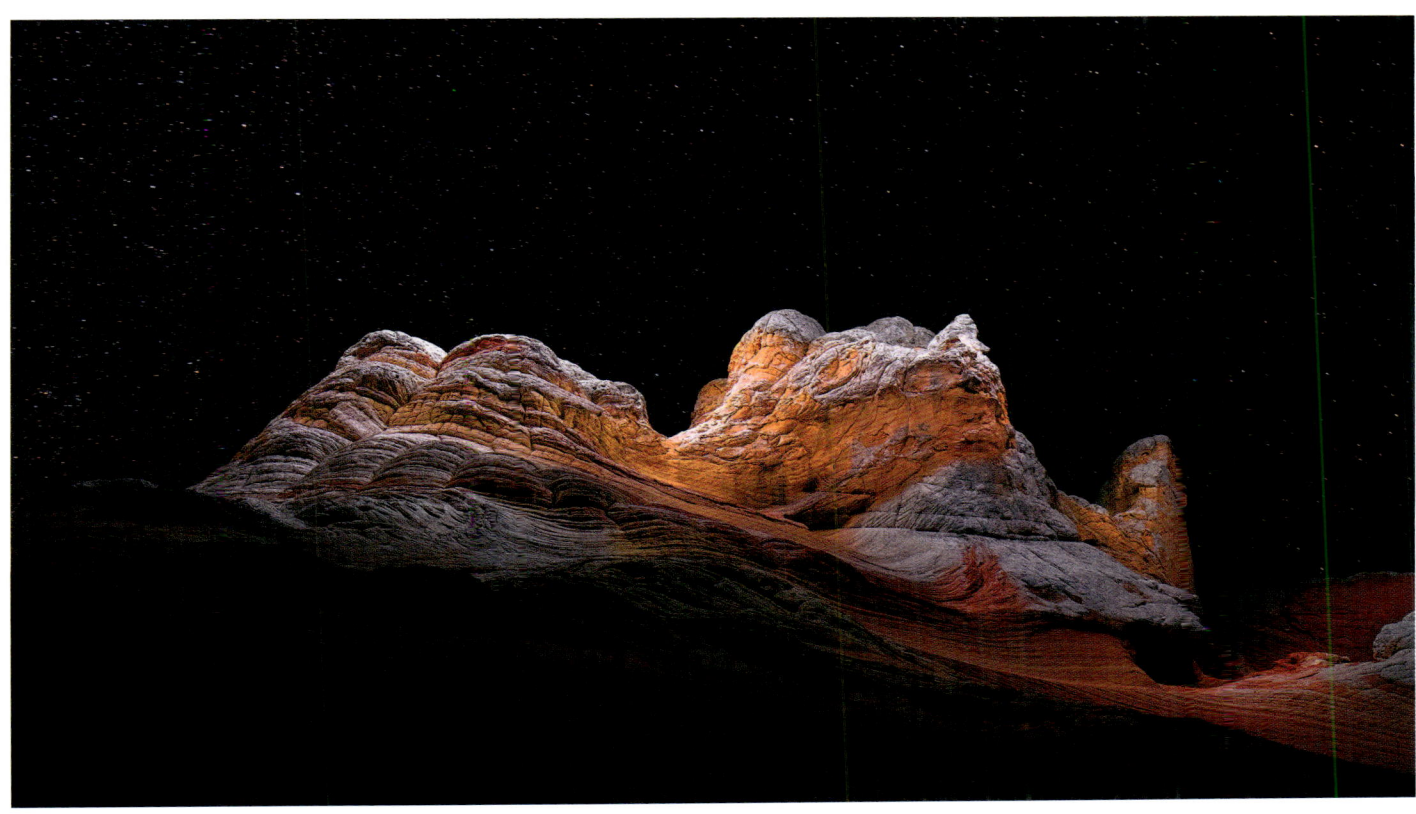

Reuben Wu
A drone bathes a Utah rock formation with light from above.

up in Salar de Uyuni, at about 10,000 feet in altitude. Some of it was flooded with a few inches of water, and some of it was completely dry and encrusted. I was fascinated by the polygons that form after it dries, so I was experimenting with placing light to create shadows. The salt is a very translucent, white surface, so it glowed in really interesting ways. And normally, if you step on these ridges, they will instantly be crushed, but because I was using a drone, I was able to fly the lights to where I needed them without any damage to the environment. I thought that was a pretty cool application of the tool.

AV: In another aeroglyph, you can see a green body of water and three vertical lines in the distance. What's the story behind that photograph?

RW: That was at Amboy in California. There's a salt-evaporation facility there, so there are all these evaporation ponds full of salt water nearby. This channel is one of the few where the water was bright, luminous green, and the bed of the channel was encrusted with salt crystals. I had this idea of using the drone to light the water and create a glowing channel of water that stretches into the distance. So I grabbed a light and put it on a fishing line, attached the fishing line to my drone, then flew the drone out to submerge the light underwater.

AV: Wow!

RW: It wasn't a very good method actually, because the light kept on bobbing up and down. What would have been better would be to have

a remote-control boat, or an inflatable dinghy I could row out there. But that was the final picture, and it looked cool.

AV: Are there new frontiers you're excited about, or new technologies you want to explore?

RW: I could see myself in the future beginning to not use the camera in my work. I feel that my role is less about making pictures and videos and more about sharing experiences. I would like to explore the world beyond using cameras, where I would be able to actually share the experience of being in the landscape itself, where you could actually interact with your surroundings. It's a pretty interesting time to be alive at the moment, with all these different technologies, like VR and AR.

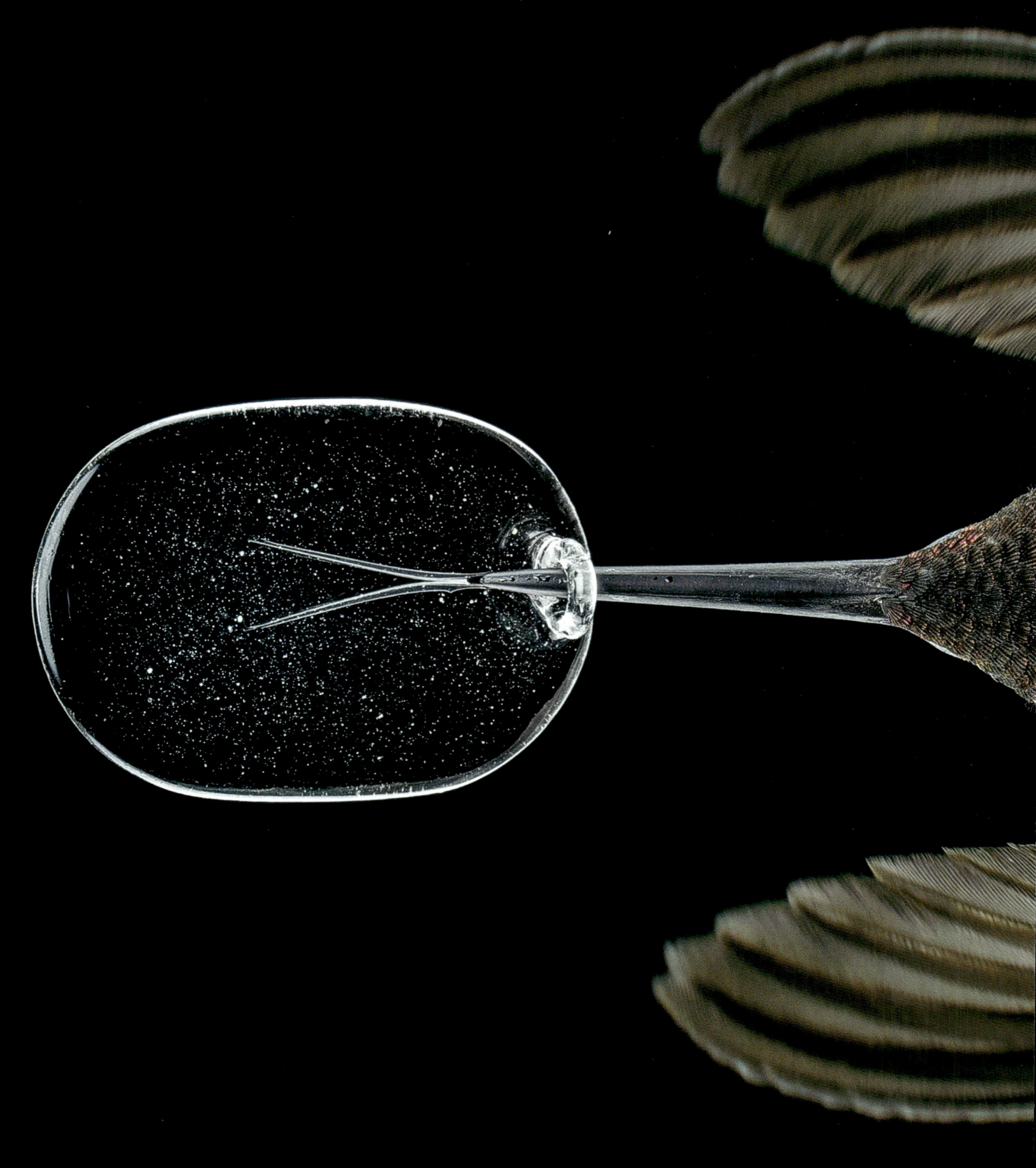

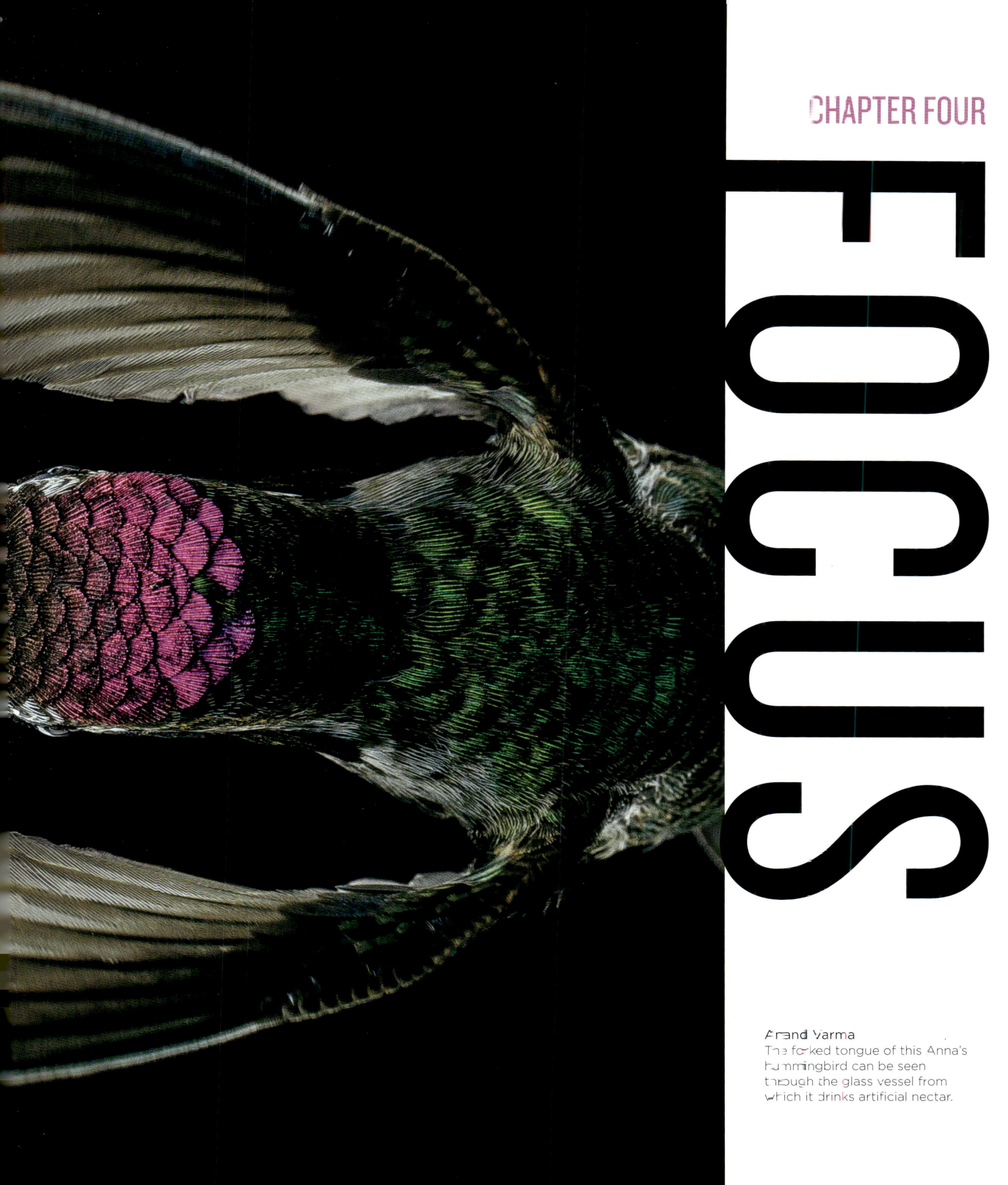

CHAPTER FOUR

FOCUS

Anand Varma
The forked tongue of this Anna's hummingbird can be seen through the glass vessel from which it drinks artificial nectar.

WHEN I PITCHED my first story idea to *National Geographic*

magazine, my editor, Susan Welchman, said: "Every photograph of a hummingbird that can be taken has already been taken." The rejection stung, but Susan had a point. Did the world need another generic photograph of a hummingbird? What new insight did I have to offer?

For inspiration, I turned to the scientists who study hummingbirds. I visited labs that were using sophisticated technology—wind tunnels and high-speed video cameras—to observe their maneuvers. I watched my friend Alejandro Rico-Guevara use transparent plastic flowers to study how hummingbirds drink. He captured video demonstrating how their flexible tongues compress and expand at incredible speed to gather nectar. Had I simply looked over his shoulder and taken a picture though, none of this magic would have been visible. I set out to re-create his experiment in a way that allowed me to photograph the wonder taking place.

I zoomed in to show the details of the hummingbird's tongue. I used a flash to freeze the bird's motion. I made sure that the light came in at just the right angle to illuminate the iridescent colors of its feathers. But controlling size, time, and light wasn't enough to clearly show what the hummingbird was doing. I needed a way to further focus the audience's attention.

I commissioned Adams & Chittenden, a scientific glassblowing company, to fabricate a "hummingbird dinner plate," and then I trained an Anna's hummingbird to drink from it. The glass dish helped me clearly define the shape of the hummingbird's tongue, thus focusing the viewer's gaze on the most fascinating element in the scene.

The photographers in this chapter use a range of strategies to direct our attention. Emanuele Biggi trains us to look carefully with his close-up of a camouflaged snake (page 311). Mark Harvey captures a golden eagle in flight (page 354). Thomas Peschak uses a shaft of light to isolate a whale shark in the open ocean (page 296). Tim Flach strips away all distractions from his frame to give us a stunning face-to-face encounter with a white tiger (page 300).

THESE IMAGES INFLUENCE WHAT WE FALL IN LOVE WITH AND WHAT WE FIGHT TO PROTECT. THIS JOURNEY BEGINS WITH WONDER.

Distance, moment, illumination, framing. These elements encompass the ideas of size, time, and light already presented in this book. But here, our definition of focus takes us beyond these technical ideas. The photographs in this chapter explore subjects that are not strictly invisible to us. Instead, each frame is a thoughtful composition that compels us to slow down and consider those layers of complexity and detail in the scene that we might miss at first glance.

By surprising us, these images make space for us to reconsider what is repulsive and what's beautiful. By holding our attention, they challenge our notion of what is boring and what is worthy of our care. By creating new connections between subject and viewer, they influence what we fall in love with and what we fight to protect.

This journey begins with wonder. ∎

FOLLOWING PAGES.
Reuben Wu
Stonehenge at sunset, illuminated from above by an LED attached to a drone

283

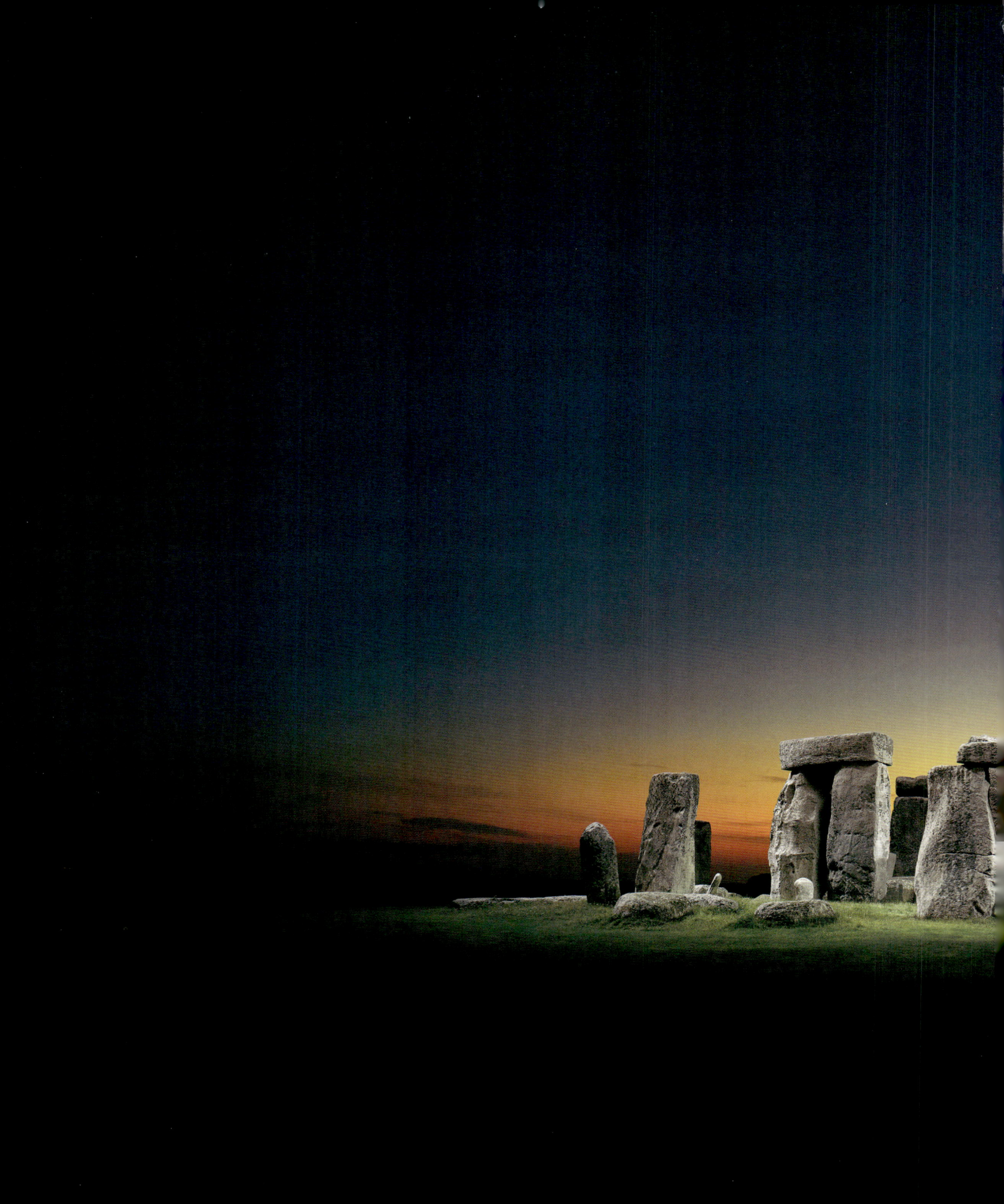

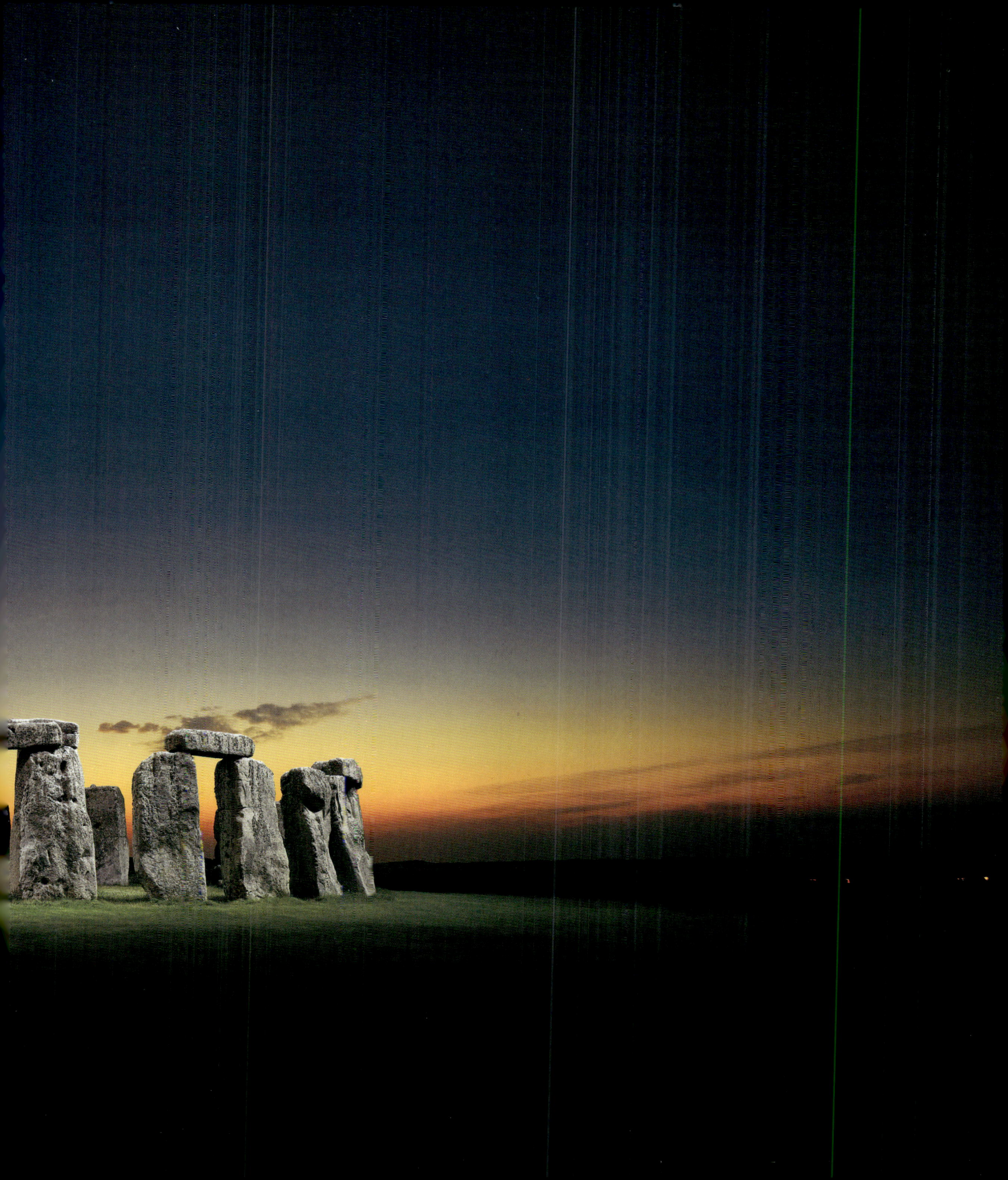

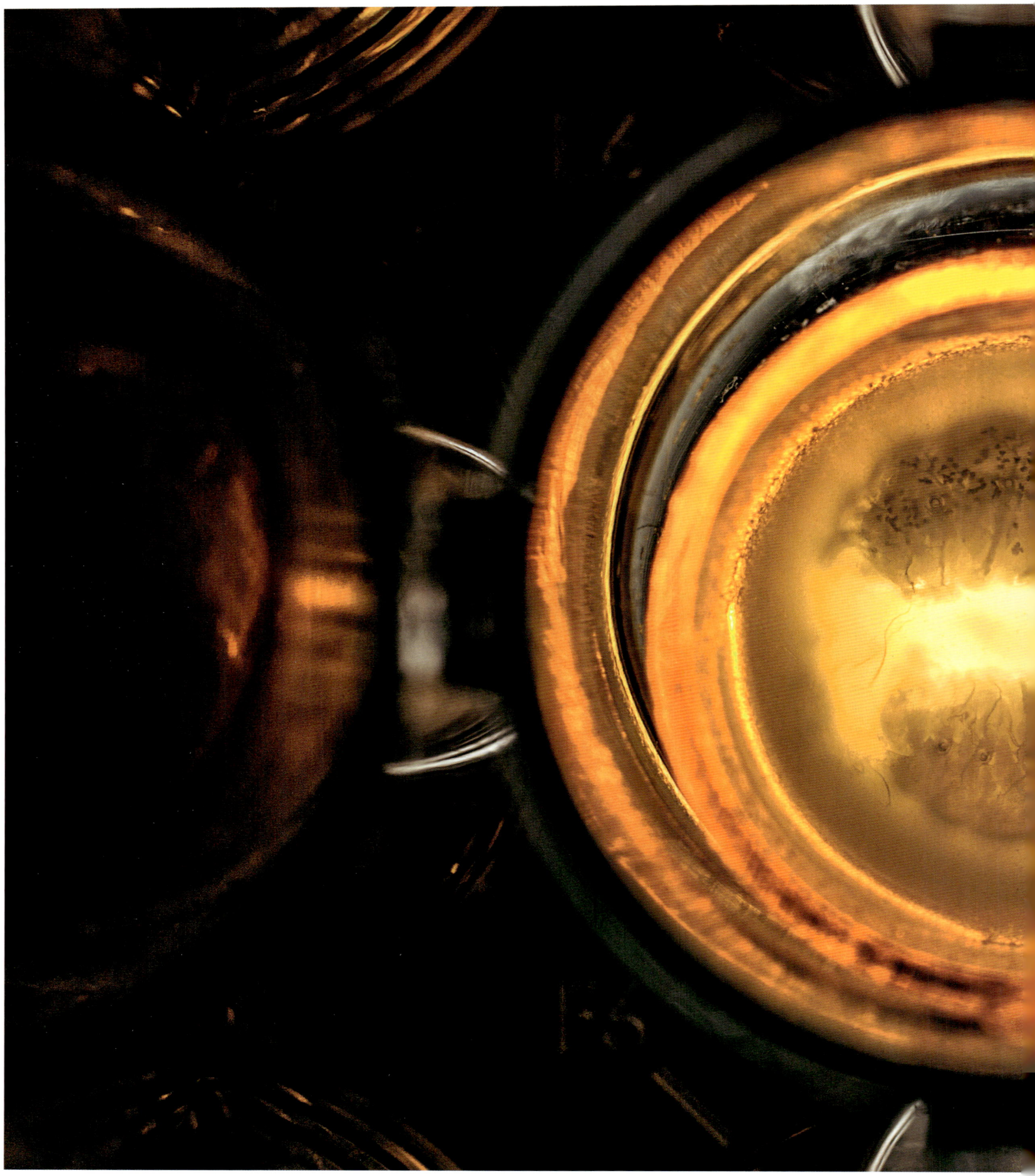

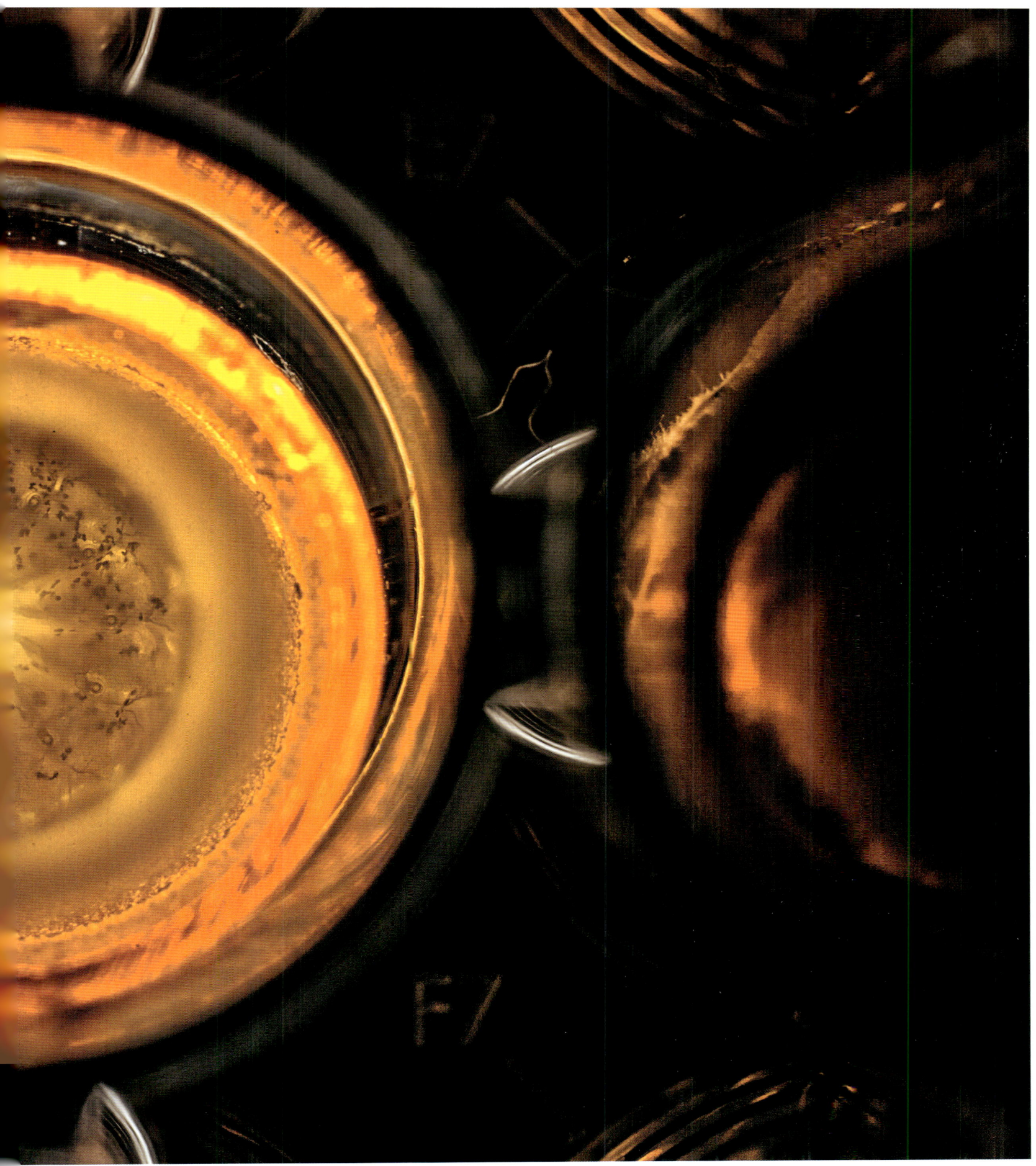

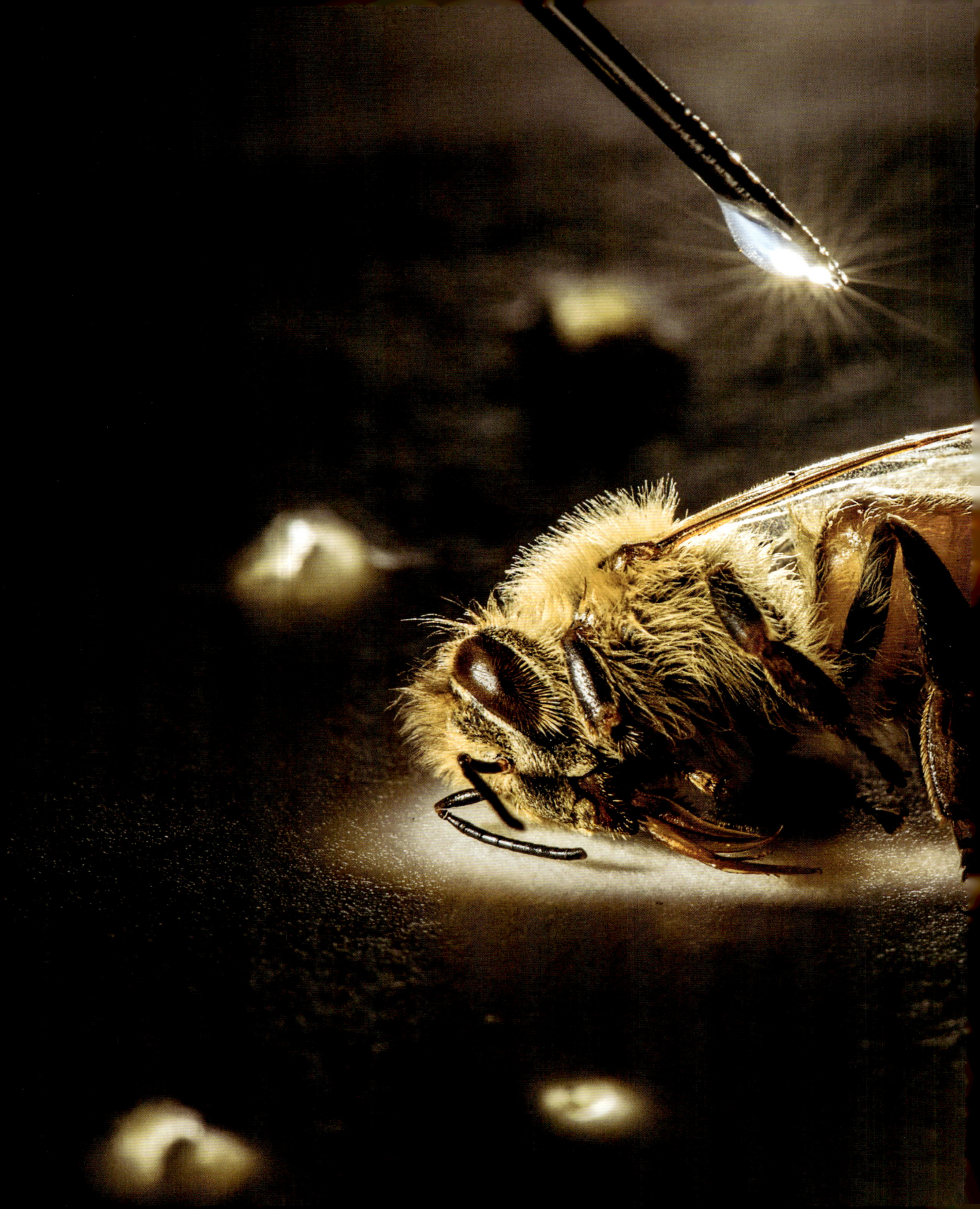

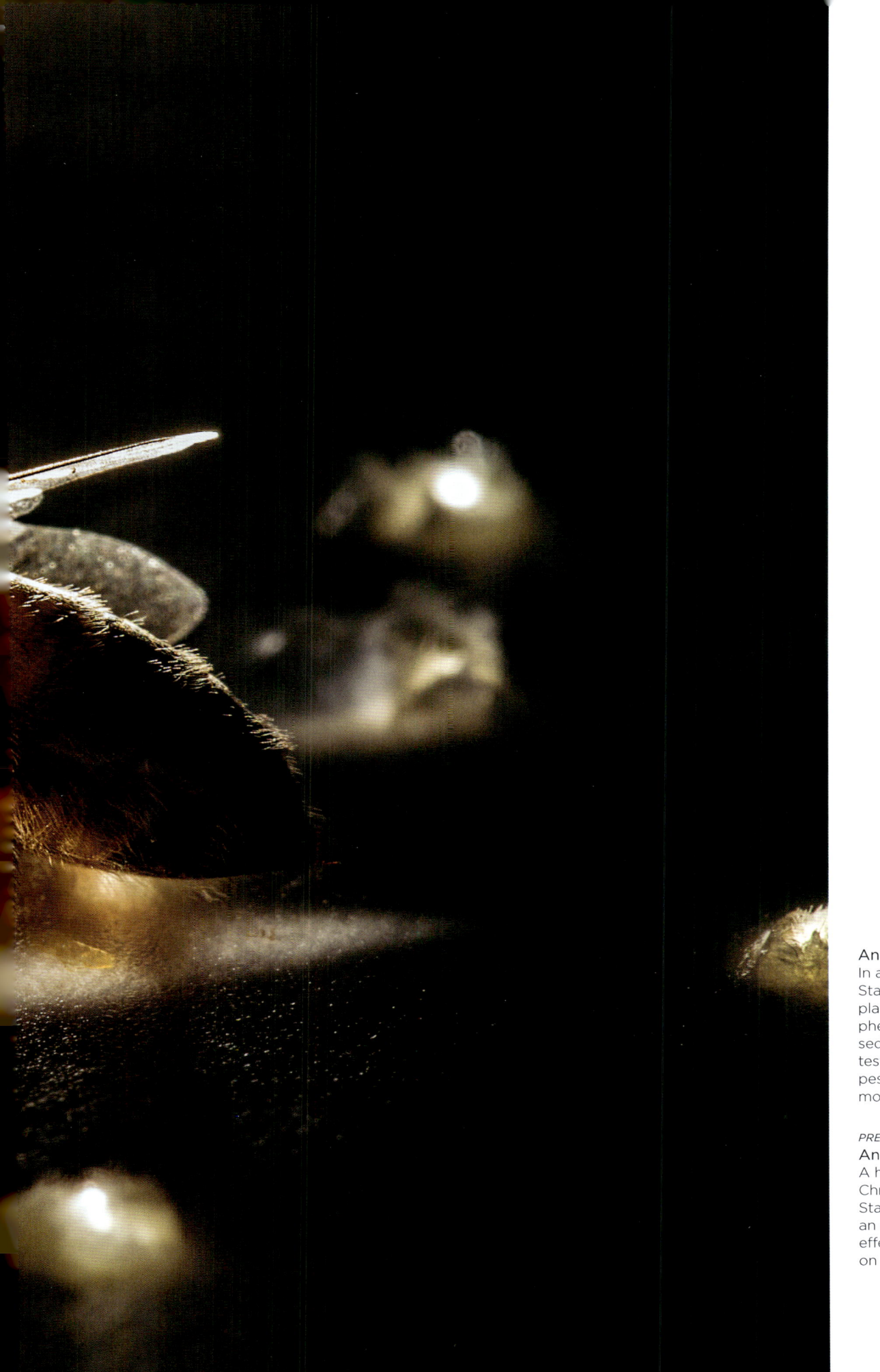

Anand Varma
In an experiment at Louisiana State University, a syringe places a minute droplet of phenothrin on a honeybee—sedated in a paper cup—to test the effects of the potent pesticide designed to kill mosquitoes.

PREVIOUS PAGES:
Anand Varma
A honeybee larva raised by Chris Mullin's lab at Penn State University as part of an experiment to study the effects of agricultural sprays on bees

Anand Varma
Bret Adee inspects one of his 72,000 beehives at sunrise on a secluded ranch in Southern California. Adee runs the largest commercial beekeeping operation in the world.

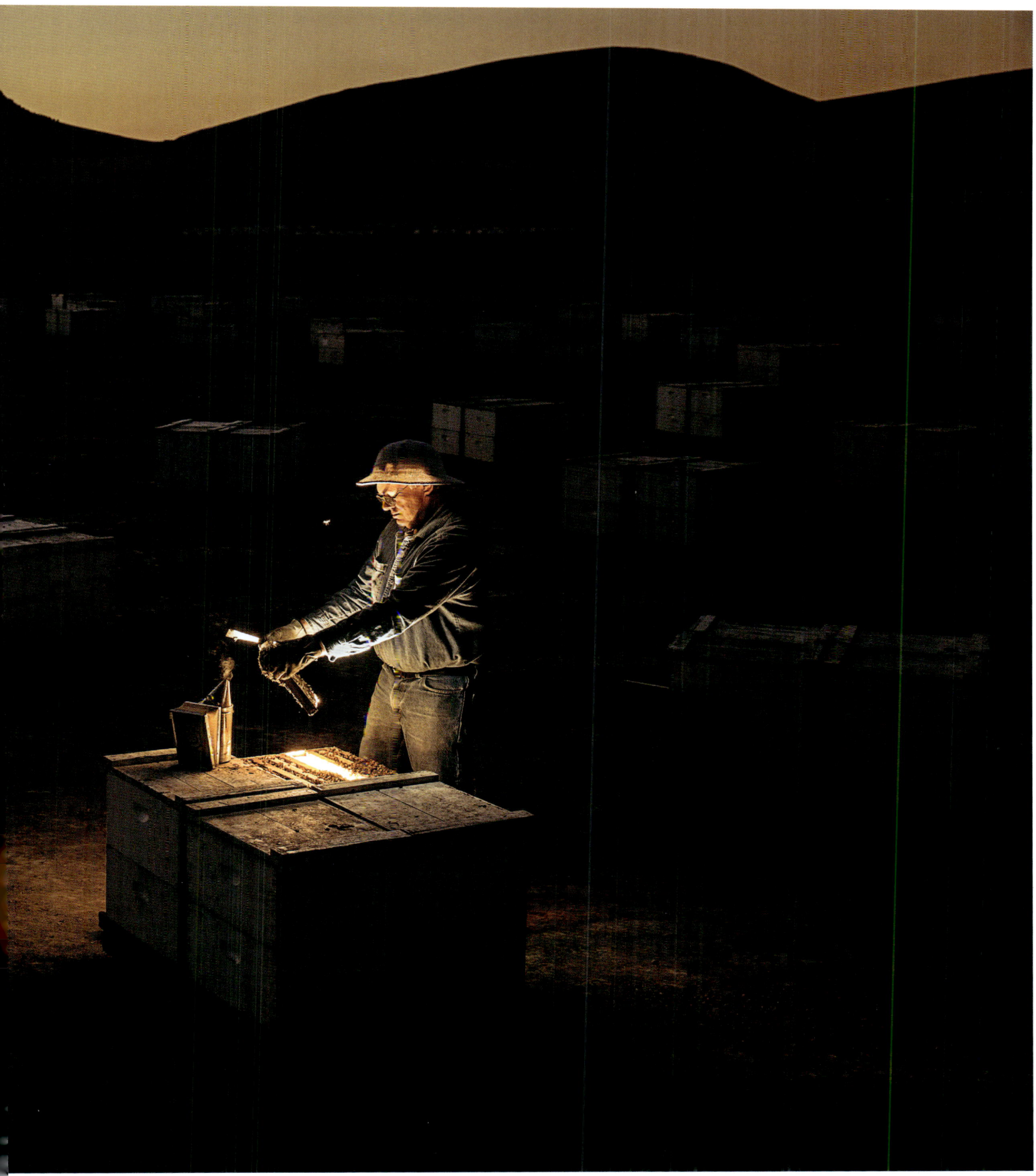

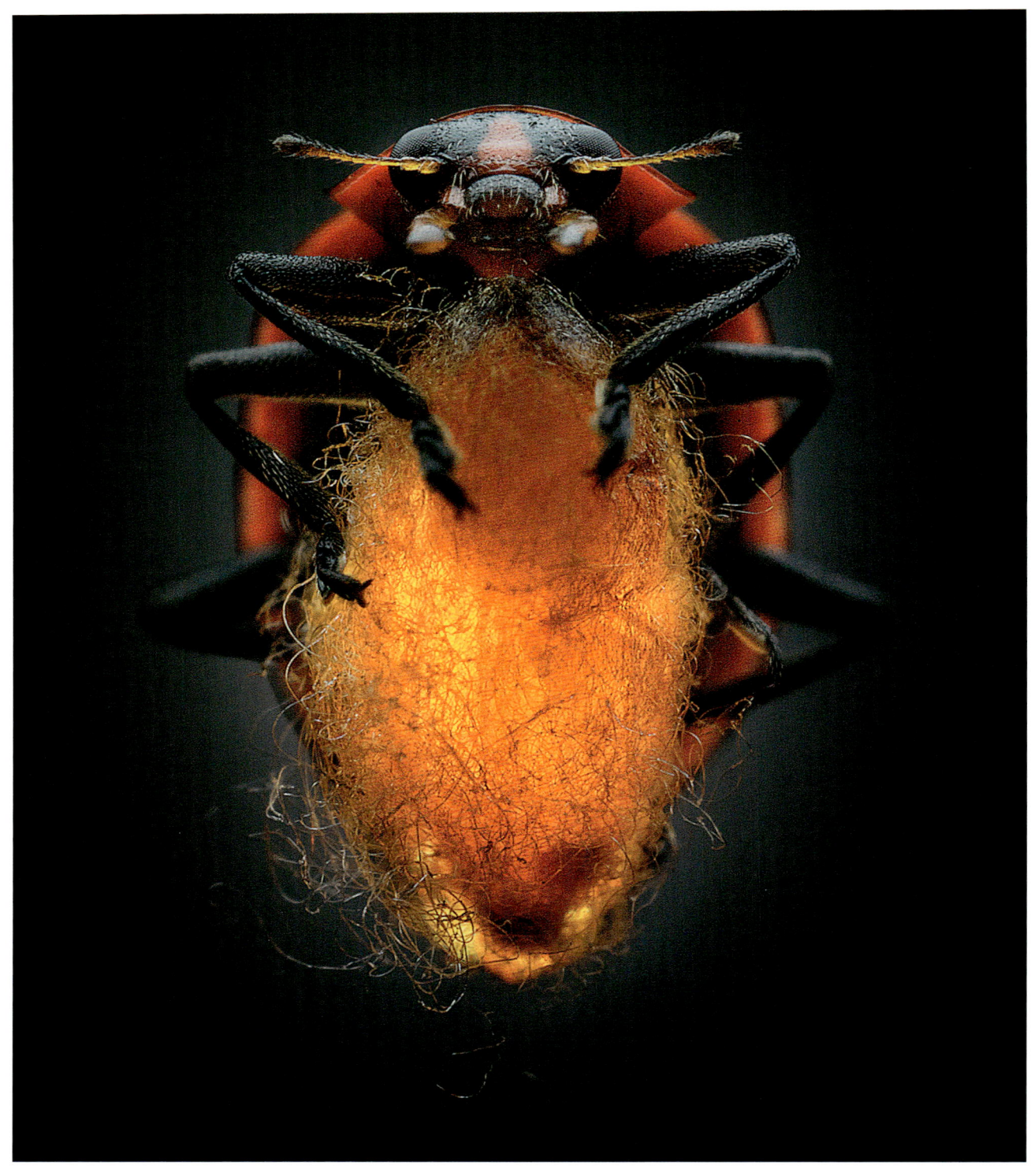

Anand Varma
A parasitic wasp spins its cocoon between the legs of a spotted lady beetle.
The wasp forces the beetle to become a bodyguard and ward off predators.

FOLLOWING PAGES:
Jonny Armstrong
A soft light beam spotlights the face of a red fox.

SURPRISING PHOTOGRAPHS
MAKE SPACE FOR US TO
RECONSIDER WHAT IS REPULSIVE
AND WHAT IS BEAUTIFUL.

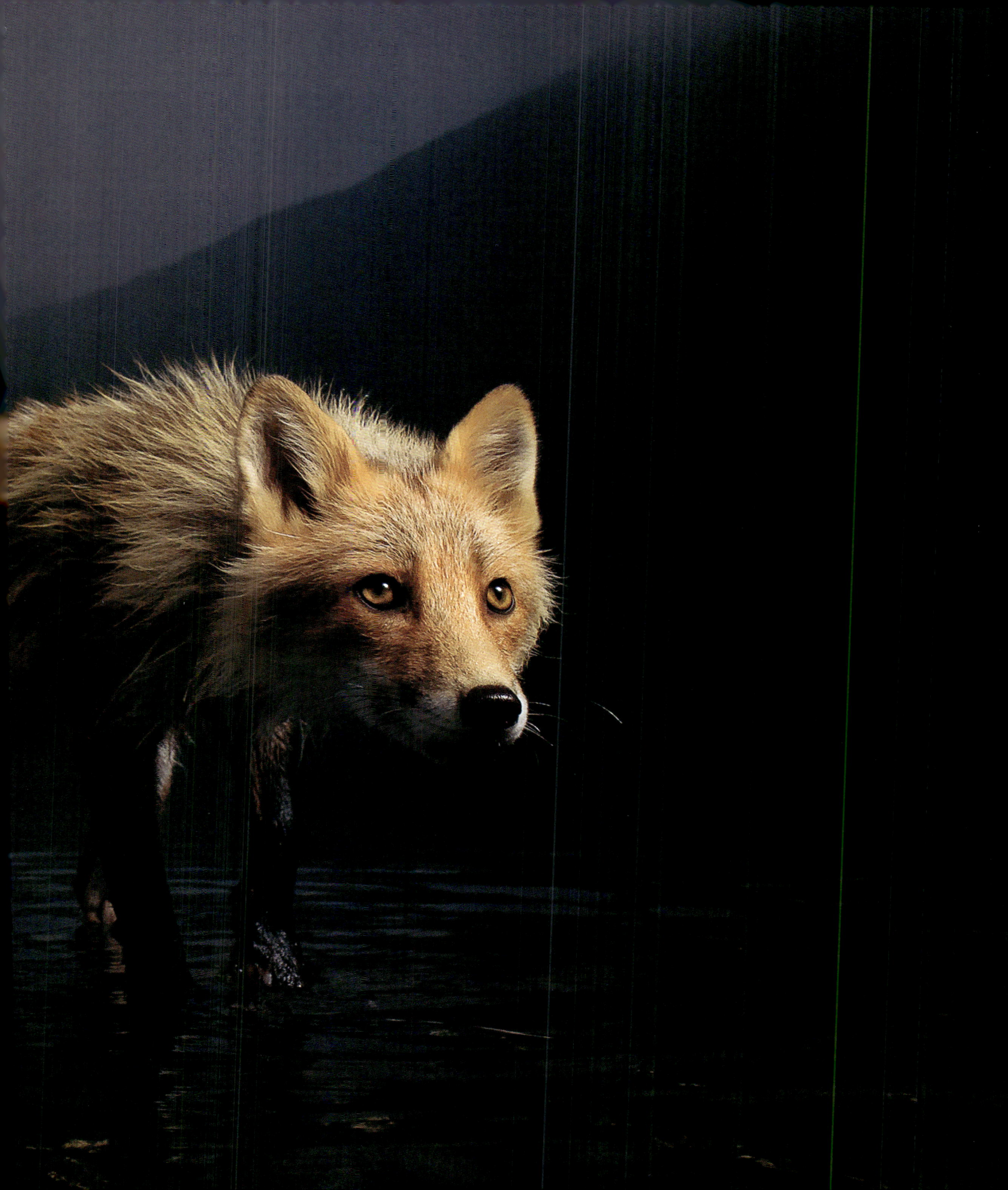

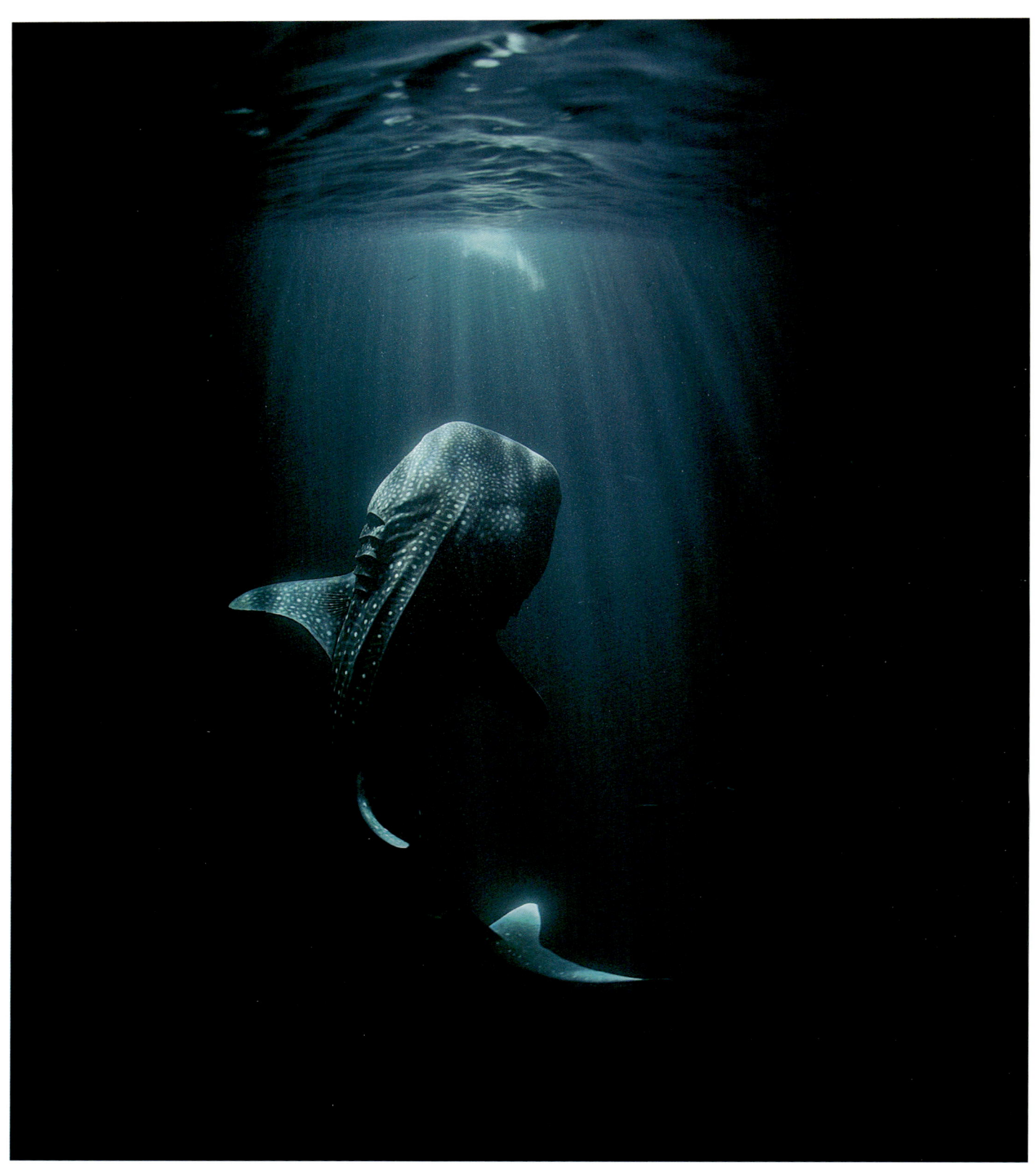

Thomas Peschak
Lights from a fishing vessel off the coast of Djibouti attract plankton,
which in turn draw the attention of a young whale shark.

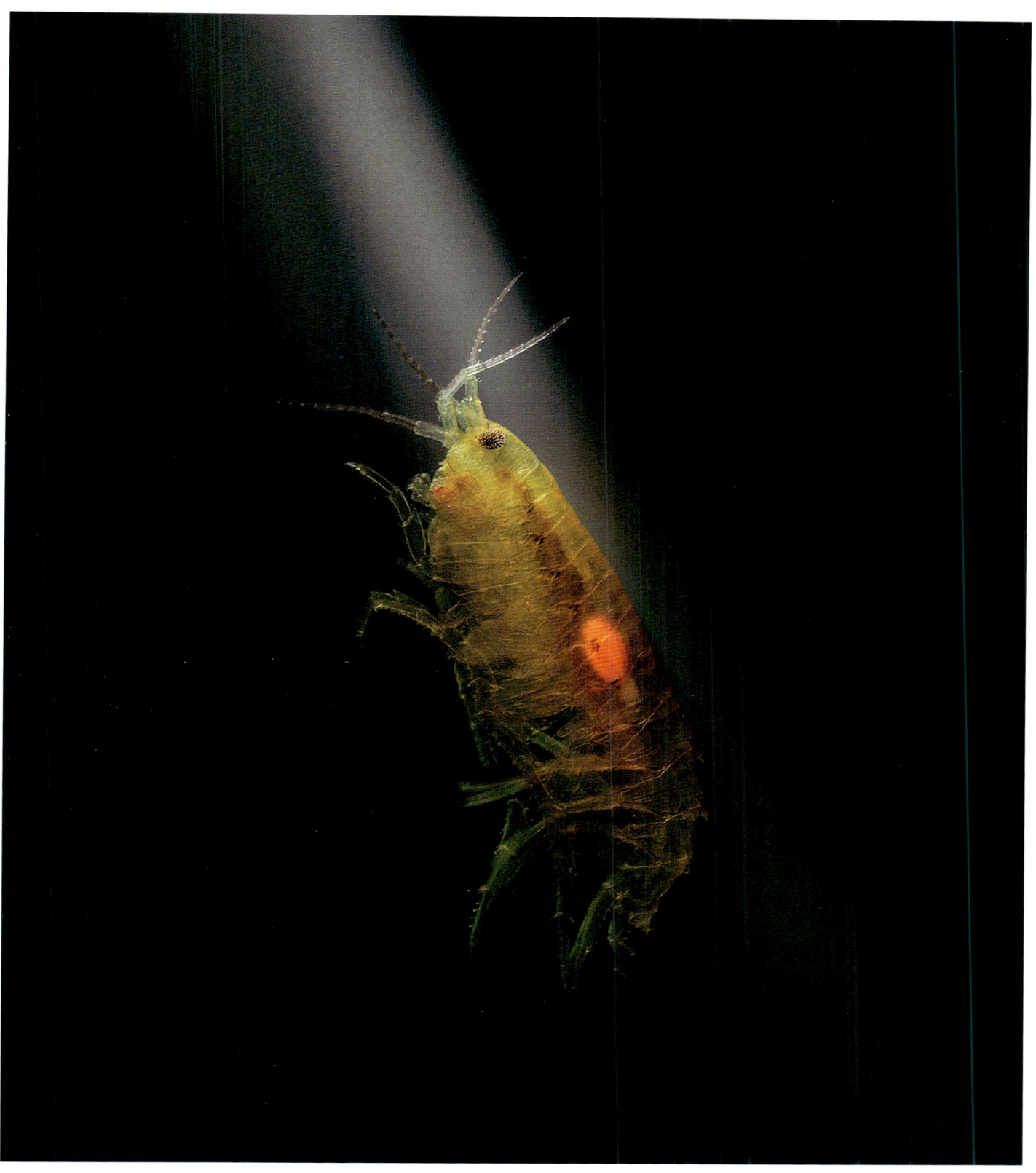

Anand Varma

The larva of a thorny-headed worm controls its host, a shrimplike amphipod crustacean, from the inside. Once invaded, the amphipod is forced to seek light, making it more likely to be eaten by a duck—the parasite's desired destination.

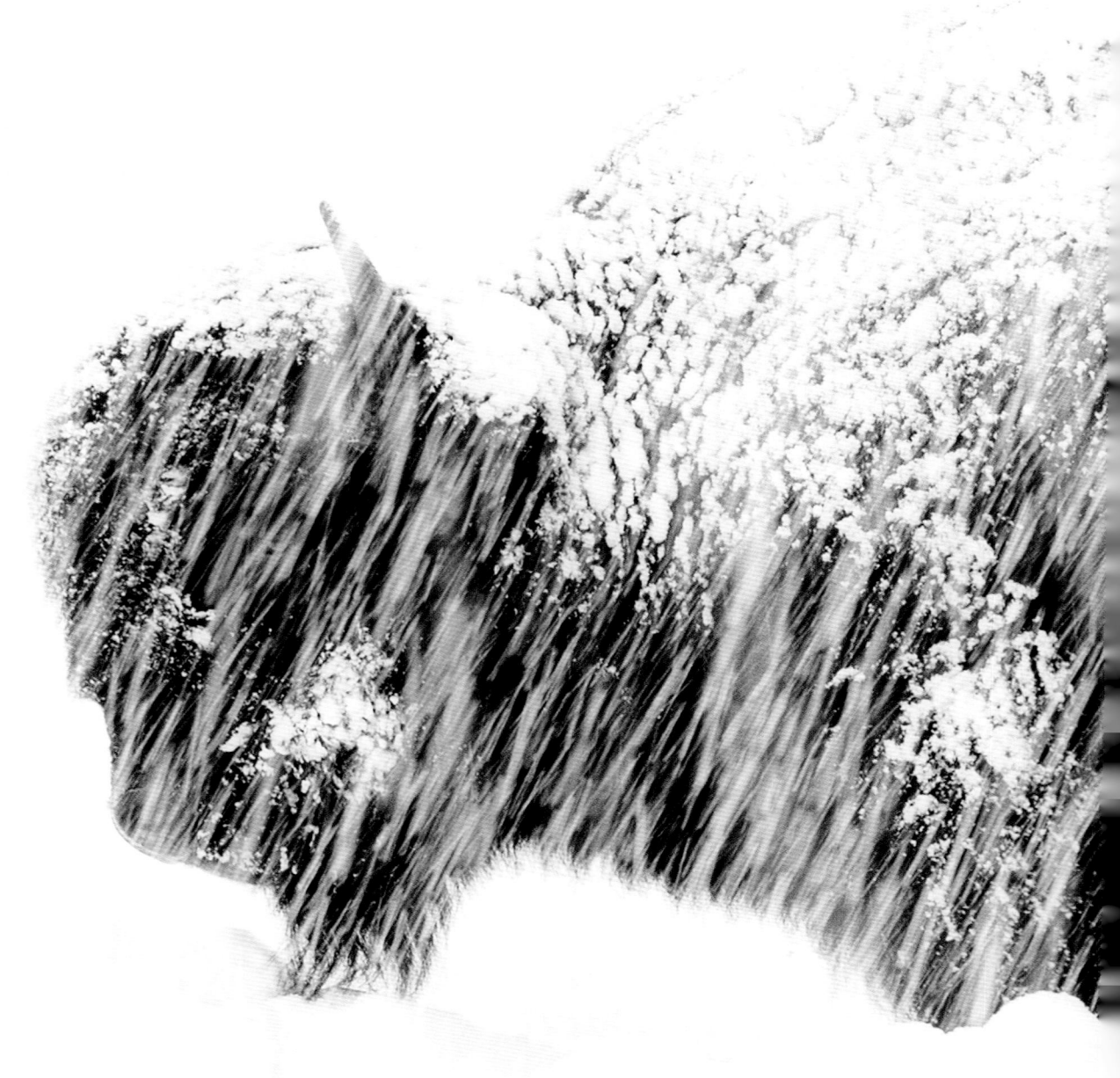

Max Waugh
A slow shutter speed blurs falling snow against the dark body of an American bison in Yellowstone National Park, Wyoming.

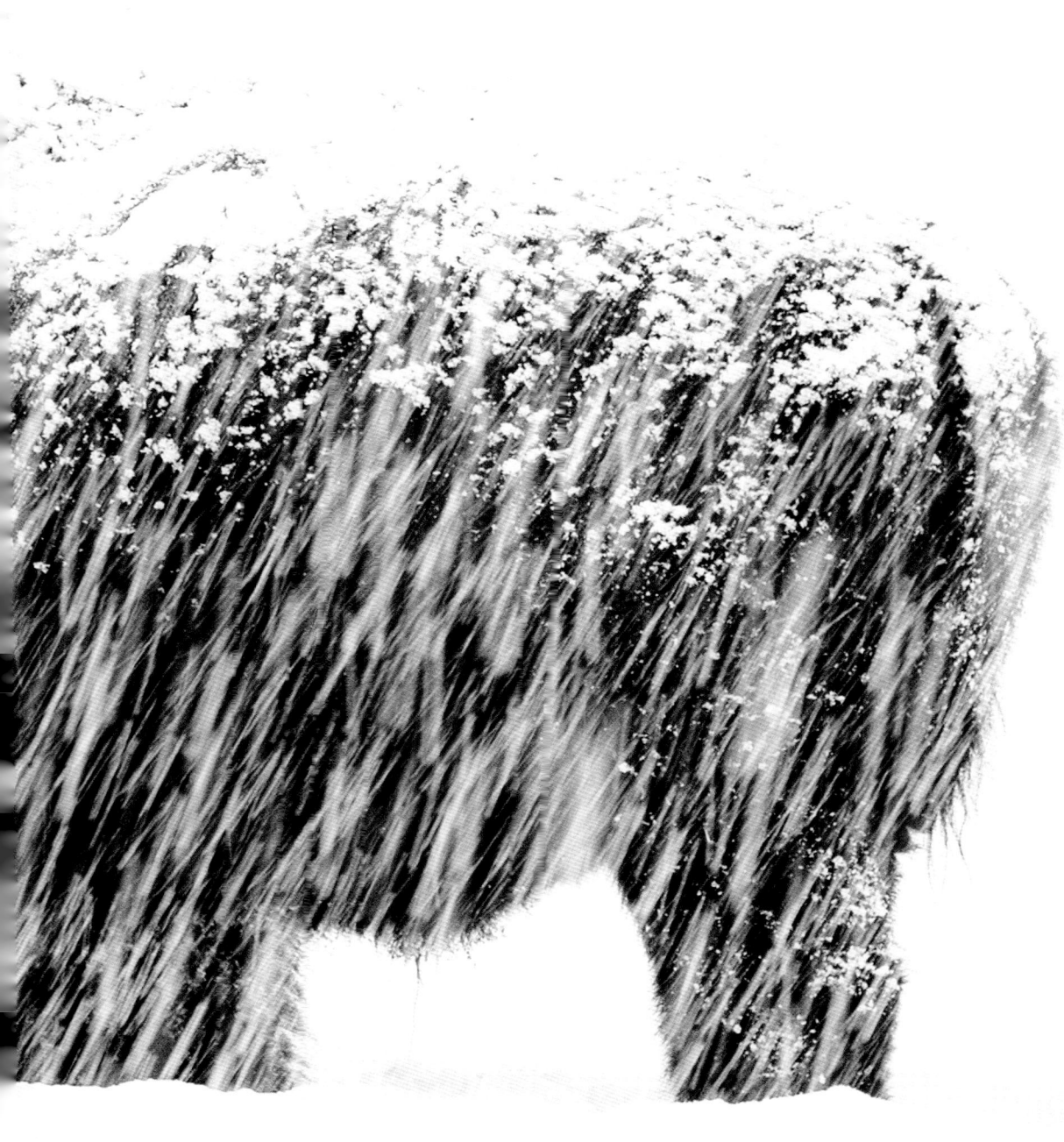

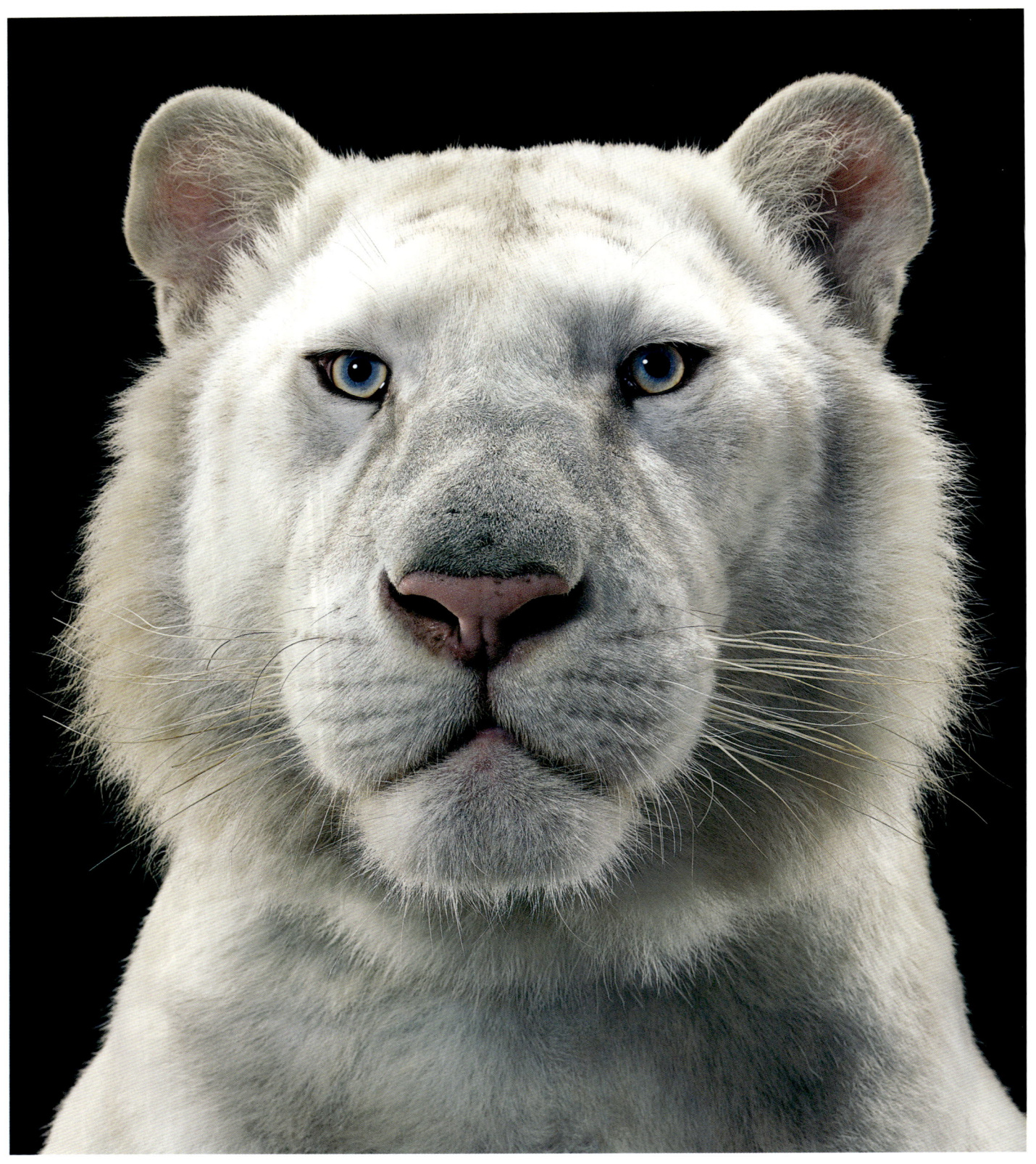

Tim Flach
Shankar, a captive-bred Bengal tiger. Its blue eyes and lack of
stripes are extremely rare, likely never found in the wild.

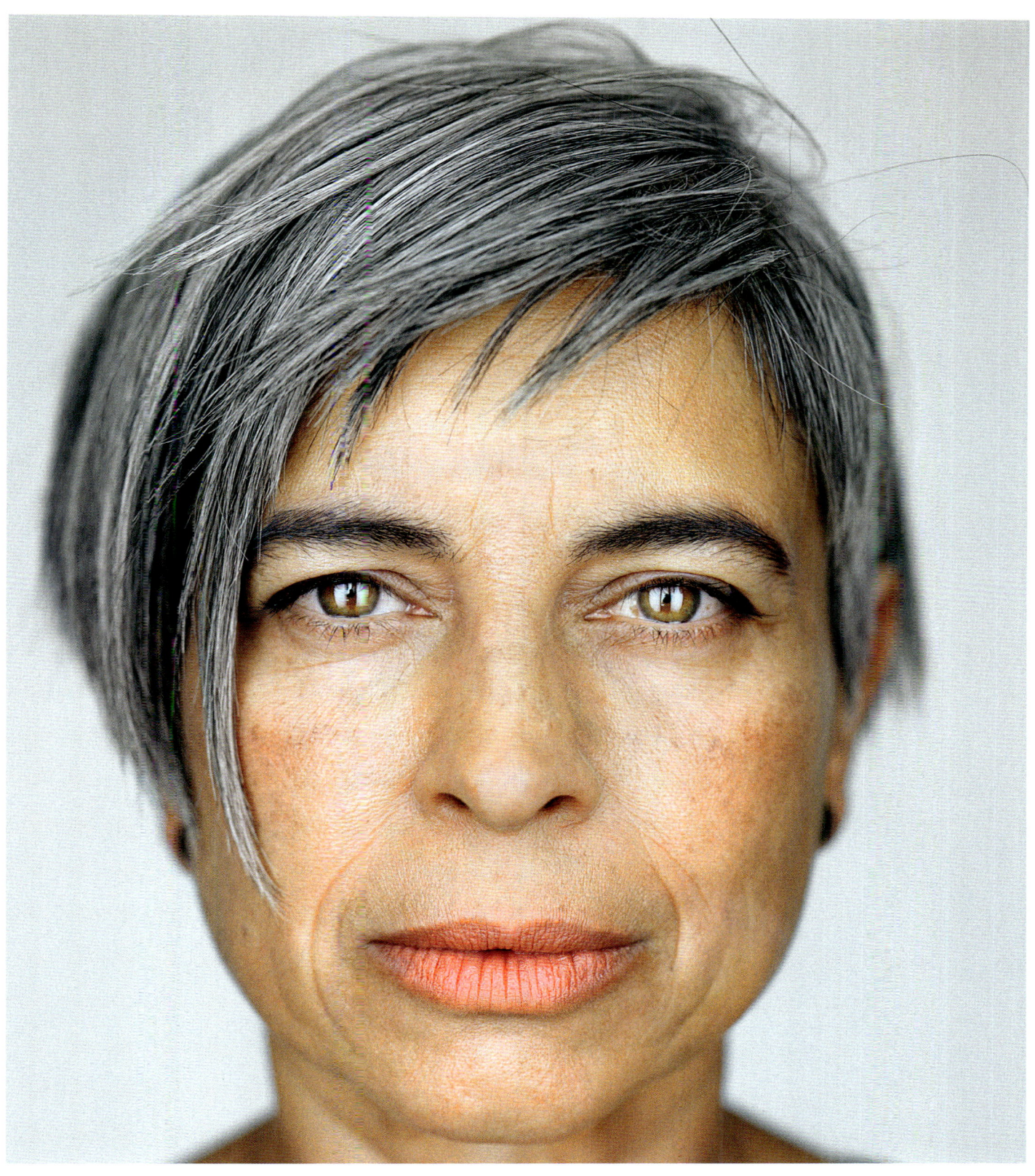

Martin Schoeller
A portrait of a woman of Asian and white descent represents multiracial families in a 2013 feature on the "changing face of America" in *National Geographic* magazine.

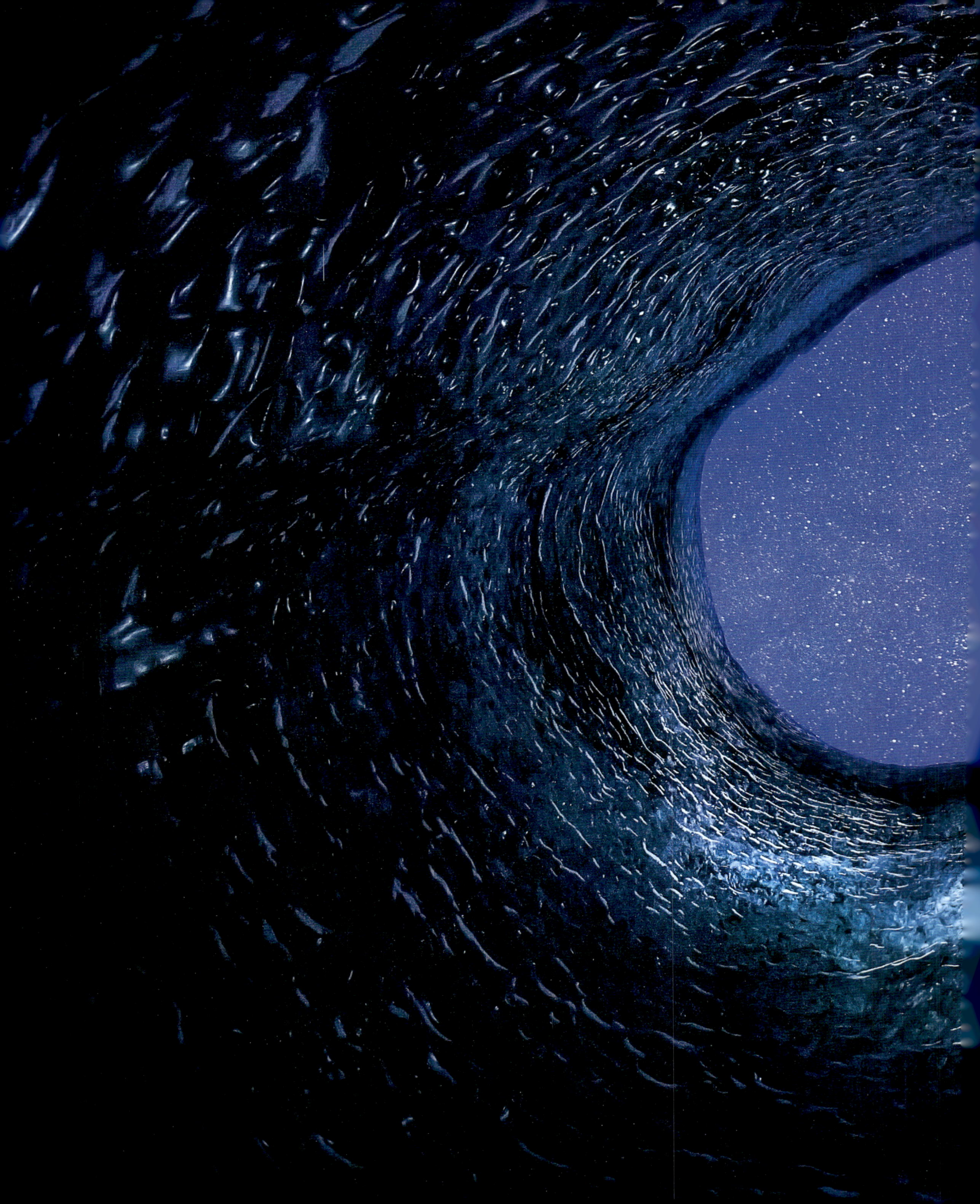

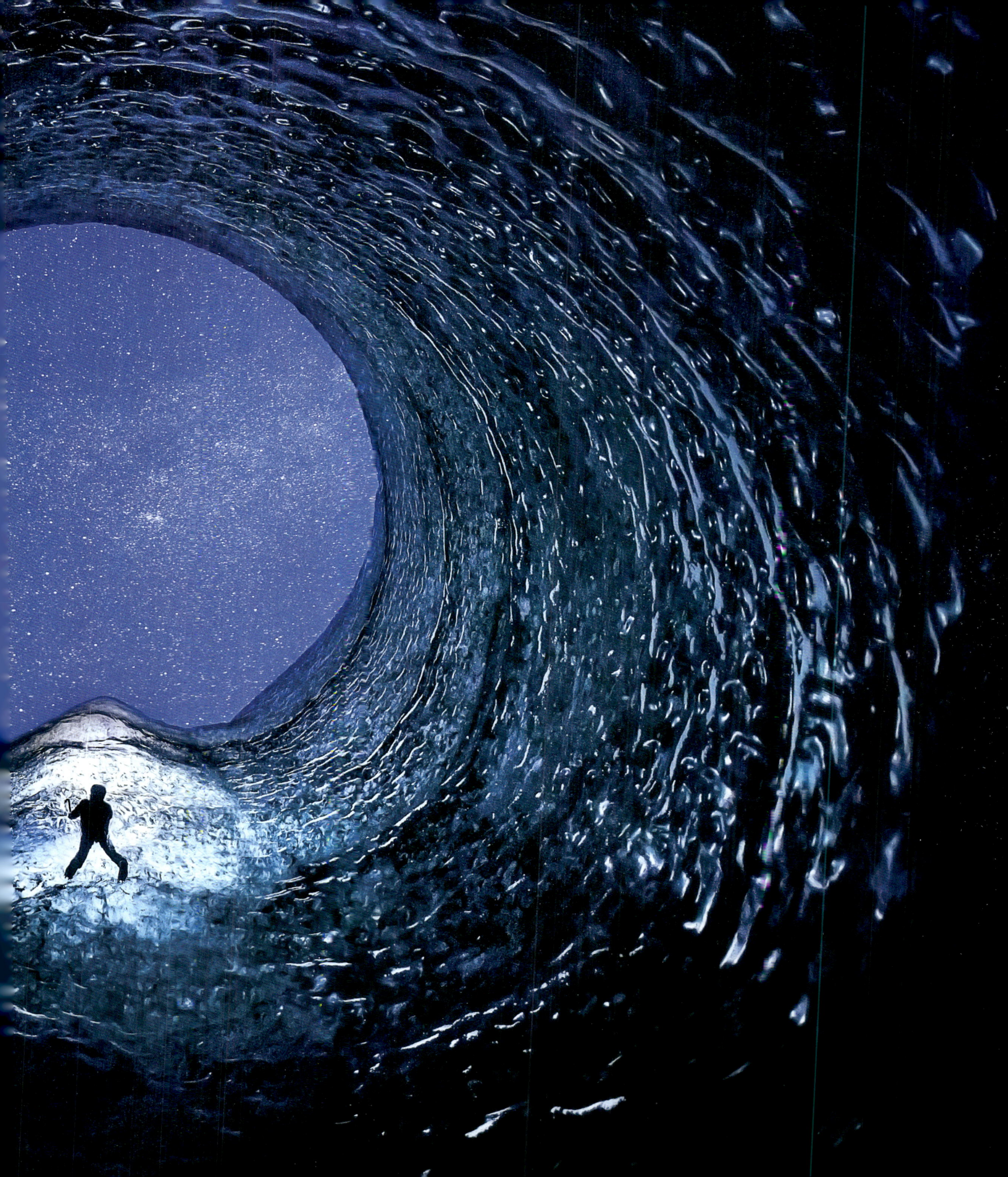

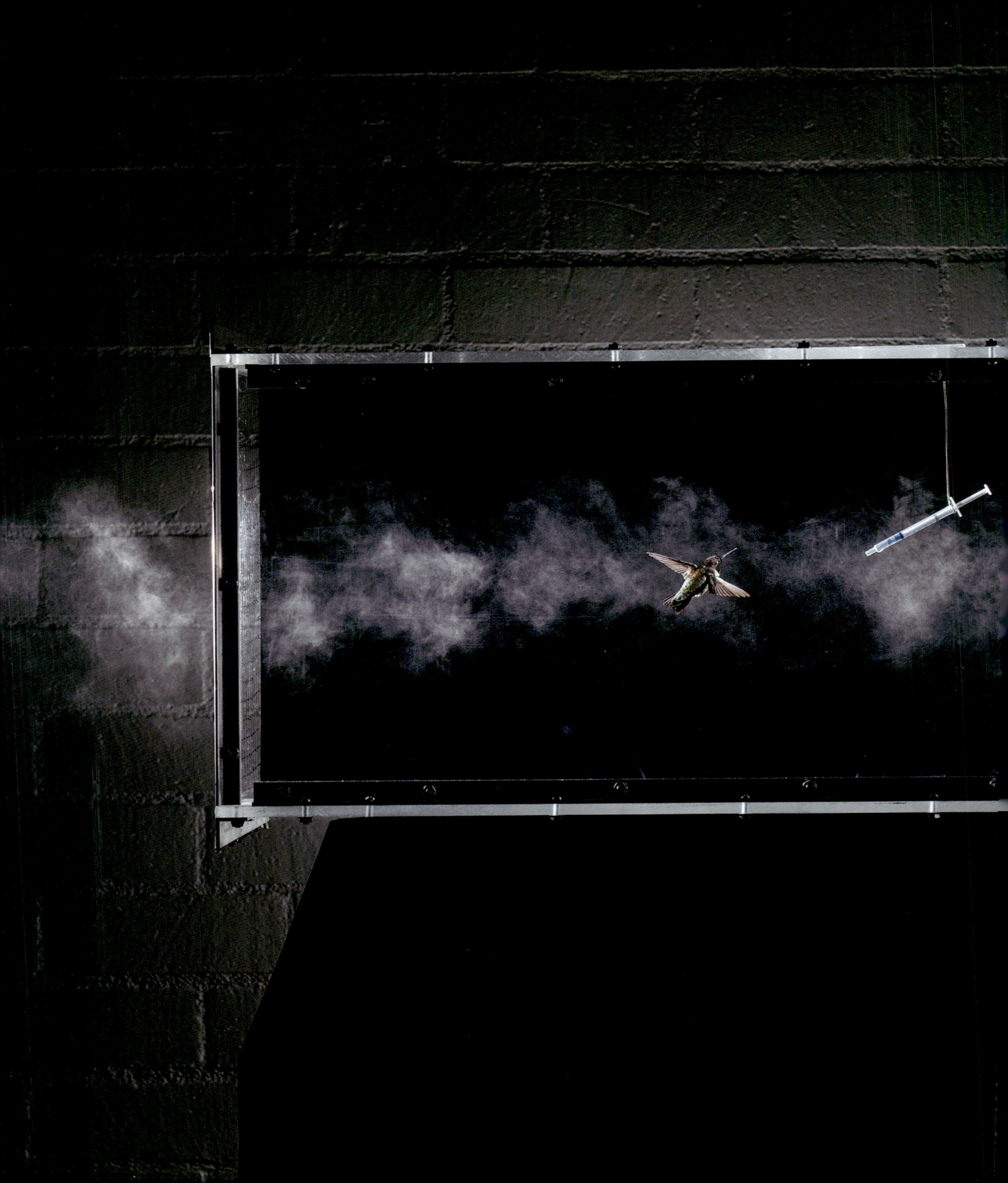

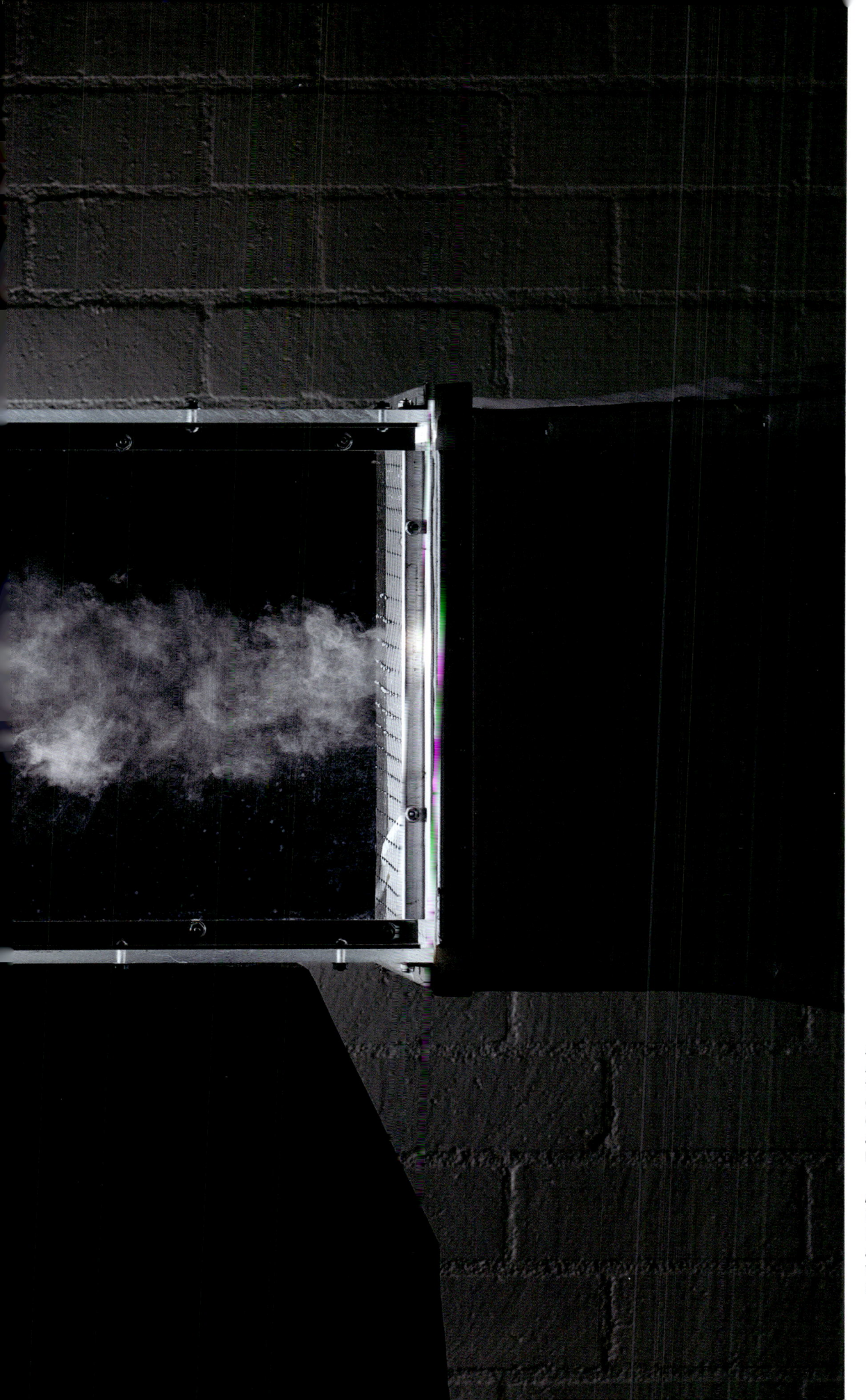

Anand Verma
A black-chinned hummingbird flies in a wind tunnel in Chris Clark's lab at the University of California, Riverside, as part of an experiment to measure the birds' flight performance.

PREVIOUS PAGES:
Paul Zizka
Ice climber Jesse Milner ascends toward the Milky Way in a shaft within Iceland's Breiðamerkurjökull glacier.

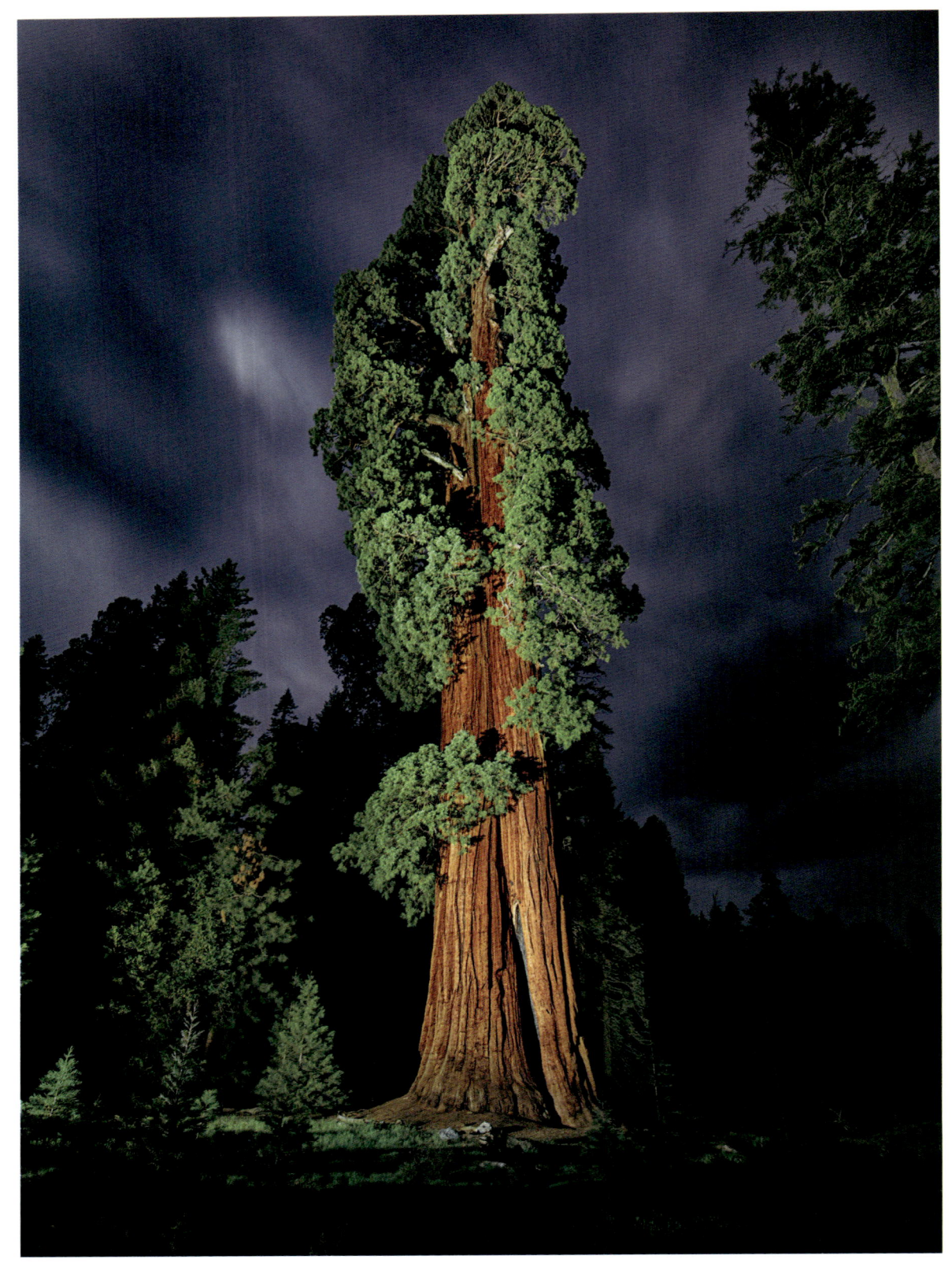

Keith Ladzinski
A giant sequoia tree isolated from its neighbors by a shaft of light
in California's Sequoia National Park

NOT EVERY SUBJECT HERE IS INVISIBLE IN THE SENSE OF BEING TOO SMALL FOR OUR EYES TO SEE. BUT EVERY IMAGE REVEALS SOMETHING WE WOULD NOT HAVE NOTICED WITHOUT THE THOUGHTFUL DIRECTION OF A PHOTOGRAPHER.

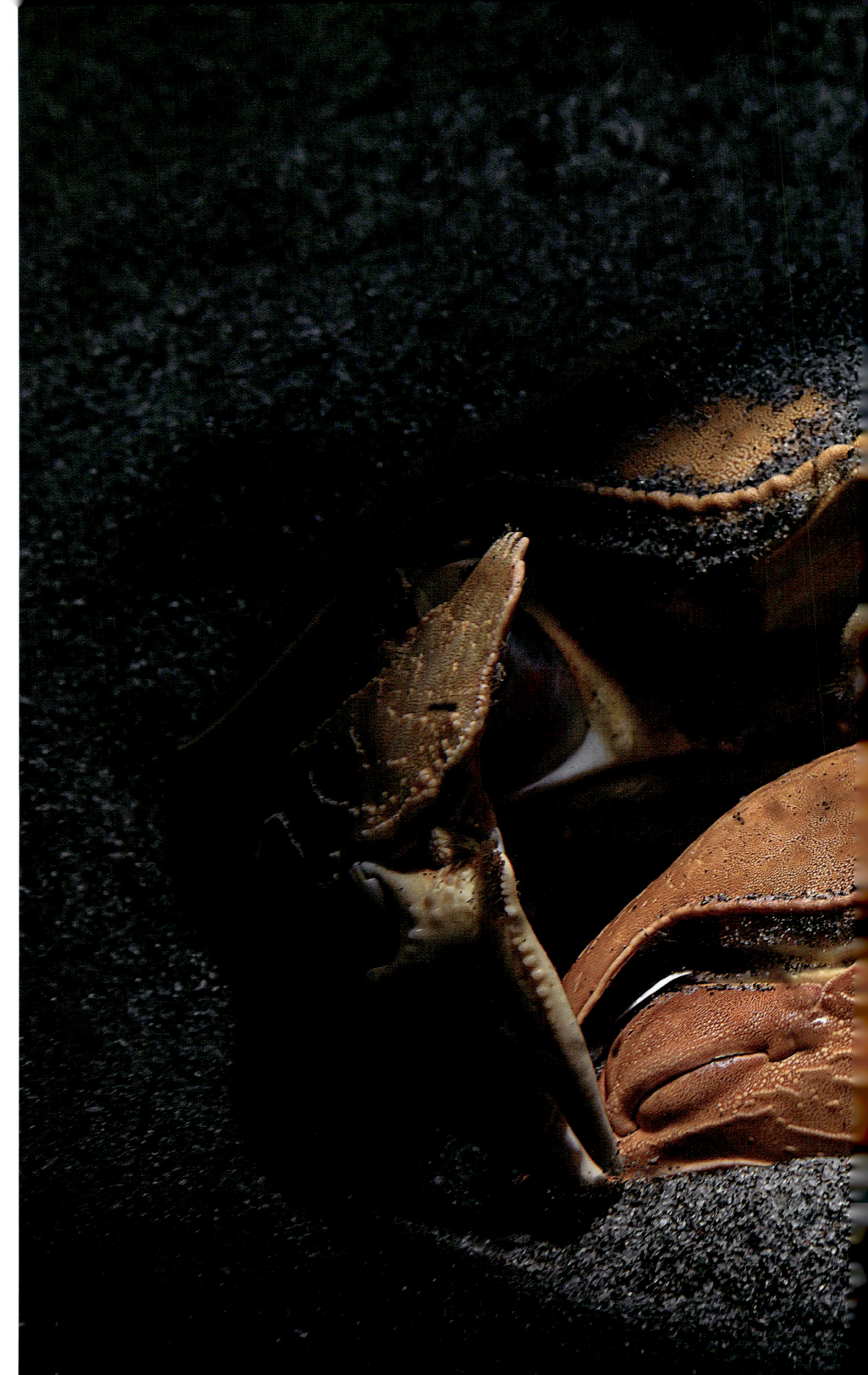

Andrey Savin
A mating pair of crabs in the Philippines

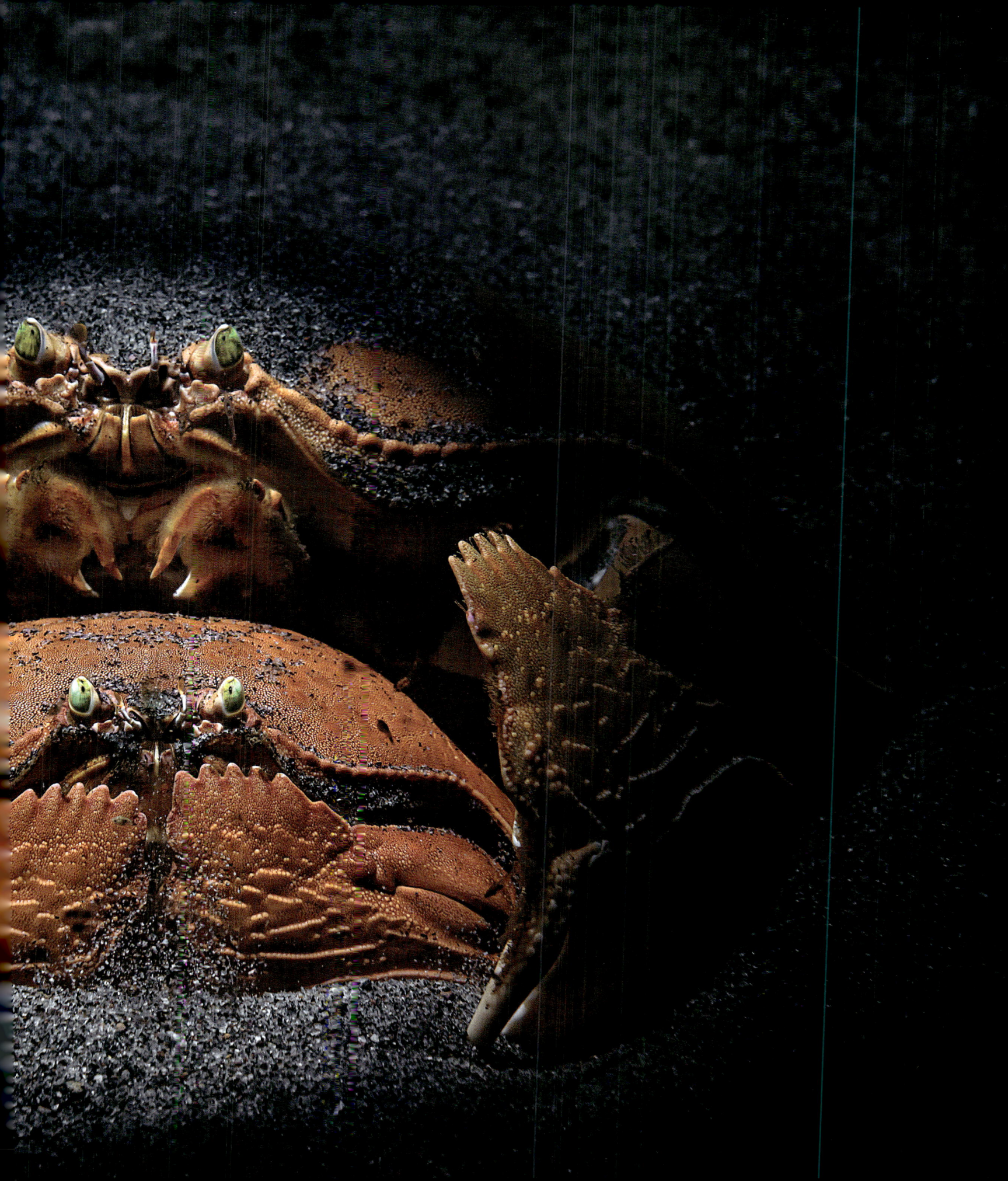

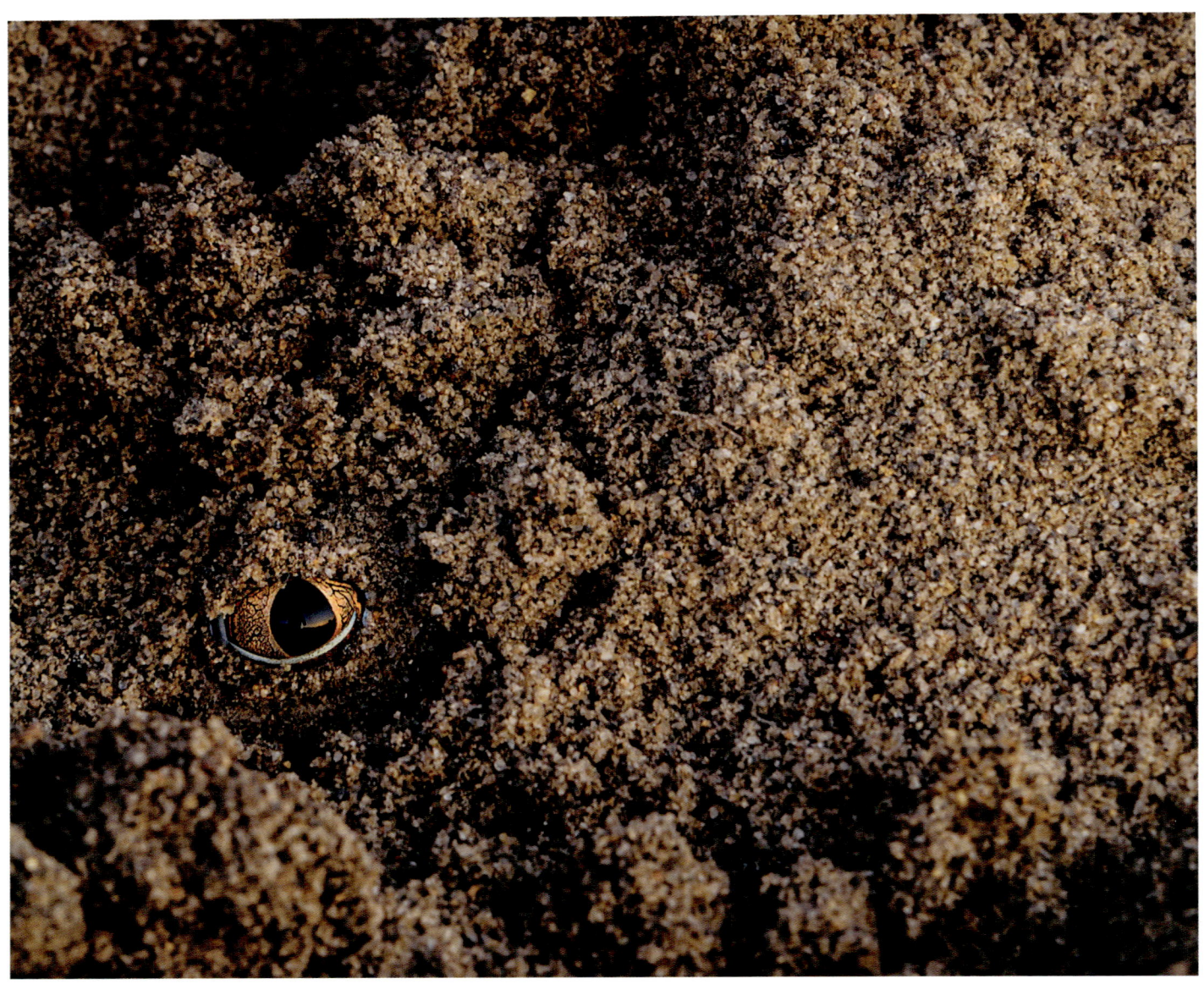

Ronan Donovan

A plains spadefoot toad takes a final look before burying itself in a sandy riverbank.
The photographer can be seen reflected in the toad's eye.

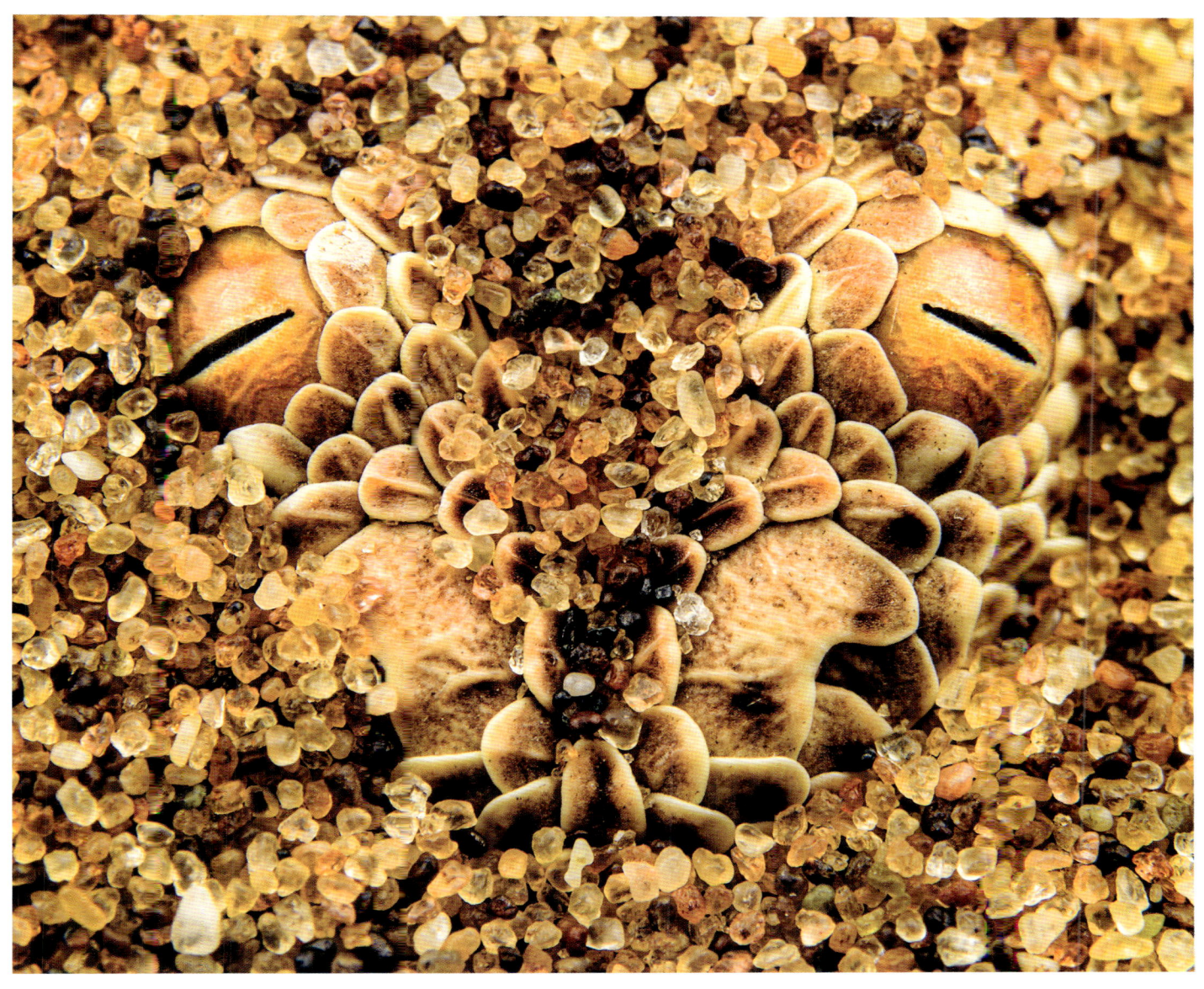

Emanuele Biggi
A venomous snake called a Peringuey's adder camouflages itself in
Namibia's desert sand.

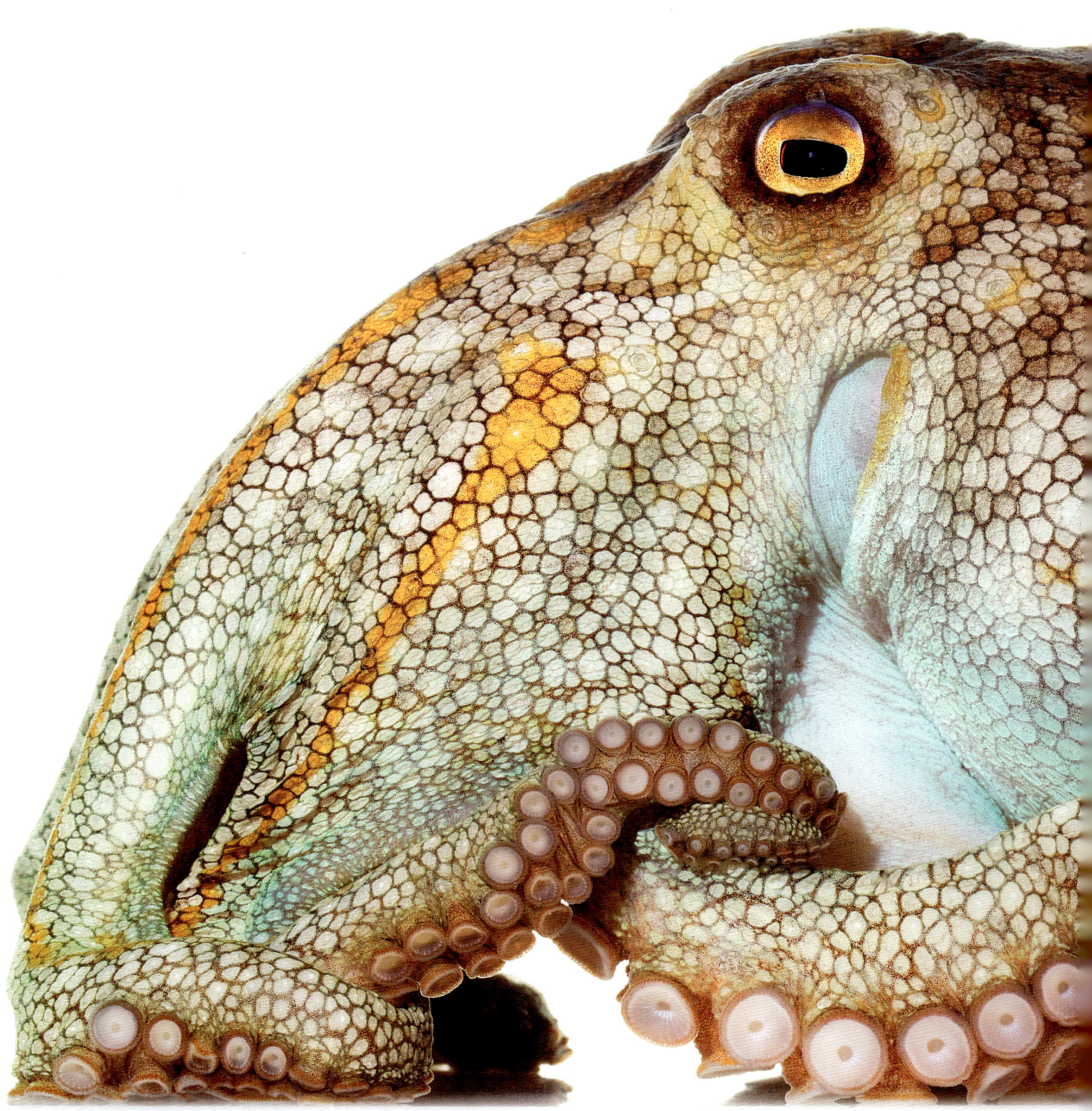

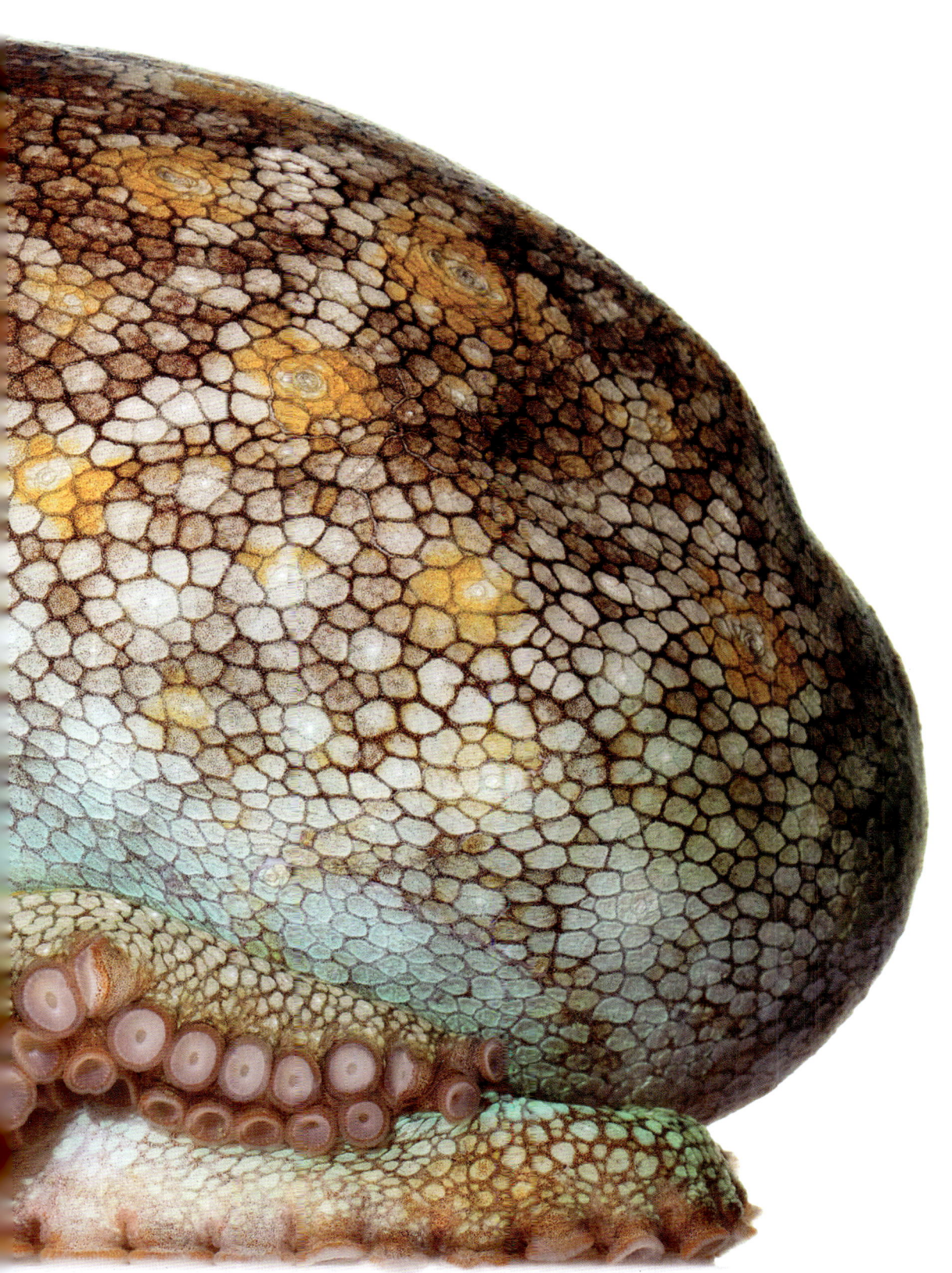

David Liittschwager
A pale octopus photographed
in an aquarium in Victoria,
Australia

INFLUENCES
FILM NOIR

Critics today still regard Orson Welles's 1941 film, *Citizen Kane,* as one of the greatest films ever made. It helped define the visual style of a genre called film noir, distinguished by its somber themes, dramatic perspectives, and stark lighting.

In the still frame of *Citizen Kane* shown here, we can only see the gesture of one man and half the face of the other. The high-contrast lighting helps guide our attention in a cluttered scene by accentuating specific details while obscuring the rest in smoke and shadow. Welles's careful use of light eliminates distractions and keeps our attention focused on the critical action.

I was introduced to this aesthetic as a kid when my dad rented classics like *Citizen Kane* and *The Maltese Falcon* for our family to watch at home. Along with graphic novels and Japanese animation, film noir left a visual imprint on my developing mind.

When I started my first assignment to photograph mind-controlling parasites, I had no solid plan for how to make compelling images—only a frantic sense that I'd better figure out some way to make these critters look cool or else my fledgling photographic career was toast.

My first trip was to visit a lab studying horsehair worms that infect crickets.

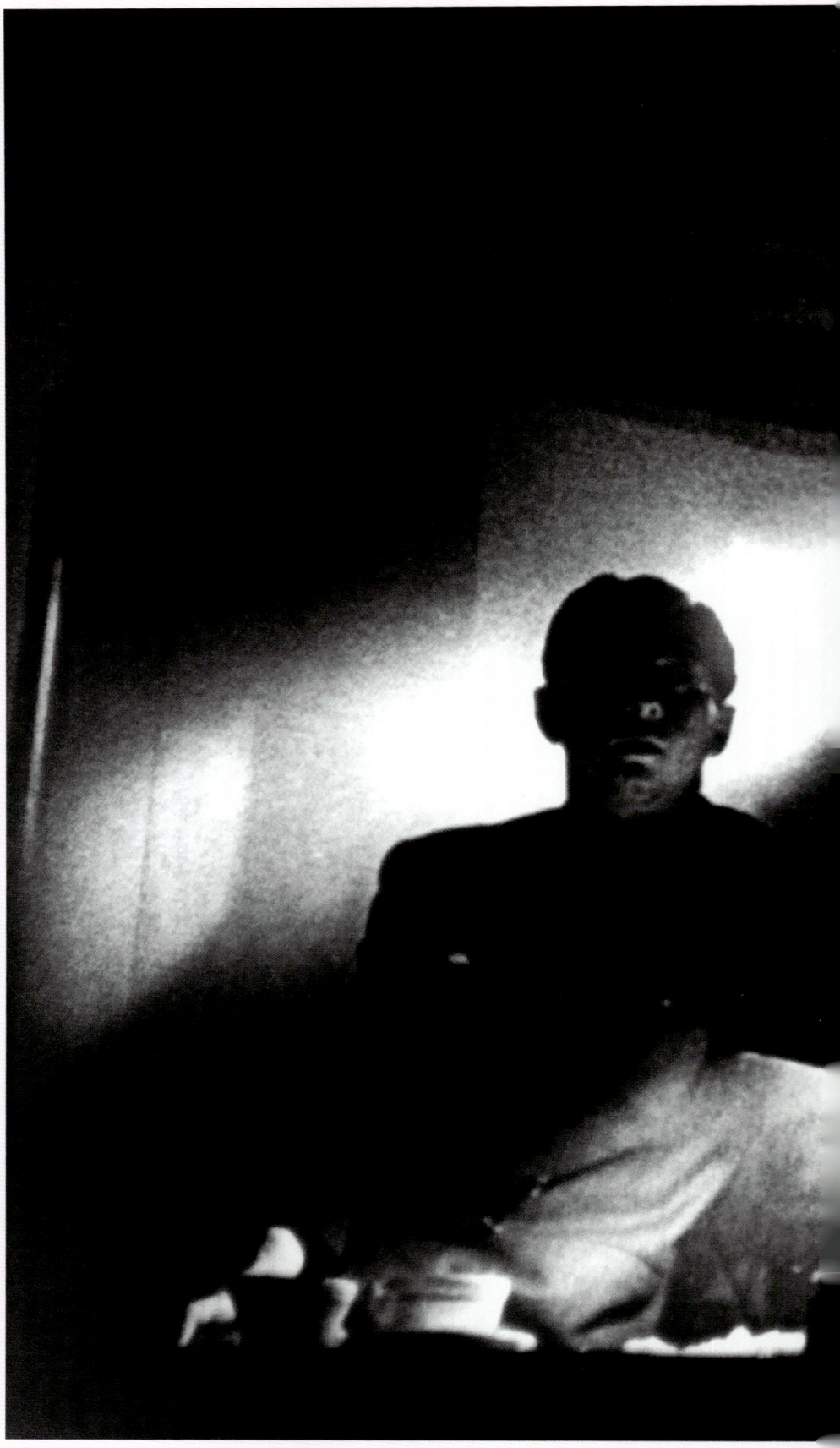

Everett Collection
A still frame from Orson Welles's 1941 film noir classic, *Citizen Kane*

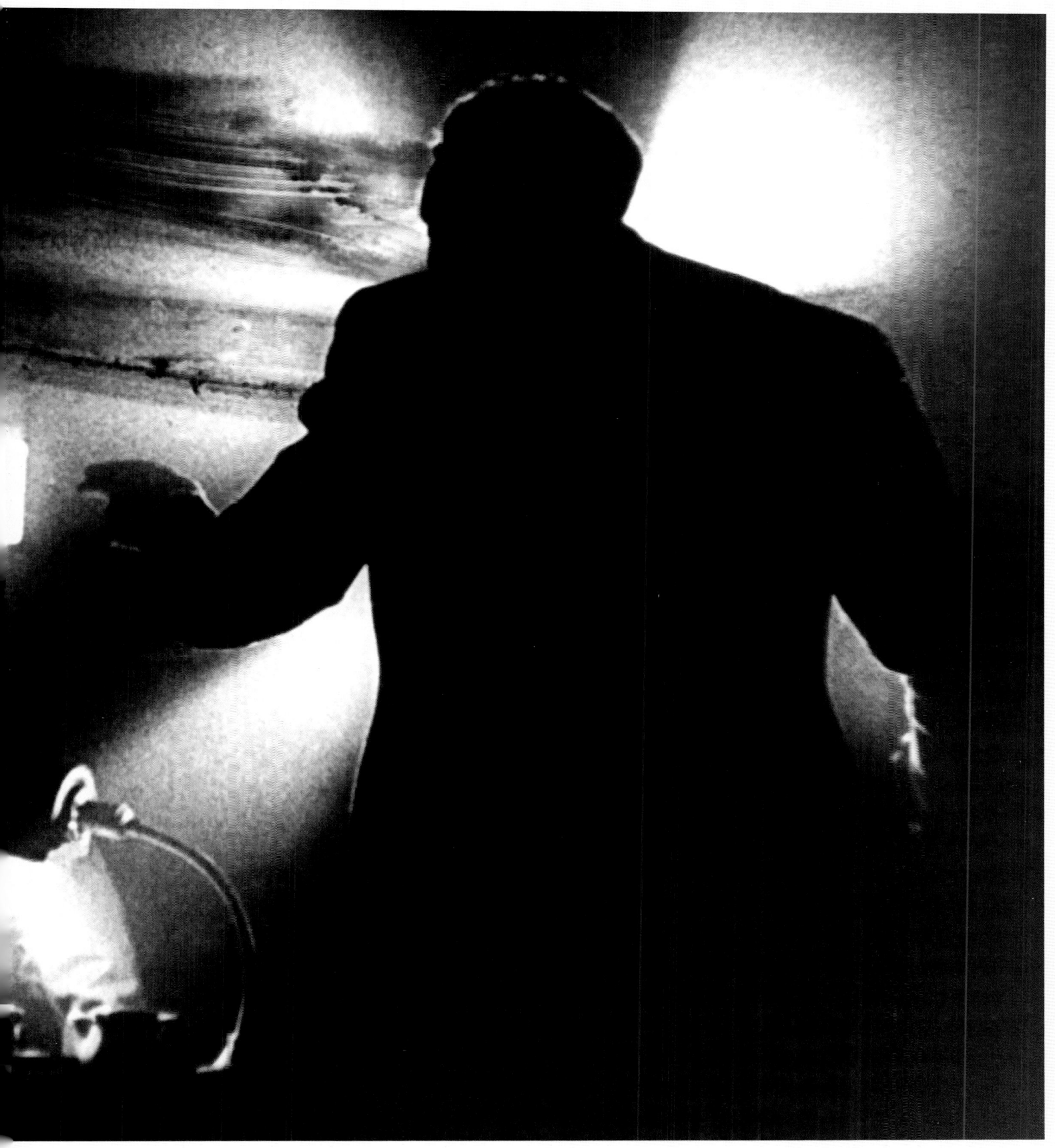

On my way, I stopped at an art-supply store to stock up on watercolors, paint, and construction paper, hoping they might help me produce a dramatic portrait of a worm. When I got to the lab, I tested every idea I could come up with.

After 28 days of trial and error and 5,585 failed photographic experiments, I finally figured out that I could take an interesting photograph by re-creating Welles's film noir style. I directed a narrow beam of light to focus on the protagonist parasite while leaving the distracting details of the cricket victim concealed in darkness.

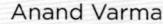

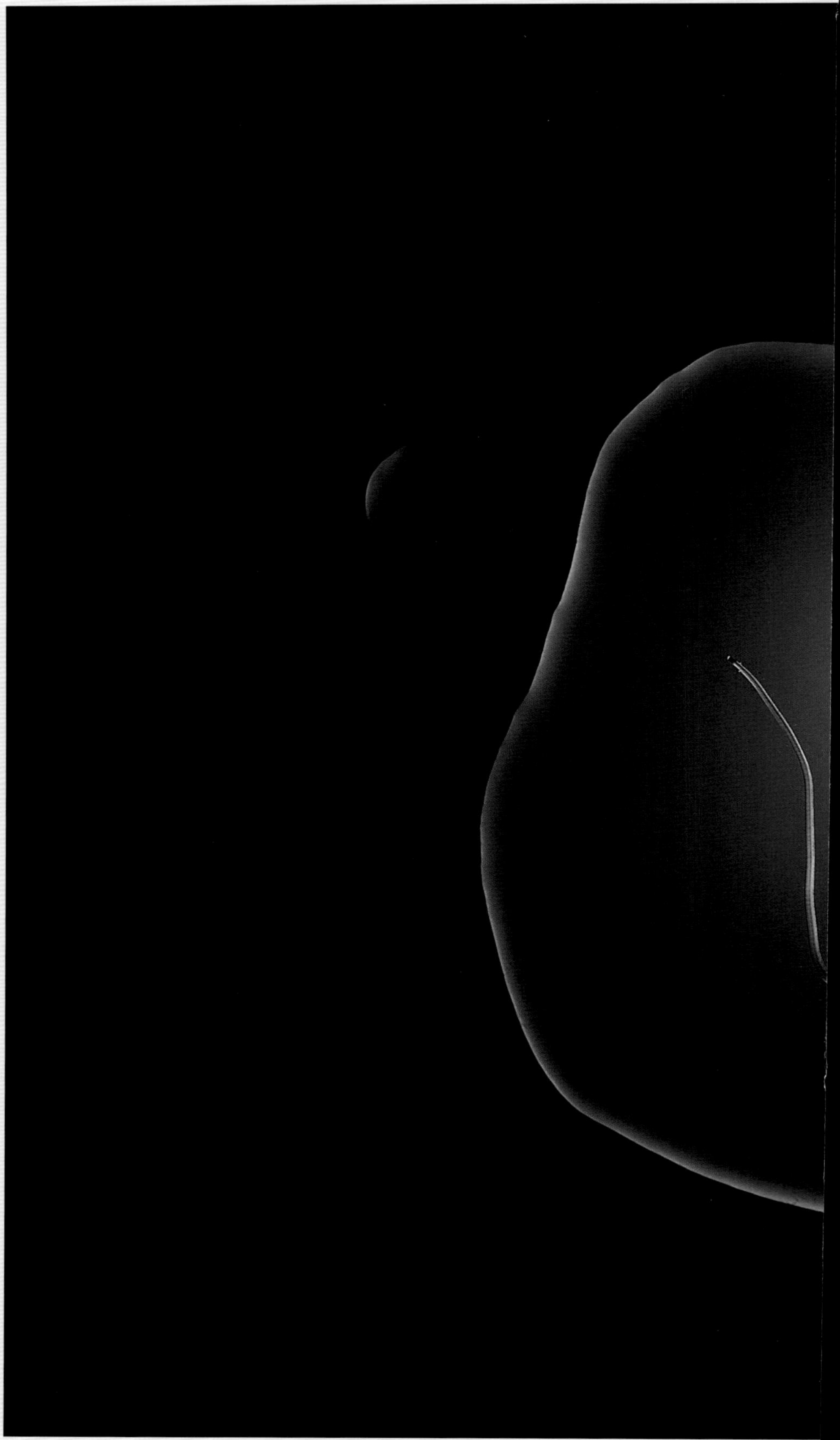

Anand Varma
A house cricket loses its will—and its life—to a horsehair worm. The cricket is terrestrial, but the adult worm is aquatic. When the mature worm is ready to leave the cricket, it alters the brain of its host and forces the cricket to leap into the nearest body of water. As the cricket drowns, an adult worm emerges, sometimes a foot in length.

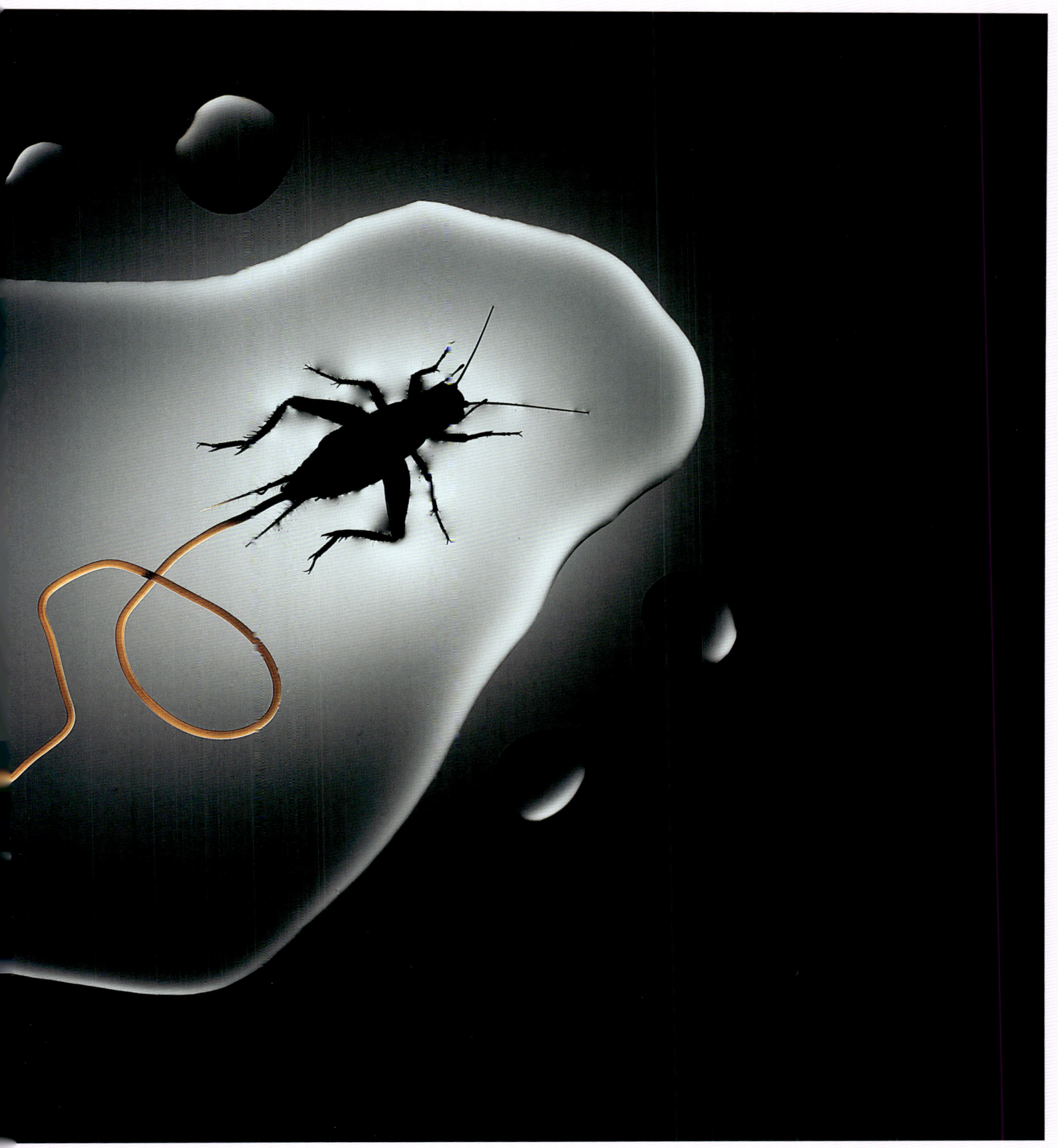

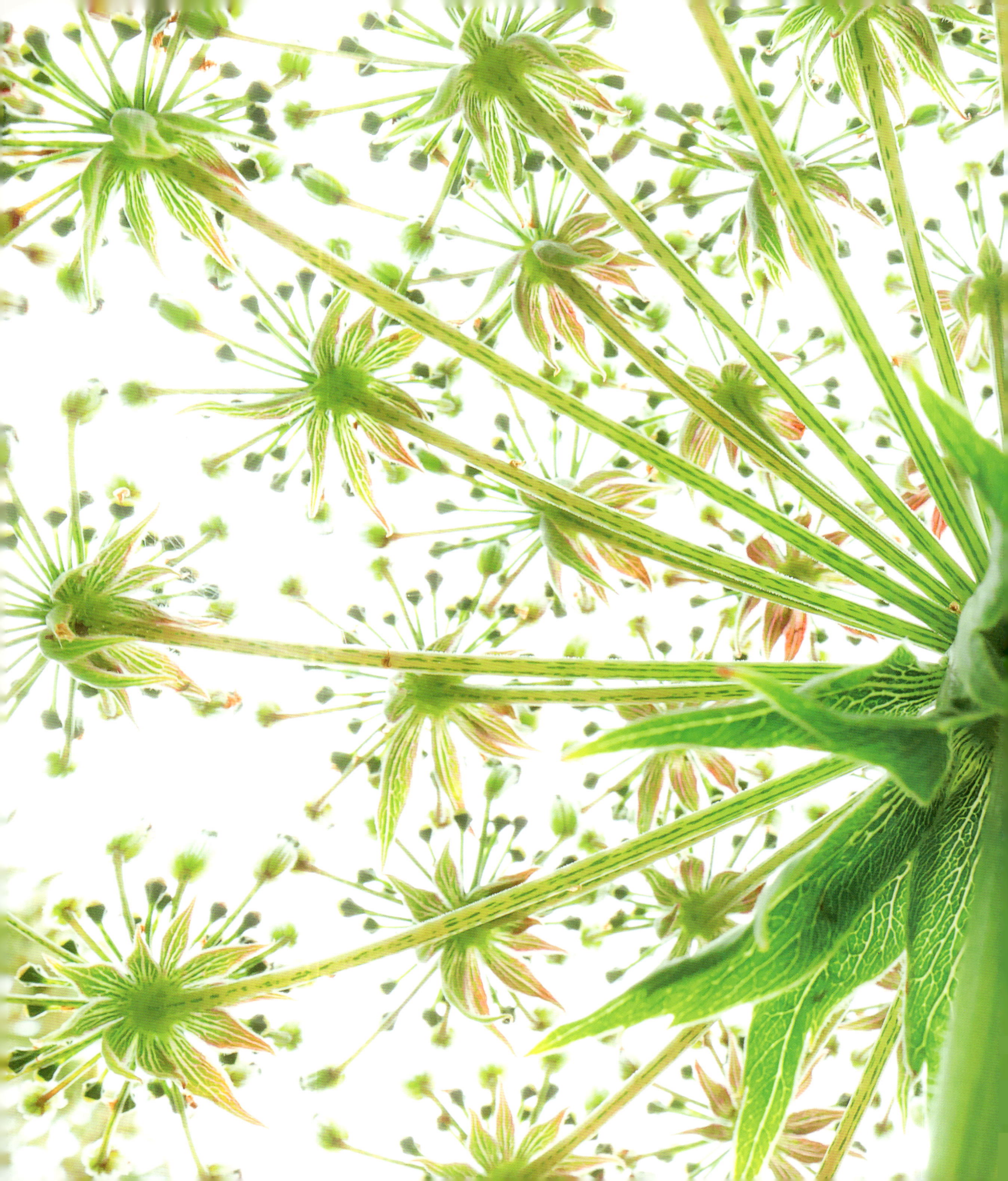

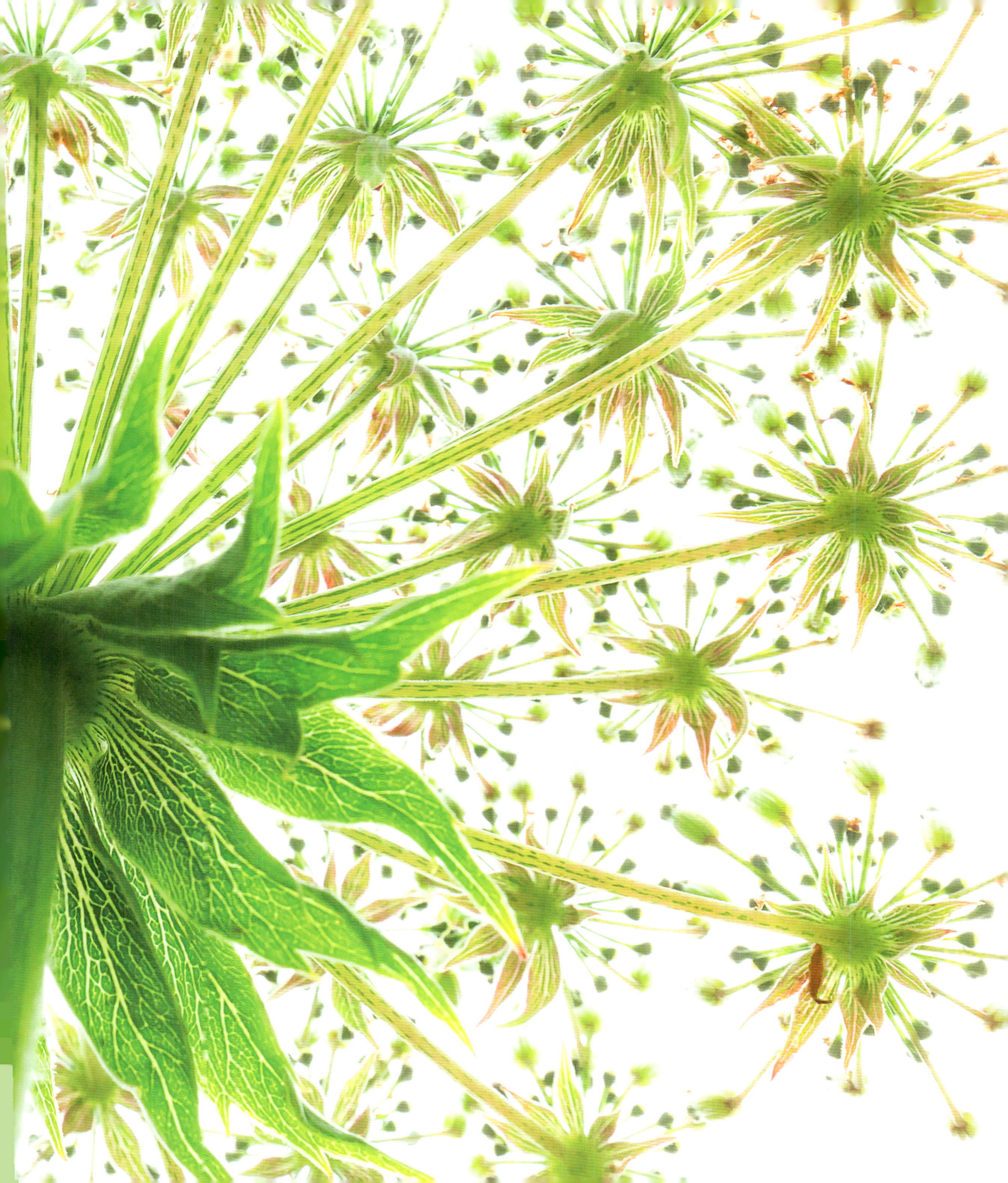

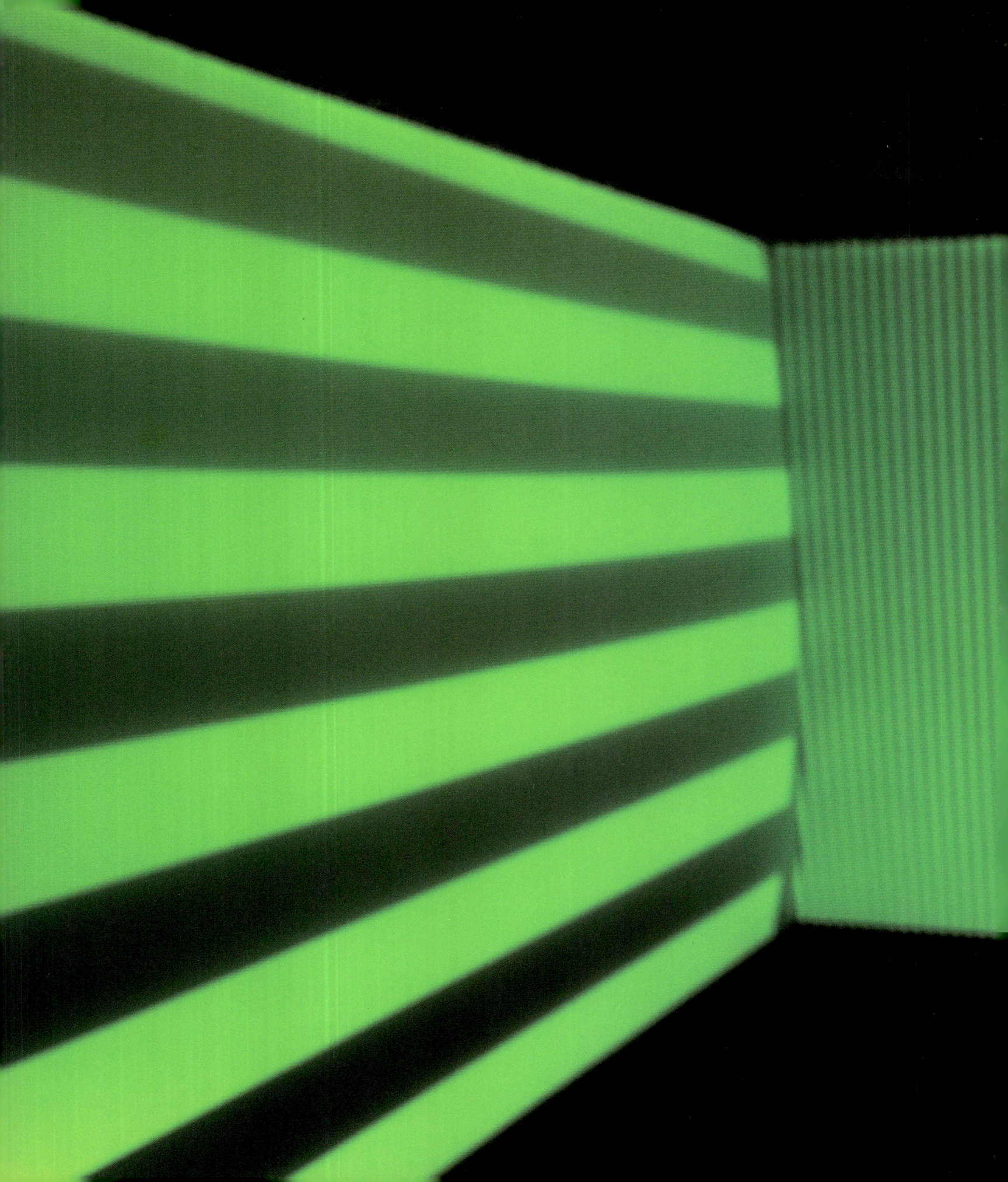

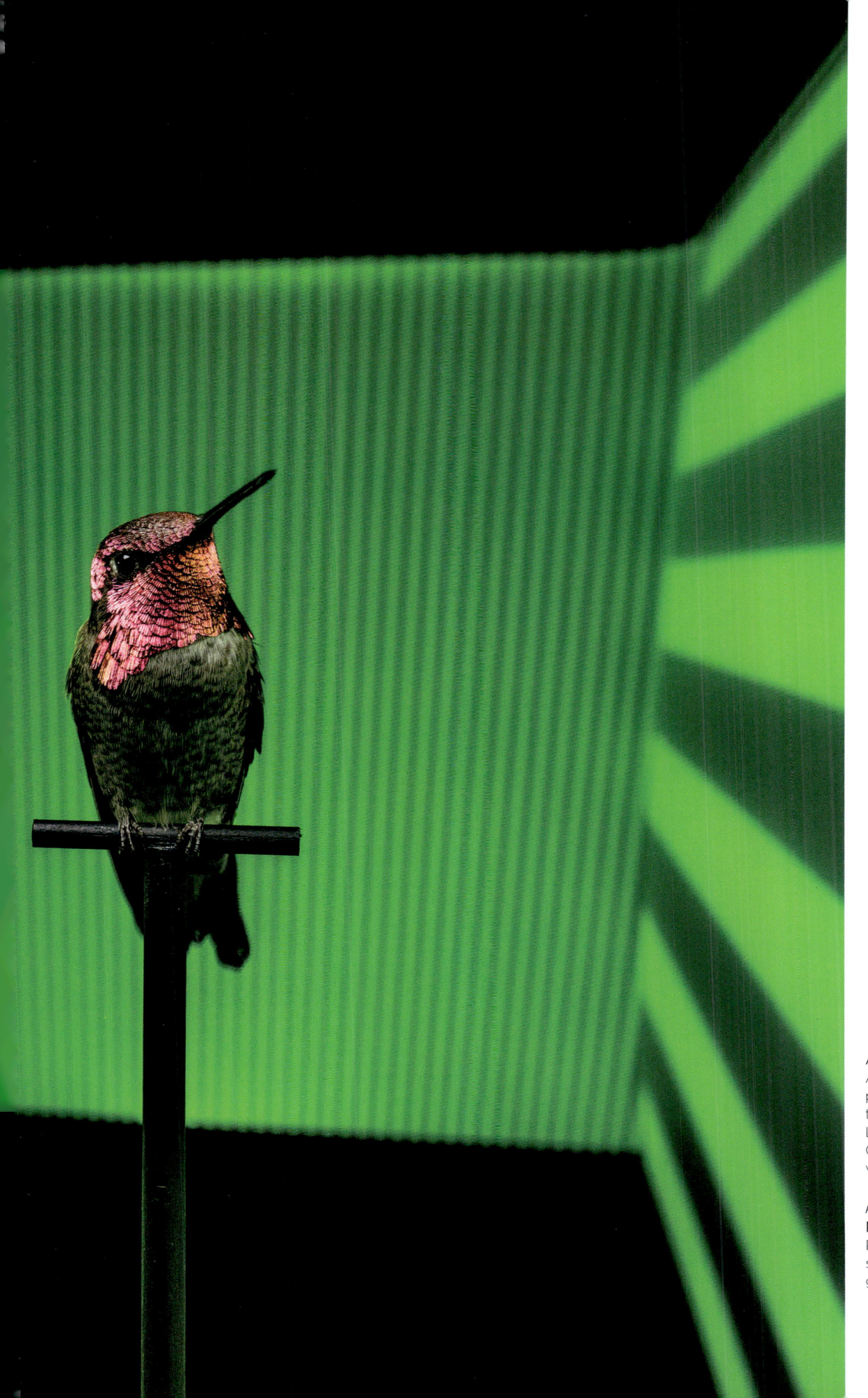

Anand Varma
An Anna's hummingbird perches inside a virtual-reality tunnel used by the Altshuler Lab at the University of British Columbia to study the birds' visual perception.

PREVIOUS PAGES:
Pere Soler
Photographed looking up the stalk, the flower of the Madeira giant black parsley plant

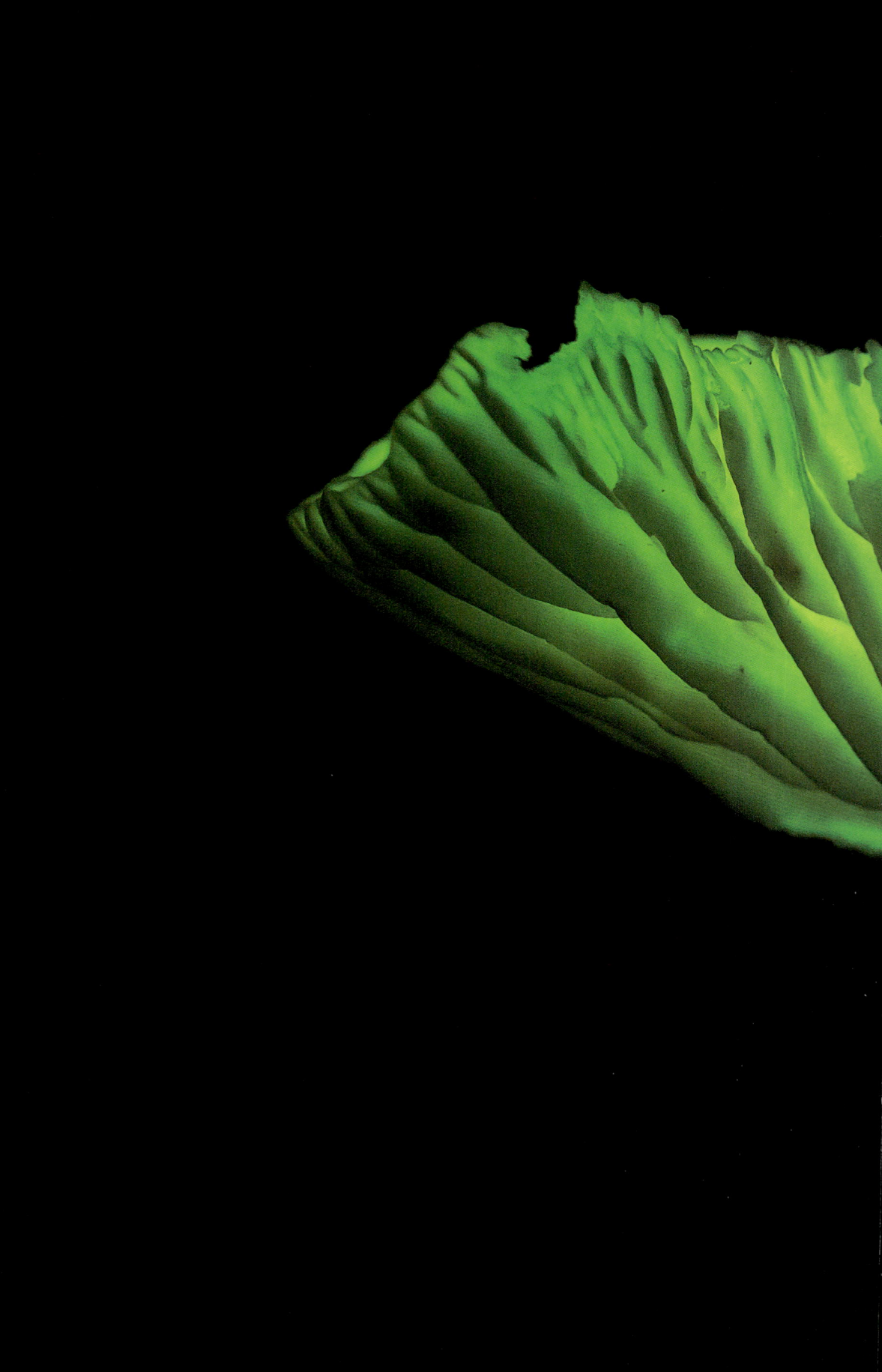

Anand Varma
A bioluminescent mushroom stands out in the palm forests of Piauí, Brazil. Scientists think this glow may attract insects that help disperse these mushrooms' spores.

FOLLOWING PAGES:
Ruedi Flück
Flares strapped on to Sammy Carlson's skis light up his nighttime path.

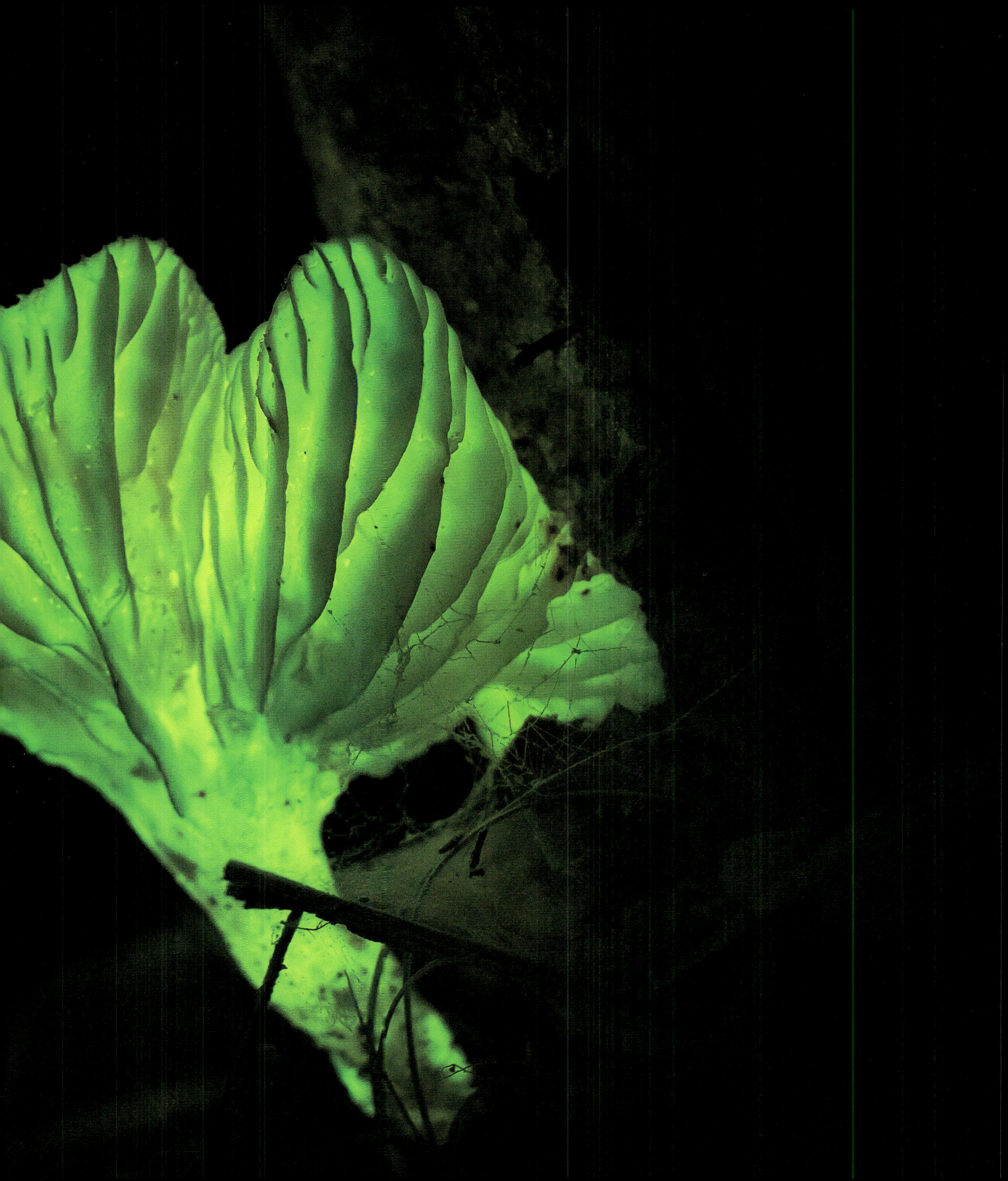

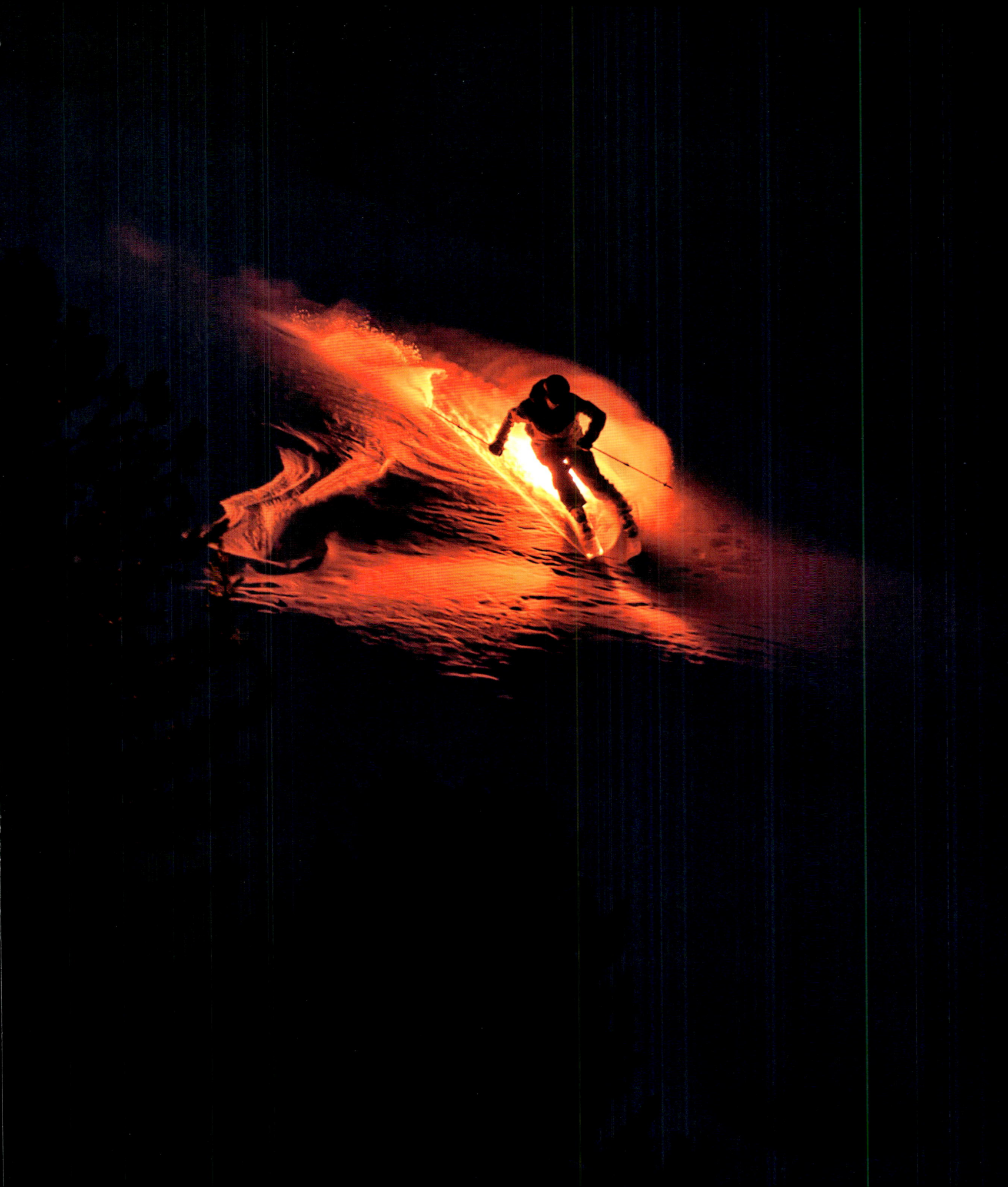

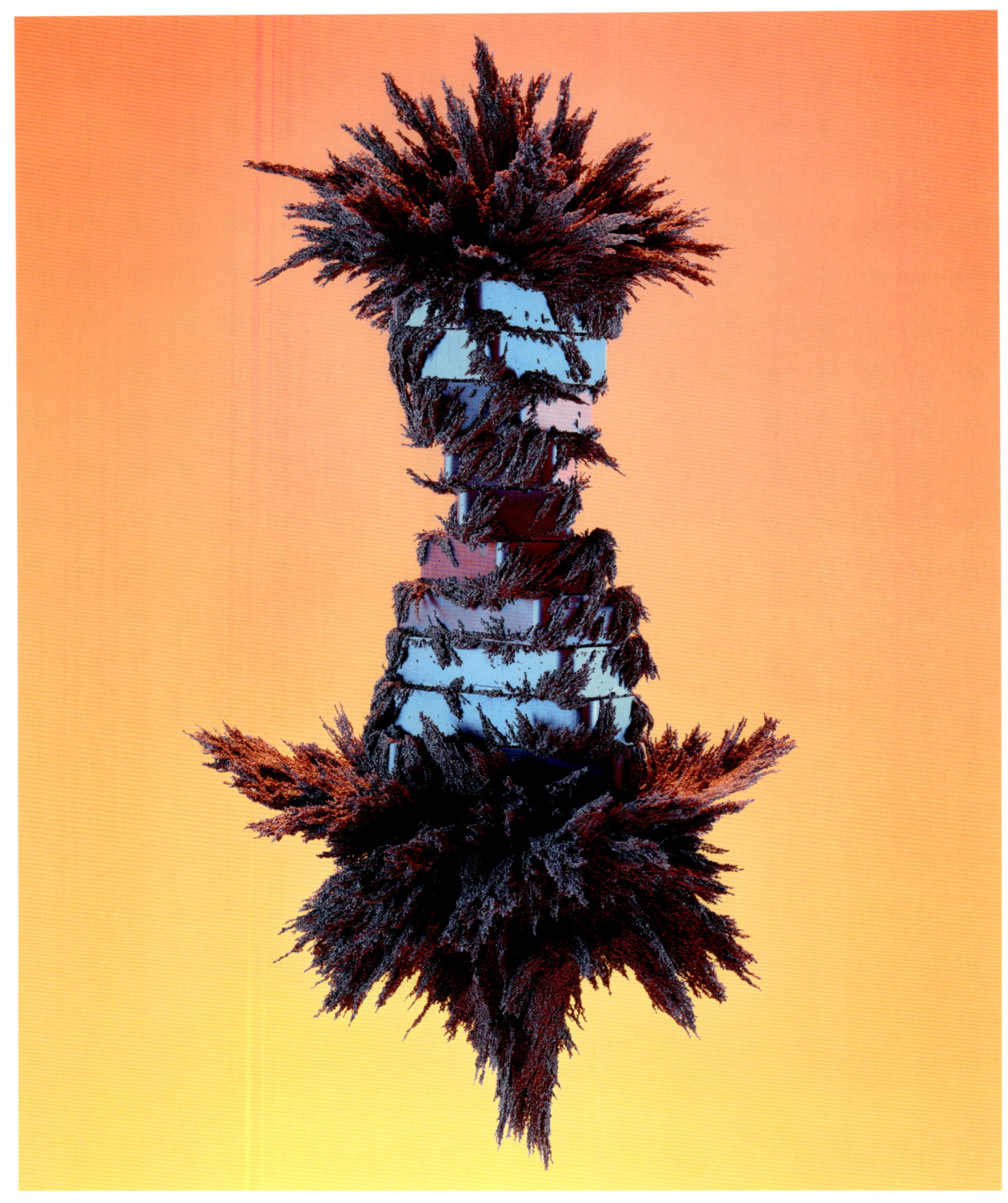

Zac Henderson
A stack of bar magnets force iron filings to take on the shape of the bars' magnetic field.

FOLLOWING PAGES:
Laurent Ballesta
Supercooled brine leaks from sea ice and forms frozen tendrils
called brinicles underwater in the Antarctic Ocean.

THE TRUE BEAUTY OF A PHOTOGRAPH
IS NOT JUST IN THE DETAILS IT CAPTURES,
BUT IN HOW IT GIVES US A FRESH WAY
TO SEE THE WORLD.

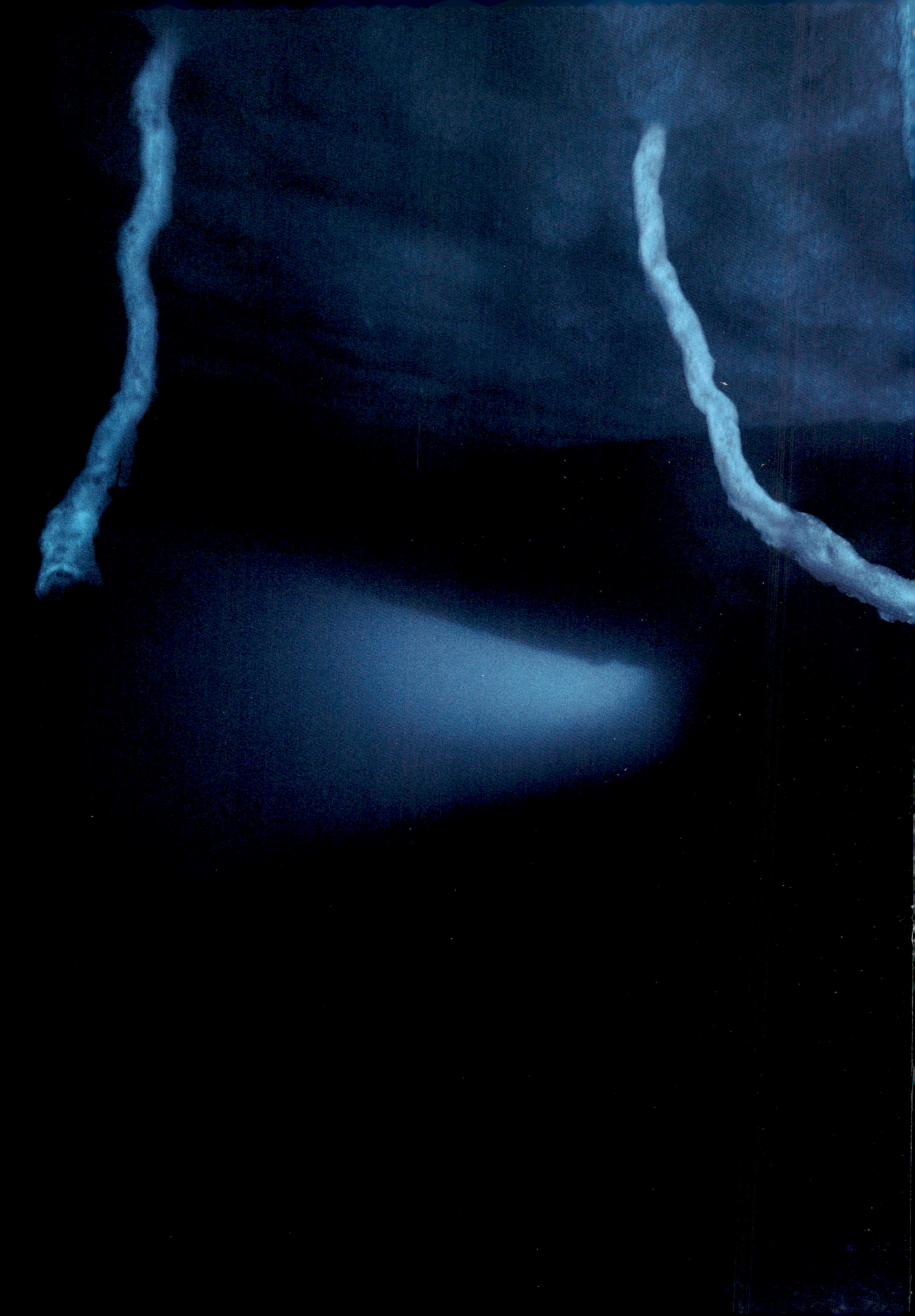

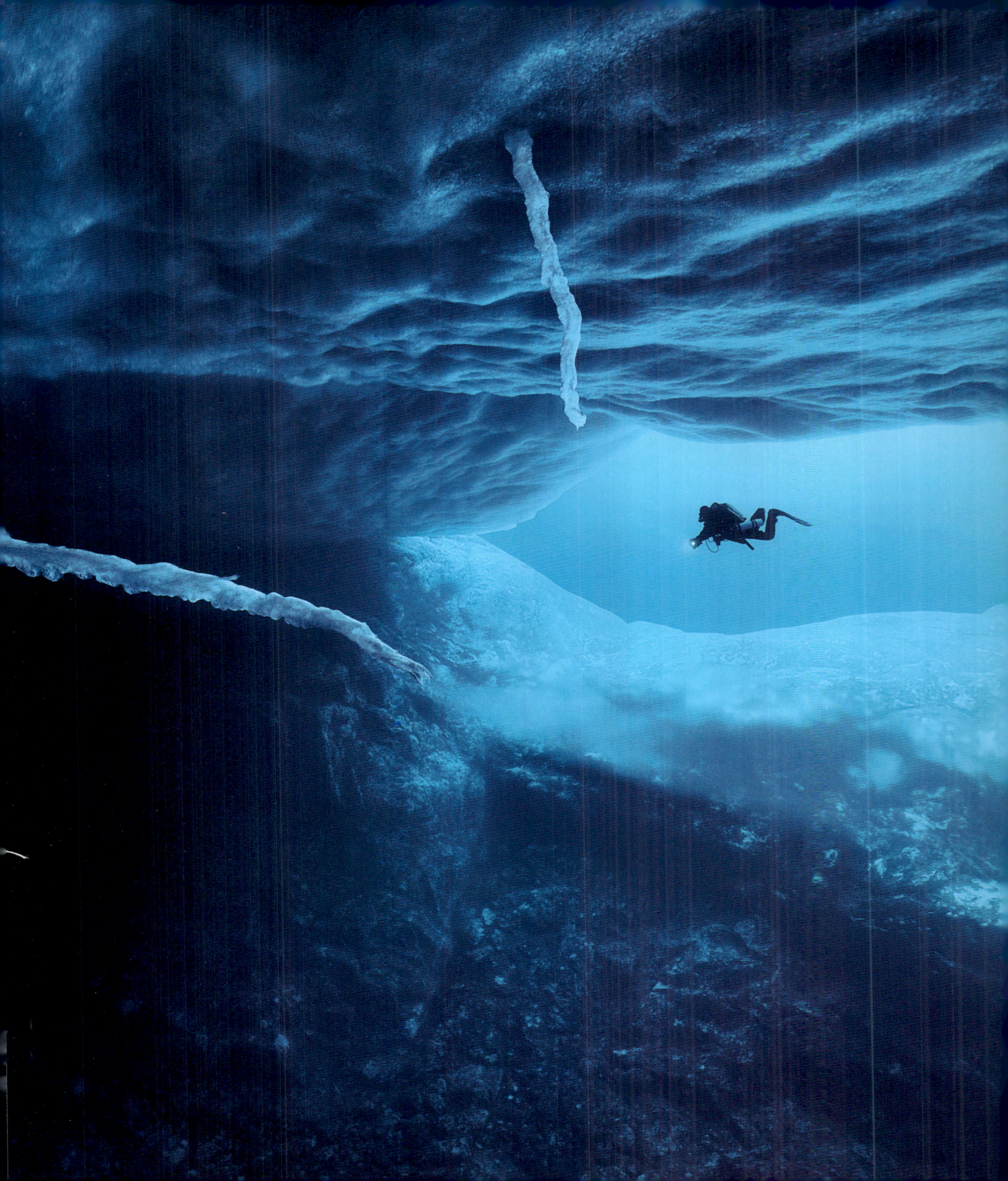

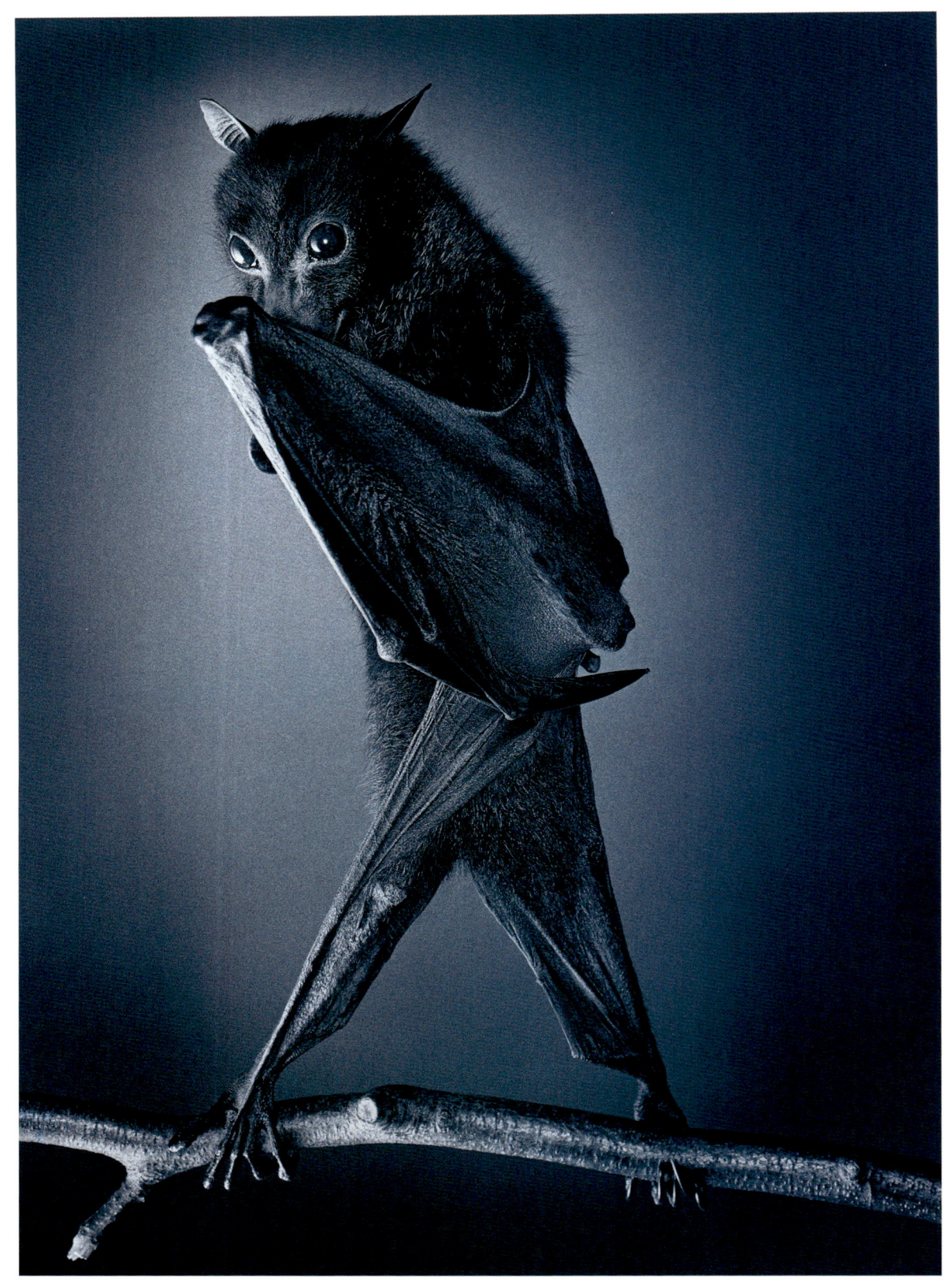

Tim Flach
An intimate view of a flying fox depicted upside down

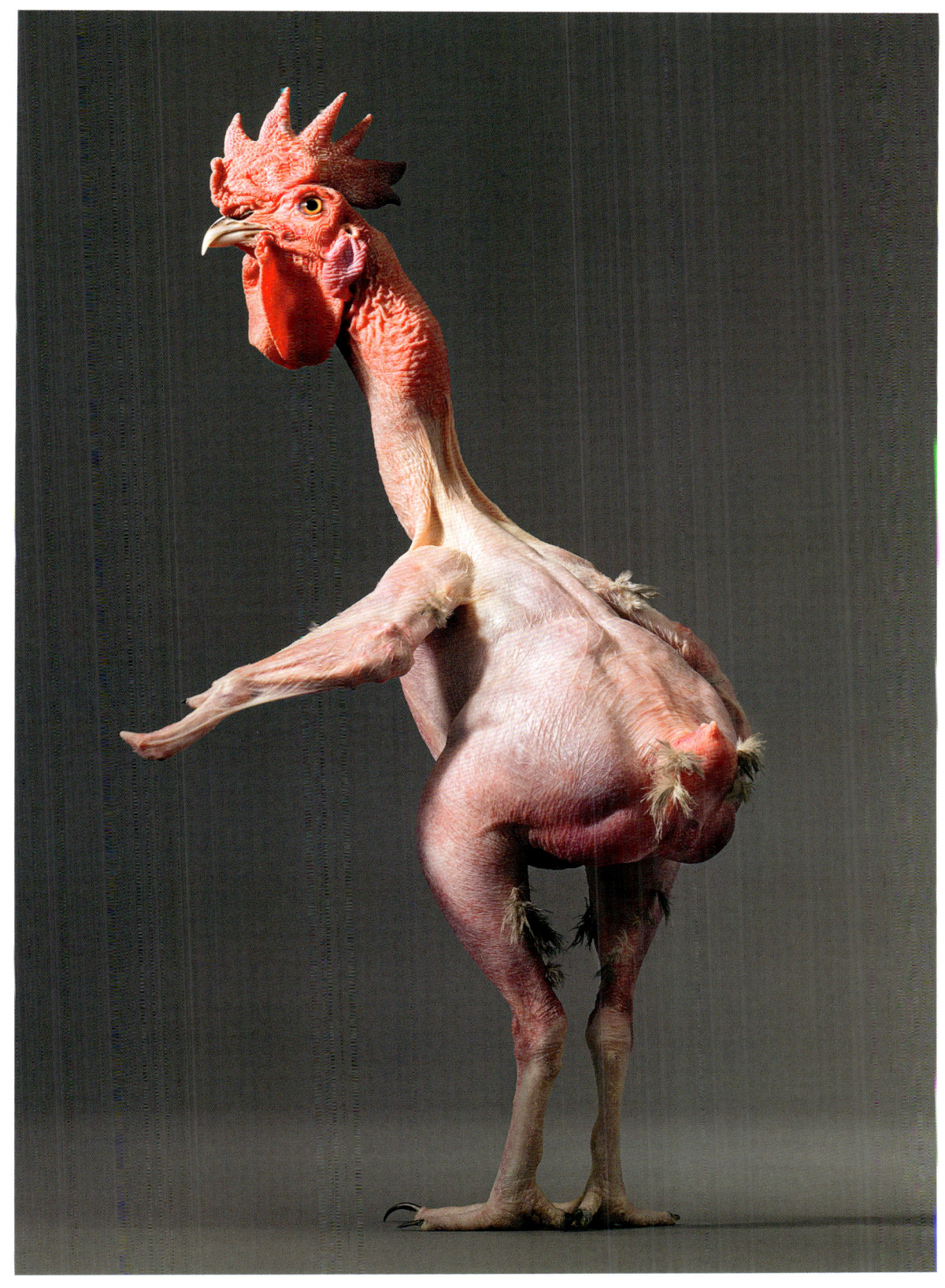

Tim Flach
A genetic mutation renders this chicken featherless, revealing a unique view of its underlying anatomy.

Javier Aznar
A katydid relies on its leaflike camouflage to feed undisturbed on a plant in Brazil.

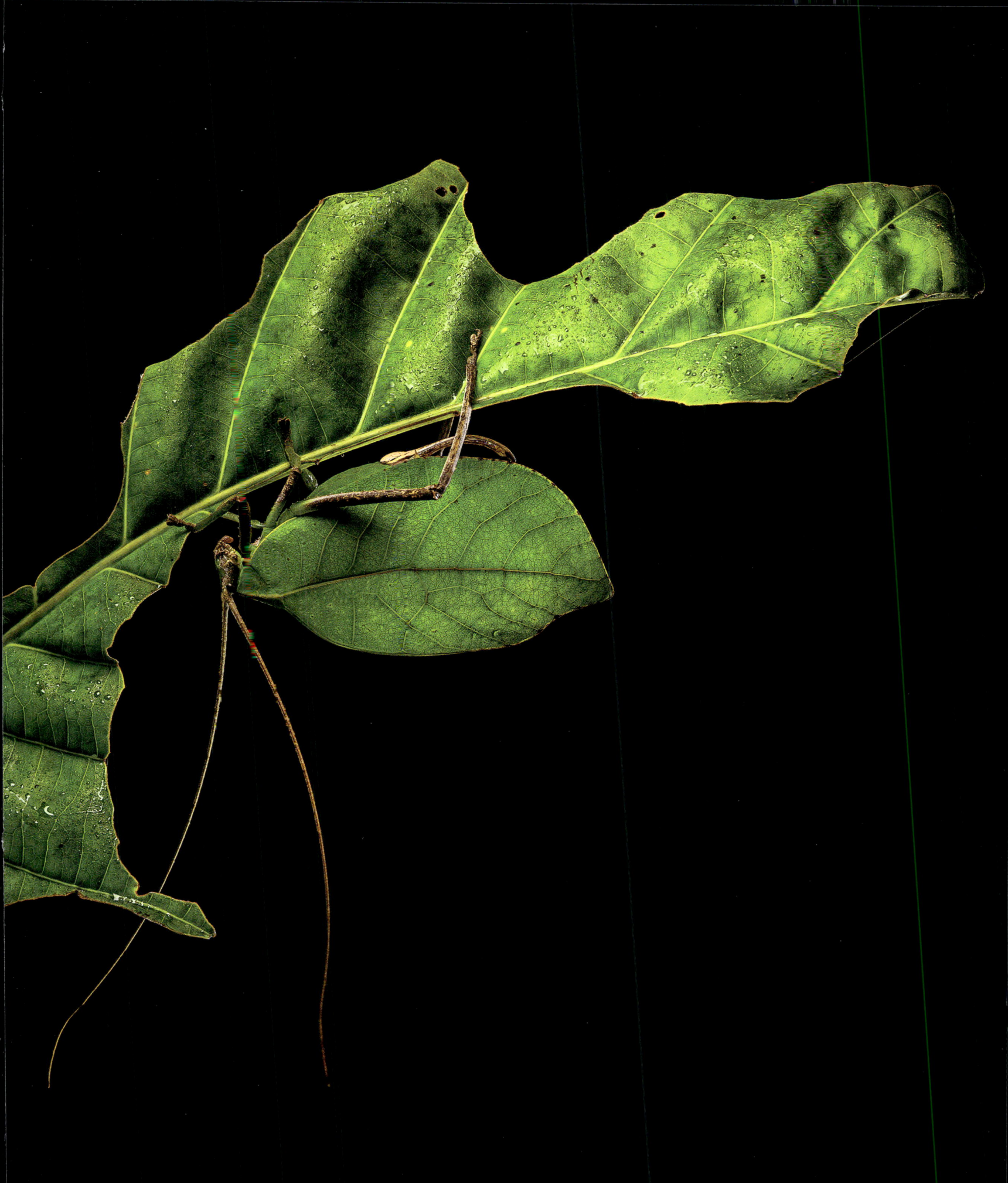

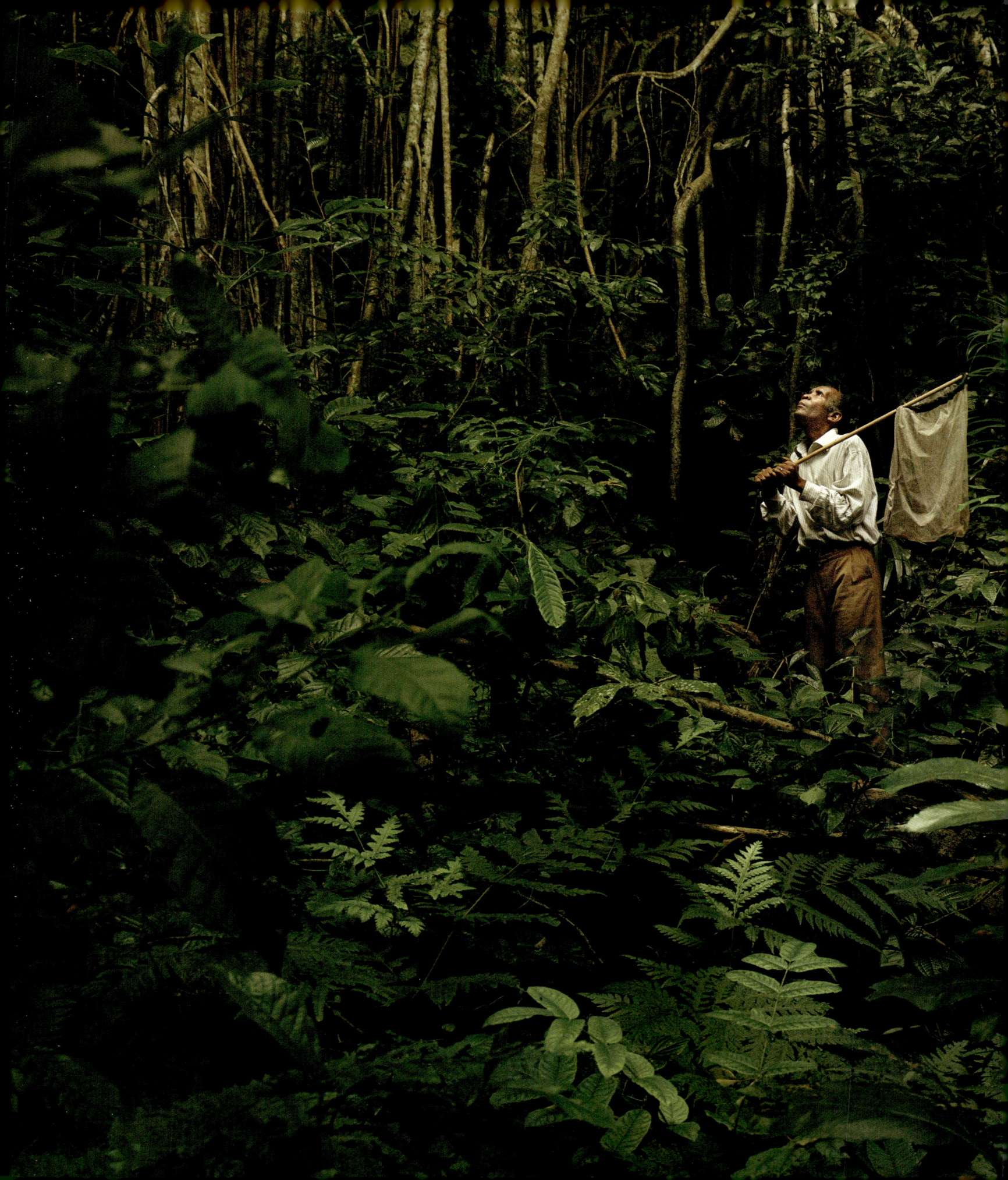

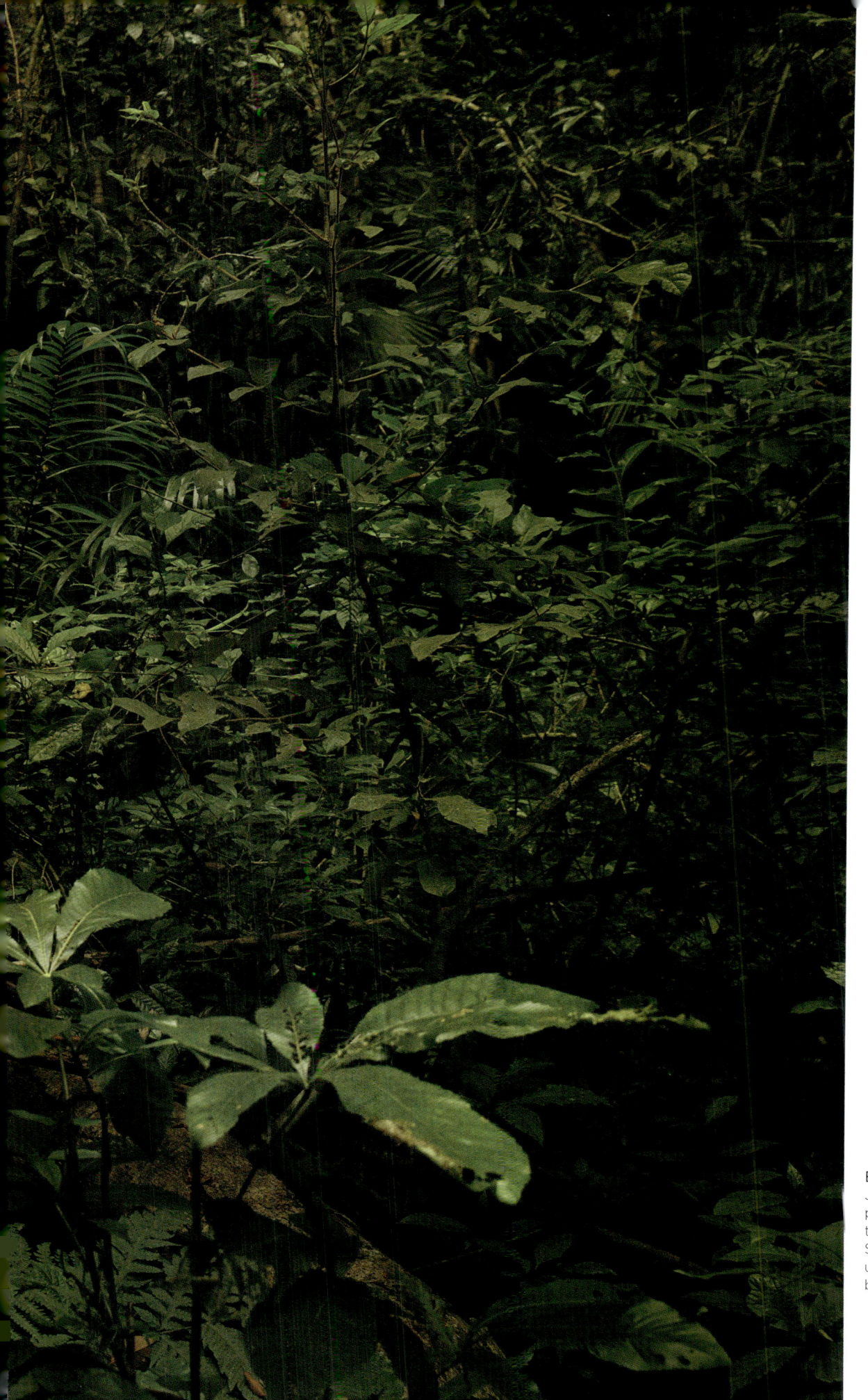

Evgenia Arbugaeva
John Mganga retraces his
path through a forest near
the Amani Hill Research
Station in Tanzania, where he
used to catch butterflies and
bugs for the lab's collections.

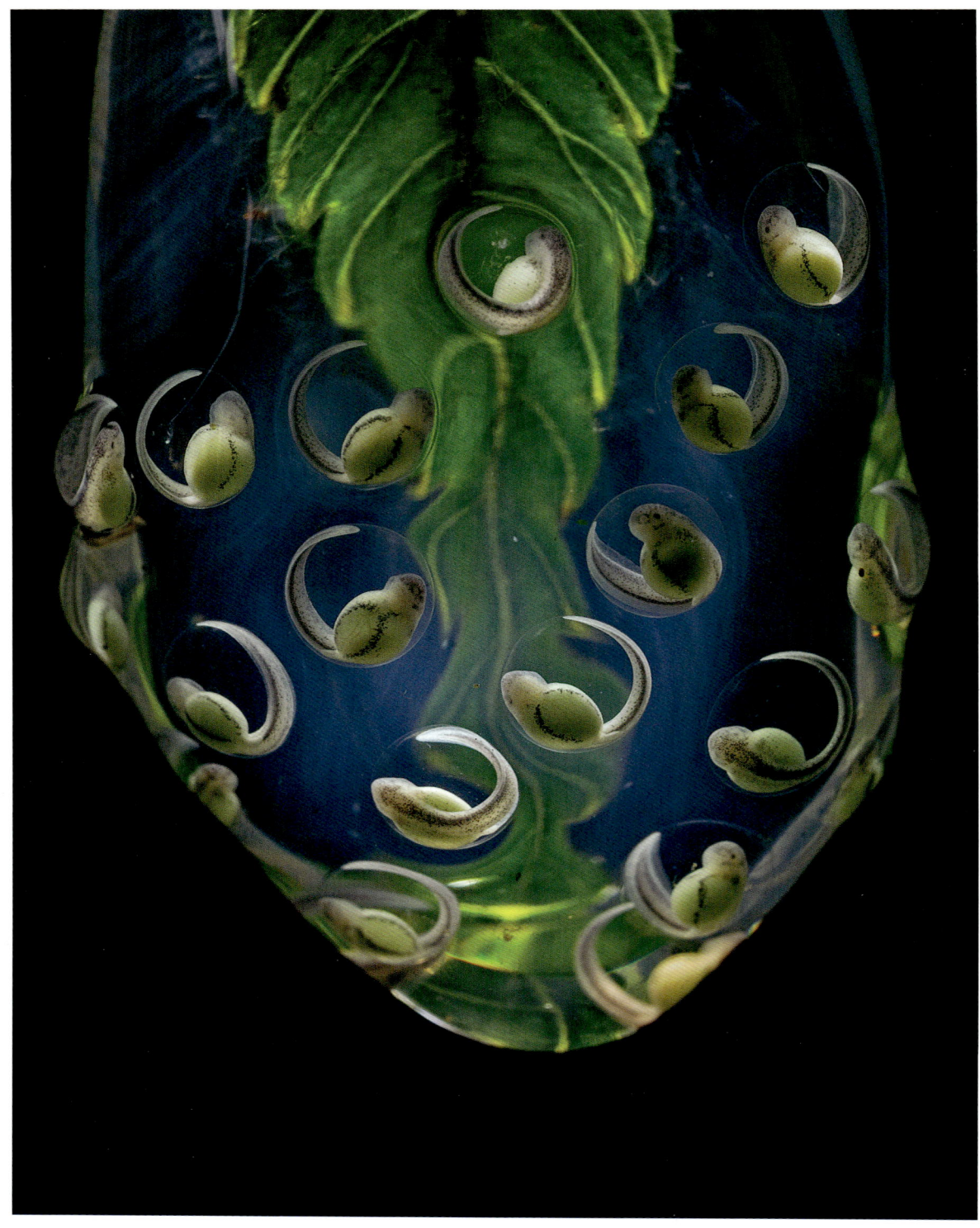

Jaime Culebras
An egg cluster of a Wiley's glass frog, endemic to the Ecuadorian Andes, hangs from
the tip of a fern. When the tiny tadpoles hatch, they drop into a stream below.

FOLLOWING PAGES:
Talal Al Rabah
Dawn light silhouettes a polar bear as it walks across the ice in Svalbard, Norway.

SCIENCE GIVES US STRUCTURED SYSTEMS FOR OBSERVING THE WORLD, WHILE PHOTOGRAPHY GIVES US TOOLS FOR EXTENDING OUR VISUAL PERCEPTION. EACH REINFORCES THE OTHER.

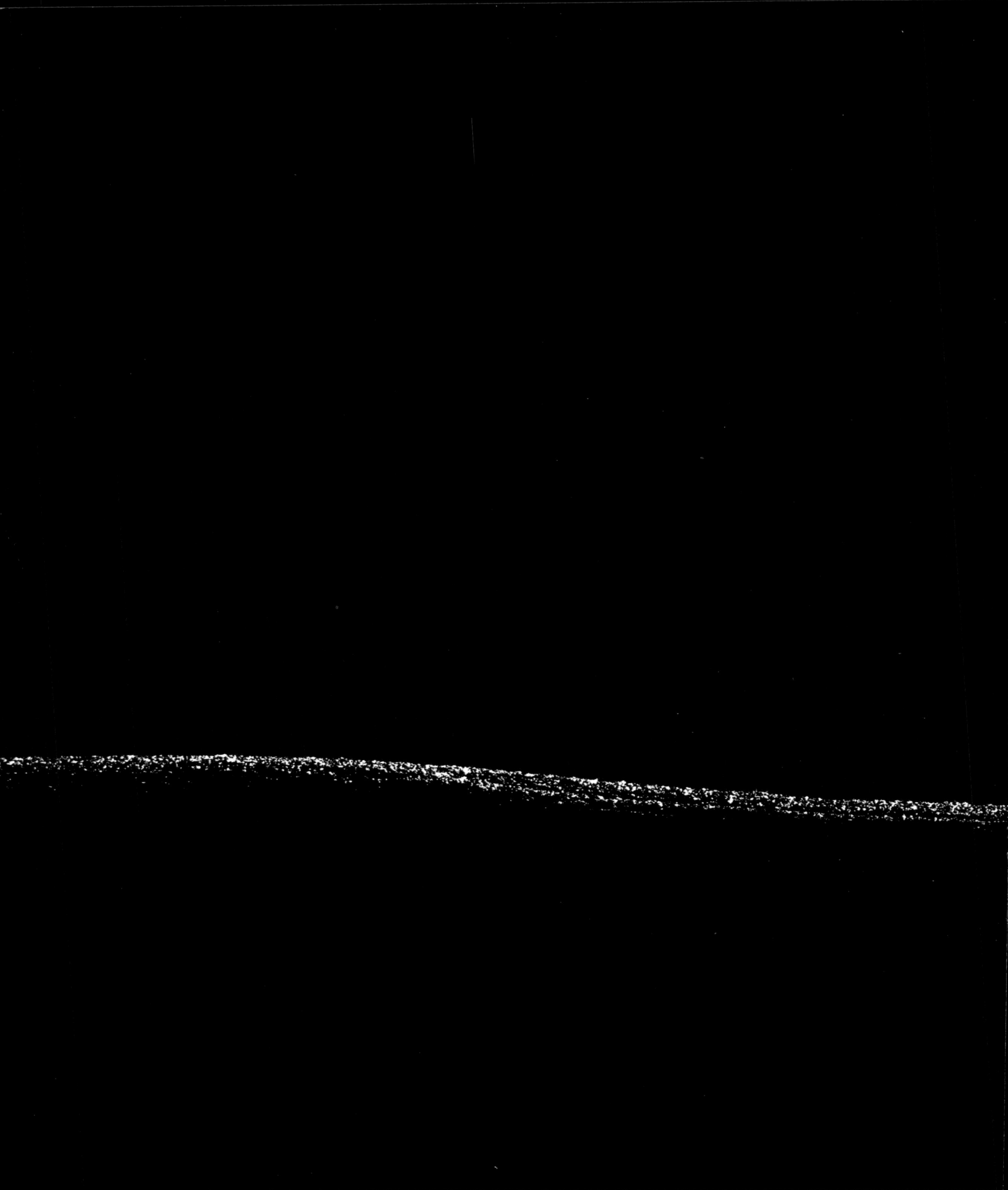

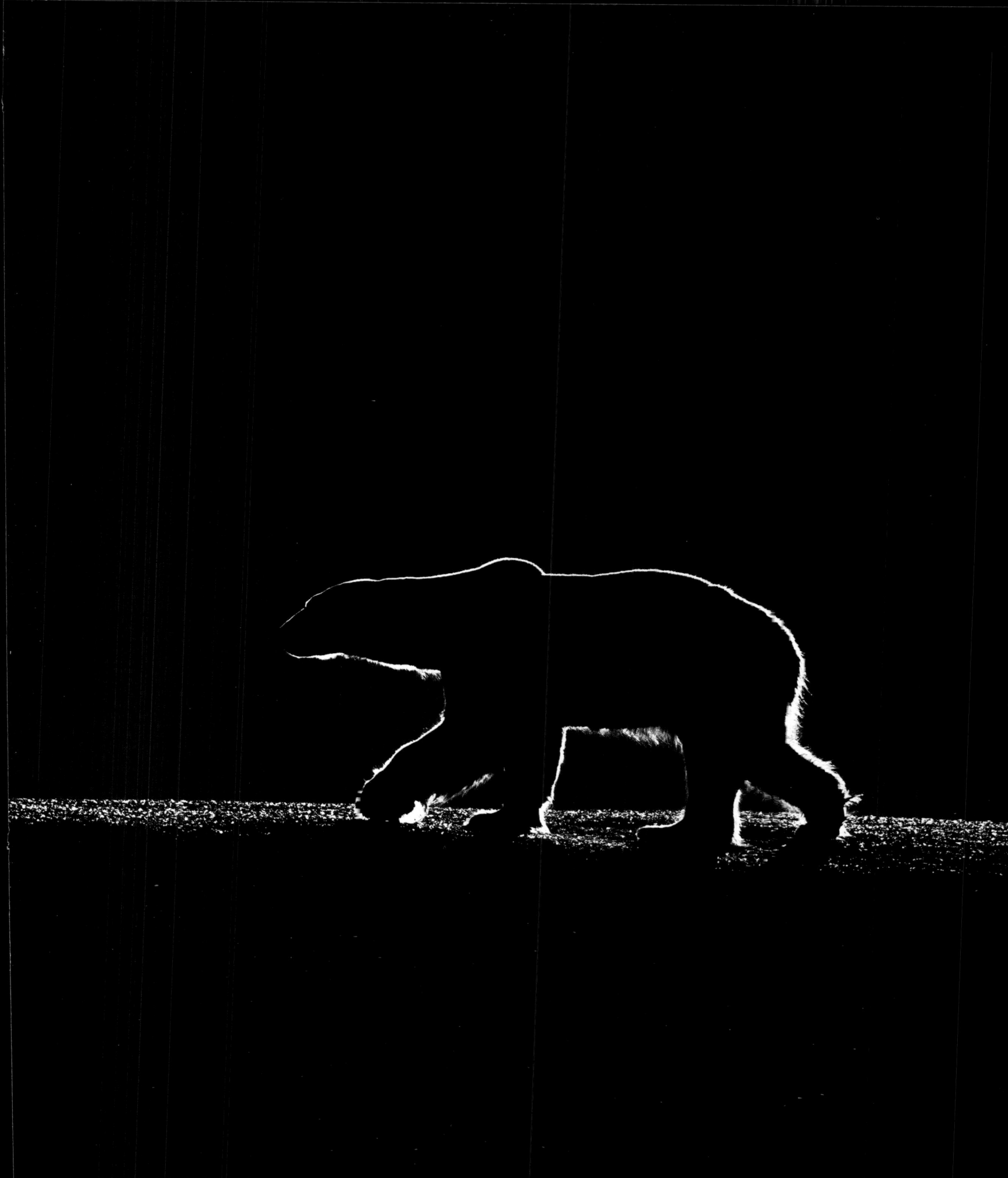

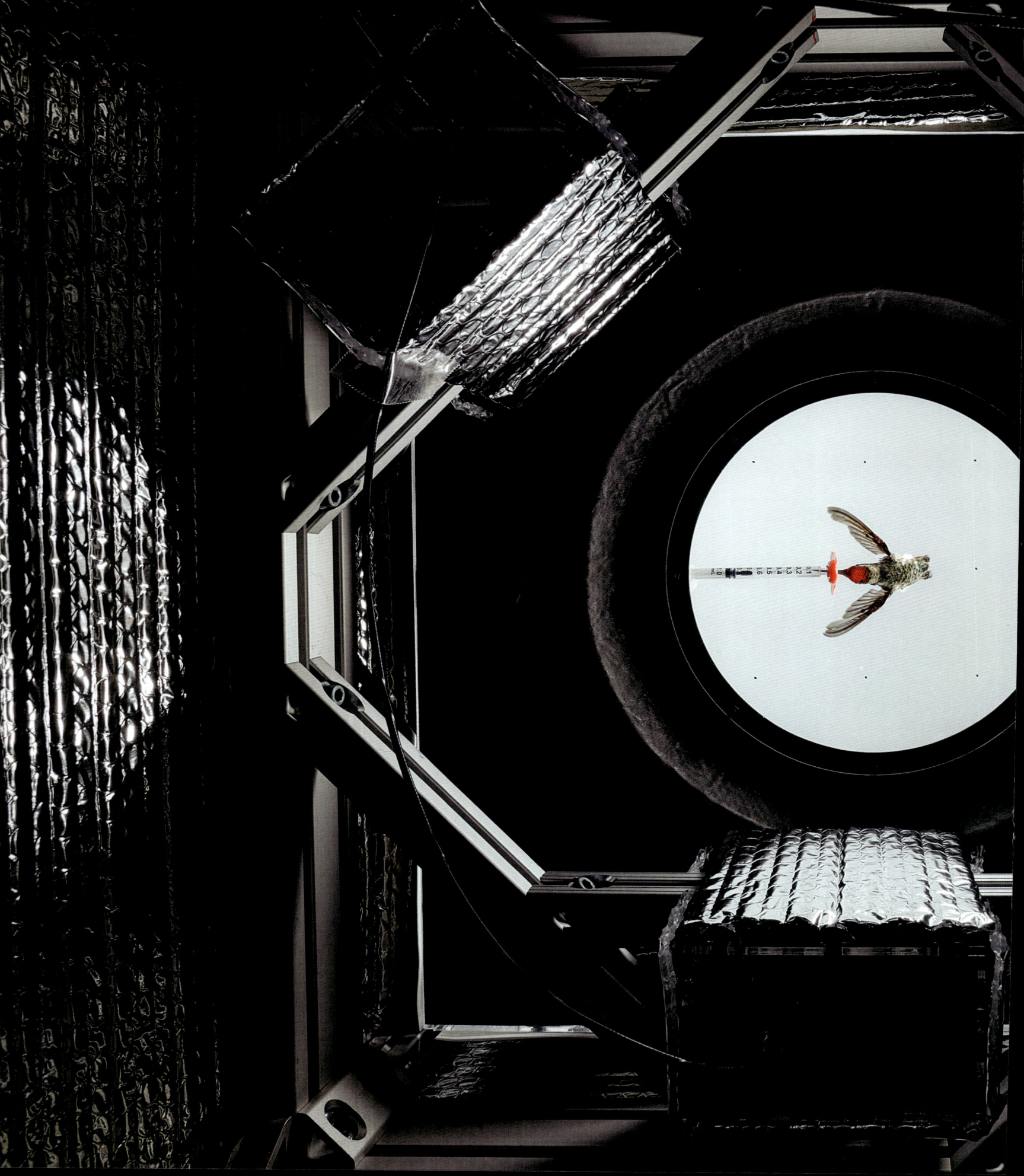

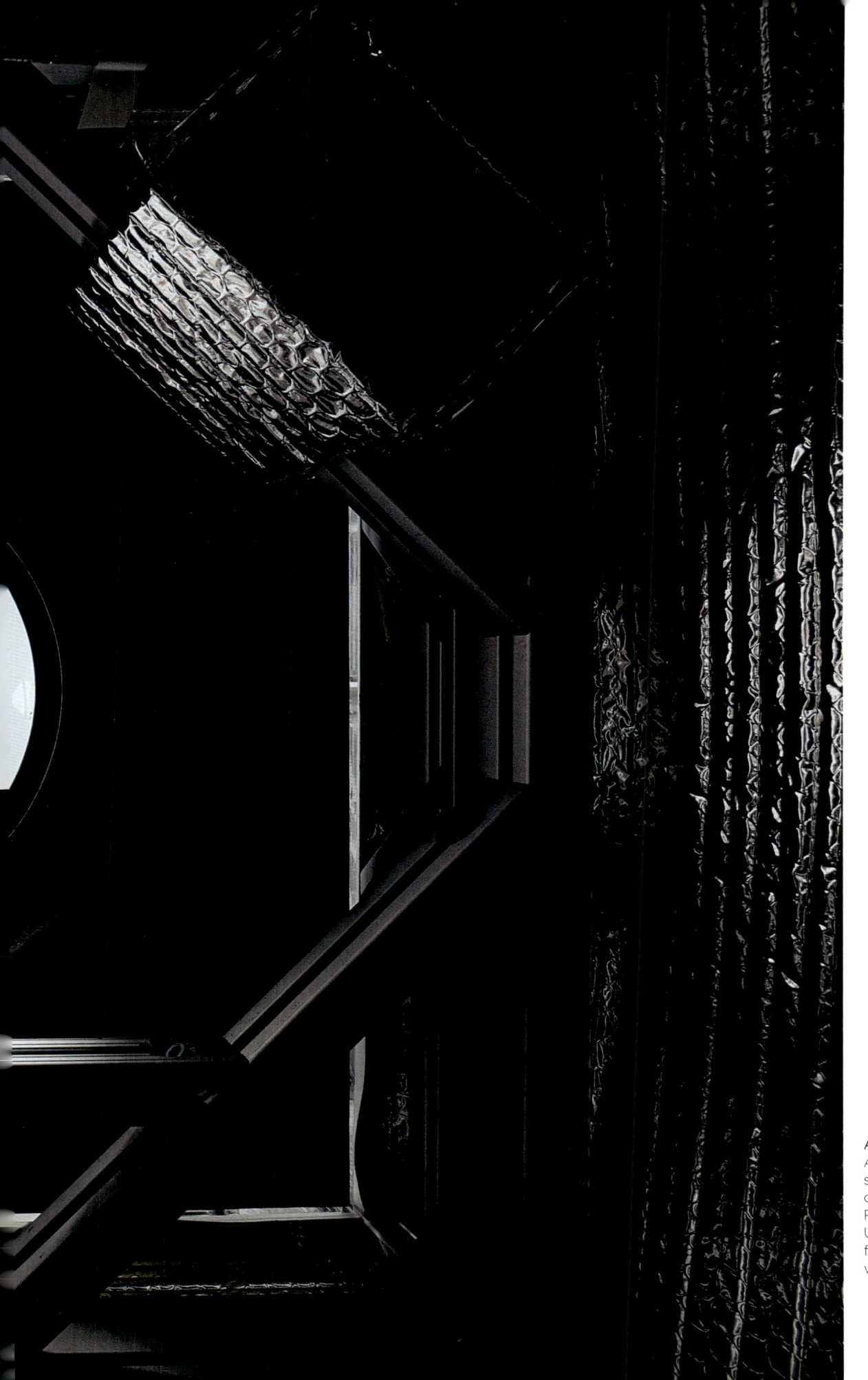

Anand Varma
An Anna's hummingbird
sips from a syringe in a
chamber designed by
Rivers Ingersoll at Stanford
University to measure the
force of these birds'
wingbeats.

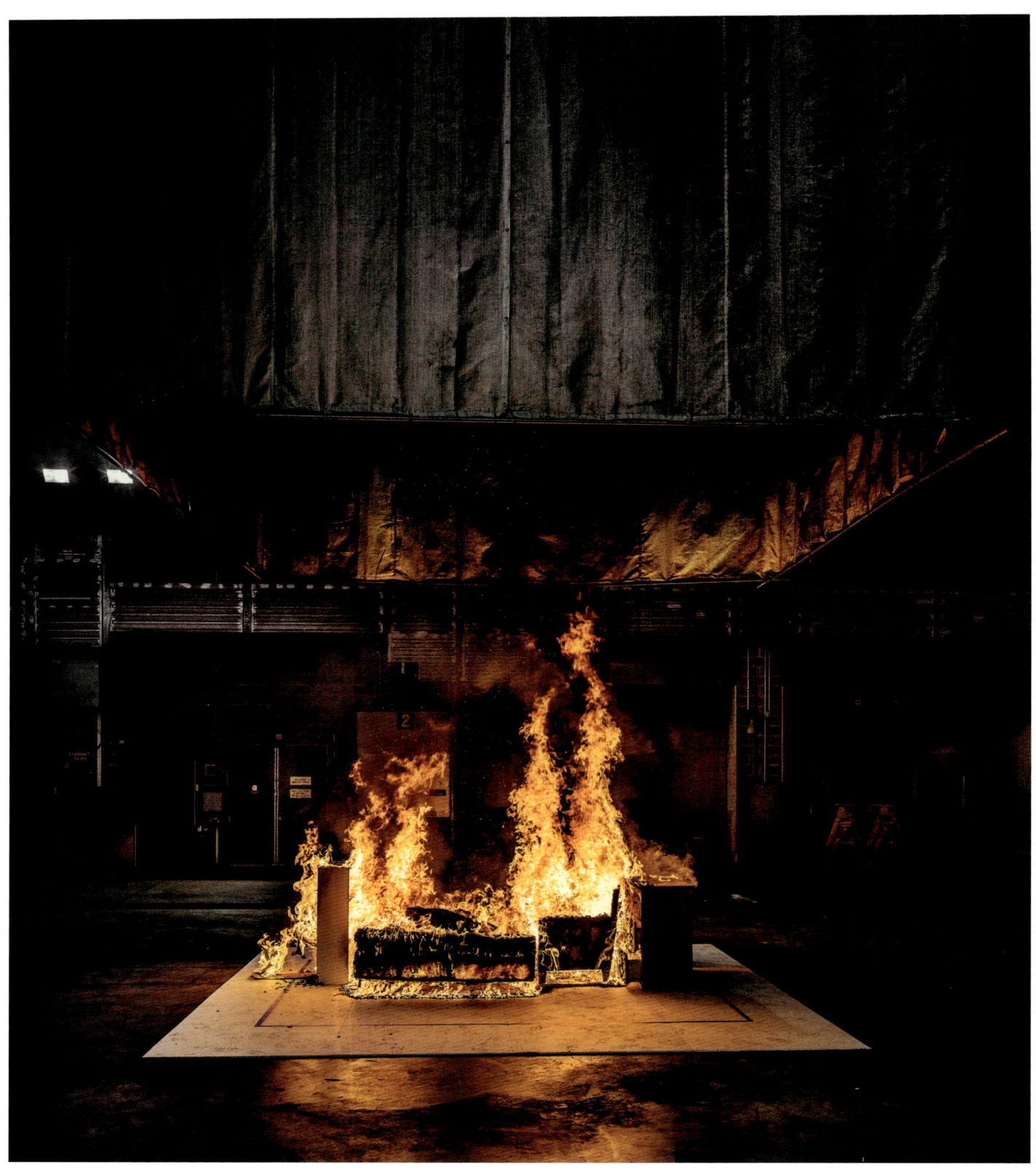

Max Aguilera-Hellweg
A model bedroom burns in an arson lab experiment run by the Bureau of Alcohol, Tobacco, Firearms and Explosives to collect data on heat, smoke, and gases released by the fire.

A THOUGHTFUL COMPOSITION COMPELS US TO CONSIDER THOSE LAYERS OF COMPLEXITY AND DETAIL THAT WE MISS AT FIRST GLANCE.

Evgenia Arbugaeva
John Mganga's hand reaches
into a box to show how
mosquitoes were fed at
Tanzania's Amani Hill Research
Station, once a center for
malaria research in Africa.

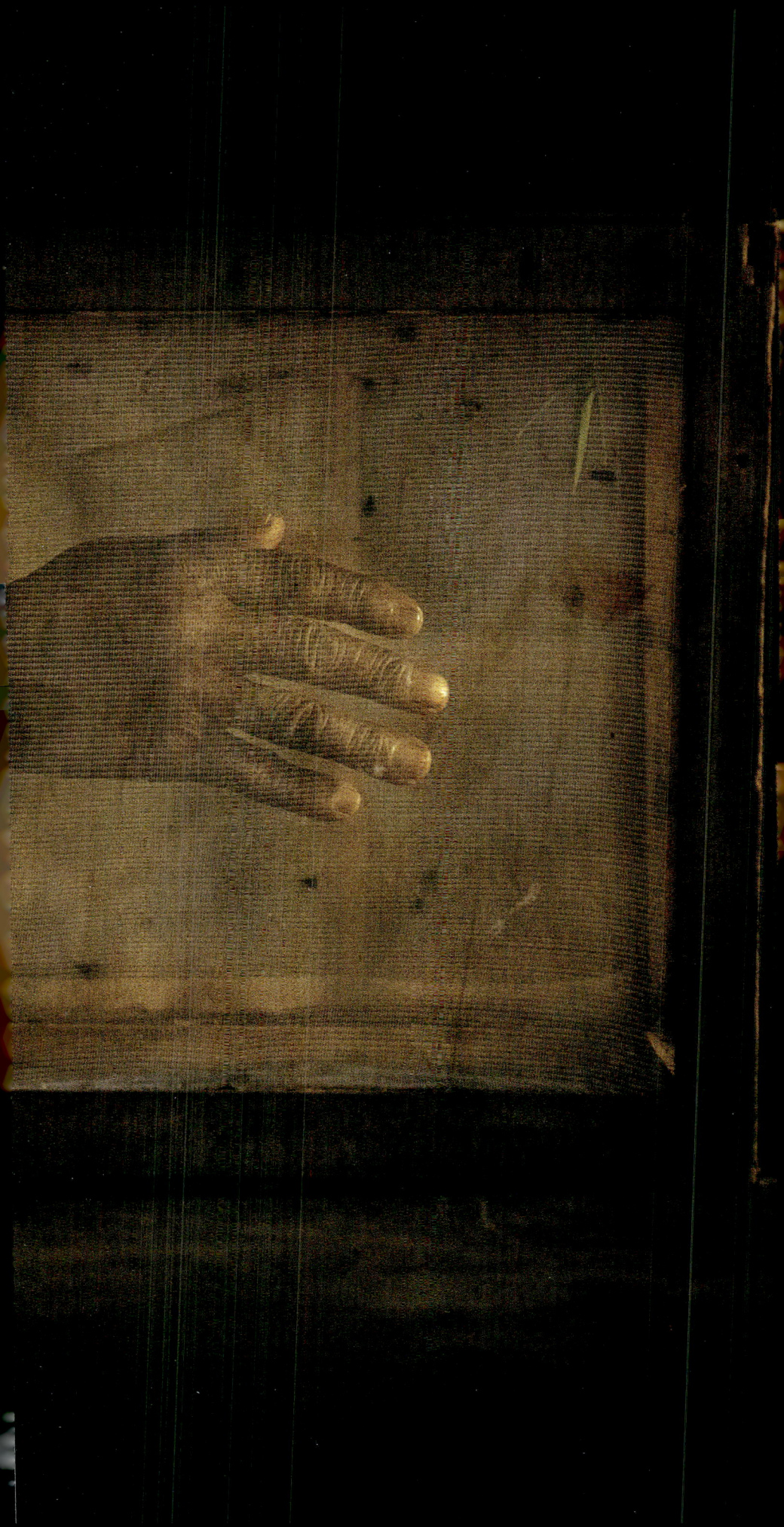

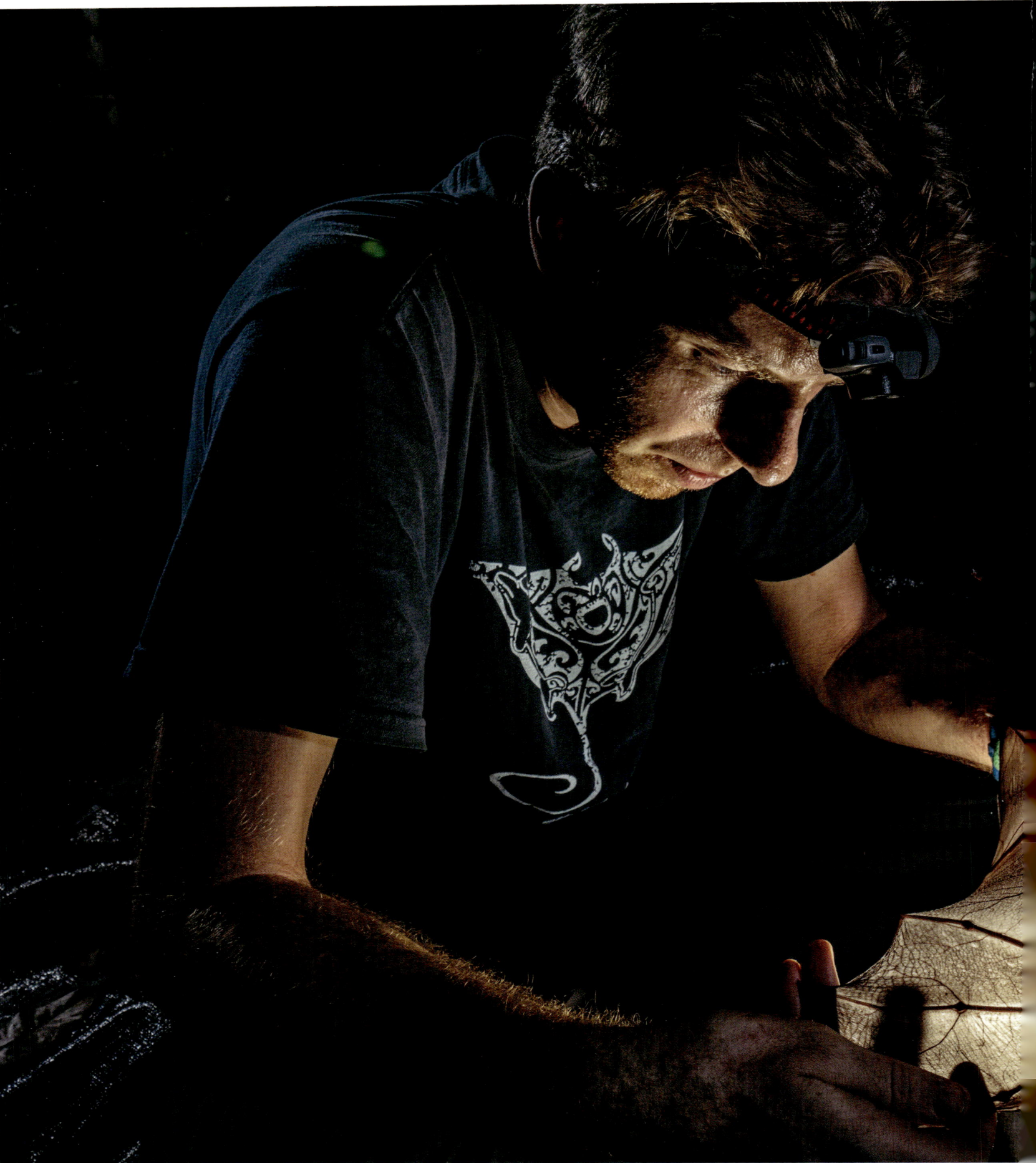

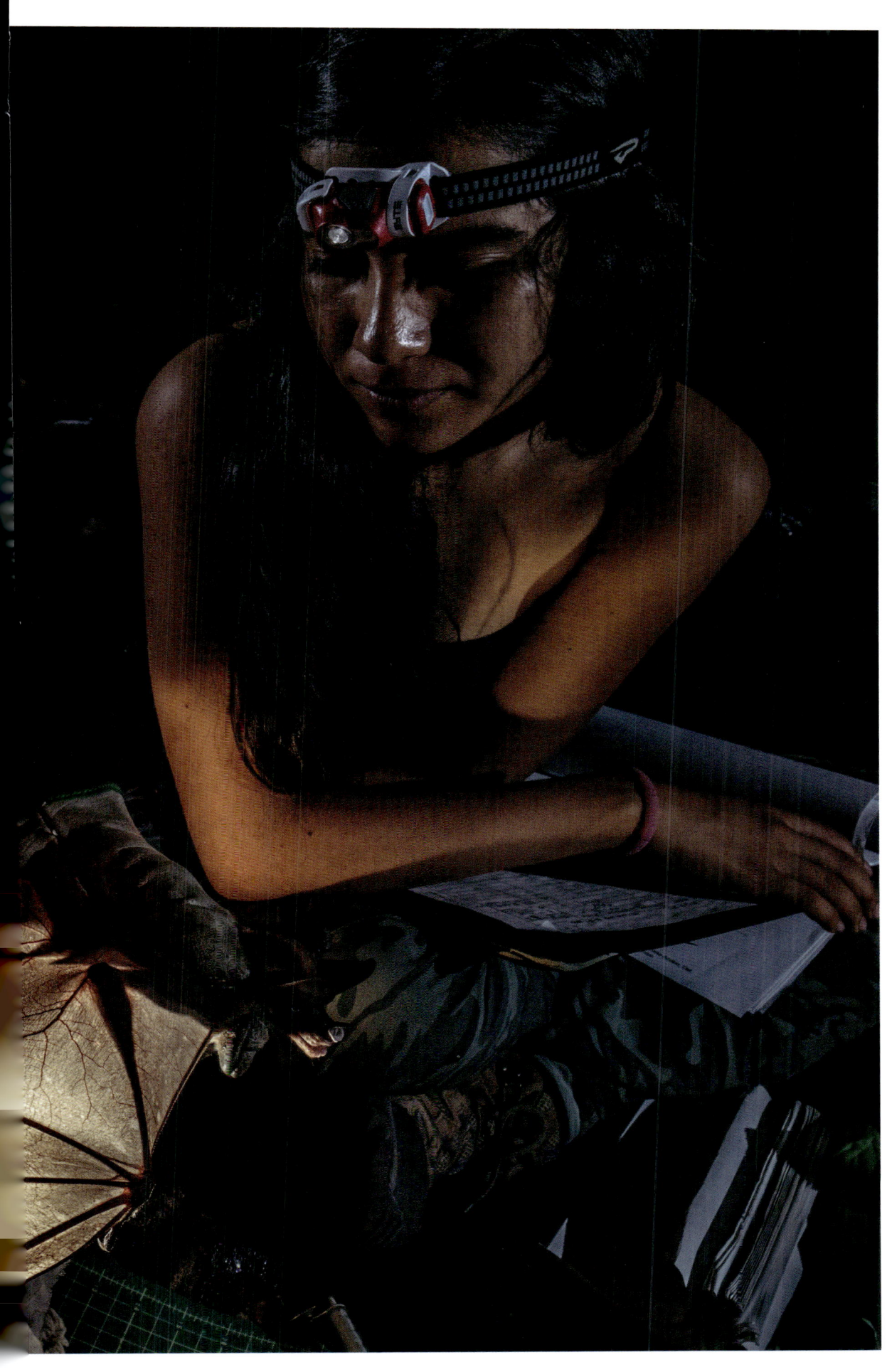

Anand Varma
Bat researcher Ivar Vleut teaches Teresa Martinez how to determine the age of a spectral bat by shining a light through its wing.

FOLLOWING PAGES:
Anand Varma
In Mexico a woolly false vampire bat casts an outsize shadow on the Maya temple called Hormiguero as it returns to its roost after a night of hunting.

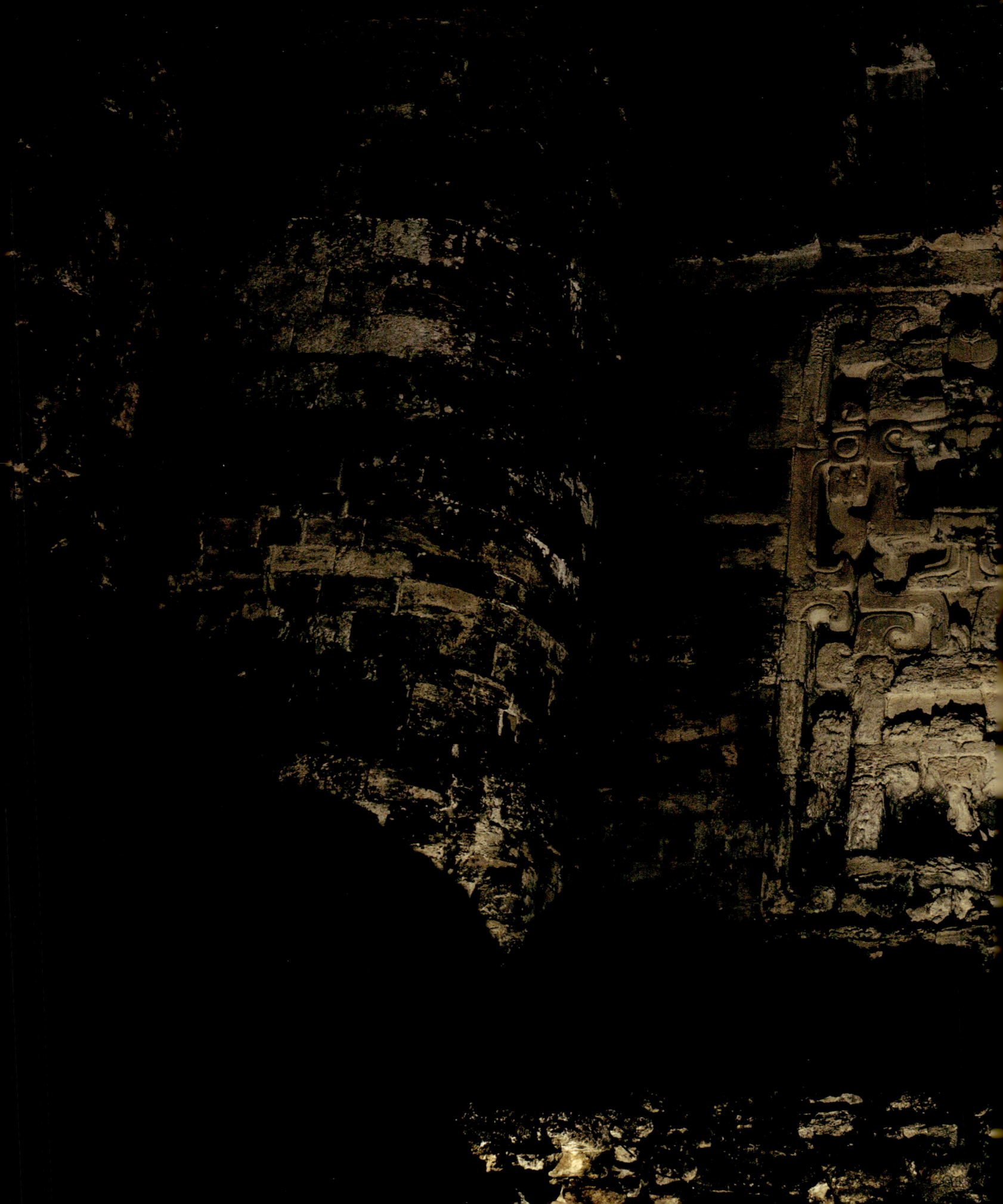

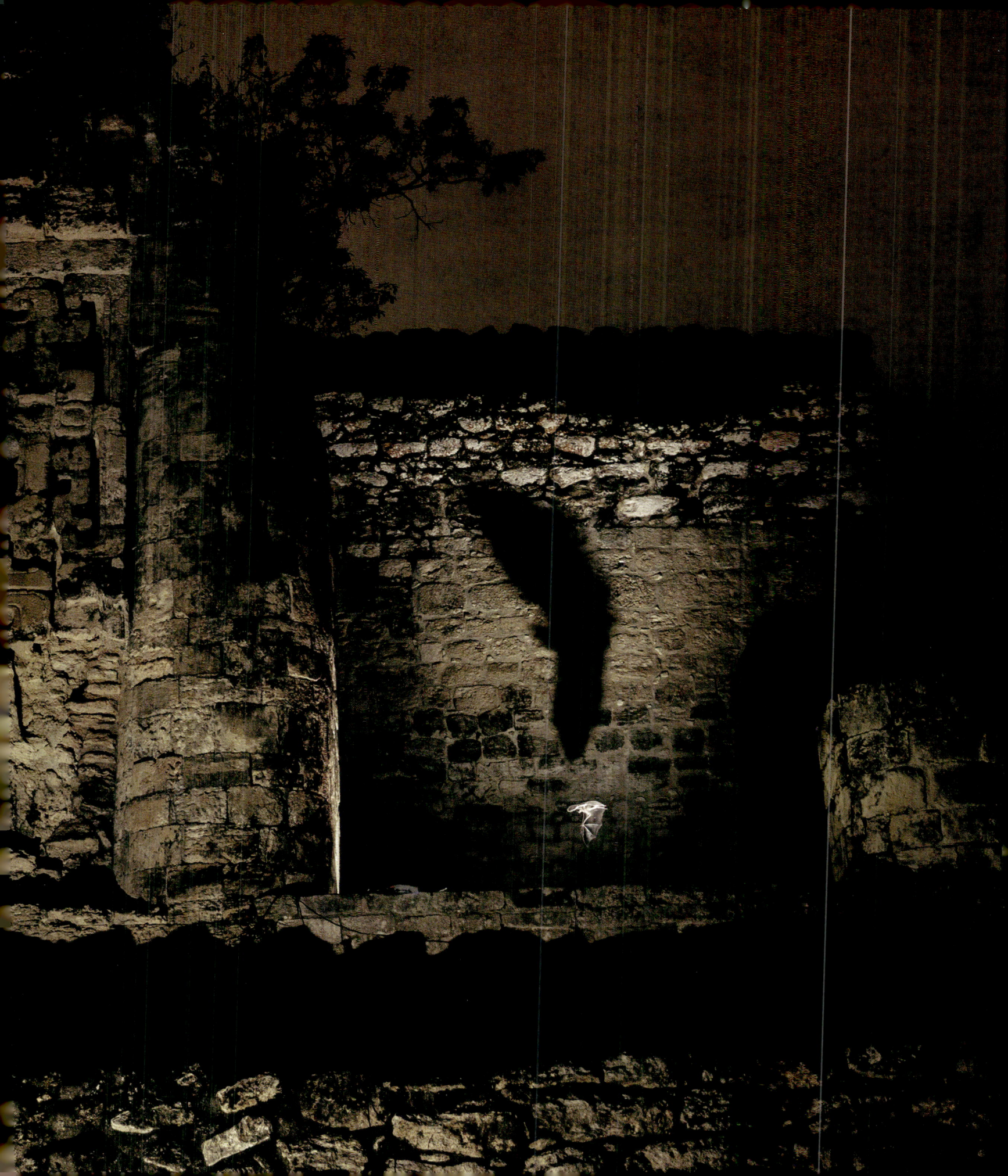

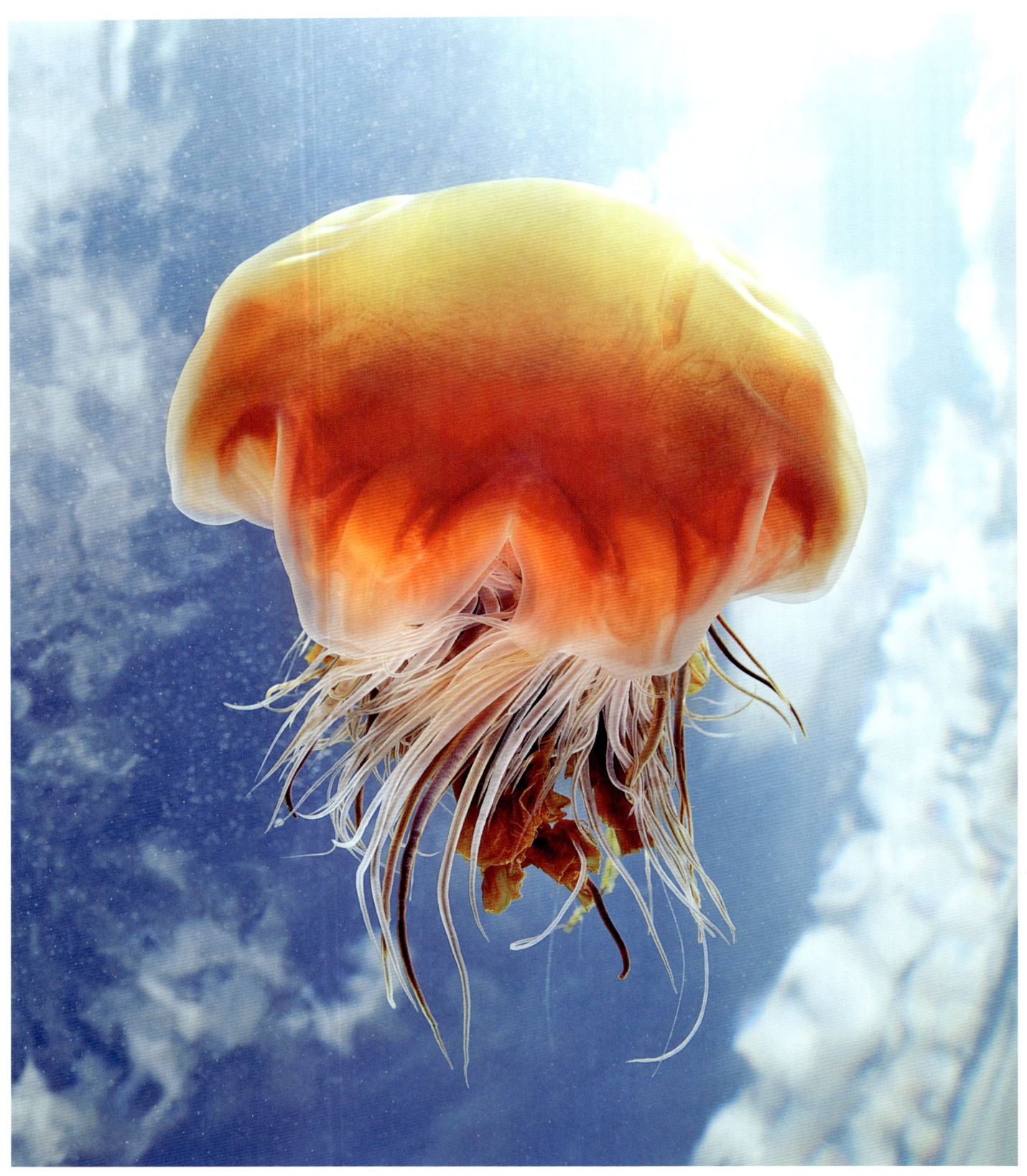

Alexander Semenov
A lion's mane jellyfish, photographed upward against a blue sky

A COMPELLING PHOTOGRAPH
EXPANDS OUR CURIOSITY
AND ENLARGES OUR CIRCLE
OF COMPASSION.

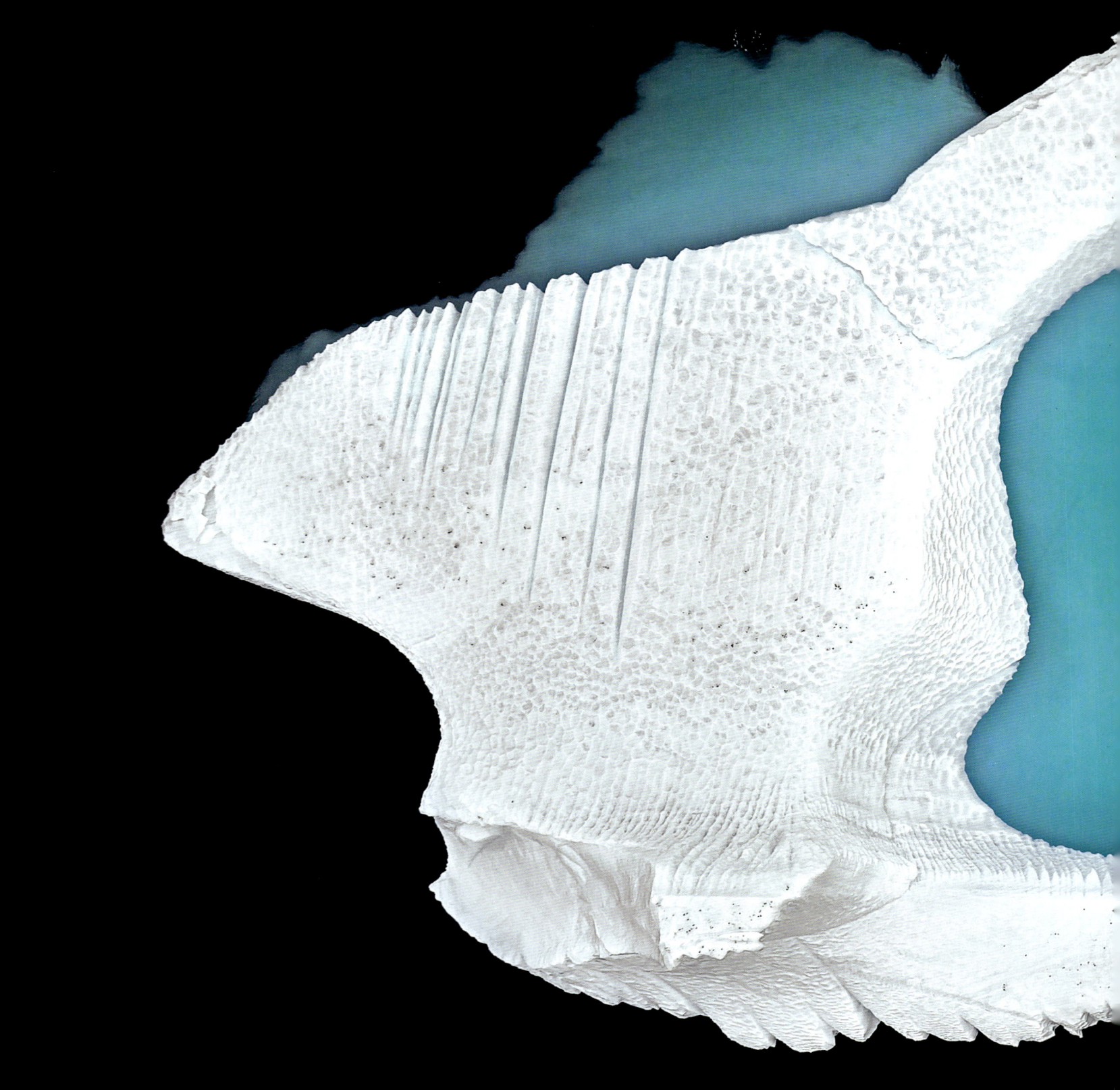

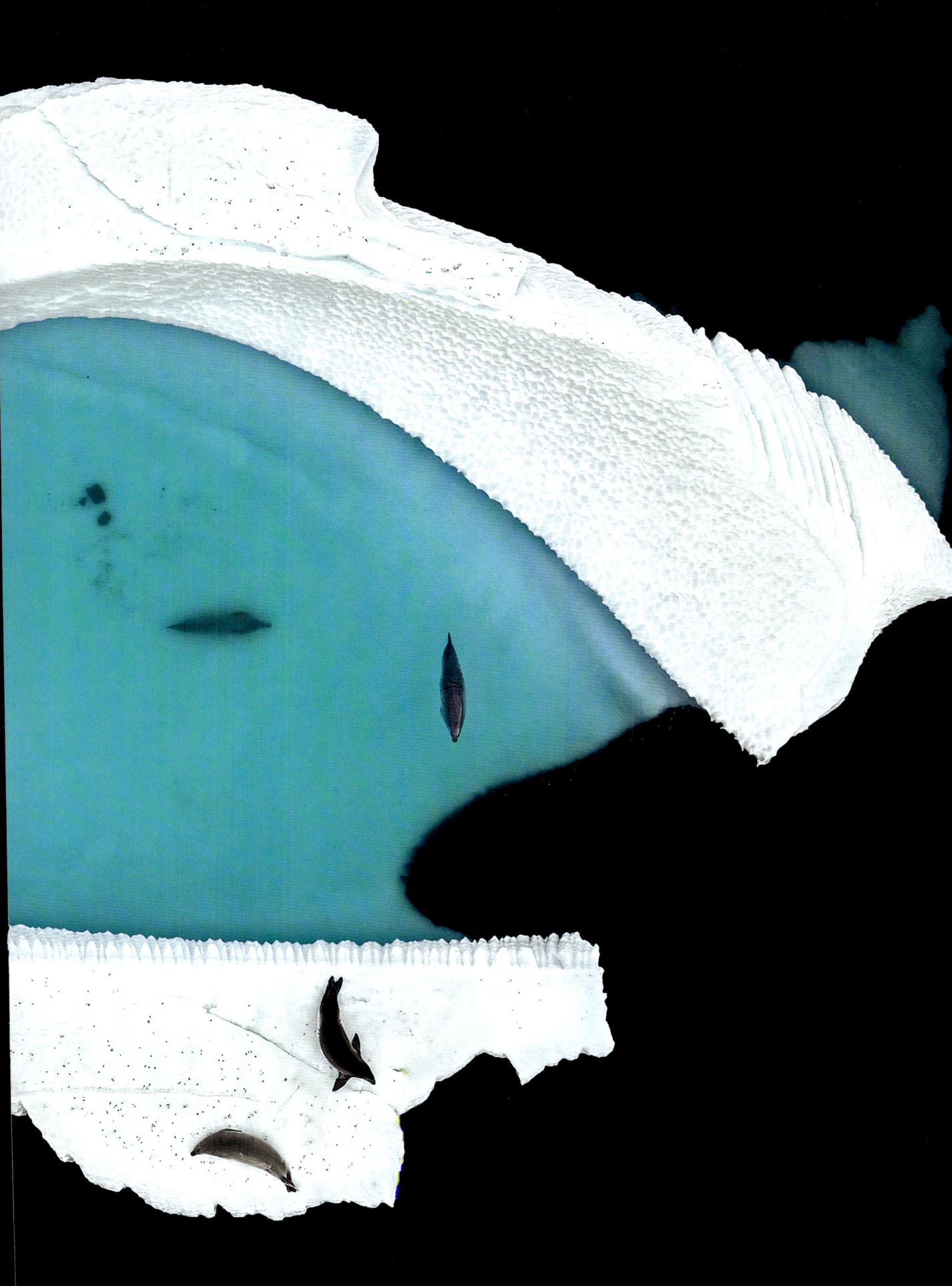

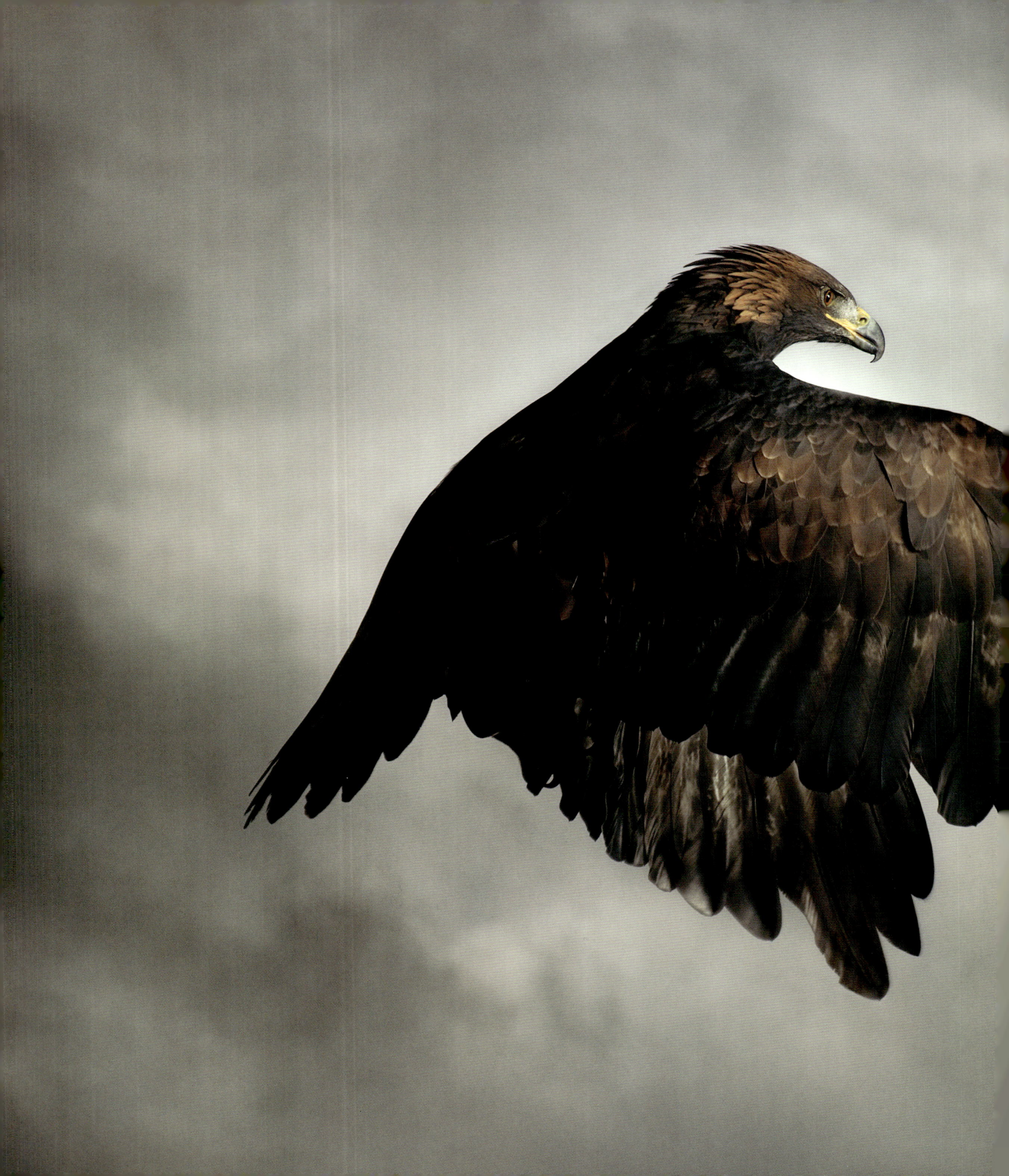

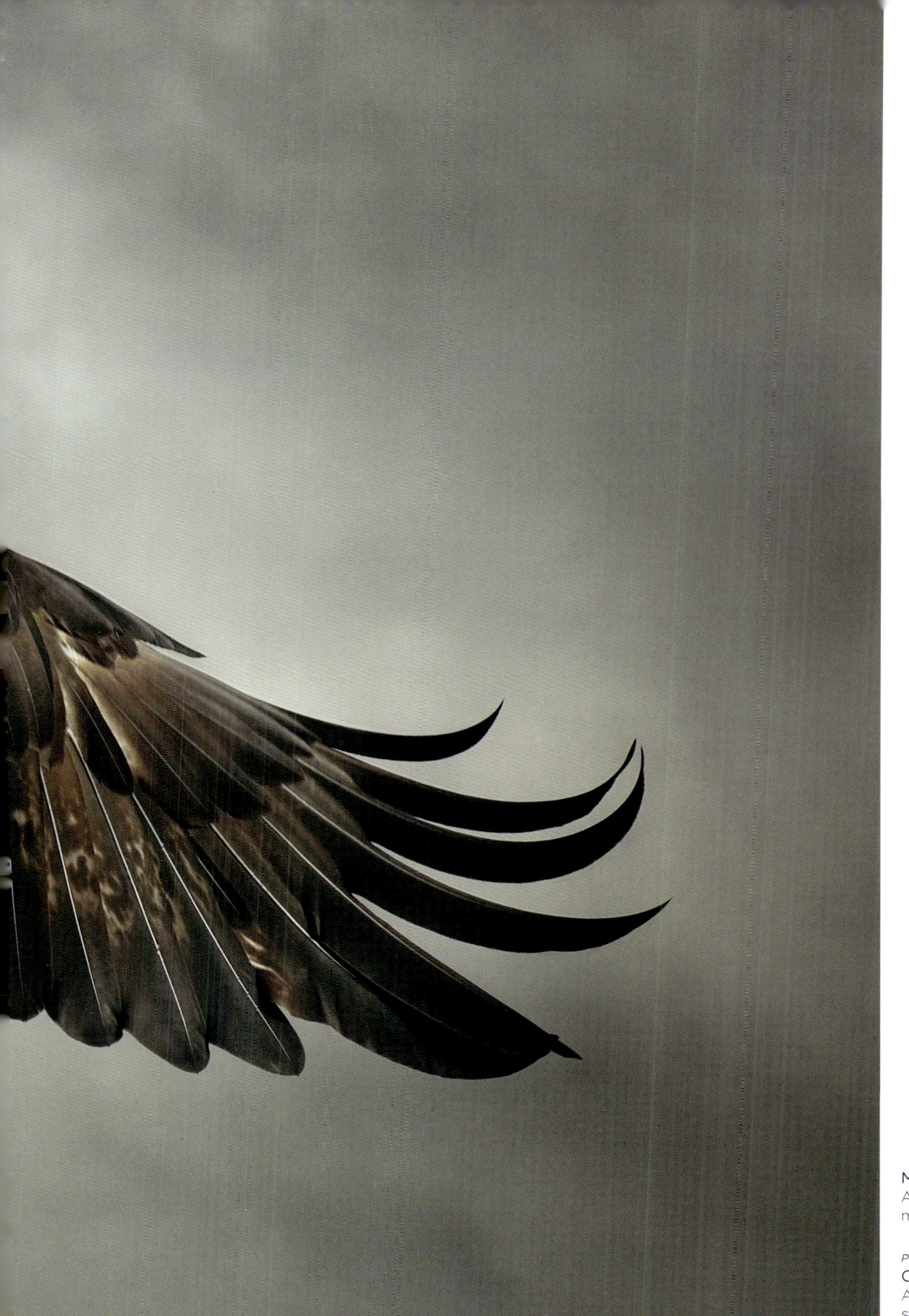

Mark Harvey
A golden eagle photographed
mid-flight

PREVIOUS PAGES:
Cristobal Serrano
An aerial view of crabeater
seals resting on an iceberg in
the Errera Channel, Antarctica

355

EVGENIA ARBUGAEVA

Evgenia Arbugaeva is a National Geographic Explorer and was one of National Geographic's four inaugural Media Innovation Fellows in 2018. Born in the Siberian port town of Tiksi on the shore of the Laptev Sea, she focuses her work most on the people, culture, and landscape of the Arctic. Her Academy Award–nominated film about the impact of climate change on the walrus population in Siberia, *Haulout,* was released in 2022.

ANAND VARMA: When did you discover your love for photography?

EVGENIA ARBUGAEVA: I feel that I've always been a photographer. When I was 15, I came from Siberia to Connecticut as an exchange student for one year, which was of course a huge cultural shock. My high school held a photography class, so I took it. Immediately, I thought it was the coolest thing ever. I couldn't believe it. After that encounter, I didn't dream of anything else. I'm so happy that I found it early in life.

AV: What draws you to photograph such remote places?

EA: My photography career is mainly focused on the northern regions of the Arctic. It's quite an aesthetic landscape. It has a very special light unlike anything I've seen anywhere in the world. One project I was working on brought me to Chukotka, the most northeastern region of Siberia along the Northern Sea Route, which connects Europe with Asia across the Arctic Ocean. I was interested in this place because I am myself from Yakutia, which is bordering with this Chukotka region. I was curious to learn about the Chukchi community of hunters. These people still live off the land and the sea.

AV: I love your photo of the walrus. Can you tell me the story behind that image?

EA: In 2017, I looked on the map for the most remote village in the Chukotka region. That brought me to Enurmino, a village on the shore of the Chukchi Sea. You can reach it only by a passenger helicopter that goes there once a month. The people there hunt walrus and whale, and one day I joined them on their hunt, and our boat landed on a strange, dark beach. The sand was almost black, and a horrendous smell was in the air. In the middle of the beach was a small hut. The people told me that every fall, thousands and thousands of walruses haul out on this beach and surround the hut. And in this hut, every autumn for the past 10 years, a marine biologist comes and studies the walruses in solitude. So the next year, I came back and joined the scientist, Maxim Chakilev, in his hut. It was an incredible experience. We were surrounded by about 100,000 walruses.

When walruses migrate, they rest on the floating ice, but because there was no ice that summer, they were forced to haul out on land. And because they arrive already exhausted, they're vulnerable to stampedes and trampling. They can't

fit on this beach, and they lay in two, sometimes three layers, so weaker animals and cubs suffocate under the pressure of bigger animals. From my perspective in the hut, it was so immediate and graphic, I was touched by it in a way that I've never been. We would open the door and see all the walruses on top of each other, and they would oftentimes be dying and we couldn't do anything about it. It was intense.

AV: How did you decide what to focus on for this picture? The choice of including the doorway is so specific.

EA: When I was in Chukotka, I listened to a lot of myths and legends about the creation of the world. And oftentimes the people's stories would come from connection with animals or human-animal transformation. Being in this place, with these animals, opened up all kinds of fantasies and imagination. In this photo, because we see the binoculars, the cigarettes in the glass jar, some screws and instruments, there is this closeness of human world and animal world. I didn't have much in terms of photography angles, so I just put my camera on the tripod and waited for perfect light. In this image, I was lucky because their breath made a mist

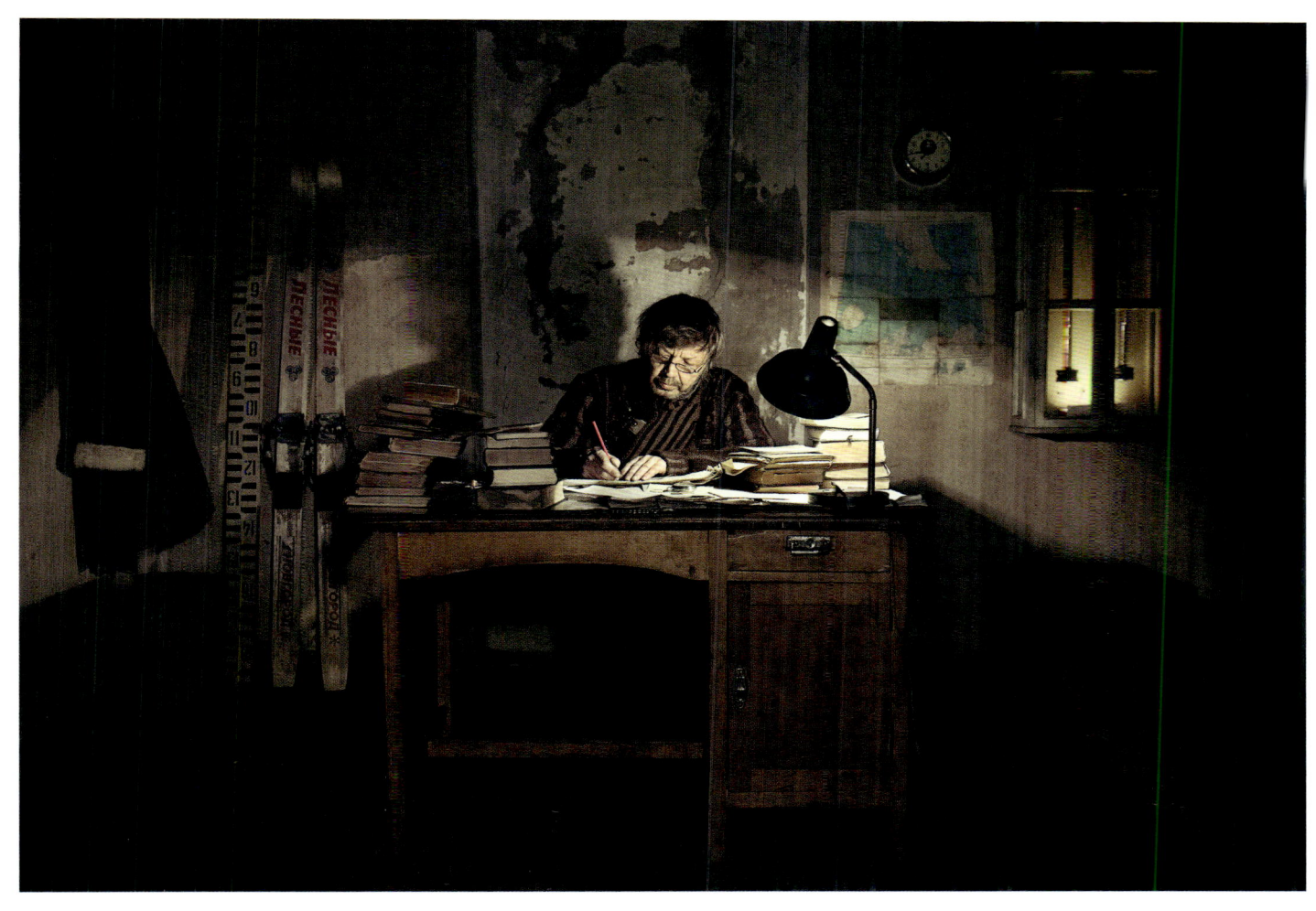

Evgenia Arbugaeva
Slava the weatherman lives alone in a meteorological station in Siberia,
surrounded by the Bering Sea.

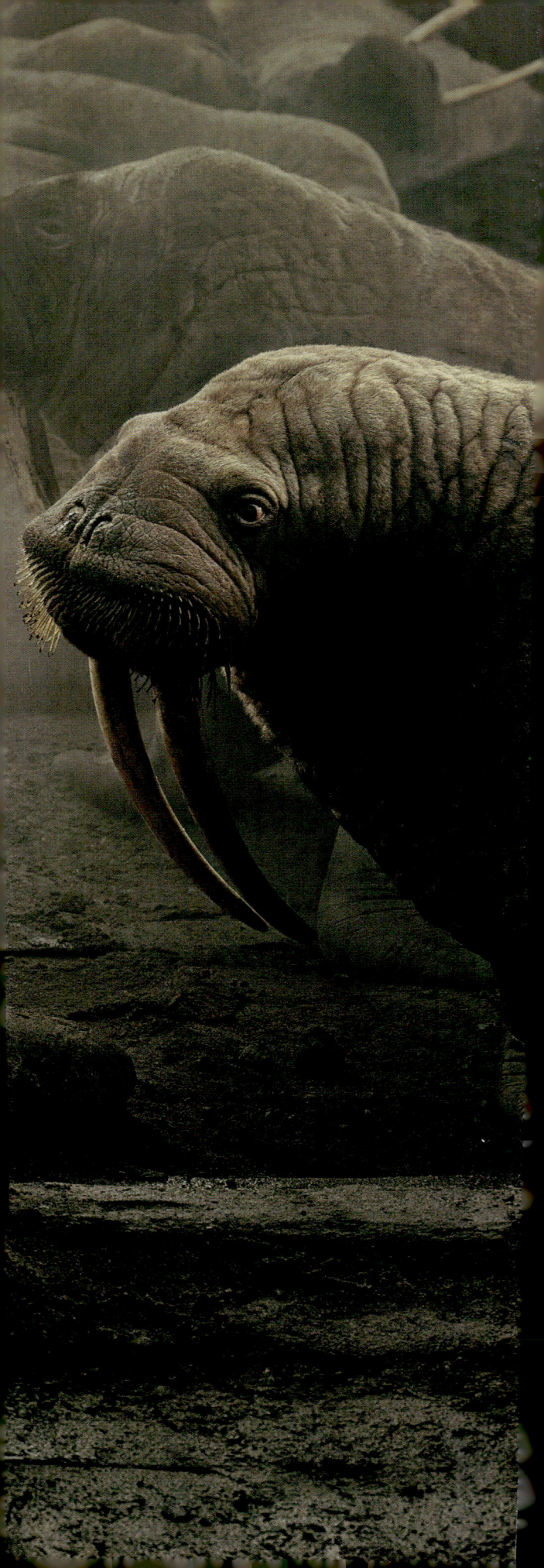

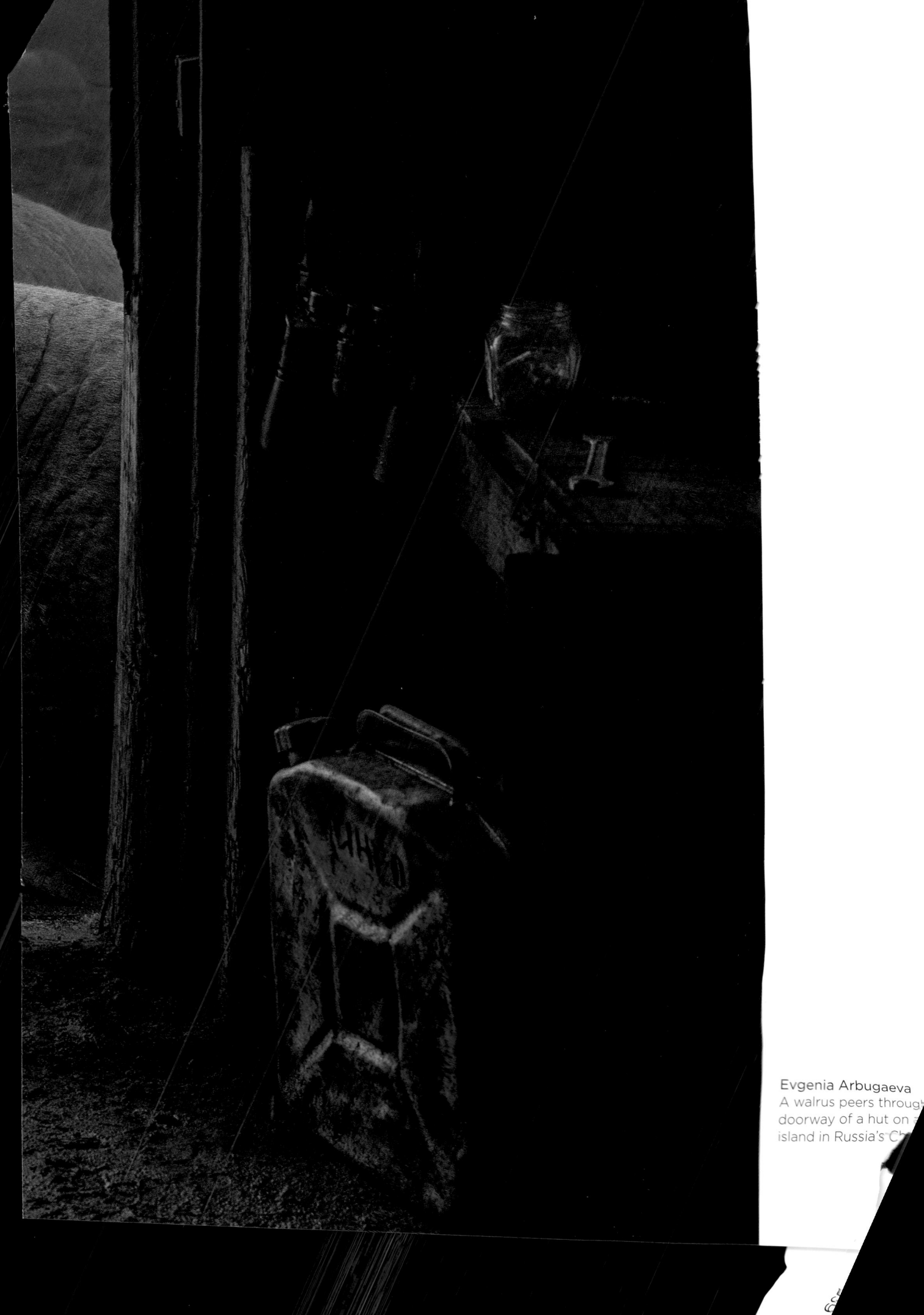

Evgenia Arbugaeva
A walrus peers throug
doorway of a hut on
island in Russia's Ch

after the sun appeared, so it created this beautiful, painterly light. I felt the walrus was looking at us, looking at me, and it looks so beautiful, but also scary, but also vulnerable. That's the magic of photography.

AV: Were there any influences or inspirations for how you took this image?

EA: In a way, I was inspired by the artist William Blake and his surreal, spiritual paintings. He has a small painting called "The Ghost of a Flea," a vision he had close to the end of his life of this half human, half insect. It was a bloodthirsty creature, but it's painted in umber colors, with a bit of gold. It's mesmerizing. I bought a postcard of this painting, and in the hut I put it next to my bunk bed. I was looking at it all the time, and I think it unconsciously informed the composition. If you put them together, you'll see some similarities about animal spirits, but also something human. The closeness of the walrus is so unusual that you start to think, Is this real? And the light creates this dreamlike quality that makes you think, Is it a painting, or is it a photo? I like to take images that are in this in-between space.

AV: I love how you're drawing inspiration from places outside of photography. I resonate with that.

EA: I think everything kind of exists already, right? It's about keeping that thread going through the generations and through different people. Many amazing artists have left his world, and to have their legacy in my work is an honor. I see my k as a continuation of the good-that existed before and has to e to exist. It's about collect-people around you, even if ad, and creating together.

AV: How did you approach telling the story of the weatherman?

EA: When I first met the weatherman, Slava, it seemed like he had already existed in my mind. He's a lone arctic wolf. We connected without many words, really, just kind of understanding each other. I was fascinated by solitude, by his solitude, how he's so comfortable in his isolation and so clearly in love with this meteorological station on a narrow peninsula in the Bering Sea. He was born on a ship, so he's been at sea all his life, and the peninsula allowed him to see the ocean on all sides, like he was living on the ship again. He's so organic. He almost disappears into the wind and the sea and the stars. I wanted to be in his company, and I wanted to be in the place that he loves so much. Because I was so fascinated by him, I was seeing everything and photographing everything from that romantic prism of him being the loner in the Arctic.

AV: Your work reveals things that we don't notice at first glance. How do you decide where to focus your camera?

EA: It's hard to answer this question, because I often feel I don't choose the places and subjects I photograph. It seems like they come to me somehow, or I'm guided to be there. For the image of the walrus, I was at this door for two weeks, every day, all day long. I like to get bored with the same situation to the point where I start to pay attention to the smallest details. I like to feel every single detail. Most of the situations and stories that I photograph, it seems that they're very quiet. There's not much happening. The images feel like a condensation of all the impressions I've had until this moment. All the things that I've

read, or the stories that I've heard, they just layer in my head, and then my eyes start to focus.

If you come to a situation without knowing anything, you focus more on things on the surface, things jumping out to you. But most of the time, these are not the most interesting things.

Oftentimes in places like the Arctic, conditions are quite harsh, so you make a story about this place to yourself. It starts to form in your head, like a poem. Experiences crystallize in your mind, and you start to pick up a certain tone, a certain mood, colors and color palettes, angles.

AV: What else are you thinking about as you work on a story?

EA: When I work on a story, I think of textures. The weatherman, for example, was a velvety story, and in terms of sound, there was a lot of cello in it. That comes from getting to know the person—when you observe the color palette of their place and things like that. I sometimes consider my work as the documentation of how my life intersects with the lives of other people. When I look back at the images, in each story I recognize very clearly that part of my life and what I was searching for.

When I was photographing the weatherman, I was trying to figure out his idea of home, because I had no home at the time. I found a lot of natural fascination in someone finding a perfect place for themselves. Now I am looking more into the spiritual connection between people and the land, especially Indigenous people. That will require a lot of conceptual thinking, or magical thinking, to photograph something that oftentimes is invisible. But I find that's a very interesting challenge.

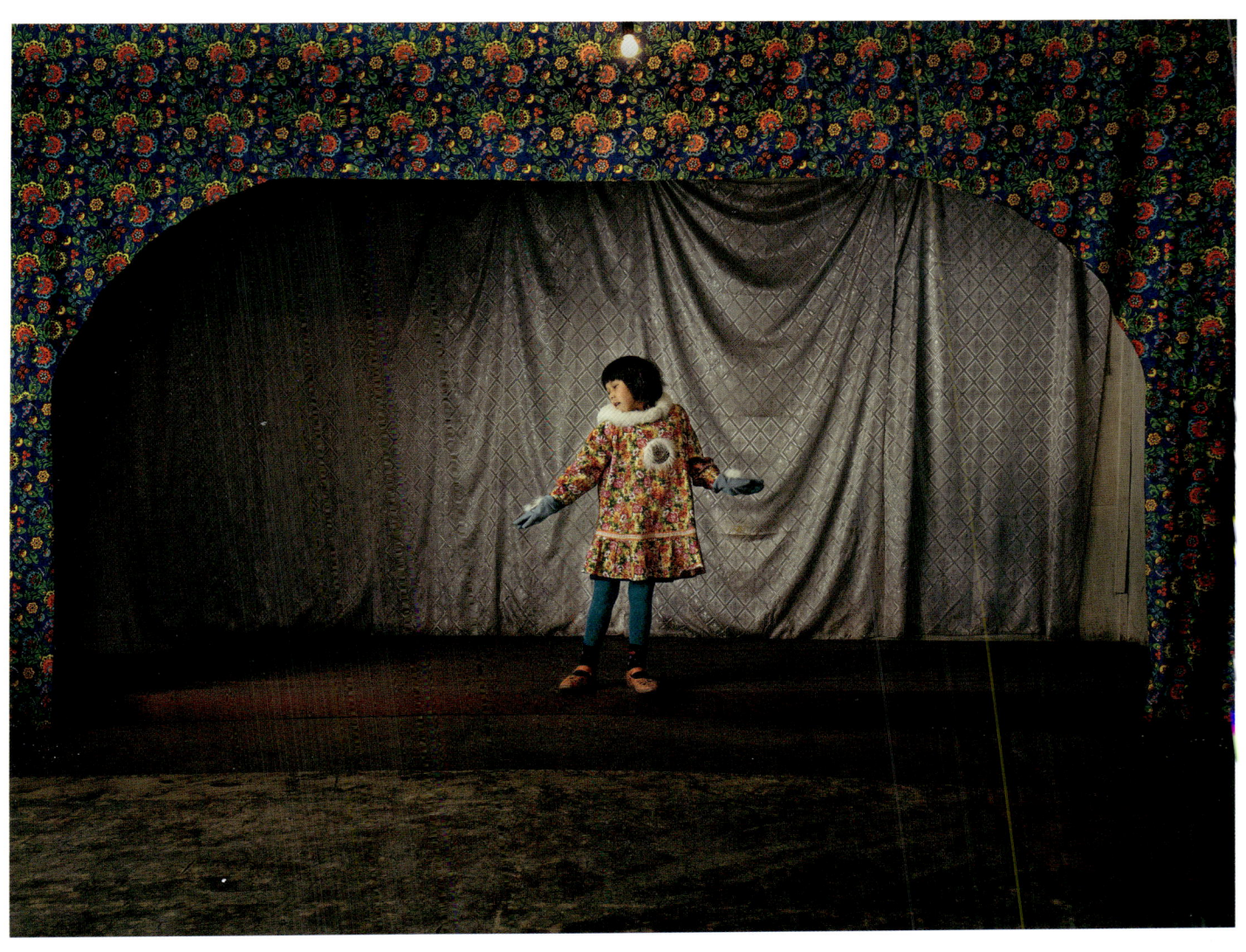

Evgenia Arbugaeva
Vika Taenom wears a customary Chukchi dress called a kamleika as she rehearses a
traditional dance in the cultural center in Enurmino, a village in far eastern Russia.

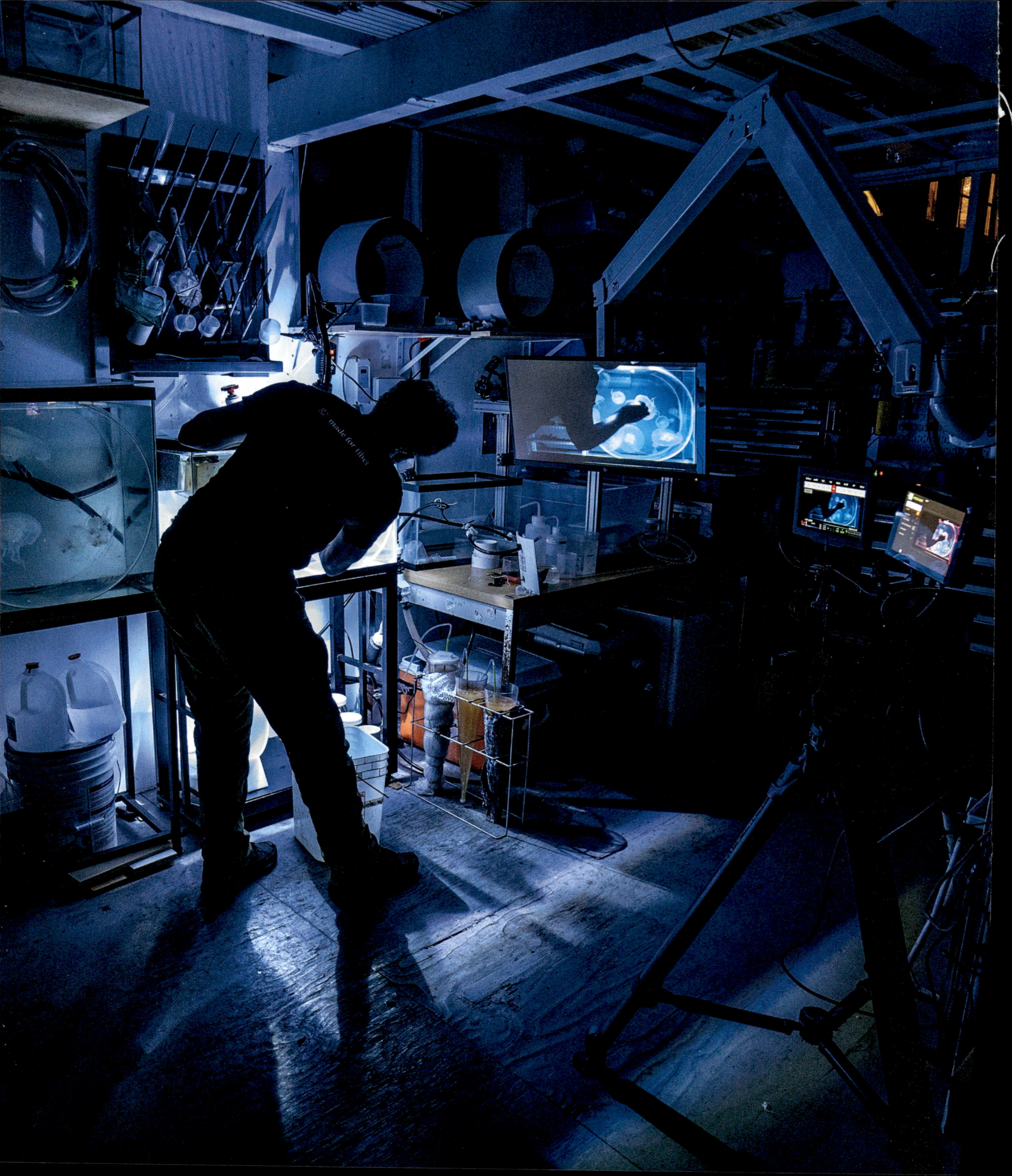

EPILOGUE

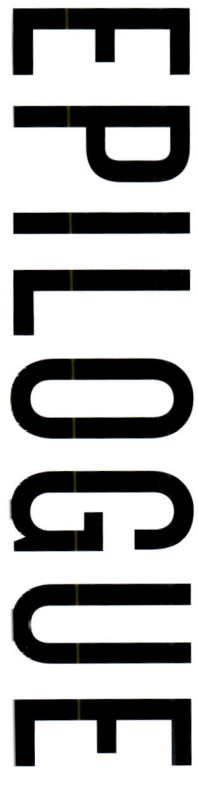

When I first decided to pursue photography, I thought I had to abandon my childhood dream of becoming a scientist. But then *National Geographic* published my story on honeybees, and a scientist told me I had captured undescribed bee behavior. In pursuit of unique images, I had stumbled onto an original observation. That's when I realized that science and photography are not mutually exclusive efforts. Science gives us structured systems for observing the world, while photography gives us tools for extending our visual perception. Each reinforces the other. Studying my subjects has inspired me to invent new photographic techniques. Experimenting with cameras has led me to capture novel scientific data.

Rather than choosing between a career in science and a career in photography, I created WonderLab in Berkeley, California, as a way to harness the combined power of both. WonderLab draws on my training as a biologist, my experience as a photographer, my fondness for tinkering, and my desire to share what I have learned with others. It is a science lab, a photography studio, a makerspace, and a classroom.

Just as each photograph in this book gives us a fresh look at our world, my mission with WonderLab is to develop new ways of visualizing science. I believe a compelling photograph can change us. It shapes what we notice, challenges our assumptions, and expands our curiosity. It motivates us to make room for new ideas, convinces us to reassess our values, and enlarges our circle of compassion. A beautiful, bewildering photograph is a wonderful place to start making sense of the staggering complexity around us.

Mark Unger
Anand Varma cleans a jellyfish tank in a garage that he converted into a film studio and biology lab—a garage that served as the precursor to WonderLab.

ACKNOWLEDGMENTS

I would like to thank the National Geographic Society and the Burroughs Wellcome Fund for supporting my work. Kaitlin Yarnall and Lou Muglia, you have been incredible champions for me and for WonderLab.

Thank you to the team at National Geographic Books, including Lisa Thomas, Elisa Gibson, Susan Hitchcock, Tyler Daswick, Becca Saltzman, Sanaa Akkach, Adrian Coakley, Libby Sander, Jill Foley, and Meredith Wilcox. Susan, I am grateful for your trust despite this being my first book. Sanaa, I learned so much from you during our marathon layout sessions.

Special thanks to Martin Oeggerli, Jen Guyton, Reuben Wu, and Evgenia Arbugaeva for agreeing to share your incredible images and talking with me for this book.

Pamela Chen, I am grateful for your insightful feedback on the manuscript. Thank you to Lucas Foglia for letting me browse your photo book collection and to Mark Unger for picking up the slack at the lab while I was focused on the book.

David Liittschwager, you inspired me to pursue photography—and Todd James, you shaped me into the photographer I am today. Thanks also to Susan Welchman for giving me my first opportunity to shoot for *National Geographic* magazine.

To my parents, Prabha and Vijay, I am grateful you believed in me as I charted this unconventional path.

Most of all, thank you, Jody, for your endless love, patience, encouragement, and support.

ILLUSTRATIONS CREDITS

42-3, Trunk Archive; 60-1, Science Source; 76-7, Biosphoto/Minden Pictures; 96-7, Eye of Science/Science Source; 148-9, Xinhua via ZUMA Wire; 152-3, elmofoto; 160, Everett Collection/Alamy Stock Photo; 160-1, © Harold Edgerton/MIT, courtesy Palm Press, Inc.; 190-1, Alamy Stock Photo; 208-9, Vitamin Sea, courtesy of The Cooking Lab, LLC; 216-7, NASA and the Hubble Heritage Team (STScI/AURA); 228, Raw data from NASA/SDO-AIA; 229 ESA & NASA/Solar Orbiter/EUI team; Data processing: E. Kraaikamp (ROB); 246-7, Eunice Kennedy Shriver National Institute of Child Health and Human Development; 248-9 and 250, GustoImages/Science Source; 251, Science Source; 256-7, Science Source.

Images copyright © Tim Flach are from titles published by Abrams Books in association with Blackwell & Ruth:
29, 64, and 168-9, from *Birds*
40-1, 300, 330, and 331, from *More Than Human*
72 and 166-7, from *Equus*

Images copyright © Martin Oeggerli were produced in cooperation with the following:
4 (2008) and 94-5 (2007) Pathology, University Basel and Prüftechnik Uri
100-1 (2012), 104 (2010), 121 (2011), and 122-3 (2013), Pathology, University Basel and School of Applied Sciences, FHNW
105 (2017), supported by Prof. D. J. Müller and D. Martinez-Martin, ETH Zurich, and D. Mathys, Nano Imaging Lab, SNI, University Basel
125 (2012), Pathology, University Basel and Bio-EM Lab, Biozentrum, University Basel

Images from the National Geographic Image Collection:
8-9, 12-3, 14, 18-9, 24-5, 52-3, 86-7, 136-7, 138-9, 144-5, 158-9, 162-3, 174, 176-7, 178, 182-3, 200-1, 222, 232-3, 264, 280-1, 284-5, 286-7, 288-9, 290-1, 292, 297, 301, 304-5, 306, 312-3, 316-7, 320-1, 340-1, 342, 346-7, 348-9

ABOUT THE AUTHOR

Anand Varma grew up exploring the woods near his childhood home in Atlanta, Georgia. As a teenager, he picked up his dad's old camera and found he could use it to feed his curiosity about nature—and to share his discoveries with others. While he was studying integrative biology at the University of California, Berkeley, he took a summer job working with National Geographic photographer David Liittschwager. The experience set him on the path of science photography.

Anand received a Young Explorers Grant from National Geographic in 2010 to document the wetlands of Patagonia. He has since photographed numerous stories for *National Geographic* magazine, including a 2014 cover story, "Mindsuckers," about parasites that control the minds of their hosts, which won the World Press Award for best nature story. He was recognized as a National Geographic Media Innovation Fellow in 2018 and a Rita Allen Foundation Civic Science Fellow in 2020. In 2022, he created the National Geographic WonderLab in Berkeley, California, to push the boundaries of how we visualize the natural world.

THE NATIONAL GEOGRAPHIC SOCIETY & WONDERLAB

The National Geographic Society is a global nonprofit that uses the power of science, exploration, education, and storytelling to illuminate and protect the wonder of our world.

Since 1888, the National Geographic Society has driven impact by identifying and investing in a global community of Explorers: leading changemakers in science, education, storytelling, conservation, and technology. National Geographic Explorers help bring our mission to life by defining some of the most critical challenges of our time, uncovering new knowledge, advancing new solutions, and inspiring transformative change throughout the world

Explorer and photographer Anand Varma sees wonder as a doorway into exploring and valuing the complexity of our world. Through WonderLab, an ambitious, state-of-the-art space, Anand is developing breakthrough photography and filmmaking techniques that capture the wonders often invisible to the naked eye. National Geographic Society supports Anand and the WonderLab's dedication to highlighting our world's hidden mysteries, demonstrating how science coupled with stunning visual storytelling sparks curiosity and creates a deeper appreciation of our natural world

To learn more about the Explorers and projects we invest in, like Anand Varma and WonderLab, as well as the efforts we support, visit natgeo.com /impact.

Since 1888, the National Geographic Society has funded more than 14,000 research, conservation, education, and storytelling projects around the world. National Geographic Partners distributes a portion of the funds it receives from your purchase to National Geographic Society to support programs including the conservation of animals and their habitats.

Get closer to National Geographic Explorers and photographers, and connect with our global community. Join us today at nationalgeographic.org/joinus

For rights or permissions inquiries, please contact National Geographic Books Subsidiary Rights: bookrights@natgeo.com

ISBN: 978-1-4262-2314-3

Printed in China

23/PPS/1